Microbial Physiology

Microbial Physiology

Editor: Dean Watson

RCALLISTO REFERENCE

www.callistoreference.com

Callisto Reference,
118-35 Queens Blvd., Suite 400,
Forest Hills, NY 11375, USA

Visit us on the World Wide Web at:
www.callistoreference.com

ISBN: 978-1-63239-809-3 (Hardback)

The publisher's policy is to use permanent paper from mills that operate a sustainable forestry policy. Furthermore, the publisher ensures that the text paper and cover boards used have met acceptable environmental accreditation standards.

Printed in the United States of America.

Cataloging-in-publication Data

Microbial physiology / edited by Dean Watson.
   p. cm.
Includes bibliographical references and index.
ISBN 978-1-63239-809-3
1. Microorganisms--Physiology. 2. Microbiology. 3. Microbial metabolism. I. Watson, Dean.
QR84 .M53 2017
579--dc23

Table of Contents

Preface...VII

Chapter 1 **Recombinant TLR5 Agonist CBLB502 Promotes NK Cell-Mediated Anti-CMV**
           Immunity in Mice..1
           Mohammad S. Hossain, Sampath Ramachandiran, Andrew T. Gewirtz,
           Edmund K. Waller

Chapter 2 **Fatty Acids from Membrane Lipids Become Incorporated into Lipid Bodies**
           during *Myxococcus xanthus* Differentiation..13
           Swapna Bhat, Tye O. Boynton, Dan Pham, Lawrence J. Shimkets

Chapter 3 **Decontamination Efficacy of Three Commercial-Off-The-Shelf (COTS)**
           Sporicidal Disinfectants on Medium-Sized Panels Contaminated with Surrogate
           Spores of *Bacillus anthracis*...23
           Jason M. Edmonds, Jonathan P. Sabol, Vipin K. Rastogi

Chapter 4 **Purification and Characterization of an Extracellular, Thermo-Alkali-Stable,**
           Metal Tolerant Laccase from *Bacillus tequilensis* SN4..31
           Sonica Sondhi, Prince Sharma, Shilpa Saini, Neena Puri, Naveen Gupta

Chapter 5 **Spatial Segregation and Aggregation of Ectomycorrhizal and Root-Endophytic**
           Fungi in the Seedlings of Two *Quercus* Species..41
           Satoshi Yamamoto, Hirotoshi Sato, Akifumi S. Tanabe, Amane Hidaka,
           Kohmei Kadowaki, Hirokazu Toju

Chapter 6 **Exo-Metabolome of *Pseudovibrio* sp. FO-BEG1 Analyzed by Ultra-High**
           Resolution Mass Spectrometry and the Effect of Phosphate Limitation...............................54
           Stefano Romano, Thorsten Dittmar, Vladimir Bondarev, Ralf J. M. Weber,
           Mark R. Viant, Heide N. Schulz-Vogt

Chapter 7 **Levels of Germination Proteins in *Bacillus subtilis* Dormant, Superdormant,**
           and Germinating Spores...65
           Yan Chen, W. Keith Ray, Richard F. Helm, Stephen B. Melville, David L. Popham

Chapter 8 **FLS2-BAK1 Extracellular Domain Interaction Sites Required for Defense**
           Signaling Activation...75
           Teresa Koller and Andrew F Bent

Chapter 9 **Genome Features of the Endophytic Actinobacterium *Micromonospora lupini***
           Strain Lupac 08: On the Process of Adaptation to an Endophytic Life Style?......................87
           Martha E. Trujillo, Rodrigo Bacigalupe, Petar Pujic, Yasuhiro Igarashi,
           Patricia Benito, Raúl Riesco, Claudine Médigue, Philippe Normand

Chapter 10 **Biological Instability in a Chlorinated Drinking Water Distribution Network**...................106
           Alina Nescerecka, Janis Rubulis, Marius Vital, Talis Juhna, Frederik Hammes

Chapter 11   Colonic Immune Suppression, Barrier Dysfunction, and Dysbiosis by Gastrointestinal *Bacillus anthracis* Infection...117
Yaíma L. Lightfoot, Tao Yang, Bikash Sahay, Mojgan Zadeh, Sam X. Cheng, Gary P. Wang, Jennifer L. Owen, Mansour Mohamadzadeh

Chapter 12   Erythrocytic Mobilization Enhanced by the Granulocyte Colony-Stimulating Factor is Associated with Reduced Anthrax-Lethal-Toxin-Induced Mortality in Mice...129
Hsin-Hou Chang, Ya-Wen Chiang, Ting-Kai Lin, Guan-Ling Lin, You-Yen Lin, Jyh-Hwa Kau, Hsin- Hsien Huang, Hui-Ling Hsu, Jen-Hung Wang, Der-Shan Sun

Chapter 13   Draft Genome Sequence Analysis of a *Pseudomonas putida* W15Oct28 Strain with Antagonistic Activity to Gram-Positive and *Pseudomonas* sp. Pathogens.......................141
Lumeng Ye, Falk Hildebrand, Jozef Dingemans, Steven Ballet, George Laus, Sandra Matthijs, Roeland Berendsen, Pierre Cornelis

Chapter 14   Architecture and Assembly of the *Bacillus subtilis* Spore Coat.................................155
Marco Plomp, Alicia Monroe Carroll, Peter Setlow, Alexander J. Malkin

Chapter 15   Loss of Cln3 Function in the Social Amoeba *Dictyostelium discoideum* Causes Pleiotropic Effects that are Rescued by Human CLN3.......................171
Robert J. Huber, Michael A. Myre, Susan L. Cotman

Chapter 16   The Mucoid Switch in *Pseudomonas aeruginosa* Represses Quorum Sensing Systems and Leads to Complex Changes to Stationary Phase Virulence Factor Regulation.......................186
Ben Ryall, Marta Carrara, James E. A. Zlosnik, Volker Behrends, Xiaoyun Lee, Zhen Wong, Kathryn E. Lougheed, Huw D. Williams

Chapter 17   Magneto-Chemotaxis in Sediment: First Insights.......................197
Xuegang Mao, Ramon Egli, Nikolai Petersen, Marianne Hanzlik, Xiuming Liu

Chapter 18   Isolate-Dependent Growth, Virulence, and Cell Wall Composition in the Human Pathogen *Aspergillus fumigatus*.......................211
Nansalmaa Amarsaikhan, Evan M. O'Dea, Angar Tsoggerel, Henry Owegi, Jordan Gillenwater, Steven P. Templeton

Permissions

List of Contributors

Index

Preface

This book unravels the recent studies in the field of microbial physiology. It also provides interesting topics for research which readers can take up. Microbial physiology refers to the biochemical examination of the microbial cell functions. It also includes an in-depth study of microbial metabolism, microbial growth, microbial cell structure, etc. While understanding the long-term perspectives of these topics, the book makes an effort in highlighting their impact as a modern tool for the growth of the discipline. It aims to shed light on some of the unexplored aspects and the recent researches in this area. Scientists and students actively engaged in the field of microbial physiology will find this book full of crucial and unexplored concepts.

This book is a comprehensive compilation of works of different researchers from varied parts of the world. It includes valuable experiences of the researchers with the sole objective of providing the readers (learners) with a proper knowledge of the concerned field. This book will be beneficial in evoking inspiration and enhancing the knowledge of the interested readers.

In the end, I would like to extend my heartiest thanks to the authors who worked with great determination on their chapters. I also appreciate the publisher's support in the course of the book. I would also like to deeply acknowledge my family who stood by me as a source of inspiration during the project.

Editor

Recombinant TLR5 Agonist CBLB502 Promotes NK Cell-Mediated Anti-CMV Immunity in Mice

Mohammad S. Hossain[1], Sampath Ramachandiran[1], Andrew T. Gewirtz[2], Edmund K. Waller[1]*

1 Department of Hematology and Medical Oncology, Division of Stem Cell and Bone Marrow Transplantation, Winship Cancer Institute, Emory University School of Medicine, Atlanta, Georgia, United States of America, **2** Department of Biology, Georgia State University, Atlanta, Georgia, United States of America

Abstract

Prior work using allogeneic bone marrow transplantation (allo-BMT) models showed that peritransplant administration of flagellin, a toll-like receptor 5 (TLR5) agonist protected murine allo-BMT recipients from CMV infection while limiting graft-vs-host disease (GvHD). However, the mechanism by which flagellin-TLR5 interaction promotes anti-CMV immunity was not defined. Here, we investigated the anti-CMV immunity of NK cells in C57BL/6 (B6) mice treated with a highly purified cGMP grade recombinant flagellin variant CBLB502 (rflagellin) followed by murine CMV (mCMV) infection. A single dose of rflagellin administered to mice between 48 to 72 hours prior to MCMV infection resulted in optimal protection from mCMV lethality. Anti-mCMV immunity in rflagellin-treated mice correlated with a significantly reduced liver viral load and increased numbers of Ly49H+ and Ly49D+ activated cytotoxic NK cells. Additionally, the increased anti-mCMV immunity of NK cells was directly correlated with increased numbers of IFN-γ, granzyme B- and CD107a producing NK cells following mCMV infection. rFlagellin-induced anti-mCMV immunity was TLR5-dependent as rflagellin-treated TLR5 KO mice had ~10-fold increased liver viral load compared with rflagellin-treated WT B6 mice. However, the increased anti-mCMV immunity of NK cells in rflagellin-treated mice is regulated indirectly as mouse NK cells do not express TLR5. Collectively, these data suggest that rflagellin treatment indirectly leads to activation of NK cells, which may be an important adjunct benefit of administering rflagellin in allo-BMT recipients.

Editor: Markus M. Heimesaat, Charité, Campus Benjamin Franklin, Germany

Funding: This work was supported by National Institutes of Health Grant R01 CA-74364-03 (to E.K.W). The funders had no role in study design, data collection and analysis, decision to publish, or preparation of the manuscript.

Competing Interests: The authors have declared that no competing interests exist.

* E-mail: ewaller@emory.edu

Introduction

CMV infection is usually asymptomatic in immune-competent healthy individuals, but may cause severe disease in immune-compromised BMT, HIV-infected AIDS, and elderly patients [1]. Interstitial pneumonitis is the most serious manifestation of CMV disease causing 30–48% patient mortality [2]. While numerous anti-viral drugs are available, the occurrence of drug-resistant CMV strains increases treatment-related complications in these patients [3]. Naturally, CMV infection is controlled by both innate and adaptive immunity [4,5]. Tabeta et al showed that innate anti-mCMV immunity is mostly controlled TLR9- and TLR3-dependent signaling during the early phase of infection, and others have shown that flagellin enhances the activation and proliferation of NK cells [6,7]. We have previously shown that prophylactic administration of native flagellin, a TLR5 agonist protein extracted from the flagella of *Salmonella typhimurium*, protected allo-BMT recipients both from GvHD and lethal CMV infection [8]. Flagellin has diverse immune-modulatory activity on both innate and adaptive immunity in mice and humans [4,9,10] [5,11,12]. The highly purified cGMP grade rflagellin variant CBLB502 is exceptionally stable, less toxic and less immunogenic than native flagellin [13,14]. Administration of rflagellin reduced radiation-induced toxicity in mice and non-human primates [13,14], but the role of flagellin-TLR5 interactions in the anti-MCMV immunity of NK cells has not been

described. The present study was undertaken to elucidate the mechanism by which rflagellin-TLR5 regulates NK cells immunity in mice infected with a lethal inoculum of mCMV.

NK cells are a major component of innate immunity, and are critical to the early immune response to mCMV infection [15–17]. The mCMV infection leads to activation of NK cells, and activated NK cells directly kill CMV-infected target cells, reducing viral replication [18]. In B6 mice, NK cells control mCMV infection through a number of activating receptors, including NKG2D, NKp46, NK1.1 (NKR-P1C), Ly49D and Ly49H. Down-stream activation via these receptors is initiated by the cytoplasmic immunoreceptor tyrosine-based activation motif (ITAM) associating with the DAP12 adaptor protein complex [19–23]. 2B4 (CD244, a non-MHC binding receptor), another surface marker of NK cells, also induces both activation and inhibitory responses depending on the phosphorylation of the cytoplasmic tyrosine motifs. The activated isoform of 2B4 induces NK cell activation via coupling with the NKG2D-DAP10 complex [19]. The killer cell lectin-like receptor G1 (KLRG1) is known to be an inhibitory surface marker for NK cells, but KLRG1 expression is also required for maturation, activation and homeostatic proliferation of NK cells [24].

In this study, we investigated how rflagellin binding to TLR5 contributes to NK cell activation and the anti-mCMV immunity of NK cells in B6 mice. We observed that a single dose of rflagellin administered 48 hours prior to mCMV infection protected mice from a lethal dose of mCMV. rFlagellin treatment led to significantly reduced viral load in the liver along with significantly increased numbers of mature, activated cytotoxic Ly49H- and Ly49D-expressing NK cells compared with the PBS-treated control mice. We found that tonic flagellin signaling through TLR5 is necessary for optimal activation of NK cells immune response to mCMV infection. Together, these data provide new mechanistic insights about the effects of rflagellin-TLR5 binding on NK cell activation against mCMV infection.

Materials and Methods

Mice

C57BL/6 (B6) mice were purchased from Jackson Laboratories (Bar Harbor, ME). TLR5$^{-/-}$ knock out (KO) mice with B6 background were bred at the Emory University animal facility. All experimental procedures conformed to *the Guide for the Care and Use of Laboratory Animals*, and were approved by the Emory University Institutional Animal Care and Use Committee (IACUC Protocol # 2001896).

Production of rflagellin and administration to mice

rFlagellin is a pharmacologically improved recombinant derivative of *Salmonella* flagellin, in which the central variable segments (domains D2 and D3) have been deleted and the structural elements required for TLR5 signaling (domains D0 and D1) are retained. The highly purified cGMP grade rflagellin variant CBLB502 is produced by Cleveland Biolabs, NY as previously described [13,25]. Briefly, the rflagellin cDNA (from *Salmonella dublin*) is overexpressed in *E. coli* and a fusion protein of flagellin with an N-terminal His$_6$-tag is purified to homogeneity by a combination of Ni-NTA chromatography and FPLC-based gel-filtration. The final product (>95% pure by SDS-PAGE) is purified from residual LPS by passing though detoxigel (Pierce, Rockford, IL). This purification process allowed us to obtain > 100 mg of pure rflagellin from 6L of bacterial culture. We obtained rflagellin from Cleveland Biolabs through a collaborative agreement between Emory University and Cleveland Biolabs. The aliquots of rflagellin were stored at −80°C and reconstituted in ice-cold 0.1% Tween-80 in PBS (PBS). A single dose of 25 µg/ 0.2 ml PBS was injected in mice i.p 48 hours before mCMV infection or otherwise stated in the experiments.

MCMV infection

rFlagellin-treated B6 or TLR5 KO mice were infected with non-lethal (1×10^5 PFU/mouse i.p) or lethal [$1 \times$LD50 (i.e., 0.5×10^6 PFU/mouse i.p) or more] doses of salivary-gland-passed Smith strain mCMV (a gift from Dr. H. Yushida, Saga University, Japan).

Liver viral load determination

Livers were aseptically harvested on days 3 and 10 post mCMV infection. The mCMV pfu per liver was determined as previously described [26]. Briefly, collected liver was homogenized and centrifuged, and serially diluted supernatants were added to confluent monolayers of 3T3 cells in 24-well tissue culture plates. After incubation for 90 minutes at 37°C, ~1 mL 2.5% methyl-cellulose in DMEM (10% FBS) was added to each well of treated 3T3 monolayers and incubated for an additional 4 days at 37°C. mCMV pfus were directly counted under a light microscope (Nikon, Melville, NY) after removing the methylcellulose and staining the 3T3 cells with methylene blue.

Isolation and measurement of leucocytes from the spleens of experimental mice

Mice were sacrificed, splenocytes were harvested, single cell suspensions were prepared and total nucleated cells per spleen were counted by using a fluorescent microscope as previously described [8].

In vivo depletion of NK cells

NK cells were depleted by using rabbit antiserum against asialo GM1 (anti-asialo GM1, Wako Chemicals) in B6 mice as previously described [26] with a slight modification. 1 vial of anti-Asialo GM1 was reconstituted in 6 ml PBS. 0.2 ml of reconstituted anti-asialo GM1 was further diluted to 0.5 ml in PBS and injected intraperitoneally in B6 mice on 4, 3 and 1 day prior to mCMV infection (5×10^5 pfu/mouse i.p). The three doses of anti-asialo GM1 selectively depleted blood CD3-NK1.1+ cells by >99% as determined by flowcytometry (Figure S2) before mCMV infection.

Measurement of NK cells cytotoxic activity

NK cell cytotoxic activity was determined by using standard 4 hour ^{51}Cr-release assay as previously described [20]. Briefly, splenocytes were harvested from rflagellin- and PBS-treated control mice on day 0, 1, 2, 3 and 8 after mCMV infection. NK-sensitive Yac-1 target cells were labeled with 37 MBq Na^{51}CrO$_4$ at 37°C for 90 min and washed three times with RPMI 1640 complete medium. The labeled target cells (1×10^4) were co-cultured with whole splenocytes (effector cells) at various effector: target (100:1, 50:1, and 25:1) ratios in a final volume of 0.2 ml fresh RPMI 1640 complete medium in 96-well U-bottomed tissue culture plates for 4 hours at 37°C. The labeled Yac-1 target cells (1×10^4) co-cultured with either only complete media or 1% Triton X were used for minimum and maximum release, respectively. The amount of ^{51}Cr released in the 0.05 ml supernatant/well was determined by a well-type gamma counter (beta liquid scintillation counter; EG&G Wallac, PerkinElmer, Ontario, Canada). Specific cytotoxicity was calculated as follows: % ^{51}Cr release = 100×(cpm experimental - cpm spontaneous release)/(cpm maximum release - cpm spontaneous release).

Measurement of TLR5 expression in FACS sorted NK cells and TLR5-transfected cell lines

To determine the TLR5 expression by NK cells, CD3-NK1.1+ NK cells were sorted by FACS from the spleens of immunologically-naïve B6 mice. Total RNAs were harvested from sorted NK cells (>4.0×10^6 sorted NK cells, >95% purity), HEK-Blue-mouse TLR5 transfected 293 cells and HEK-Blue-Null2-k 293 Cells (InvivoGen, CA) using RNeasy kit (Qiagen). Single strand cDNA was synthesized using 1 µg of total RNA, and QuantiTect Reverse Transcription Kit (Qiagen) in a total volume of 20 µL. 5 µl of cDNA reaction was used in the PCR reaction with primers specific for TLR5 (forward primers 5 -GGA CAC TGA AGG ATT TGA AGA TG-3 and reverse primers 5 -GGA CCA TCT GTA TGC TTG GAA TA-3) [27] or GAPDH as a control. Samples were amplified by 35 cycles and loaded on to a 1.5% agarose gel and subjected to electrophoresis. Specific bands were detected by staining with ethidium bromide.

Serum cytokines measured by Luminex assay

Serum was harvested on day 0, 2 and 3 after MCMV infection from rflagellin- and PBS-treated B6 mice. Mouse 26-plex and IFN-α/β kits were purchased from Affymetrix Inc (Santa Clara, CA) and the Luminex assay (Luminex Corp., Austin, Texas) was performed in a blinded fashion by the Immunology Core

Laboratory at Stanford University (Stanford, CA) according to the manufacturer's recommendations. All samples were assayed in a single batch, and each sample was measured in duplicate. Plates were read using a Luminex 200 instrument (Luminex Corp) as previously described [28].

Flow cytometry

The NK cells (CD3-NK1.1+) were determined by staining of splenocytes harvested from B6 mice with anti-mouse mAbs to CD3 and NK1.1. The activation status of NK cells was determined by staining the cells with mAbs to ICOS-1, CD69, KLRG1, 2B4, Ly49G2, Ly49C/H, Ly49D, Ly49H, CD122, CD11b, CD27, CD107a, etc. All antibodies were purchased from either BD Pharmingen (San Jose, CA) or eBioscience. The frequencies of granzyme B and IFN-γ producing NK cells were determined by staining for intracellular cytokines after 4 hours of *in vitro* stimulation of whole splenocytes with PMA-calcium ionomycin with Golgi Plug cocktail (BD Pharmingen) as described before [8]. The stained cells were acquired by FACScanto (Becton Dickinson, San Jose, CA) and analyzed by FlowJo software.

Statistical analyses

Student's *t*-test and Log Rank test were used to determine the statistical significance of the acquired data. Differences between groups were considered statistically significant when p value<0.05 was obtained.

Results

Prophylactic rflagellin administration induced strong anti-MCMV immunity

We have previously shown that prophylactic administration of two doses of native flagellin (50 μg/mouse i.p) 3 hours before irradiation and 24 hours after transplant protected allo-BMT recipients from GvHD by inducing transient immunosuppression of donor T cells. Paradoxically, flagellin treatment also protected allo-BMT recipients from lethal mCMV infection [8]. Like native flagellin, rflagellin also protected allo-BMT recipients from GvHD in a similar fashion, with the optimal i.p dose between 25 μg to 50 μg/mouse (our unpublished data). To study the mechanism by which flagellin confers protection from mCMV infection without the immunological complexity created by allo-transplantation, we studied the effects of rflagellin-treatment in non-transplanted WT B6 mice infected with lethal dose of mCMV. First, to confirm whether prophylactic rflagellin administration could enhance innate and adaptive immune responses to mCMV, WT B6 mice were treated with 25 μg rflagellin i.p or PBS 96, 72, 48, 24, 12 or 0 hours prior to infection with a lethal i.p dose (1×10^6 pfu, 2×LD50) of mCMV. All mice that received rflagellin 72 or 48 hours prior to MCMV infection survived (p<0.05 compared with PBS-treated control mice) to 17 days post-infection, a time at which CMV-induced pathogenicity had resolved. In contrast, mice receiving rflagellin at earlier (96 hours) or later times (24 or 12 hours before mCMV infection) had 40%, 80% and 80% survival, respectively (Figure 1A). Interestingly, all mice receiving rflagellin at the same time as mCMV infection (0 hour) died within 5 days (Figure 1A). Control mice treated with PBS 48 hours before mCMV infection had 37.5% survival (Figure 1A).

We used the weight-loss of individual mice 5 days after infection as a measure of mCMV pathogenicity. Weight losses were similar among all rflagellin- and PBS-treated control groups, except the mice that received rflagellin simultaneously with mCMV infection (0 hour) had all died before day 5 post mCMV infection and the data were not available (Figure 1B). Next, we determined the

relative effectiveness of prophylactic rflagellin administration by infecting rflagellin-treated mice 48 hours later with a range of mCMV doses. Both rflagellin- and PBS-treated control mice receiving a very high dose, 2.5×10^6 pfu/mouse (i.e., 5×LD50) died within 7 days after mCMV infection (Figure 1C). The rflagellin-treated mice receiving 0.5×10^6 pfu/mouse (i.e., 1×LD50) or 1×10^6 pfu/mouse (i.e., 2×LD50) had 100% and 90% survival (p<0.05 compared with the corresponding PBS-treated groups), respectively (Figure 1C). In contrast, only 40% and 20% of PBS-treated mice survived after receiving 0.5×10^6 pfu/mouse or 1×10^6 pfu/mouse of mCMV, respectively (Figure 1C). Second, we determined the effect of 25 μg rflagellin/ mouse i.p administered 24 or 48 hours *after* a lethal dose of mCMV (0.5×10^6 pfu/mouse i.p) in WT B6 mice. Similar to PBS-treated control mice, WT B6 mice receiving rflagellin 24 or 48 hours after mCMV infection had less than 40% survival (Figure 1D). These data suggest that a single intraperitoneal dose of rflagellin administered 48 hours before mCMV infection yields the maximal effect on anti-mCMV immunity. Additionally, administration of rflagellin 25 μg/mouse i.p did not cause any noticeable toxicity as determined by weight loss within 48 hours (Figure S1A) and prevented weight loss typically seen following low dose mCMV infection (1×10^5 pfu/mouse i.p) (Figure S1B). In contrast, PBS-treated control mice had significant weight lost by day 3 after mCMV infection compared with the weight prior to infection (Figure S1C).

Since the liver is one of the primary target organs for mCMV infection in mice [29] and anti-mCMV immunity is inversely correlated with viral load, we next determined the viral load in the liver of rflagellin- and PBS-treated mice on days 3 and 10 following a non-lethal (i.e., 1×10^5 pfu/mouse i.p) dose of mCMV infection. Mice treated with rflagellin 48 hours before mCMV infection had significantly reduced viral load (**p<0.005) in the liver on day 3 and had faster liver viral clearance (not detectable, ND) on day 10 after mCMV infection compared with the PBS-treated control mice (Figure 2A).

Since flagellin is the only known ligand for TLR5 and rflagellin avidly binds TLR5 [13], we next confirmed the requirement for rflagellin-TLR5 immune interaction in anti-mCMV immunity by using TLR5 KO mice. TLR5 KO B6 mice had increased susceptibility to mCMV infection compared with the WT B6 mice (Figure S2A and S2B) with a LD50 of mCMV ~2-fold less than in WT mice (Figure S2C). TLR5 KO mice were treated with rflagellin or PBS 48 hours before mCMV infection (i.e., 1×10^5 pfu/mouse i.p) and viral load was determined on day 3 and 10 after mCMV infection. Both rflagellin- and PBS-treated TLR5 KO mice had similar liver virus titers on day 3 after mCMV infection (Figure 2B), but viral titers were ~10-fold higher in TLR5 KO mice compared with PBS-treated WT mice (Figure 2A). These data suggest that endogenous signaling through TLR5 is important in protecting WT mice from mCMV infection. Additionally, mice treated with rflagellin prior to mCMV infection had less mCMV-induced pathogenicity (weight loss) compared with PBS-treated WT mice (Figure 2C). WT mice that received rflagellin at the same time as a sub lethal mCMV infection (i.e., 1×10^5 pfu/mouse i.p) had more weight loss, appeared sick, and had to be sacrificed by day 3 post mCMV infection. Mice treated simultaneously with rflagellin and mCMV had higher liver viral loads on day 3 after mCMV infection compared with PBS-treated control mice (Figure 2D). Taken together, these data suggest that rflagellin initiates immune responses that require 1–2 days to become fully active in protecting mice from mCMV infection.

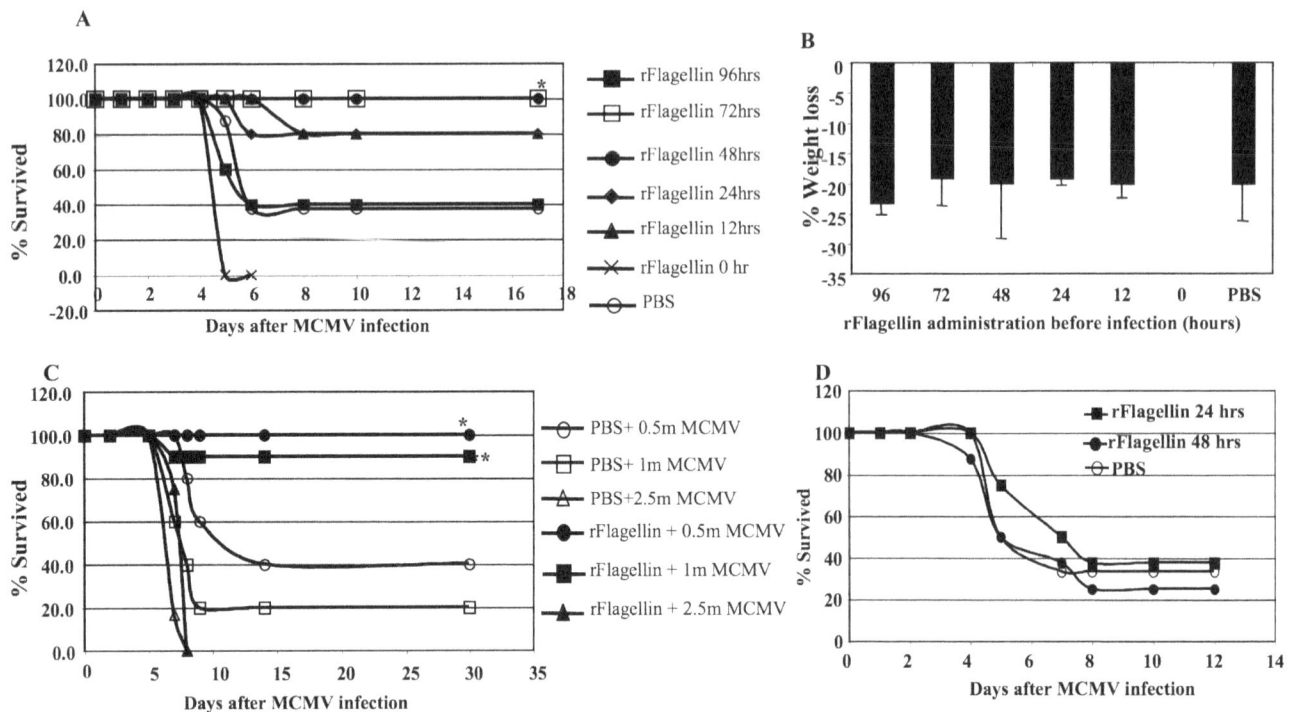

Figure 1. Prophylactic rflagellin administration protected mice from lethal mCMV infection. A. A total of 5 groups WT B6 mice were treated with 25 µg rflagellin/mouse i.p 96, 72, 48, 24 or 0 hours before a lethal dose ($2 \times LD50$, 1×10^6 pfu/mouse i.p) of mCMV infection. Control mice were treated with PBS only 48 hours before the same lethal dose of mCMV infection. Infected mice were monitored every day to record for mortality. The % survival recorded until day 17 after mCMV infection is shown. The symbol "*" indicates the p value<0.05, Log Rank Test of groups rflagellin 48 hrs or 72 hrs vs PBS-treated group. B. The % weight loss measured on day 5 after mCMV infection. All mice receiving rflagellin at 0 hours after mCMV infection died before day 5 after mCMV infection and the % weight loss data of this group was not available. 5–10 mice were used per group. C. WT B6 mice were treated with a single dose of rflagellin 25 µg/mouse or 0.2 ml PBS i.p. 48 hours later mice were infected either with 0.5×10^6, 1×10^6 or 2.5×10^6 mCMV pfu/mouse i.p. The % survival on day 28 after mCMV infection is presented. 10 mice were used per group. The % survival recorded until day 17 after mCMV infection is shown. The symbol "*" indicates the p value<0.05, Log Rank Test of while compared with the survival data of rflagellin-treated mice vs corresponding mCMV infection dose in PBS-treated mice. 5–10 mice were used per group. D. WT B6 mice were infected with 0.5×10^6 mCMV pfu/mouse i.p. A single dose of rflagellin (25 µg/mouse) was injected i.p 24 or 48 hours after mCMV infection. Control mice were injected with 0.2 ml PBS i.p 24 hours after mCMV infection. Survival of mice after mCMV infection was monitored each day and % survived mice until 12 days after infection is presented. 6–8 mice were used per group.

Anti-mCMV immunity in rflagellin-treated mice mostly mediated by NK cells

NK cells are the major component of innate immunity, and they play a key role in controlling mCMV infection [26]. To confirm the anti-mCMV immunity in rflagellin-treated mice is mediated by NK cell, we next depleted NK cells *in vivo* by administering anti-asialo GM1 before and after rflagellin treatment and infecting mice with a lethal dose (5×10^5 pfu i.p) mCMV 48 hours after rflagellin treatment (Figure 3A). Anti-asialo-GM1 treated mice had >99% NK depletion (Figure S3). Interestingly, all mice treated with anti-asialo GM1 alone or with rflagellin died within 8 days following mCMV infection while rflagellin- and PBS-treated control mice had 100% and 80% survival, respectively (p<0.001 comparing PBS-treated group to treated groups) (Figure 3B). These data indicate that anti-mCMV immunity in rflgellin-treated mice is dependent upon the presence of NK cells.

Prophylactic rflagellin administration enhanced anti-mCMV immunity by increasing the numbers of activated cytotoxic NK cells

The peak number of activated cytotoxic NK cells is typically seen in the spleen 2–3 days after infection [26]. To explore the role

of NK cells against mCMV infection in rflagellin-treated mice, we next analyzed the anti-mCMV immunity of NK cells in the spleen of rflagellin-treated WT mice on day 0 and 3 after mCMV infection (2 and 5 days after rflagellin administration). To confirm the role of TLR5-signaling in anti-mCMV immunity of NK cells after rflagellin administration, we studied anti-mCMV immunity of NK cells in rflagellin-treated and mCMV-infected TLR5 KO mice. Total numbers of splenocytes were significantly increased 2 days after rflagellin-treatment (day 0 after mCMV infection) in WT mice compared with PBS-treated control mice, with the greatest effect seen 3 days following mCMV infection (Figure 4A). Surprisingly, rflagellin-treated TLR5 KO mice had significantly increased numbers of splenocytes 2 days (day 0 mCMV infection) after rflagellin-treatment compared to PBS-treated TLR5 KO mice, but no differences in splenocyte numbers were detected 3 days after mCMV infection (Figure 4B). The numbers of splenic CD3-NK1.1+ NK cells, KLRG1+, ICOS-1+ and CD69+ activated NK cells were significantly higher on both days 0 and 3 after mCMV infection in rflagellin-treated WT B6 mice compared with PBS-treated control mice (Figure 4C, 4E, 4G and 4I). While rflagellin treatment resulted in an increase in the numbers of total splenic NK cells and KLRG1+ NK cells 48 hours later in TLR5 KO mice, there was no significant effect on

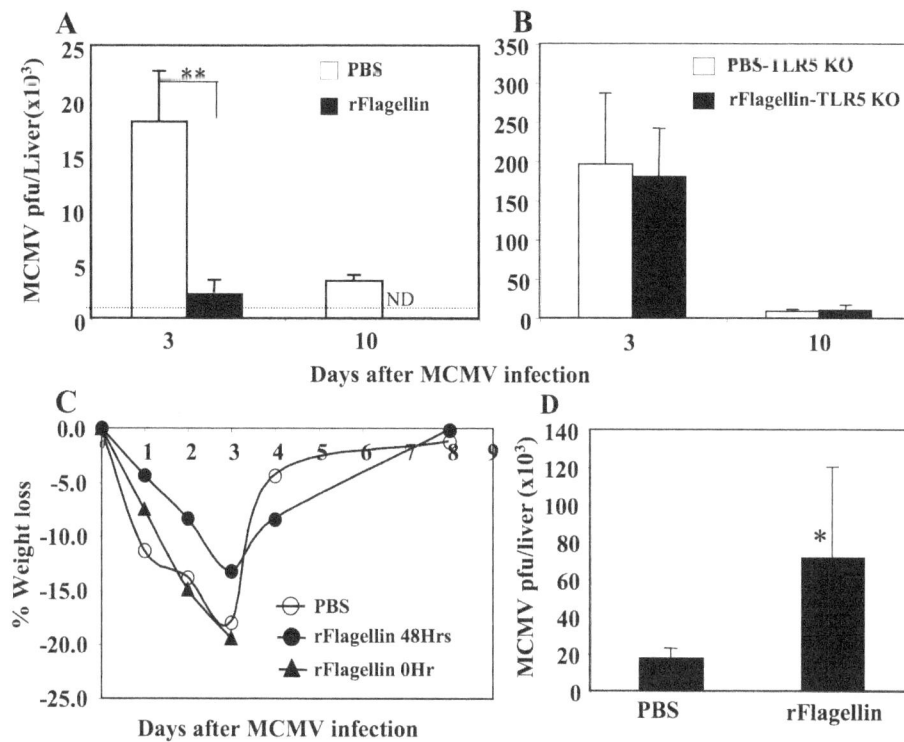

Figure 2. Prophylactic rflagellin administration reduced liver mCMV load in WT B6 mice. WT B6 and TLR5 KO B6 mice were given 25 µg rflagellin or 0.2 ml PBS i.p 48 hours before a sub-lethal dose (1×10^5 pfu/mouse i.p) of mCMV. Mice were sacrificed on day 3 and 10 after mCMV infection and viral load per liver was determined as described in Materials and Methods. A. Virus titer in livers of rflagellin- and PBS-treated WT B6 mice. B. Virus titer in livers of rflagellin- and PBS-treated TLR5 KO B6 mice. The data are the representative of three similar experiments using 5 mice per group at each time point. C. WT B6 mice were given 25 µg rflagellin/mouse (rFlagellin 48 Hrs, closed circle) or 0.2 ml PBS (open circle) i.p 48 hours before or at the same time as (rFlagellin 0 Hr, closed triangle) a sub-lethal mCMV infection (1×10^5 pfu/mouse i.p). Weights of individual mice were measured on day 0, 1, 2, 3, 4 and 8 days after mCMV infection. Percent weight changes per group of experimental mice are presented. D. Mice receiving rflagellin (rFlagellin 0 Hr) during mCMV infection became dehydrated and hunched and were sacrificed on day 3 after mCMV infection and liver viral load was determined. The data are the representative of two similar experiments using 5 mice per group. The symbols "*" and "**"represent the p values <0.05 and <0.005, respectively, Students t-Test.

numbers of NK cell subsets or ICOS-1+ or CD69+ NK cells by day 3 after MCMV infection compared with the PBS-treated TLR5 KO control mice (Figures 4D, 4F, 4H and 4J). However, administration of 25 µg highly purified native flagellin 2 days prior to mCMV infection in TLR5 KO mice did not have any effect on NK cells in spleen in contrast to significantly increased numbers of splenic NK cells and KLRG1+ NK cells in WT mice (Figure S4). We next determined the expression of other activation and/or inhibitory markers on NK cells harvested from the spleen of rflagellin-treated and mCMV-infected WT B6 mice [22,30–32]. The numbers of CD11b+, CD122+, 2B4+, Ly49G2+, Ly49C/H+ and Ly49D+ NK cells were significantly increased in the spleen 2 days after rflagellin treatment with these differences persisting (with lower absolute numbers of NK cells) 3 days later following mCMV infection compared with the PBS-treated control mice (Figure 4K to 4P). To determine whether rflagellin activated NK cells in mice through direct or indirect pathways, we next examined the TLR5 expression on NK cells by RT-PCR. FACS-sorted CD3-NK1.1+ NK cells harvested from naïve B6 mice did not express TLR5 (Figure 4Q lane 3), validated by using the TLR5 transfected cell lines (Figure 4S lane 1) and TLR5 negative cell lines (Figure 4S lane 2). These data suggest that rflagellin-TLR5 interactions indirectly activate NK cells.

Quantitative anti-mCMV activity of NK cells against mCMV infection is generally determined by measuring the cytolytic activity of NK cells *ex vivo* using ^{51}Cr-pulsed Yac-1 target cells [26,33,34]. We therefore determined the cytolytic activity of NK cells against ^{51}Cr-pulsed Yac-1 target cells in splenocytes harvested from the rflagellin- and PBS-treated WT or TLR5 KO mice on days 0 and 3 after mCMV infection. As expected, NK-cell cytolytic activity was significantly increased in the spleen of rflagellin-treated WT mice 48 hours after rflagellin treatment and also on day 3 after mCMV infection compared with the PBS-treated control mice (Figure 5A and 5B). There was no difference in NK cell lytic activity in rflagellin-treated TLR5 KO mice compared with PBS-treated TLR5 KO mice (Figure 5C and 5D).

To explore the time to peak NK lytic activity in rflagellin-treated WT mice, we next determined the kinetics of NK cell lytic activity in the spleens of both rflagellin- and PBS-treated WT mice on days 0, 1, 2, 3 and 8 after mCMV infection. As expected, significantly higher levels of NK lytic activity were detected 48 hours after rflagellin-treatment, and 1 and 3 days after mCMV infection in rflagellin-treated mice compared with the PBS-treated mice, while no cytolytic activity was detected on day 8 after mCMV infection in either group of mice (Figure 5E). Collectively, these data suggest that rflagellin transiently enhances the activation and cytolytic activity of NK cells, and that pre-treatment with rflagellin 2 days prior to mCMV infection results in optimal anti-mCMV activity of NK cells.

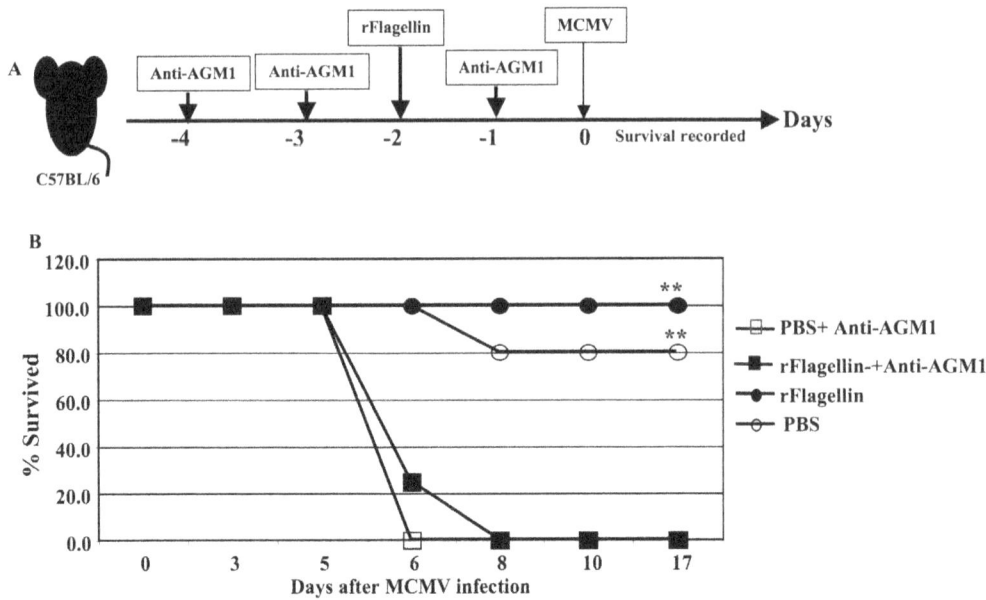

Figure 3. NK cells are required to induce enhanced early anti-mCMV immunity in rflagellin-treated mice. A). Experimental design of anti-asialo GM1 administration in WT B6 mice to deplete NK cells *in vivo*. Reconstituted anti-asialo GM1 in PBS and 0.5 ml was injected i.p to B6 mice on −4, −3 and −1 days of mCMV infection as described in Materials and Methods. Control WT B6 mice were injected with 0.5 ml PBS. 25 μg rflagellin was injected per mouse i.p 48 hours before mCMV infection in anti-asialo GM1-treated and or PBS treated WT B6 mice. All groups of treated mice were infected with a lethal dose (5×10^5 pfu/mouse) of mCMV i.p on day 0. B). Survival data were recorded by observing mice every day or mice were euthanized having weight loss >25% following mCMV infection and percent survived mice of each group are presented. This experiments was performed once using 8 to 10 mice per group. The symbol "**"represents the *p* value<0.005, Log Rank Test (Kaplan-Meier estimator).

The increased killing of NK sensitive Yac-1 target cells by splenocytes harvested from the rflagellin-treated WT mice could be due to either increased numbers of activated NK cells or increased cytolytic activity per cell. In absence of mCMV infection rflagellin-treated splenocytes harvested from the WT mice had < 2-folds increase numbers of NK cells per spleen compare with the PBS-treated splenocytes (Figure 4C). But the rflagellin-treated splenocytes showed Yac-1 target cells killing activity (13.6%±3.1% at 100:1 effector/target ratio) whereas PBS-treated splenocytes showed (−2.6%±2.1% at 100:1 effector/target ratio) which is at least >13 times more (even considering PBS-treated splenocytes killing effect 0–1% at 100:1 effector/target ratio) (Figure 5A). These data indicated that rflagellin treatment enhanced the lytic activity of NK cells in addition to increasing the total numbers of NK cells in the spleen. A number of previously published studies have shown that cytotoxic activity of NK cells is directly related to the degranulation of lysosomal-associated membrane protein-1 (LAMP-1 or CD107a) by NK cells [18,35,36]. To further confirm the increased lytic activity of NK cells in rflagellin-treated mice, we next measured the degranulation of CD107a in NK cells harvested from the spleens of rflagellin- and PBS-treated mice 2 days after rflagellin treatment (day 0 after mCMV infection) and 3 days after mCMV infection. NK cells from rflagellin-treated mice expressed increased levels of surface CD107a 2 days following rflagellin treatment (Figure 6A). In contrast, by day 3 after mCMV infection a larger proportion of NK cells had degranulated, and differences were not seen in comparing NK cells from PBS- to rflagellin-treated mice (Figure 6A). However, significantly higher numbers of CD107a+ NK cells per spleen were determined on both day 0 and 3 after mCMV infection in rflagellin-treated mice compared with the PBS-treated mice (Figure 6B). These data suggest the direct

evidence of increased cytolytic activity of NK cells following rflagellin treatment.

rFlagellin enhanced NK cytolytic activity through increasing the numbers of mature Ly49H+ NK cells

A 4-stage model of NK cell maturation pathways associated with increased effector function has been described based on the expression of CD11b and CD27. The suggested sequence of maturation stages of NK cells is: CD11b$^-$CD27$^-$ (double negative, DN)→CD11b$^-$CD27$^+$→CD11b$^+$CD27$^+$ (double positive, DP)→CD11b$^+$CD27$^-$ [30]. Additionally, mCMV infection increases the expression of Ly49H on activated cytotoxic effector NK cells and specifically enhances killing of mCMV-infected target cells *in vivo* [35]. We next investigated the effect of rflagellin treatment on the numbers of Ly49H-expressing NK cells and the frequencies of the 4 maturation stages of Ly49H-expressing NK cell subpopulations on day 3 after mCMV infection. Although the percentage of CD3-NK1.1+ NK cells was higher (but statistically insignificant) in rflagellin-treated mice compared with PBS-treated mice (Figure 7A), the percentages of Ly49H+ NK cells increased significantly on day 3 after MCMV infection in rflagellin-treated mice compared with the PBS-treated control mice (Figure 7B). rFlagellin treatment increased the frequencies of CD11b$^-$CD27$^+$, DP and CD11b$^+$CD27$^-$ NK cell subpopulations on day 3 after mCMV infection compared with the PBS-treated mice (Figure 7C). The percentages of cells expressing Ly49H increased significantly in the DN and CD11b-CD27+ subsets of NK cells, but not the more mature DP and CD11b+ CD27− subsets, in rflagellin-treated mice on day 3 after mCMV infection compared with the PBS-treated mice (Figure 7D). However, the absolute numbers of Ly49H+NK cells and all 4 maturation subsets of Ly49H+NK cells per spleen were significantly increased in rflagellin-treated mice compared with PBS-treated mice on day 3 following mCMV

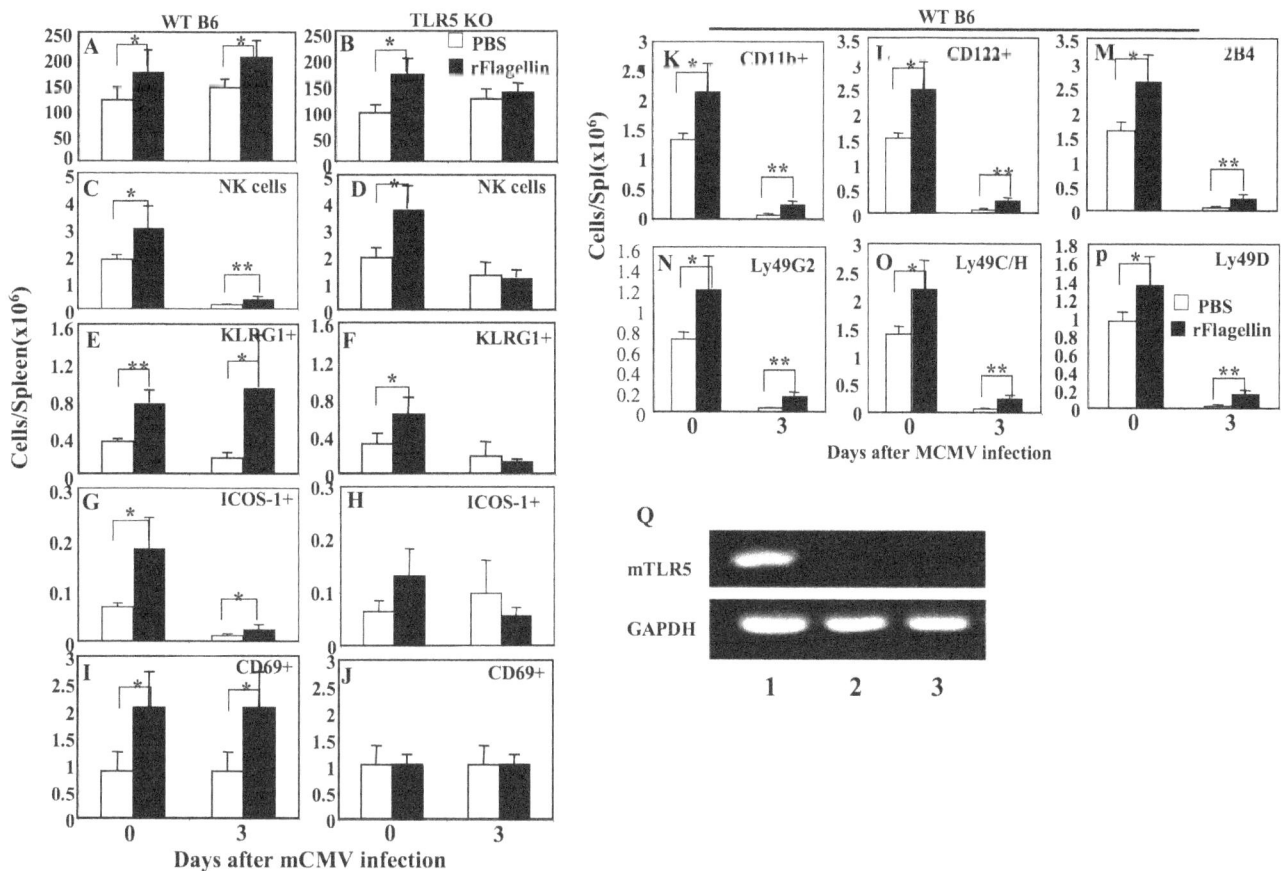

Figure 4. rFlagellin treatment increased NK cell activation in the absence and presence of mCMV infection. Splenocytes were harvested from rflagellin- and PBS-treated WT B6 and TLR5 KO B6 mice on day 0 and 3 after mCMV infection (1×10^5 pfu/mouse i.p). A and B. Nucleated cells per spleen were determined from WT B6 and TLR5 KO B6 mice. C–J. Numbers of: CD3-NK1.1+ NK cells (C and D); KLRG2+ NK cells (E and F); ICOS-1+ NK cells (G and H); and CD69+ NK cells (I and J) per spleen were measured from WT B6 and TLR5 KO B6 mice. K–P. Numbers of: CD11b+ NK cells (K); CD122+ NK cells (L); 2B4+ NK cells (M); Ly49G2+ NK cells (N); Ly49C/H+ NK cells (O); and Ly49D+ NK cells (P) per spleen were determined from rflagellin- and PBS-treated WT B6 mice on day 0 and 3 after mCMV infection. The "*" and "**" represent p values<0.05 and <0.005, respectively, Students t-test. Q. mRNAs were harvested from the TLR5-transfected and TLR5-ve Null cells and FACS-sorted splenic CD3-NK1.1+ NK cells from naïve WT B6 mice as described in Materials and Methods. The cDNA bands specific for TLR5 and GAPDH were measured by RT-PCR and were visualized by ethidium bromide staining. Lane 1 = TLR5 expressing cells, Lane 2 = FACS sorted NK cells, and Lane 3 = TLR5-ve Null cells.

infection (Figure 7E). These data suggest that rflagellin treatment enhanced NK cell maturation, and upregulated Ly49H expression on all NK cell subsets following mCMV infection.

rFlagellin enhanced IFN-γ and granzyme B producing NK cells

We previously showed that highly purified native flagellin reduced GvHD in allo-BMT recipients through reduced production of IFN-γ, TNF-α and IL-6 during the first 10 days post-transplant [8]. Since the anti-MCMV immunity of NK cells is mostly controlled by a set of cytokines/chemokines induced by mCMV infection [33,34,37,38], we first measured the numbers of IFN-γ and granzyme B-producing splenic NK cells in rflagellin-treated mice following in vitro culture in media with brefeldin A alone or following stimulation with PMA-ionomycin plus brefeldin A. Compared with the PBS-treated control mice, the numbers of IFN-γ producing NK cells in rflagellin-treated mice were significantly higher on day 2 after mCMV infection without stimulation (Figure 8A) and on both days 1 and 2 after mCMV infection following PMA-ionomycin stimulation (Figure 8B). The numbers of splenic granzyme B+ NK cells were significantly

higher in rflagellin-treated mice on day 0 post mCMV infection without stimulation, (Figure 8C and 8E) and on day 1, 2 and 3 post mCMV infection after PMA-ionomycin stimulation compared with the PBS-treated control mice (Figure 8D and 8F). These data suggest that increased anti-mCMV activity of NK cells in rflagellin-treated mice is mediated by increased number of IFN-γ and granzyme B-producing NK cells.

rFlagellin reduced production of pro-inflammatory cytokines in the absence of CMV infection

Besides NK cells, antigen-presenting cells (APCs), epithelial cells, and endothelial cells produce cytokines/chemokines which directly or indirectly control anti-mCMV immunity of NK cells [37]. We next compared the levels of serum cytokines and chemokines in rflagellin-treated versus PBS-treated mice. Serum harvested from the rflagellin-treated mice had significantly reduced levels of IFN-α (but not IFN-β) 48 hours after treatment compared with the PBS-treated control mice, but similar levels of IFN-α (and IFN- on days 2 and 3 after MCMV infection (Figure 8G)). Moreover, we measured significantly reduced serum levels of IL-1, IL-5, IL-12p40 (not IL-12p70) and IL-10 on

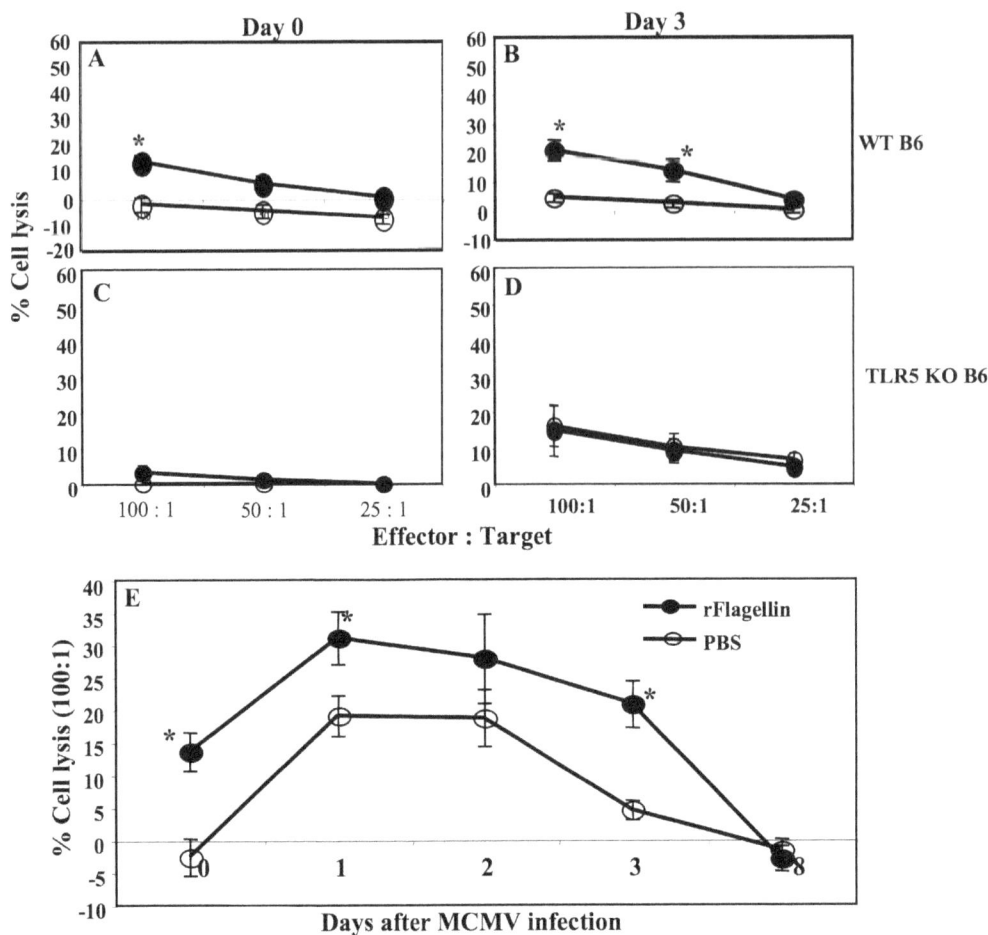

Figure 5. rFlagellin increased NK cell lytic activity in the presence and absence of mCMV infection. Splenocytes were harvested on days 0 and 3 after mCMV infection and NK cell lytic activity was measured using standard 4-hour ^{51}Cr-release assay by Yac-1 target cells as described in Materials and Methods. A and B. The % cell lytic activity of NK cells of rflagellin- and PBS-treated WT B6 mice on days 0 and 3 after infection. C and D. The % cell lytic activity by NK cells harvested from rflagellin- and PBS-treated TLR5 KO B6 mice on day 0 and 3 post infection. E. Kinetics of NK cells lytic activity from the splenocytes harvested from rflagellin and PBS-treated WT B6 mice on day 0, 1, 2, 3 and 8 after mCMV infection. The data shown in A–D are representative of three independent experiments and data shown in E are from one experiment. 5 mice were used per group per time point. The "*" indicates p value<0.05, Student's T-test.

48 hours after rflagellin treatment compared with the PBS-treated control mice (Figure 8H). In contrast, there were no differences in levels of any of the 26 cytokines/chemokines tested on day 3 after mCMV infection between rflagellin- and PBS-treated mice (only representative pro-inflammatory cytokines are shown in Figure 8I). Collectively, these data suggest that rflagellin administration reduces production of pro-inflammatory cytokines and IFN-α without decreasing the activation and maturation of NK cells (Figure 4 and 5A).

Discussion

Using an established mouse model of mCMV infection, we have explored the mechanism by which rflagellin enhances anti-mCMV immunity of NK cells. This study is an important extension of our previously reported work focused on the pre-clinical use of rflagellin to reduce GvHD and opportunistic infections. We have shown that administration of highly purified native flagellin reduced GvHD in murine allo-BMT recipients and protects against mCMV infection [8] and that peritransplant administration of either 25 μg or 50 μg rflagellin resulted in a comparable reduction in the severity of GvHD (unpublished data). In the

current studies, we therefore used a single intraperitoneal dose of 25 μg rflagellin to elucidate the mechanism by which rflagellin enhances anti-mCMV immunity. We observed that prophylactic administration of rflagellin protected WT B6 mice from lethal mCMV infection, and that optimization of anti-mCMV immunity depends on the timing of rflagellin administration. Administration of rflagellin 48 hours before mCMV infection led to significantly decreased viral load that was associated with increased numbers of mature, activated cytotoxic NK cells without a concomitant increase in pro-inflammatory cytokines.

These data indicate that tonic signaling by rflagellin through TLR5 is required for optimal activation of NK cells in response to mCMV infection. The activation of NK cells by rflagellin in TLR5 KO mice was functionally incomplete as increased cytolytic activity was not seen compared with NK cells from PBS-treated TLR5 KO mice (Figure 4C & 4D) and the NK cells in TLR5 KO mice did not upregulate the activation markers CD69 and ICOS-1 following rflagellin treatment and MCMV infection (Figure 4H and 4J). While mCMV-infected TLR5 KO mice had >10-fold higher titers of virus in the liver compared with WT B6 mice on day 3 after mCMV infection, TLR5 KO mice successfully

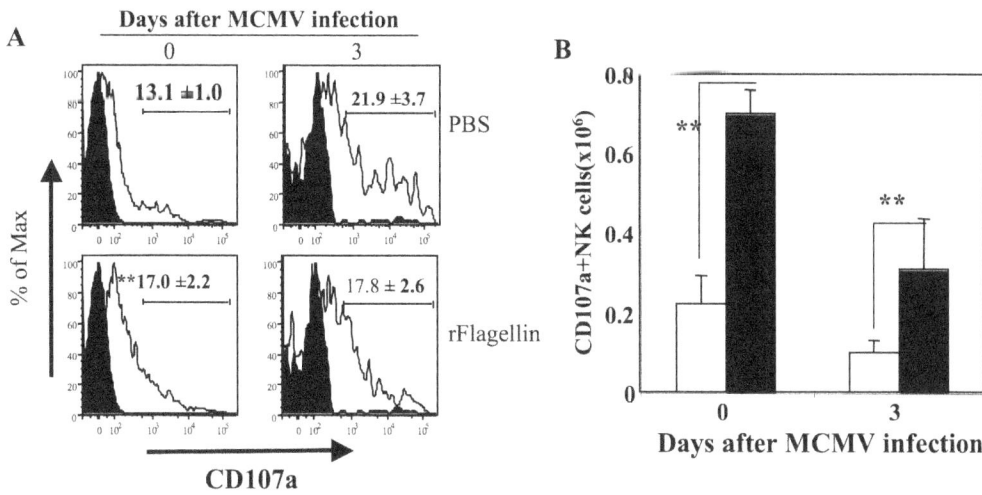

Figure 6. rFlagellin enhanced NK cells cytotoxicity by increasing CD107a degranulation in NK cell. Splenocytes were harvested from rflagellin- and PBS-treated mice on days 0 (48 hours after rflagellin treatment) and 3 after mCMV infection (1×10^5 pfu/mouse i.p). Flowcytometric analysis for CD107a expressed by NK cells was performed by both regular surface stainings as described in Materials and Method. A. The representative FACS data of CD3-NK1.1+ NK cells expressing CD107a in the spleen of rflagellin- and PBS-treated mice determined by regular surface staining. B. The absolute numbers of CD107a+NK cells per spleen were determined from rflagellin- and PBS-treated mice on day 0 (48 hours after rflagellin treatment) and 3 after mCMV infection by using the flowcytometric data. 5 mice were used per group per time point. The "*" and "**" represent the p values<0.05 and <0.005, respectively, Student's T-test.

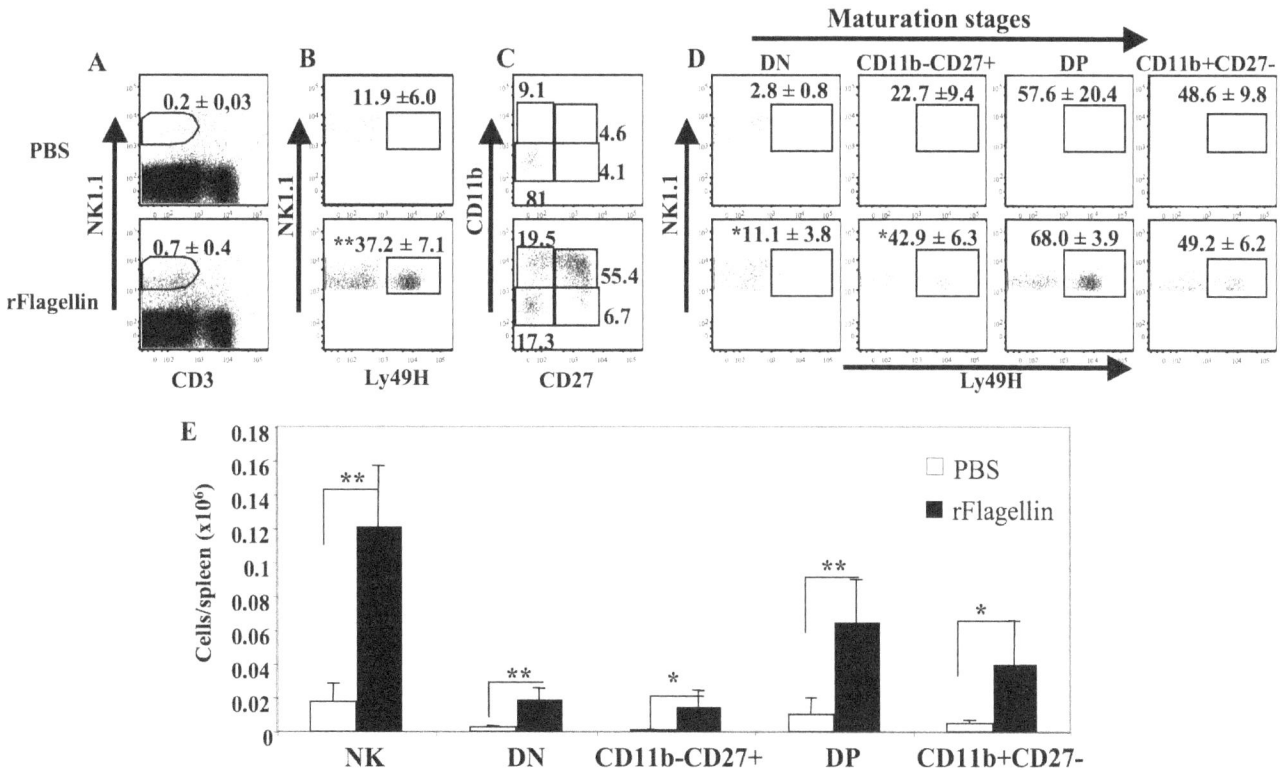

Figure 7. rFlagellin treatment enhanced maturation and increased expression of Ly49H on NK cells after mCMV infection. Harvested splenocytes on day 3 after mCMV infection from rflagellin- and PBS-treated control mice were stained with mAbs to NK1.1, CD27, CD11b along with Ly49H as described in Materials and Methods. A. FACS plots of % CD3-NK1.1+ NK cells of lymphocyte-gated populations. B. % Ly49H expressed by NK cells. C. CD11b−CD27−(DN), CD11b−CD27+, CD11b+CD27+ (DP) and CD11b+CD27− NK cell populations. D. Ly49H+ NK cells of 4 subsets gated populations described in C. E. The total numbers of Ly49H+ NK cells and all 4 subsets of NK cells (as described in D) per spleen expressed Ly49H on day 3 after mCMV infection. The "*" represents p value<0.05, Student's T-test. 5 mice were used per group.

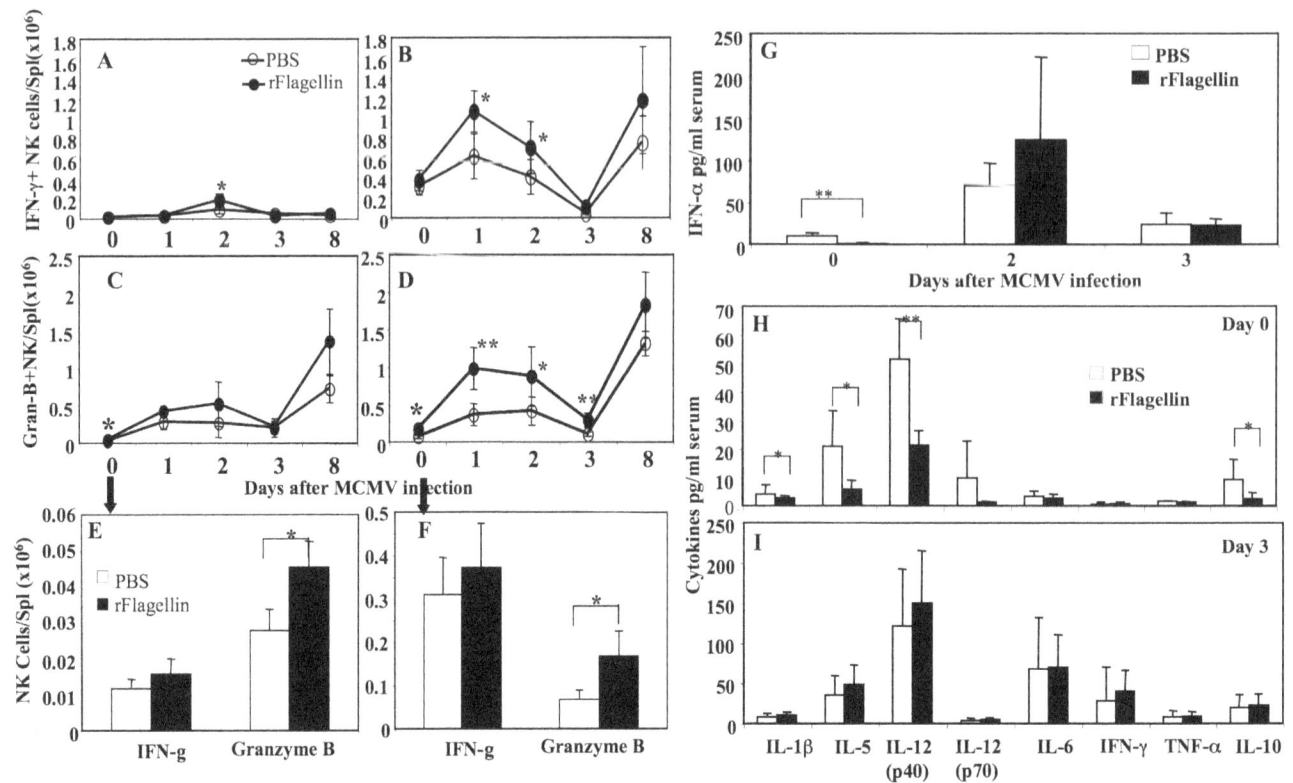

Figure 8. The effect of rflagellin treatment on cytokine production in the presence and absence of mCMV infection. Splenocytes harvested from rflagellin- and PBS-treated control mice on day 0, 1, 2, 3 and 8 after mCMV infection were stimulated with PMA ionomycin for 4 hours at 37°C as described in Materials and Methods. Cells were stained for intracellular expression of IFN-γ and granzyme B along with NK cell surface markers. A and B represent the numbers of IFN-γ producing NK cells per spleen in the absence and presence of PMA-ionomycin stimulation. C and D represent the numbers of granzyme B producing NK cells per spleen in the absence and presence of PMA ionomycin stimulation. E and F represent the numbers of IFN-γ and granzyme B producing NK cells per spleen in the absence and presence of PMA ionomycin stimulation, harvested on day 0 after mCMV infection. G. Serum IFN-α on day 0, 2 and 3 after mCMV infection determined by Luminex assay. H and I represent serum cytokines determined on day 0 and 3 after mCMV infection, determined by Luminex assay. 5 mice were used per group in each time point. The "*" and "**" represent p values<0.05 and <0.005, respectively, Student's T-test.

recovered from mCMV infection (Figure 2), indicating that other non-TLR5-dependent pathways activate and initiate anti-mCMV innate immunity [6].

This study also indicates a role for non-TLR5 dependent signaling in response to rflagellin. We observed increased numbers of NK cells and KLRG1+ NK cells in the spleen of rflagellin-treated TLR5 KO mice 48 hours after treatment (Figure 4B, 4D and 4F). These data suggest that rflagellin may activate NK cells through a TLR5-independent pathway, consistent with prior reports of both TLR5-dependent and independent pathways of flagellin-induced immune responses [39,40]. In addition to direct flagellin-TLR5 interactions, intracellular flagellin also binds with the cytosolic immunosurveillance proteins NLR (nucleotide-binding domain, leucine-rich repeat)-containing apoptosis inhibitory proteins, NAIPs, in a TLR5-independent pathway [39,40].

Ly49H is a CMV-1 encoded NK cell-activating receptor that specifically recognizes the m157 viral protein on the surface of mCMV-infected cells in association with DAP12 adaptor protein complex [19–23], and *in vivo* depletion of Ly49H by mAb in mCMV infection has been reported to increase viral titers in infected organs [35]. 2B4, also known as CD244, is a non-MHC binding receptor that also activates NK cells against mCMV infection by coupling with NKG2D-DAP10 adaptor complex molecules [19].

Our data indicate that up-regulation of the Ly49H and 2B4 surface proteins are the predominant mechanism underlying rflagellin-enhanced NK cell immunity against mCMV infection (Figure 4M and 4O) [19,41]. Additionally, rflagellin treatment enhanced the expression of activation markers on NK cells and also increased the numbers of NK cells in the spleens of WT B6 mice (Figure 4). Cytotoxic activity of NK cells is directly related to the degranulation of lysosomal-associated membrane protein-1 (LAMP-1 or CD107a) by NK cells [18,35,36] and we also measured significantly increased numbers of CD107a degranulation in NK cells in rflagellin- treated mice compared with PBS-treated mice both in absence and presence of mCMV infection (Figure 6). Therefore, the increased killing of NK sensitive Yac-1 target cells by splenocytes from rflagellin-treated mice is due to increased numbers of activated NK cells as well as increased cytotoxic activity of individual NK cells.

A relevant question related to this work is whether the administration of rflagellin could have clinical utility in patients at risk for opportunistic viral infections such as allo-BMT recipients. We have shown that a single dose of rflagellin administered 48 hours before mCMV infection enhanced anti-mCMV immunity, and administration of rflagellin at other times relative to mCMV infection had a reduced positive effect on anti-mCMV immunity (Figure 1A and 1D). Thus, the optimal schedule of rflagellin administration, 2 days before viral infection, might preclude its clinical application as treatment or prophylaxis for mCMV infection. However, administration of rflagellin significantly reduced production of inflammatory cytokines on day 0 of mCMV infection (Figure 8G and H), and peritransplant administration of native flagellin reduced GvHD and also reduced inflammatory cytokines in allo-BMT recipients [8]. Since inflammatory cytokines have been directly correlated with GvHD pathogenesis [42–44], early post-transplant administration of rflagellin in allo-BMT could provide clinical benefit in allo-BMT recipients by enhancing NK cell activity [45] without increasing the risk of GvHD. Alternatively, ex vivo treatment of NK cells with rflagellin might be useful to generate activated NK cells that could be used as adoptive cellular therapy in patients, as human NK cells express TLR5 [27]. Recently published data showed that repeated subcutaneous administration of very low dose rflagellin (e.g, 2 µg/mouse) in allo-BMT recipients enhanced anti-tumor immunity of CD8+ T cells without increasing GvHD toxicity [46]. The use of repeated administration of much lower doses of rflagellin during mCMV infection in non-transplant settings as well as in the peritransplant period in allo-BMT recipients may enhance anti-mCMV immunity, and these experimental approaches are currently under investigation in our lab.

In summary, we have shown that administration of a single dose of rflagellin significantly enhanced innate immunity by increasing the activation status and cytotoxic activity of NK cells against mCMV-infected targets. The rflagellin used in this study is a pharmacologically optimized TLR5 agonist that is less toxic and less immunogenic than native flagellin, and is currently being evaluated in clinical trials as a vaccine adjuvant. The results from this study provide mechanistic insights that may be exploited for clinical benefit using rflagellin to reduce opportunistic infections in immune-compromised patients.

Supporting Information

Figure S1 rFlagellin did not induce noticeable toxicity in mice. WT B6 mice were treated with 25 µg rflagellin/mouse i.p 48 hours before a sub-lethal dose (1×10^5 pfu/mouse i.p) of mCMV infection. Weight and overall physical activity of individual mouse were recorded as parameters of rflagellin toxicity. A). Weight loss of rflagellin-treated mice (n = 20) was determined by measuring weight on 0 and 48 hours after rflagellin treatment. No signs of physical sickness in rflagellin-treated mice after 48 hours of injection. B). Weight loss of rflagellin-treated individual mouse (n-9) was determined by measuring weight on 0 and 3 days after MCMV infection. C). Weight loss of rflagellin-treated mice (n-8) was determined by measuring weight of individual mouse on 0 and 3 days after MCMV infection. The "*" represents p values<0.05, Student's T-test.

Figure S2 TLR5 KO B6 mice are more susceptible to mCMV infection than WT B6 mice. Four groups of WT B6 and TLR5 KO B6 mice were infected with 2.5×10^5 pfu/mouse, 5×10^5 pfu/mouse, 1×10^6 pfu/mouse or 2.5×10^6 pfu/mouse i.p mCMV. Survival of infected mice was monitored by recording and weight every day. Mice having >25% weight loss were euthanized and included in the list of mortality. A. Percent survival of WT B6 mice data. B. Percent survival of TLR5 KO B6 mice data. 5–10 mice were used per group. C. The LD50 of WT B6 mice and TLR5 KO B6 mice against mCMV infection were calculated from the survival data of Figure A and B.

Figure S3 Treatment of anti-asialo GM1 caused >99% *in vivo* NK cell depletion. 0.5 ml of reconstituted anti-asialo GM1 in PBS were injected to B6 mice on −4, −3 and −1 days of mCMV infection as described in Materials and Methods and in Figure 3. Control WT B6 mice were injected with 0.5 ml PBS. 25 µg rflagellin/mouse i.p was injected 48 hours before mCMV infection in anti-asialo GM1-treated and or PBS treated WT B6 mice. Representative two mice from PBS-treated control group, two mice from anti-asialo GM1-treated group and one mouse from anti-asialo GM1 and rflagellin-treated group were bled before mCMV infection. Depletion of NK cells in blood was determined by flowcytometry.

Figure S4 Administration of native flagellin had no effect on NK cells in TLR5 KO mice. WT B6 and TLR5 KO B6 mice were treated with highly purified native flagellin (25 µg/mouse i.p) extracted from the *S. typhimurium*. Control mice were injected with 0.2 ml PBS i.p. 48 hours later both native flagellin- and PBS-treated mice were sacrificed and splenocytes were harvested. The numbers of nucleated cells per spleen were determined by counting the cells under microscope. The numbers of NK cell and KLRG1+ NK cells were determined by FACS. 5 mice were used per group. The "**" represents p values<0.005, Student's T-test.

Acknowledgments

The authors thank Cleveland Biolabs, NY for providing cGMP grade recombinant flagellin variant CBLB502 and also thank Dr. Hiroki Yoshida for providing the pathogenic strain salivary gland passed murine CMV.

Author Contributions

Conceived and designed the experiments: MSH SR ATG EKW. Performed the experiments: MSH SR. Analyzed the data: MSH SR EKW. Contributed reagents/materials/analysis tools: MSH SR ATG EKW. Wrote the paper: MSH SR ATG EKW.

References

1. Paar DP, Pollard RB (1996) Immunotherapy of CMV infections. Advances in experimental medicine and biology 394: 145–151.
2. Nomura F, Shimokata K, Sakai S, Yamauchi T, Kodera Y, et al. (1990) Cytomegalovirus pneumonitis occurring after allogeneic bone marrow transplantation: a study of 106 recipients. Japanese journal of medicine 29: 595–602.
3. Langston AA, Redei I, Caliendo AM, Somani J, Hutcherson D, et al. (2002) Development of drug-resistant herpes simplex virus infection after haploidentical hematopoietic progenitor cell transplantation. Blood 99: 1085–1088.
4. Lenac T, Arapovic J, Traven L, Krmpotic A, Jonjic S (2008) Murine cytomegalovirus regulation of NKG2D ligands. Med Microbiol Immunol 197: 159–166.
5. Babic M, Krmpotic A, Jonjic S (2011) All is fair in virus-host interactions: NK cells and cytomegalovirus. Trends Mol Med 17: 677–685.
6. Tabeta K, Georgel P, Janssen E, Du X, Hoebe K, et al. (2004) Toll-like receptors 9 and 3 as essential components of innate immune defense against

mouse cytomegalovirus infection. Proceedings of the National Academy of Sciences of the United States of America 101: 3516–3521.
7. Tsujimoto H, Uchida T, Efron PA, Scumpia PO, Verma A, et al. (2005) Flagellin enhances NK cell proliferation and activation directly and through dendritic cell-NK cell interactions. Journal of leukocyte biology 78: 888–897.
8. Hossain MS, Jaye DL, Pollack BP, Farris AB, Tselanyane ML, et al. (2011) Flagellin, a TLR5 agonist, reduces graft-versus-host disease in allogeneic hematopoietic stem cell transplantation recipients while enhancing antiviral immunity. Journal of immunology 187: 5130–5140.
9. Zeng H, Wu H, Sloane V, Jones R, Yu Y, et al. (2006) Flagellin/TLR5 responses in epithelia reveal intertwined activation of inflammatory and apoptotic pathways. American journal of physiology Gastrointestinal and liver physiology 290: G96–G108.

10. Salamone GV, Petracca Y, Fuxman Bass JI, Rumbo M, Nahmod KA, et al. (2010) Flagellin delays spontaneous human neutrophil apoptosis. Lab Invest 90: 1049–1059.

11. Zhang Y, Joe G, Hexner E, Zhu J, Emerson SG (2005) Host-reactive CD8+ memory stem cells in graft-versus-host disease. Nat Med 11: 1299–1305.

12. Zhang Y, Joe G, Hexner E, Zhu J, Emerson SG (2005) Alloreactive memory T cells are responsible for the persistence of graft-versus-host disease. J Immunol 174: 3051–3058.

13. Burdelya LG, Krivokrysenko VI, Tallant TC, Strom E, Gleiberman AS, et al. (2008) An agonist of toll-like receptor 5 has radioprotective activity in mouse and primate models. Science 320: 226–230.

14. Eaves-Pyles T, Murthy K, Liaudet L, Virag L, Ross G, et al. (2001) Flagellin, a novel mediator of Salmonella-induced epithelial activation and systemic inflammation: I kappa B alpha degradation, induction of nitric oxide synthase, induction of proinflammatory mediators, and cardiovascular dysfunction. Journal of immunology 166: 1248–1260.

15. Brautigam AR, Dutko FJ, Olding LB, Oldstone MB (1979) Pathogenesis of murine cytomegalovirus infection: the macrophage as a permissive cell for cytomegalovirus infection, replication and latency. The Journal of general virology 44: 349–359.

16. Dalod M, Hamilton T, Salomon R, Salazar-Mather TP, Henry SC, et al. (2003) Dendritic cell responses to early murine cytomegalovirus infection: subset functional specialization and differential regulation by interferon alpha/beta. The Journal of experimental medicine 197: 885–898.

17. Biron CA, Su HC, Orange JS (1996) Function and Regulation of Natural Killer (NK) Cells during Viral Infections: Characterization of Responses in Vivo. Methods 9: 379–393.

18. Vahlne G, Becker S, Brodin P, Johansson MH (2008) IFN-gamma production and degranulation are differentially regulated in response to stimulation in murine natural killer cells. Scand J Immunol 67: 1–11.

19. Lanier LL (2008) Up on the tightrope: natural killer cell activation and inhibition. Nature immunology 9: 495–502.

20. Li JM, Southerland L, Hossain MS, Giver CR, Wang Y, et al. (2011) Absence of vasoactive intestinal peptide expression in hematopoietic cells enhances Th1 polarization and antiviral immunity in mice. Journal of immunology 187: 1057–1065.

21. Wu X, Chen Y, Wei H, Sun R, Tian Z (2012) Development of murine hepatic NK cells during ontogeny: comparison with spleen NK cells. Clinical & developmental immunology 2012: 759765.

22. Pyzik M, Gendron-Pontbriand EM, Vidal SM (2011) The impact of Ly49-NK cell-dependent recognition of MCMV infection on innate and adaptive immune responses. Journal of biomedicine & biotechnology 2011: 641702.

23. Daeron M, Latour S, Malbec O, Espinosa E, Pina P, et al. (1995) The same tyrosine-based inhibition motif, in the intracytoplasmic domain of Fc gamma RIIB, regulates negatively BCR-, TCR-, and FcR-dependent cell activation. Immunity 3: 635–646.

24. Huntington ND, Tabarias H, Fairfax K, Brady J, Hayakawa Y, et al. (2007) NK cell maturation and peripheral homeostasis is associated with KLRG1 up-regulation. Journal of immunology 178: 4764–4770.

25. Yoon SI, Kurnasov O, Natarajan V, Hong M, Gudkov AV, et al. (2012) Structural basis of TLR5-flagellin recognition and signaling. Science 335: 859–864.

26. Hossain MS, Takimoto H, Hamano S, Yoshida H, Ninomiya T, et al. (1999) Protective effects of hochu-ekki-to, a Chinese traditional herbal medicine against murine cytomegalovirus infection. Immunopharmacology 41: 169–181.

27. Lauzon NM, Mian F, MacKenzie R, Ashkar AA (2006) The direct effects of Toll-like receptor ligands on human NK cell cytokine production and cytotoxicity. Cell Immunol 241: 102–112.

28. Li JM, Hossain MS, Southerland L, Waller EK (2013) Pharmacological inhibition of VIP signaling enhances antiviral immunity and improves survival in murine cytomegalovirus-infected allogeneic bone marrow transplant recipients. Blood 121: 2347–2351.

29. Salem ML, Hossain MS (2000) Protective effect of black seed oil from Nigella sativa against murine cytomegalovirus infection. International journal of immunopharmacology 22: 729–740.

30. Chiossone L, Chaix J, Fuseri N, Roth C, Vivier E, et al. (2009) Maturation of mouse NK cells is a 4-stage developmental program. Blood 113: 5488–5496.

31. Fu B, Wang F, Sun R, Ling B, Tian Z, et al. (2011) CD11b and CD27 reflect distinct population and functional specialization in human natural killer cells. Immunology 133: 350–359.

32. Kim S, Iizuka K, Kang HS, Dokun A, French AR, et al. (2002) In vivo developmental stages in murine natural killer cell maturation. Nature immunology 3: 523–528.

33. Orange JS, Biron CA (1996) Characterization of early IL-12, IFN-alphabeta, and TNF effects on antiviral state and NK cell responses during murine cytomegalovirus infection. Journal of immunology 156: 4746–4756.

34. Orange JS, Biron CA (1996) An absolute and restricted requirement for IL-12 in natural killer cell IFN-gamma production and antiviral defense. Studies of natural killer and T cell responses in contrasting viral infections. Journal of immunology 156: 1138–1142.

35. Alter G, Malenfant JM, Altfeld M (2004) CD107a as a functional marker for the identification of natural killer cell activity. J Immunol Methods 294: 15–22.

36. Graubardt N, Fahrner R, Trochsler M, Keogh A, Breu K, et al. (2013) Promotion of liver regeneration by natural killer cells in a murine model is dependent on extracellular adenosine triphosphate phosphohydrolysis. Hepatology 57: 1969–1979.

37. Biron CA (1997) Activation and function of natural killer cell responses during viral infections. Current opinion in immunology 9: 24–34.

38. Biron CA (1998) Role of early cytokines, including alpha and beta interferons (IFN-alpha/beta), in innate and adaptive immune responses to viral infections. Seminars in immunology 10: 383–390.

39. Honko AN, Mizel SB (2005) Effects of flagellin on innate and adaptive immunity. Immunologic research 33: 83–101.

40. Kofoed EM, Vance RE (2012) NAIPs: building an innate immune barrier against bacterial pathogens. NAIPs function as sensors that initiate innate immunity by detection of bacterial proteins in the host cell cytosol. BioEssays : news and reviews in molecular, cellular and developmental biology 34: 589–598.

41. Dokun AO, Kim S, Smith HR, Kang HS, Chu DT, et al. (2001) Specific and nonspecific NK cell activation during virus infection. Nature immunology 2: 951–956.

42. Tanaka J, Imamura M, Kasai M, Sakurada K (1995) Transplantation-related complications predicted by cytokine gene expression in the mixed lymphocyte culture in allogeneic bone marrow transplants. Leukemia & lymphoma 19: 27–32.

43. Ohta M, Tateishi K, Kanai F, Ueha S, Guleng B, et al. (2005) Reduced p38 mitogen-activated protein kinase in donor grafts accelerates acute intestinal graft-versus-host disease in mice. European journal of immunology 35: 2210–2221.

44. Antin JH, Ferrara JL (1992) Cytokine dysregulation and acute graft-versus-host disease. Blood 80: 2964–2968.

45. Hokland M, Jacobsen N, Ellegaard J, Hokland P (1988) Natural killer function following allogeneic bone marrow transplantation. Very early reemergence but strong dependence of cytomegalovirus infection. Transplantation 45: 1080–1084.

46. Ding X, Bian G, Leigh ND, Qiu J, McCarthy PL, et al. (2012) A TLR5 agonist enhances CD8(+) T cell-mediated graft-versus-tumor effect without exacerbating graft-versus-host disease. Journal of immunology 189: 4719–4727.

Fatty Acids from Membrane Lipids Become Incorporated into Lipid Bodies during *Myxococcus xanthus* Differentiation

Swapna Bhat, Tye O. Boynton, Dan Pham, Lawrence J. Shimkets*

Department of Microbiology, University of Georgia, Athens, Georgia, United States of America

Abstract

Myxococcus xanthus responds to amino acid limitation by producing fruiting bodies containing dormant spores. During development, cells produce triacylglycerides in lipid bodies that become consumed during spore maturation. As the cells are starved to induce development, the production of triglycerides represents a counterintuitive metabolic switch. In this paper, lipid bodies were quantified in wild-type strain DK1622 and 33 developmental mutants at the cellular level by measuring the cross sectional area of the cell stained with the lipophilic dye Nile red. We provide five lines of evidence that triacylglycerides are derived from membrane phospholipids as cells shorten in length and then differentiate into myxospores. First, in wild type cells, lipid bodies appear early in development and their size increases concurrent with an 87% decline in membrane surface area. Second, developmental mutants blocked at different stages of shortening and differentiation accumulated lipid bodies proportionate with their cell length with a Pearson's correlation coefficient of 0.76. Third, peripheral rods, developing cells that do not produce lipid bodies, fail to shorten. Fourth, genes for fatty acid synthesis are down-regulated while genes for fatty acid degradation are up regulated. Finally, direct movement of fatty acids from membrane lipids in growing cells to lipid bodies in developing cells was observed by pulse labeling cells with palmitate. Recycling of lipids released by Programmed Cell Death appears not to be necessary for lipid body production as a *fadL* mutant was defective in fatty acid uptake but proficient in lipid body production. The lipid body regulon involves many developmental genes that are not specifically involved in fatty acid synthesis or degradation. MazF RNA interferase and its target, enhancer-binding protein Nla6, appear to negatively regulate cell shortening and TAG accumulation whereas most cell-cell signals activate these processes.

Editor: Dirk-Jan Scheffers, University of Groningen, Groningen Institute for Biomolecular Sciences and Biotechnology, Netherlands

Funding: Research in this publication was supported by the National Science Foundation under award number MCB 0742976 and by the National Institute of General Medical Sciences of the National Institutes of Health under award number 1RO1GM095826. The funders had no role in study design, data collection and analysis, decision to publish, or preparation of the manuscript.

Competing Interests: The authors have declared that no competing interests exist.

* E-mail: shimkets@uga.edu

Introduction

Lipid bodies are carbon storage organelles found in most eukaryotic organisms. They consist of triacylglycerides (TAGs) surrounded by a single phospholipid layer and associated proteins. In eukaryotes, lipid bodies are dynamic organelles that regulate lipid metabolism, membrane trafficking, signaling, and protein degradation. Alterations in TAG metabolism influence the risk of developing diabetes and other metabolic diseases in humans [1]. Therefore, biogenesis and regulation of lipid bodies has become an area of intense research. TAGs are rarely found in prokaryotes, where the most prevalent carbon storage molecules include polyhydroxyalkanoates, trehalose, and glycogen. Nevertheless, lipid bodies occur in several Actinomycetes and a few Proteobacteria species such as *Acinetobacter calcoaceticus* ADP1 [2]. In these prokaryotes, TAG synthesis generally occurs in response to high carbon to nitrogen ratio [3].

Recently, lipid bodies were discovered in *Myxococcus xanthus*, a member of the δ-Proteobacteria that forms fruiting bodies containing spores [4]. *M. xanthus* development is induced by carbon limitation [5,6] suggesting that TAGs are produced from existing cellular material. Lipid bodies first appear 6 h after development is induced by amino acid deprivation, attain their largest size at 18 h, and disappear in the mature spore. It is not clear how *M. xanthus* lipid body synthesis is mediated or regulated. Disruption of genes encoding proteins associated with mature lipid bodies did not compromise lipid body production [4].

DK1622 (wild type) cells observed after nutrient deprivation on a solid surface contain lipid bodies of various sizes that can be visualized by Nile red staining. These lipid bodies comprise a substantial portion of the cell volume. The chemical composition of the lipid body lipids is known in detail and consists primarily of TAGs, some containing ether-linked fatty alcohols instead of ester-linked fatty acids [4]. As development is induced by carbon limitation, where does the carbon for lipid body production originate? Unlike *Bacillus* endospore formation where the spore is formed inside a mother cell, *Myxococcus* sporulation is an encystment in which the long, thin rod-shaped cells shorten then round up to become spherical myxospores [7]. Cylindrical cells about 7 µm in length and 1 µm in diameter produce spherical spores roughly 1.8 µm in diameter. Excluding the thick cortex and spore coat layers, the diameter of the membrane-bound spore interior is about 1 μm [8]. Thus, the membrane surface area declines from 23.6 μm^2 to 3.1 μm^2. On theoretical grounds, membrane phospholipids could serve as the principle carbon source for TAGs within lipid bodies with little biochemical complexity to the conversion. Two alternate possibilities also exist. Fatty acids could be salvaged from cells undergoing Programmed Cell Death (PCD) or they could be synthesized *de novo* using carbon reserves from elsewhere in the cell.

In this work, we show that lipid body production is closely coupled with reduction in cell length during development. When growing cells were pulse labeled with palmitate, the label appeared in the membrane during growth then shifted into lipid bodies during development. The results point to a novel method of producing TAGs from phospholipids.

Results and Discussion

Lipid bodies appear as cell length diminishes in WT cells

Lipid body size was quantified using Nile red stained wild type DK1622 cells [4]. Vegetative cells grown in a rich medium (0 h) have no lipid bodies (Figure 1A). 6 h after starvation initiates development, several small lipid bodies appear close to the membrane. Over the next 12 h, lipid bodies increase in size and number. By 18 h, the peak of lipid body production, the area of the cell stained with Nile red is roughly 20% the cross-sectional area of the cell. Cell shortening begins soon after initiation of development in WT cells and coincides temporally with the appearance of lipid bodies. By 18 h, a few cells have become spherical while the cylindrical cells are about 40% shorter (Figure 1B). Lipid bodies then decline in size and number after 18 h (Figure 1B) before finally disappearing completely in mature spores. The coupling of diminishing cell length with lipid body accumulation suggests the use of internal carbon sources as a reservoir for TAG building blocks. Another possibility is the recycling of lipids released by cells undergoing PCD. A small amount of lysis occurs by 18 h [9].

We attempted to distinguish between these possibilities by examining the phenotype of a *fadL* mutant. FadL, a porin that facilitates the movement of fatty acids across the outer membrane, is essential for fatty acid catabolism in bacteria [10]. In *Escherichia coli*, long chain fatty acid uptake is abolished in mutants lacking *fadL* [11]. The phenotype of *M. xanthus fadL* mutant LS3125 (MXAN7040) is strikingly similar to that of the wild type, proficient in both development and lipid body formation. Thus, it appears as if the majority of carbon used to synthesize TAGs is produced from internal reservoirs generated from the shrinking cell.

Peripheral rods remain long and are devoid of lipid bodies

Within the fruiting body, both cell shortening and lipid body production are observed in 80% of cells destined for PCD, and 10% of cells that sporulate [9]. Peripheral rods comprise the remaining 10% of developing cells and rarely enter the fruiting body [12]. While the function of peripheral rods remains unknown, they appear to express different proteins compared with cells inside the fruiting body [12]. Peripheral rods do not make lipid bodies [4]. If lipid bodies are derived from the existing cell membrane and thus coupled to the shrinking cell, then peripheral rods might be expected to remain long. We determined the lengths of cells lacking lipid bodies. The average length of 30 cells remained the same through three different time points, 18 h (7.0 μm±1.4), 30 h (7.2 μm±1.4), and 48 h (7.3 μm±1.7). Thus, developing cells devoid of lipid bodies remain long throughout

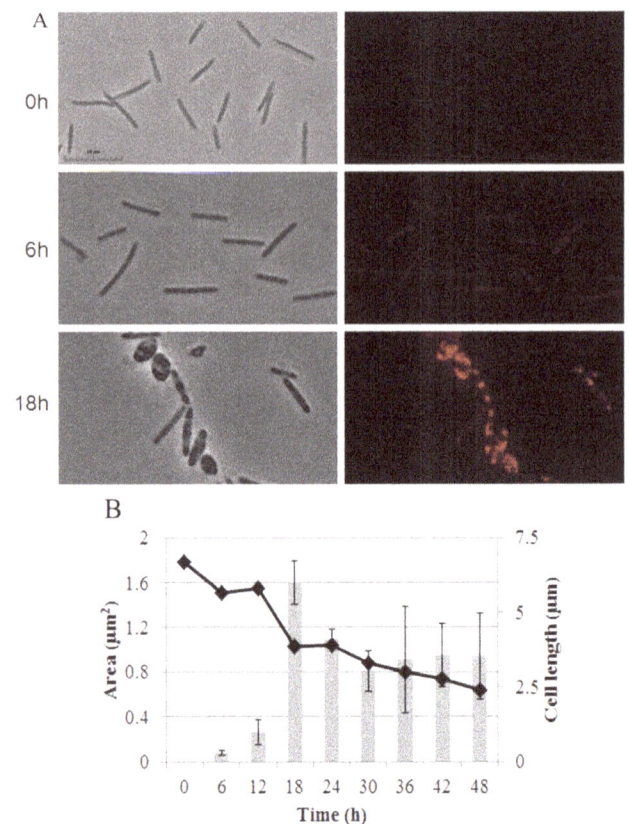

Figure 1. Lipid body production in WT cells during development. (A) DK1622 cells stained with the lipophilic dye Nile red at times indicated during development. Phase (Left), fluorescence (Right). Bar is 10 μm. (B) Lipid bodies were quantified by measuring the average cross sectional area stained with Nile red using at least 30 cells (grey bars). Cell length was measured using phase contrast images of 30 randomly chosen cells (filled diamonds). At 48 h, the cells are a nearly equal mixture of long, peripheral rods and spherical myxospores.

development. This observation is consistent with the idea that cell shortening and lipid body production are coupled processes.

Genes for fatty acid synthesis are down regulated during development

To determine whether the internal carbon reservoir for lipid body production involves existing fatty acids or instead utilizes *de novo* fatty acid synthesis, we examined the expression of fatty acid biosynthetic genes during development. If TAGs are derived from membrane phospholipids, one might expect a decline in transcription of fatty acid biosynthesis genes. Gene expression was quantified using published microarray data from developing wild type cells (Accession number GSE9477) [13]. The *M. xanthus* fatty acid pathways assigned by the Kyoto Encyclopedia of Genes and Genomes (KEGG) Pathways database are shown in figure 2 [14,15]. Proteins proposed to mediate each reaction are shown next to the arrows as both MXAN numbers and, where known, 4-letter protein names.

M. xanthus straight chain fatty acid synthesis begins with an activation sequence that produces malonyl acyl carrier protein (Malonyl ACP) (Figure 2, upper right). All genes required for straight chain primer synthesis are down-regulated with respect to vegetative cells (MXAN4039, 2.1-fold; MXAN2704, 5.7-fold; MXAN5768, 1.5-fold, MXAN0082, 1.3-fold, and MXAN4771,

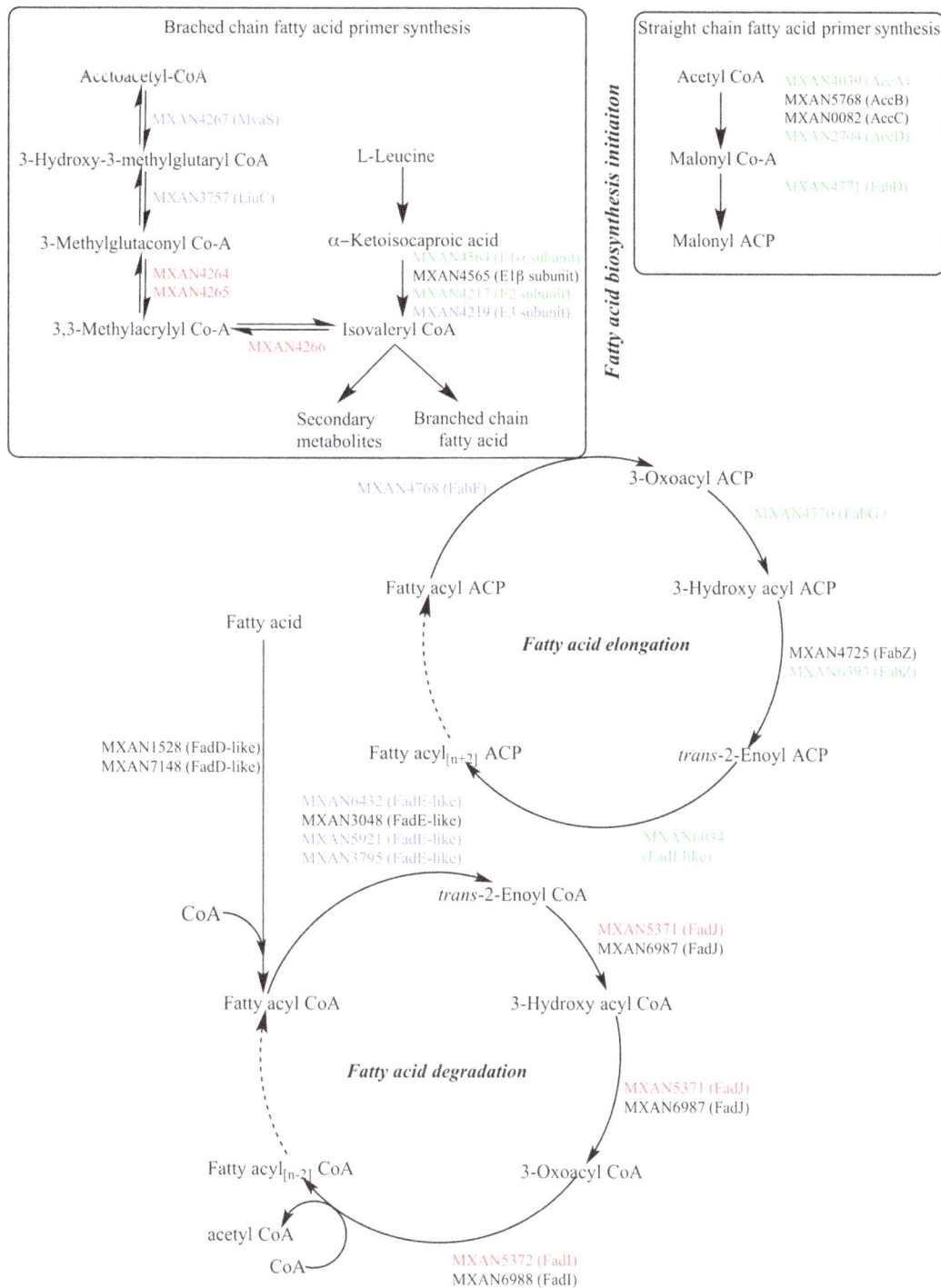

Figure 2. *M. xanthus* fatty acid metabolic pathways were analyzed at the transcriptional level using available microarray data from developing cells [13]. The pathway information and gene annotations were obtained from the Kyoto Encyclopedia of Genes and Genomes (KEGG) pathway database. Genes that are down-regulated (≥1.5 fold), up-regulated (≥1.5 fold) or unchanged during development are shown in green, red, and black, respectively. Blue denotes genes that were not on the microarray. A dashed arrow between two compound names implies that the two names represent the same compound in different stages of polymerization or depolymerization.

1.9-fold). For branch chain fatty acids, the enzyme branched chain keto acid dehydrogenase (BCKAD) makes primers that are then elongated. Here, L-leucine is converted to isovaleryl CoA, the primer for the *iso* odd family of fatty acids including *iso*15:0 the major fatty acid. MXAN4564 encodes the E1α subunit of BCKAD and is down-regulated 12.7-fold while MXAN4565

(E1β subunit) and MXAN4217 (E2 subunit) are down-regulated 1.4-fold and 1.5-fold, respectively (Figure 2, upper left). The E3 subunit of BCKAD was not present on the microarray and thus undetermined. *M. xanthus* has an additional, novel pathway for making isovaleryl CoA [16–18]. Three genes identified in this pathway are up-regulated during development (MXAN4264, 2.9

fold; MXAN4265, 2.0-fold; MXAN4266, 3.8-fold). While the upregulation of the latter pathway is counter to the results with BCKAD, this pathway is thought to primarily function in secondary metabolite production. For elongation, both straight and branch chain primers are extended two carbons at a time using the fatty acid synthase cycle until they reach full length (Figure 2, center circle). Genes within the cycle are down-regulated (MXAN4770; 1.5-fold, MXAN4725, 1.4-fold, MXAN6393, 1.6-fold, and MXAN6034, 1.4-fold).

The most common and energetically effective pathway for fatty acid catabolism is β-oxidation. Here, two carbon atoms are removed from the fatty acid with each turn of the cycle. In general, genes involved in β-oxidation are up-regulated during development. In *E. coli*, the first step involves esterification of a fatty acid molecule to a CoA moiety by FadD. The KEGG database predicts MXAN1528 as a possible FadD homolog. However, a BLASTP search using the *E. coli* K12 FadD sequence suggested MXAN7148 as the closest homolog. Whereas transcription of MXAN1528 does not change, transcript levels of MXAN7148 increased 1.3-fold (Figure 2, bottom circle). In the next step, acyl CoA is oxidized to enoyl CoA by FadE. The KEGG database identified 3 homologs. A fourth, MXAN3795, is the closest homolog identified by BLASTP using *E. coli* K12 FadE. Only MXAN3048 is on the array and expression increases 1.3-fold. The subsequent steps in the fatty acid degradation cycle are catalyzed by FadI and FadJ. *M. xanthus* contains two homologs of each. MXAN5371 (FadJ) and MXAN5372 (FadI) are up-regulated 3.0-fold and 2.3-fold respectively whereas the other pair is not up-regulated. Thus, it appears that genes involved in the degradation of fatty acids are expressed during development, consistent with the observation that lipid bodies disappear in mature spores.

In summary, the data call into question the hypothesis that the fatty acid content of lipid bodies is synthesized *de novo* during development. Microarray data show that fatty acid biosynthesis genes are down regulated. These data are consistent with the idea that preformed fatty acids are used to make TAGs. The largest source of fatty acids is the pool of membrane phospholipids, which are of limited utility in a shrinking cell. TAGs provide an uncharged, non-toxic intermediate that can be stored until needed for carbon and energy during spore maturation.

Most developmental mutants have reduced lipid bodies

If cell shortening and lipid body production are obligately coupled, then mutants defective in cell shortening should also be defective in lipid body production. We examined over 30 developmental mutants at 18 h, the peak of lipid body production in wild type cells. The mutant set includes most of the commonly studied developmental mutants known to have defects in fruiting body morphogenesis, myxospore differentiation, or both. Lipid body area per cell was averaged from 30 cells at 18 hours of development. The mutants show a wide range of variability from 0% to nearly 140% of wild type levels (Table 1). Average cell length was also calculated for these cells (Table 1). The mutant set reflects a continuum in shape change including those that fail to decrease cell length, those that initiate cell shortening, and those that shorten and ultimately sporulate despite delayed timing. A plot of lipid body area vs. cell length at 18 h shows that lipid body area increases as cell length declines over much of the mutant set (Figure 3). The line describing the best fit to the entire mutant collection passes through the standard deviations for WT cells. The Pearson's correlation coefficient between cell length and lipid body area for the entire mutant collection is 0.76. These results argue that for most mutants, cell length is proportional to lipid

body content regardless of the stage at which the mutants are blocked.

The mutant collection defines at least two stages in the cell shortening/lipid body production process. The first stage is represented by mutants whose lipid body content is 40% or less of wild type and whose average cell length is 6 μm or longer at 18 hours. Most of the known signaling mutants cluster in this group including *asgA* (A-signal), *bsgA* (B-signal), *csgA* (C-signal), *dsgA* (D-signal), and *esg MXAN4265* (E-signal). All are blocked within the first six hours of development and these represent their terminal developmental phenotypes [19,20]. The second mutant cluster shortens substantially (4–6 μm), though not quite as much as wild type, and produces fewer TAGs (40–80% of WT). Some, like *fibA* and *mrpA*, ultimately sporulate [21] though *fibA* produces fewer spores than wild type and has a reduced rate of germination [22]. Others in this cluster, like *difA* and *mrpC*, fail to sporulate and represent terminal phenotypes [23,24].

In conclusion, most developmental mutants are defective in producing lipid bodies even though the products of the mutant genes have no direct role in lipid metabolism. Furthermore, lipid body content is proportional with cell length as if lipid bodies are derived from some portion of the shrinking cell. As the correlation coefficient is <1, there may be several different carbon sources for TAG production that vary among mutant strains. Alternately, fatty acid oxidation could also reduce the correlation coefficient in some strains because TAGs are just a temporary residence for fatty acids. The microarray results indicate that fatty acids are consumed during development, consistent with the observation that mature spores lack lipid bodies.

Lipid body production is regulated by branched chain fatty acids and an RNA interferase

Several mutants have an altered relationship between lipid body content and cell length that differs substantially from the correlation coefficient. The genes represented by these mutants may play direct roles in regulating cell shortening and/or lipid body production. A mutant that under produces lipid bodies relative to cell length is MXAN6704. Mutants that over produce lipid bodies relative to cell length include *asgB*, *epsA*, *nla6*, and *fabH*. *mazF* is unique in that while lipid body content is proportional to cell length, the cells are unusually short at this stage of development.

mazF encodes an RNA interferase recognizing an 8-nucleotide sequence [25]. MazF appears to be a negative regulator of shortening since *mazF* cells are about half as long as WT cells and have a corresponding increase in lipid content. Among the MazF targets is enhancer-binding protein *nla6* mRNA [26]. The *nla6* mutant has similar though not quite as dramatic properties (Figure 3). The *nla6* mutant overproduces lipid bodies and is slightly shorter than wild type. No other genes in this mutant collection are known to be MazF targets [26]. *nla28* and *actB* transcription is activated by Nla6, but inactivation of these genes diminishes lipid accumulation rather than enhancing it so they appear not to be negative regulators (Table 1 and Figure 3) [27].

MXAN6704 under produces lipid bodies relative to cell length and seems to encode a Gcn5-related N-acetyltransferase (GNAT). GNATs use acetyl coenzyme A to transfer an acetyl group to the primary amine of an acceptor molecule such as lysine [28]. *MXAN6704* mutants make neither fruiting bodies nor spores. While the *MXAN6704* mutant produces only 5% of the wild type level of lipid bodies, the cells shorten more than most mutants that fail to make lipid bodies (Figure 3) indicating a partial uncoupling of cell shortening from lipid body production. Protein acetylation is thought to regulate the activity of enzymes controlling carbon

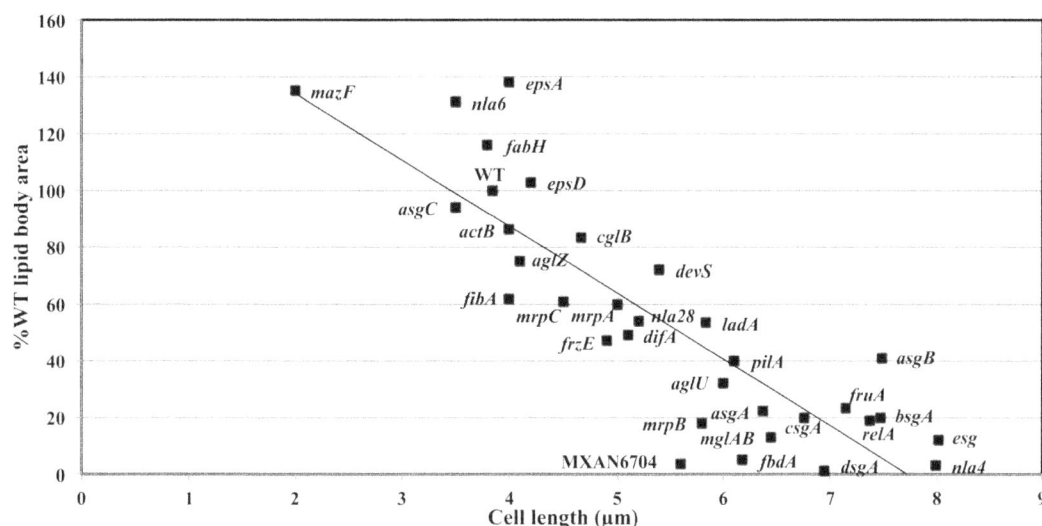

Figure 3. Lipid body area correlates with cell length in developmental mutants. Lipid body production at 18 h is compared with average cell length, each derived from the average of 30 cells. Adjacent to the plotted point is the name of the mutated gene. Compiled from the data in table 1.

flow across metabolic pathways, especially the flow of acetate as it relates to fatty acid synthesis and degradation [29,30]. One might predict that this mutant has enhanced fatty acid degradation relative to wild type.

The *fabH* mutant is one of several mutants that over produce TAGs. *fabH* has high levels of the principle fatty acid in TAGs, branched chain fatty acid *iso*15:0, due to a defect in producing straight chain fatty acids [31]. The opposite extreme is seen for *esg* *MXAN4265*, which is unable to produce *iso*15:0 and fails to shorten or make lipid bodies. When *iso*15:0 or a TAG called TG1 that contains an ether-linked *iso*15:0 fatty alcohol are added to *esg* *MXAN4265* cells, either molecule restores lipid body production at physiological concentrations concomitantly with cell shortening and sporulation [9]. TG1 and *iso*15:0 are the twin components of the E-signal whose sensory mechanisms remain unknown [9]. These results suggest an autocatalytic cycle in which cell shortening stimulates the production of E-signal to further stimulate cell shortening.

Fatty acids in membrane lipids of growing cells become incorporated into lipid bodies during development

Since fatty acid biosynthetic genes are down regulated during development, we directly investigated the possibility that membrane phospholipids serve as the source of preformed fatty acids for lipid body lipids. Growing *Myxococcus* cells were pulse labeled with the fatty acid palmitic acid alkyne (PA[alk]). Using click chemistry [32], Alexa Fluor 488 azide was attached to the PA[alk] in permeabilized cells prior to visualization with fluorescence microscopy. Incorporation of the fatty acid was examined during growth in CYE complex media [33], M1 defined media [34], and A1 minimal media [35]. Fatty acid incorporation as measured by fluorescence was only detected in A1. To determine the positions of membranes and lipid bodies in the micrographs, cells were also stained with Nile red.

In wild-type cells (Figure 4, DK1622), PA[alk] is detected in the membranes of vegetative cells grown in A1 (0 h) as a thin layer of fluorescence that outlines each cell. The chemical form(s) was not identified, but a likely possibility is phosphatidylethanolamine, which comprises 70% of the phospholipid in *Myxococcus* [36]. By

24 h of development, PA[alk] becomes prominent within DK1622 lipid bodies demonstrating that the TAGs are at least partially derived from the fatty acids in membrane lipids. PA[alk] is also incorporated into membranes of a strain deficient in making lipid bodies (Figure 4, LS3931), but fully formed lipid bodies are rarely observed. No detectable signal is seen in lipid bodies from wild-type cells grown without PA[alk] fatty acid indicating that there is no bleed through between fluorescence channels (Figure 4, last line).

Detection of exogenous PA[alk] incorporation when cells are grown in minimal media demonstrates fatty acid uptake. As such, the possibility exists that lipid bodies are formed from exogenous material during development. Lipid bodies first arise during the early stages of PCD that ultimately claims 80% of the initial cell population. Lipids scavenged from dead cells would provide an excellent source of preformed fatty acids. To examine whether fatty acids are recycled from dead cells, incorporation of PA[alk] was examined in LS3125, a mutant strain lacking *fadL*, an outer membrane protein necessary for fatty acid uptake. No detectable signal was seen in this mutant, though lipid bodies are clearly visible (Figure 4, LS3125). These results indicate that incorporation of exogenous cellular material from dead cells is not required for mature lipid body formation.

In summary, PA[alk] fluorescence is always found concurrent with lipids stained with Nile red, as shown in each of the merged images. In growing cells, these are membrane lipids that surround the cell. In developing wild type cells, lipid bodies are strongly labeled from preformed fatty acids consistent with the idea that membrane phospholipids are mobilized to make TAGs as the membrane surface area declines. As the results are not quantitative, we cannot rule out the possibility that the lipid body fatty acids supplied by membrane phospholipids are augmented by some *de novo* fatty acid synthesis.

The developmental regulon regulates the lipid body regulon

Figure 5 provides a simplistic model of the lipid body/ developmental regulon in *M. xanthus* illustrating the regulatory points for lipid body production. Development begins with amino acid deprivation. Nla4, an enhancer binding protein, activates

Table 1. Lipid body production and average cell length in developmental mutants.

Strain	%WT lipid body area/cell[a]	Cell length (μm)[b]
actB	86.0±27.3	4.0±0.8
aglU	32.0±0.8	6.0±1.0
aglZ	76.0±33.9	4.1±2.0
asgA	22.0±10.1	6.4±1.4
asgB	41.0±10.1	7.5±1.1
asgC	94.0±11.1	3.5±1.5
bsgA	20.0±13.2	7.5±2.5
cglB	76.0±19.5	4.7±1.2
csgA	20.0±1.0	6.8±1.4
devS	72.0±19.3	5.4±0.8
difA	49.0±9.6	5.1±0.9
dsgA	1.0±0.8	7.0±1.2
epsA	138.0±37.3	4.0±0.8
epsD	111.0±25.3	4.2±1.2
esg MXAN4265	11.0±9.6	8.0±2.1
fabH	123.0±22.6	3.8±0.5
fbdA	5.0±4.7	6.2±0.9
fibA	61.0±10.0	4.3±0.7
fruA	23.0±6.0	7.2±1.7
frzE	47.0±10.0	5.0±1.7
ladA	53.3±27.0	5.8±1.2
mazF	135.0±11.1	2.0±0.5
mglAB	13.0±11.0	6.4±1.5
mrpA	60.0±9.0	5.0±0.9
mrpB	18.0±8.0	5.8±1.4
mrpC	61.0±23.1	4.5±0.9
MXAN6704	3.6±3.5	5.6±1.4
nla28	54.0±10.7	5.2±1.4
nla4	2.0±2.3	8.0±2.0
nla6	133.0±40.0	3.5±0.6
pilA	40.0±10.0	6.1±2.1
relA	19.0±13.0	7.4±1.7
Wild type	100.0±10.8	3.6±1.2

[a]The average lipid body area at 18 h, the peak of lipid body production in wild type cells, was calculated from 30 cells and compared with WT plus or minus the standard deviation.
[b]Average cell length at 18 h plus of minus the standard deviation.

expression of relA (synthesis of (p)ppGpp) and initiates the stringent response [37–39]. Both mutants exhibit little shortening and little TAG accumulation (Figure 3). Sensing starvation initiates production of the A-signal, a quorum signal composed of specific amino acids designed to determine whether the cell density is sufficient for development [40]. E-signaling also begins about this time [23,41].

Five genes are required for A-signal production. While asgA (22%) and asgB (41%) deficient strains produce few lipid bodies relative to wild type levels, the asgC (94%) mutant produces nearly normal levels. These results argue that lipid body production is not dependent on the A-signal, but requires another function provided by asgA and asgB. AsgA is a hybrid histidine kinase/response

regulator [42]. AsgB is a putative DNA binding protein [43]. Perhaps AsgA and AsgB directly or indirectly activate genes required for lipid body production irrespective of their role in A-signal production.

A-signaling, or one of the Asg proteins [23], and the E-signal, a mixture of fatty acid iso15:0 and ether triacylglyceride TG1 [9], initiates expression of the mrpAB operon encoding a two-component system [44]. MrpA encodes a histidine protein kinase that presumably phosphorylates response regulator MrpB (Figure 5). The situation may be more complex since inactivation of each gene yields different phenotypes. The mrpB mutation is more severe, blocking aggregation, sporulation, and lipid body production (Figure 3). The mrpA mutation blocks sporulation but has a more modest impact on cell shortening and lipid body production.

MrpB-P goes on to activate expression of the transcription factor mrpC, essential for development [23]. MrpC autoactivates its own expression and induces expression of mazF and fruA. MrpC binding sites are observed in the promoter regions of all three genes [25,45,46]. MrpC and FruA regulate transcription of many genes involved in sporulation [23,46,47]. Loss of Lon protease, referred to as BsgA for its role in B-signal production [48,49], leads to dramatic reduction in lipid bodies. BsgA is thought to activate MrpC by removing 33 amino acids from the N-terminus, though the reaction has never been demonstrated in vitro [45]. mrpC, bsgA, and fruA mutants under produce lipid bodies, though mrpC produces more TAGs than the other two (Figures 3 and 5). MazF is an RNA interferase that degrades enhancer binding protein nla6 mRNA [26]. Nla4 activates transcription of nla6 and can be considered a positive regulator of shortening and TAG production since nla4 mutants fail to do either (Figure 3) [27].

actB encodes a transcriptional activator of the act operon leading to increased expression of csgA [50]. csgA produces the C-signal, which is essential for both aggregation and sporulation [51]. C-signaling activates response regulator FruA, a response regulator and transcriptional activator of many morphogenesis genes [52]. The method of FruA activation by C-signaling is presumably phosphorylation, though the cognate histidine kinase has not been identified. Curiously, csgA mutants have a more severe phenotype than actB mutants, even though csgA appears to lie downstream of actB (Figures 3 and 5). A csgA mutant fails to aggregate, sporulate, or produce lipid bodies whereas the actB mutant aggregates and produces nearly normal levels of lipid bodies (84%; Figure 3 and Table 1), but doesn't sporulate well [53]. Aggregates are formed as the C-signal levels rise, but even higher C-signal levels are required to induce sporulation [54,55]. Although ActB increases C-signaling, actB mutants already produce one-fourth of the WT level of CsgA, which is sufficient to induce aggregation [56]. This level of C-signaling in actB cells also seems sufficient for the initial phase of cell shortening and lipid body production. The mglAB mutant is nonmotile and consequently defective in C-signaling, which requires cell alignment for efficient transmission of C-signal [57]. mglAB is also severely defective in lipid body production, comparable to csgA (Table 1).

Sporulation within fruiting bodies is triggered by DevTRS, proteins of unknown function whose production is activated in part by FruA and in part by a second LysR-type activator, LadA, which also binds to the dev promoter region [58–60]. ladA and devS mutants produce comparable and relatively high levels of lipid bodies suggesting that mutations in late developmental genes have only limited reduction of lipid bodies.

In summary, there is no single point in the developmental program where lipid body production becomes activated to the full extent. TAG synthesis is inhibited by mutations in most

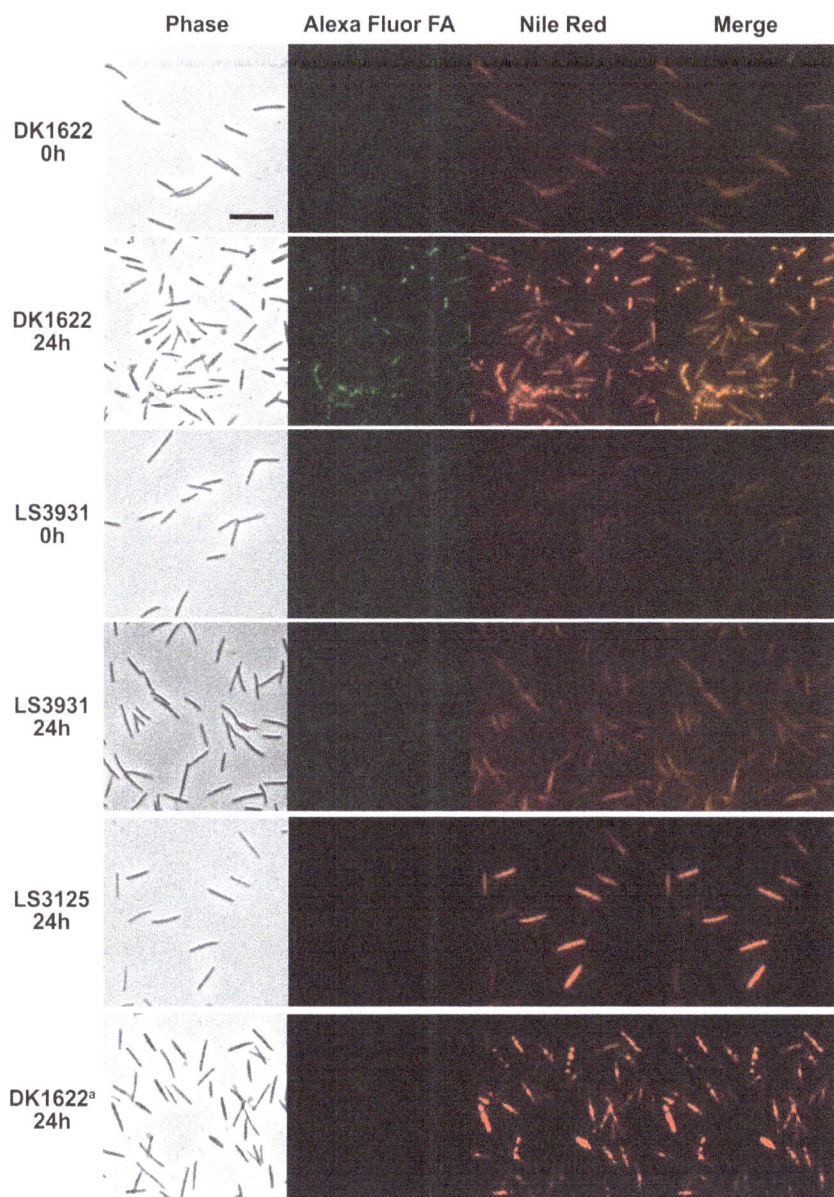

Figure 4. Lipid bodies are derived from membrane phospholipids. *Myxococcus* cells were grown in the presence of palmitic acid alkyne (PA^{alk}) in A1 minimal medium and stained with Nile red. Click chemistry was used to attach Alexa Fluor 488 azide to PA^{alk} prior to visualization. Both Alex Fluor fatty acid and the Nile red stain are visible in membranes during vegetative growth (DK1622 0 h). The PA^{alk} was removed and cells were allowed to develop on TPM agar (DK1622 24 h). Labeled membrane lipids were incorporated into Nile red stained lipid bodies, seen in both channels and the merged images. In a strain deficient in lipid body production (LS3931), incorporation occurs into membrane lipids (LS3931 0 h) but not lipid bodies (LS3931 24 h). A *fadL* mutant (LS3125), defective in fatty acid uptake, is unable to incorporate PA^{alk} altogether. Wild-type cells grown in the absence of PA^{alk} then allowed to develop (DK1622^a 24 h) exhibit no fluorescence (bottom row, second panel) indicating little bleed through between channels. Scale bar is 10 μm.

developmental genes even though these genes are not directly related to lipid metabolism. These include most of the early cell signal-producing genes, which abolish lipid body production. Taken together, these results suggest that development consists of a series of checkpoints designed to precisely couple shortening with TAG production.

Conclusions

M. xanthus development is initiated by carbon limitation. Unexpectedly, lipid bodies containing TAGs become a major,

though temporary, development-specific product that is consumed during spore maturation. Where does the carbon for TAG synthesis originate? Two major reservoirs of carbon may be available to sporulating cells, an extracellular one generated by PCD and an internal one generated by cell shrinkage. PCD eliminates 80% of the cells, and fatty acids could be recycled by extracellular phospholipases and incorporated into prespores. *M. xanthus* produces many lipolytic enzymes that could facilitate recycling, but deletion of their genes had little effect on spore yield [61]. Furthermore, mutation of *fadL*, which encodes a porin required for fatty acid assimilation, did not disrupt development or

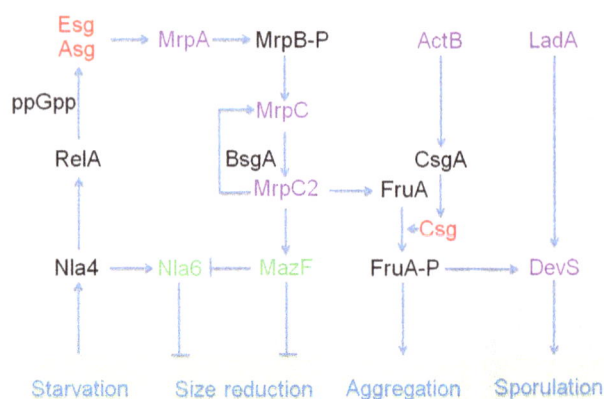

Figure 5. Regulation of lipid body production during *M. xanthus* fruiting body development. Mutants blocked in synthesis of proteins labeled in in black are deficient in lipid body synthesis (> 40%). Those labeled in purple produce intermediate levels of lipid bodies (40–80%). Those in green produce near WT levels or higher (> 80%). Red letters indicate developmental signals. Asg, A-signal; Csg, C-signal; Esg, E-signal.

diminish lipid body production. Clearly, lipid bodies can be produced without recycling extracellular lipids.

As there is a >80% decrease in cell volume and membrane surface area during differentiation, carbon reserves from the shrinking cell could easily be mobilized into TAGs. This possibility is supported by a temporal relationship between diminishing cell length and lipid body production in wild type cells. Furthermore, there is a strong correlation between cell length and lipid body content among mutants blocked during various stages of development. The shrinking cell could produce TAGs by either *de novo* fatty acid synthesis or by recycling of fatty acids in membrane phospholipids. Genes involved in fatty acid biosynthesis are down regulated, so we instead focused our attention on lipid conversion. Membrane lipids that were pulse labeled with palmitate during growth served as a source of fatty acids for lipid bodies. While these results illustrate the movement of fatty acids from membrane lipids to TAGs, the experiments were not quantitative and we cannot rule out the possibility that mobilization is supplemented with some fatty acid synthesis and/or some recycling of extracellular lipids. We predict the existence of complex machinery designed to shorten the cell and mobilize membrane phospholipids to TAGs as part of a pathway in which fatty acids are eventually degraded to produce mature spores. The enzymes involved in synthesizing TAGs from phospholipids remain unknown. *M. xanthus* contains a single diacylglycerol acyltransferase gene, but deletion of this gene did not eliminate TAG production ([4] and unpublished).

Materials and Methods

Bacterial strains and growth condition

Table 2 lists the bacterial strains used in this study and their sources. *M. xanthus* strains were grown in CYE broth [1% Bacto casitone (Difco), 0.5% yeast extract (Difco), 10 mM 4-morpholinepropanesulfonic acid (MOPS) (pH 7.6), and 0.1% $MgSO_4$)] at 32°C with vigorous shaking. Kanamycin and Bacto agar (Difco) were added to CYE at final concentrations of 50 µg/ml and 1.5% respectively. Development was induced on TPM agar plates

Table 2. *M. xanthus* strains used in this study.

Bacterial strain	Genotype	Reference or source
DK1622	WT	[65]
DK3260	dsgA429	[66]
DK4398	asgB480	[67]
DK5057	asgA476	[68]
DK5061	asgC767	[69]
DK5614	fabH	[31]
DK6204	ΔmglAB	[70]
DK10410	ΔpilA	[71]
DK10603	ΔactB	[50]
DK11063	fruA	[52]
DK11209	ΔdevS	[72]
LS313	cglB	Lawrence Shimkets
LS1191	esg MXAN4265	[9]
LS1193	mazF	Lawrence Shimkets
LS2208	ΔfibA	Lawrence Shimkets
LS2225	fbdA	[73]
LS2442	ΔcsgA	Lawrence Shimkets
LS2702	MXAN6704	This study
LS3931	ΔcsgA, socA	Lawrence Shimkets
MS10	relA	[74]
AG328	nla28	[37]
AG304	nla4	[37]
AG306	nla6	[37]
MPVlysR	ladA	[60]
MXH1777	ΔaglU	[75]
MXH2265	ΔaglZ	[76]
RGM252	bsgA302	[48]
SW810	ΔepsA	[77]
SW813	ΔepsD	[77]
SW2802	ΔmrpB	[78]
SW2807	ΔmrpA	[78]
SW2808	ΔmrpC	[78]
SW600	ΔfrzE	[79]
YZ601	ΔdifA	[24]

[10 mM Tris HCl, pH 7.6, 1 mM $KH(H_2)PO_4$, pH 7.6, 10 mM $MgSO_4$, 1.5% agar (Difco)].

Strain construction

magellan-4 transposon mutagenesis was used for isolating new fruiting body deficient mutants [62]. Fruiting body deficient strains containing a *magellan*-4 insertion were then backcrossed to *M. xanthus* DK1622 by electroporation of 1 µg genomic DNA or by generalized transduction with phage Mx4 [63]. Insertion regions were identified by cloning and sequencing [64]. LS2702 contains a *magellan*-4 insertion within *MXAN6704*.

Nile red staining

M. xanthus strains were grown to a density of 5×10^8 cells/ml then resuspended in 100 µl dH_2O to a final density of 5×10^9 cells/ml. Aliquots of 10 µl were spotted on TPM agar and incubated for various times. Lipid body staining was carried out as described by Hoiczyk et al with modifications [4]. A 0.5 mg/ml stock solution of Nile red (Sigma Aldrich) prepared in 100% ethanol was diluted in dH$_2$O to a final concentration of 1.25 µg/ml and added directly on top of cells developing on TPM agar. The plates were incubated for 2 h at 32°C. Cells were resuspended in a drop of TPMF buffer [TPM buffer containing 10% ficoll], and examined with a fluorescence microscope (Leica Microsystems, DM5500B). Digital images were obtained using a QICAM FAST 1394 camera (Q Imaging systems, Compix Inc.).

The fluorescence images were digitally altered using Simple PCI (Hamamatsu Corporation) to remove background noise, improve contrast, and measure cell length and fluorescence intensity. Average lipid body area and cell length were calculated from 30 randomly chosen cells.

Fatty acid incorporation

For incorporation of labeled fatty acids, Myxococcus xanthus cells were grown to a density of 1×10^8 cells/ml in A1 minimal media [0.5% potassium aspartate, 0.5% sodium pyruvate, 0.05% (NH$_4$)$_2$SO$_4$, 0.2% MgSO$_4$, 0.125 mg/ml spermidine, 0.1 mg/ml each asparagine, isoleucine, phenylalanine and valine, 0.05 mg/ml leucine, 0.01 mg/ml methionine, 1 µg/ml cobalamin, 10 µM FeCl$_3$, 10 µM CaCl$_2$, 1 mM KH(H$_2$)PO$_4$ pH 7.6, and 10 mM Tris HCl pH 7.6] [35] containing 100 µM palmitic acid alkyne (Cayman Chemical).

For visualization, click chemistry was used to attach Alexa Fluor 488 azide to the alkyne moiety of the fatty acid after growth or development. 1×10^9 cells were permeablized by resuspension in 1 ml TPM containing 2% paraformaldehyde and incubated with gentle shaking for 1 h at room temperature. Cells were then harvested at 13,000 rpm for 1 min and resuspended in 1 ml phosphate buffered saline (PBS), pH 7.4, with 0.02% Tween 20. Cells were next harvested and washed in 1 ml TGE buffer (50 mM glucose, 20 mM Tris pH 7.6, 10 mM EDTA), resuspended in 1 ml TGE containing 10 µg/ml lysozyme, and incubated for 10 min at room temperature. Cells were washed in 1 ml PBS containing 2% BSA, then resuspended in 500 µl Click-iT reaction (Life Technologies) containing 440 µl Click-iT buffer, 10 µl 100 mM CuSO$_4$, and 50 µl 1 mg/ml Alexa Fluor 488 azide. Cells were incubated for 30 min at room temperature in the absence of light. After the reaction was complete, cells were harvested and resuspended in 1 ml TPM containing 2% BSA, stained with Nile red as previously described in this paper, and visualized using a fluorescence microscope (Leica Microsystems, DM5500B).

Acknowledgments

We would like to thank James Ward for help with fluorescence microscopy.

Author Contributions

Conceived and designed the experiments: SB LJS. Performed the experiments: SB TOB DP. Analyzed the data: SB TOB DP LJS. Contributed reagents/materials/analysis tools: LJS. Wrote the paper: SB TOB LJS.

References

1. Greenberg AS, Coleman RA, Kraemer FB, McManaman JL, Obin MS, et al. (2011) The role of lipid droplets in metabolic disease in rodents and humans. J Clin Invest 121: 2102–2110.
2. Alvarez HM, Steinbuchel A (2002) Triacylglycerols in prokaryotic microorganisms. Appl Microbiol Biotechnol 60: 367–376.
3. Waltermann M, Hinz A, Robenek H, Troyer D, Reichelt R, et al. (2005) Mechanism of lipid-body formation in prokaryotes: how bacteria fatten up. Mol Microbiol 55: 750–763.
4. Hoiczyk E, Ring MW, McHugh CA, Schwar G, Bode E, et al. (2009) Lipid body formation plays a central role in cell fate determination during developmental differentiation of Myxococcus xanthus. Mol Microbiol 74: 497–512.
5. Manoil C, Kaiser D (1980) Guanosine pentaphosphate and guanosine tetraphosphate accumulation and induction of Myxococcus xanthus fruiting body development. J Bacteriol 141: 305–315.
6. Manoil C, Kaiser D (1980) Accumulation of guanosine tetraphosphate and guanosine pentaphosphate in Myxococcus xanthus during starvation and myxospore formation. J Bacteriol 141: 297–304.
7. Shimkets L, Seale TW (1975) Fruiting-body formation and myxospore differentiation and germination in Myxococcus xanthus viewed by scanning electron microscopy. J Bacteriol 121: 711–720.
8. Voelz H, Dworkin M (1962) Fine structure of Myxococcus xanthus during morphogenesis. J Bacteriol 84: 943–952.
9. Bhat S, Ahrendt T, Dauth C, Bode HB, Shimkets LJ (2014) Two lipid signals guide fruiting body development of Myxococcus xanthus. MBio 5: e00939-00913.
10. van den Berg B (2005) The FadL family: unusual transporters for unusual substrates. Curr Opin Struct Biol 15: 401–407.
11. Nunn WD, Simons RW (1978) Transport of long-chain fatty acids by Escherichia coli: mapping and characterization of mutants in the fadL gene. Proc Natl Acad Sci U S A 75: 3377–3381.
12. O'Connor KA, Zusman DR (1991) Development in Myxococcus xanthus involves differentiation into two cell types, peripheral rods and spores. J Bacteriol 173: 3318–3333.
13. Shi X, Wegener-Feldbrugge S, Huntley S, Hamann N, Hedderich R, et al. (2008) Bioinformatics and experimental analysis of proteins of two-component systems in Myxococcus xanthus. J Bacteriol 190: 613–624.
14. Kanehisa M, Araki M, Goto S, Hattori M, Hirakawa M, et al. (2008) KEGG for linking genomes to life and the environment. Nucleic Acids Res 36: D480–484.
15. Okuda S, Yamada T, Hamajima M, Itoh M, Katayama T, et al. (2008) KEGG Atlas mapping for global analysis of metabolic pathways. Nucleic Acids Res 36: W423–426.
16. Bode HB, Ring MW, Schwar G, Altmeyer MO, Kegler C, et al. (2009) Identification of additional players in the alternative biosynthesis pathway to isovaleryl-CoA in the myxobacterium Myxococcus xanthus. Chembiochem 10: 128–140.
17. Dickschat JS, Bode HB, Kroppenstedt RM, Muller R, Schulz S (2005) Biosynthesis of iso-fatty acids in myxobacteria. Org Biomol Chem 3: 2824–2831.
18. Li Y, Luxenburger E, Muller R (2013) An alternative isovaleryl CoA biosynthetic pathway involving a previously unknown 3-methylglutaconyl CoA decarboxylase. Angew Chem Int Ed Engl 52: 1304–1308.
19. Kroos L, Kaiser D (1987) Expression of many developmentally regulated genes in Myxococcus depends on a sequence of cell interactions. Genes Dev 1: 840–854.
20. Li SF, Shimkets LJ (1993) Effect of dsp mutations on the cell-to-cell transmission of CsgA in Myxococcus xanthus. J Bacteriol 175: 3648–3652.
21. Bonner PJ, Black WP, Yang Z, Shimkets LJ (2006) FibA and PilA act cooperatively during fruiting body formation of Myxococcus xanthus. Mol Microbiol 61: 1283–1293.
22. Lee B, Mann P, Grover V, Treuner-Lange A, Kahnt J, et al. (2011) The Myxococcus xanthus spore cuticula protein C is a fragment of FibA, an extracellular metalloprotease produced exclusively in aggregated cells. PLoS One 6: e28968.
23. Sun H, Shi W (2001) Analyses of mrp genes during Myxococcus xanthus development. J Bacteriol 183: 6733–6739.
24. Yang Z, Ma X, Tong L, Kaplan HB, Shimkets LJ, et al. (2000) Myxococcus xanthus dif genes are required for biogenesis of cell surface fibrils essential for social gliding motility. J Bacteriol 182: 5793–5798.
25. Nariya H, Inouye M (2008) MazF, an mRNA interferase, mediates programmed cell death during multicellular Myxococcus development. Cell 132: 55–66.
26. Boynton TO, McMurry JL, Shimkets LJ (2013) Characterization of Myxococcus xanthus MazF and implications for a new point of regulation. Mol Microbiol 87: 1267–1276.
27. Giglio KM, Caberoy N, Suen G, Kaiser D, Garza AG (2011) A cascade of coregulating enhancer binding proteins initiates and propagates a multicellular developmental program. Proc Natl Acad Sci U S A 108: E431–439.
28. Zhang J, Sprung R, Pei J, Tan X, Kim S, et al. (2009) Lysine acetylation is a highly abundant and evolutionarily conserved modification in Escherichia coli. Mol Cell Proteomics 8: 215–225.
29. Tucker AC, Escalante-Semerena JC (2013) Acetoacetyl-CoA synthetase activity is controlled by a protein acetyltransferase with unique domain organization in Streptomyces lividans. Mol Microbiol 87: 152–167.
30. Newman JC, He W, Verdin E (2012) Mitochondrial protein acylation and intermediary metabolism: regulation by sirtuins and implications for metabolic disease. J Biol Chem 287: 42436–42443.
31. Bode HB, Ring MW, Kaiser D, David AC, Kroppenstedt RM, et al. (2006) Straight-chain fatty acids are dispensable in the myxobacterium Myxococcus xanthus for vegetative growth and fruiting body formation. J Bacteriol 188: 5632–5634.
32. Thiele C, Papan C, Hoelper D, Kusserow K, Gaebler A, et al. (2012) Tracing fatty acid metabolism by click chemistry. ACS Chem Biol 7: 2004–2011.

33. Kearns DB, Shimkets LJ (1998) Chemotaxis in a gliding bacterium. Proc Natl Acad Sci U S A 95: 11957–11962.

34. Witkin SS, Rosenberg E (1970) Induction of morphogenesis by methionine starvation in *Myxococcus xanthus*: polyamine control. J Bacteriol 103: 641–649.

35. Bretscher AP, Kaiser D (1978) Nutrition of *Myxococcus xanthus*, a fruiting myxobacterium. J Bacteriol 133: 763–768.

36. Orndorff PE, Dworkin M (1980) Separation and properties of the cytoplasmic and outer membranes of vegetative cells of *Myxococcus xanthus*. J Bacteriol 141. 914–927.

37. Caberoy NB, Welch RD, Jakobsen JS, Slater SC, Garza AG (2003) Global mutational analysis of NtrC-like activators in *Myxococcus xanthus*: identifying activator mutants defective for motility and fruiting body development. J Bacteriol 185: 6083–6094.

38. Singer M, Kaiser D (1995) Ectopic production of guanosine penta- and tetraphosphate can initiate early developmental gene expression in *Myxococcus xanthus*. Genes Dev 9: 1633–1644.

39. Ossa F, Diodati ME, Caberoy NB, Giglio KM, Edmonds M, et al. (2007) The *Myxococcus xanthus* Nla4 protein is important for expression of stringent response-associated genes, ppGpp accumulation, and fruiting body development. J Bacteriol 189: 8474–8483.

40. Kuspa A, Plamann L, Kaiser D (1992) A-signalling and the cell density requirement for *Myxococcus xanthus* development. J Bacteriol 174: 7360–7369.

41. Downard J, Ramaswamy SV, Kil KS (1993) Identification of *esg*, a genetic locus involved in cell-cell signaling during *Myxococcus xanthus* development. J Bacteriol 175: 7762–7770.

42. Plamann L, Li Y, Cantwell B, Mayor J (1995) The *Myxococcus xanthus asgA* gene encodes a novel signal transduction protein required for multicellular development. J Bacteriol 177: 2014–2020.

43. Plamann L, Davis JM, Cantwell B, Mayor J (1994) Evidence that *asgB* encodes a DNA-binding protein essential for growth and development of *Myxococcus xanthus*. J Bacteriol 176: 2013–2020.

44. Ellehauge E, Norregaard-Madsen M, Sogaard-Andersen L (1998) The FruA signal transduction protein provides a checkpoint for the temporal co-ordination of intercellular signals in *Myxococcus xanthus* development. Mol Microbiol 30: 807–817.

45. Nariya H, Inouye S (2006) A protein Ser/Thr kinase cascade negatively regulates the DNA-binding activity of MrpC, a smaller form of which may be necessary for the *Myxococcus xanthus* development. Mol Microbiol 60: 1205–1217.

46. Mittal S, Kroos L (2009) Combinatorial regulation by a novel arrangement of FruA and MrpC2 transcription factors during *Myxococcus xanthus* development. J Bacteriol 191: 2753–2763.

47. Ueki T, Inouye S (2003) Identification of an activator protein required for the induction of *fruA*, a gene essential for fruiting body development in *Myxococcus xanthus*. Proc Natl Acad Sci U S A 100: 8782–8787.

48. Gill RE, Bornemann MC (1988) Identification and characterization of the *Myxococcus xanthus bsgA* gene product. J Bacteriol 170: 5289–5297.

49. Gill RE, Cull M, Fly S (1988) Genetic Identification and cloning of a gene required for developmental cell interactions in *Myxococcus xanthus*. J Bacteriol 170: 5279–5288.

50. Gronewold TM, Kaiser D (2007) Mutations of the act promoter in *Myxococcus xanthus*. J Bacteriol 189: 1836–1844.

51. Shimkets LJ, Gill RE, Kaiser D (1983) Developmental cell interactions in *Myxococcus xanthus* and the *spoC* locus. Proc Natl Acad Sci USA 80: 1406–1410.

52. Sogaard-Andersen L, Slack FJ, Kimsey H, Kaiser D (1996) Intercellular C-signaling in *Myxococcus xanthus* involves a branched signal transduction pathway. Genes Dev 10: 740–754.

53. Gronewold TM, Kaiser D (2001) The *act* operon controls the level and time of C-signal production for *Myxococcus xanthus* development. Mol Microbiol 40: 744–756.

54. Kim SK, Kaiser D (1991) C-factor has distinct aggregation and sporulation thresholds during *Myxococcus* development. J Bacteriol 173: 1722–1728.

55. Li S, Lee BU, Shimkets LJ (1992) *csgA* expression entrains *Myxococcus xanthus* development. Genes Dev 6: 401–410.

56. Gronewold TM, Kaiser D (2002) *act* operon control of developmental gene expression in *Myxococcus xanthus*. J Bacteriol 184: 1172–1179.

57. Kim SK, Kaiser D (1990) C-factor: a cell-cell signaling protein required for fruiting body morphogenesis of *M. xanthus*. Cell 61: 19–26.

58. Julien B, Kaiser AD, Garza A (2000) Spatial control of cell differentiation in *Myxococcus xanthus*. Proc Natl Acad Sci U S A 97: 9098–9103.

59. Thony-Meyer L, Kaiser D (1993) *devRS*, an autoregulated and essential genetic locus for fruiting body development in *Myxococcus xanthus*. J Bacteriol 175: 7450–7462.

60. Viswanathan P, Ueki T, Inouye S, Kroos L (2007) Combinatorial regulation of genes essential for *Myxococcus xanthus* development involves a response regulator and a LysR-type regulator. Proc Natl Acad Sci U S A 104: 7969–7974.

61. Moraleda-Munoz A, Shimkets LJ (2007) Lipolytic enzymes in *Myxococcus xanthus*. J Bacteriol 189: 3072–3080.

62. Rubin EJ, Akerley BJ, Novik VN, Lampe DJ, Husson RN, et al. (1999) *In vivo* transposition of mariner-based elements in enteric bacteria and mycobacteria. Proc Natl Acad Sci U S A 96: 1645–1650.

63. Campos JM, Geisselsoder J, Zusman DR (1978) Isolation of bacteriophage MX4, a generalized transducing phage for *Myxococcus xanthus*. J Mol Biol 119: 167–178.

64. Youderian P, Burke N, White DJ, Hartzell PL (2003) Identification of genes required for adventurous gliding motility in *Myxococcus xanthus* with the transposable element mariner. Mol Microbiol 49: 555–570.

65. Kaiser D (1979) Social gliding is correlated with the presence of pili in *Myxococcus xanthus*. Proc Natl Acad Sci U S A 76: 5952–5956.

66. Cheng Y, Kaiser D (1989) *dsg*, a gene required for *Myxococcus* development, is necessary for cell viability. J Bacteriol 171: 3727–3731.

67. Kuspa A, Kroos L, Kaiser D (1986) Intercellular signaling is required for developmental gene expression in *Myxococcus xanthus*. Dev Biol 117: 267–276.

68. Hagen DC, Bretscher AP, Kaiser D (1978) Synergism between morphogenetic mutants of *Myxococcus xanthus*. Dev Biol 64: 284–296.

69. Kuspa A, Kaiser D (1989) Genes required for developmental signalling in *Myxococcus xanthus*: three *asg* loci. J Bacteriol 171: 2762–2772.

70. Hartzell P, Kaiser D (1991) Function of MglA, a 22-kilodalton protein essential for gliding in *Myxococcus xanthus*. J Bacteriol 173: 7615–7624.

71. Wu SS, Kaiser D (1997) Regulation of expression of the *pilA* gene in *Myxococcus xanthus*. J Bacteriol 179: 7748–7758.

72. Viswanathan P, Murphy K, Julien B, Garza AG, Kroos L (2007) Regulation of *dev*, an operon that includes genes essential for *Myxococcus xanthus* development and CRISPR-associated genes and repeats. J Bacteriol 189: 3738–3750.

73. Bonner PJ, Shimkets LJ (2006) Cohesion-defective mutants of *Myxococcus xanthus*. J Bacteriol 188: 4585–4588.

74. Harris BZ, Kaiser D, Singer M (1998) The guanosine nucleotide (p)ppGpp initiates development and A-factor production in *Myxococcus xanthus*. Genes Dev 12: 1022–1035.

75. White DJ, Hartzell PL (2000) AglU, a protein required for gliding motility and spore maturation of *Myxococcus xanthus*, is related to WD-repeat proteins. Mol Microbiol 36: 662–678.

76. Yang R, Bartle S, Otto R, Stassinopoulos A, Rogers M, et al. (2004) AglZ is a filament-forming coiled-coil protein required for adventurous gliding motility of *Myxococcus xanthus*. J Bacteriol 186: 6168–6178.

77. Lu A, Cho K, Black WP, Duan XY, Lux R, et al. (2005) Exopolysaccharide biosynthesis genes required for social motility in *Myxococcus xanthus*. Mol Microbiol 55: 206–220.

78. Sun H, Shi W (2001) Genetic studies of *mrp*, a locus essential for cellular aggregation and sporulation of *Myxococcus xanthus*. J Bacteriol 183: 4786–4795.

79. Shi W, Yang Z, Sun H, Lancero H, Tong L (2000) Phenotypic analyses of *frz* and *dif* double mutants of *Myxococcus xanthus*. FEMS Microbiol Lett 192: 211–215.

Decontamination Efficacy of Three Commercial-Off-The-Shelf (COTS) Sporicidal Disinfectants on Medium-Sized Panels Contaminated with Surrogate Spores of *Bacillus anthracis*

Jason M. Edmonds[1]⁹, Jonathan P. Sabol[2], Vipin K. Rastogi[1]*⁹

1 U.S. Army - Edgewood Chemical Biological Center, Research, Development and Engineering Command, Aberdeen Proving Ground, Maryland, United States of America,
2 EXCET, Inc., Springfield, Virginia, United States of America

Abstract

In the event of a wide area release and contamination of a biological agent in an outdoor environment and to building exteriors, decontamination is likely to consume the Nation's remediation capacity, requiring years to cleanup, and leading to incalculable economic losses. This is in part due to scant body of efficacy data on surface areas larger than those studied in a typical laboratory (5×10-cm), resulting in low confidence for operational considerations in sampling and quantitative measurements of prospective technologies recruited in effective cleanup and restoration response. In addition to well-documented fumigation-based cleanup efforts, agencies responsible for mitigation of contaminated sites are exploring alternative methods for decontamination including combinations of disposal of contaminated items, source reduction by vacuuming, mechanical scrubbing, and low-technology alternatives such as pH-adjusted bleach pressure wash. If proven effective, a pressure wash-based removal of *Bacillus anthracis* spores from building surfaces with readily available equipment will significantly increase the readiness of Federal agencies to meet the daunting challenge of restoration and cleanup effort following a wide-area biological release. In this inter-agency study, the efficacy of commercial-of-the-shelf sporicidal disinfectants applied using backpack sprayers was evaluated in decontamination of spores on the surfaces of medium-sized (~1.2 m²) panels of steel, pressure-treated (PT) lumber, and brick veneer. Of the three disinfectants, pH-amended bleach, Peridox, and CASCAD evaluated; CASCAD was found to be the most effective in decontamination of spores from all three panel surface types.

Editor: Srinand Sreevatsan, University of Minnesota, United States of America

Funding: Funding for this program was provided by the Joint Science and Technology Office of the Defense Threat Reduction Agency. The funders had no role in study design, data collection and analysis, decision to publish, or preparation of the manuscript.

Competing Interests: The authors have the following interests: Jon Sabol is employed by EXCET, Inc.

* Email: vipin.k.rastogi.civ@mail.mil

⁹ These authors contributed equally to this work.

Introduction

In 2001, a number of letters containing *Bacillus anthracis* spores, the causative agent for the deadly anthrax disease, were processed and delivered to their respective recipients by the United States Postal Service resulting in contamination of several building interiors, including U.S. Postal & Distribution Centers in Brentwood, DC, Trenton, NJ, Hart Senate Office Building, Washington DC, and American Media Inc., Boca Raton, FL [1]. Despite heavy contamination levels of several building interiors, remediation of building interiors was achieved successfully by fumigation with chlorine dioxide (CD) or vaporous hydrogen peroxide (VHP) [2,3,4]. A number of solution based sporicidal disinfectants have been approved by U.S. EPA's Office of Pesticides Programs, but were not used to great extent because their efficacy has been proven in laboratory-scale studies only.

Three standardized test methods (ASTM 2197-02, ASTM 2414-05, AOAC *Official Method* 2008–05) are available for determining the efficacy of sporicidal disinfectants and gaseous fumigants under pristine laboratory conditions [5,6,7]. These standardized test methods often use small (<1-cm) size test carriers of smooth, non-porous, and hard surfaces, such as steel and glass. Consequently, such methods are not suited for conducting efficacy studies of biologically contaminated wide area urban environments which include building structures composed of a vast array of both porous and non-porous materials. Furthermore, the current test methodologies rely on complete submersion of the inoculated coupons in test chemical, or use of adequate volume of test chemical to completely cover the contaminated surface during the test of liquid disinfectants. In the field, neither the test conditions are idealized, nor is the immersion of contaminated vertical (5, 6) or complex surfaces a possibility. Although a number of gaseous decontamination technologies have been investigated within large volumes of air within buildings [7,8,9,10,11], the application of gaseous technologies to outside surfaces of buildings over a large urban area would likely be challenging. Therefore, the assessment of liquid based decontamination technologies over large areas is relevant to large scale biological remediation.

Since adequate test methodologies [7] are lacking for field assessment, efficacy data on commercial of-the-shelf (COTS) sporicidal agents in decontamination of complex scale surfaces of mitigation and remediation potential is limited. Under the auspices of Interagency Biological Restoration Demonstration (IBRD), a federally funded program with the goal to reduce the time and resources required to recover and restore wide urban areas post-environmental incident, this study was initiated to generate quantitative efficacy data which could be extrapolated to large wide area decontamination attempts. Specific aspects of appropriate methodology included use of mid-size panel assembly (1.2 m²), spore inoculation and their sampling, decon application, sample concentration and spore enumeration.

Depending on the surface composition and the decontamination technology tested, the ability to recover viable spores from the panels after decontamination trials varied. Even though, a correlation between lab-scale assessment and field remediation is lacking with respect to post-decontamination spore recovery, related approaches and data is highly desirable. Restoration of buildings for occupancy is a very complex issue and requires a high degree of public trust in federal agencies authorized in declaring areas decontaminated with near-zero risk of infection. Buildings are composed of highly complex and porous surfaces which potentially pose a long-term and/or recurring threat to occupants and passers-by. The ability to efficiently and effectively decontaminate porous materials to safe levels is of great concern to officials charged with deeming a structure safe for re-occupancy after a contamination has occurred. Method flow charts and efficacy data from experiments using Peridox, pH-amended Ultra Clorox Germicidal bleach, and CASCAD, in decontamination of mid-size panels are presented in this paper.

Methods and Materials

Test Inoculum

Ten gram of *Bacillus atrophaeus* subspecies *globigii* Dugway 1088 batch 040 spores (BG spores) were washed with cold water (4+ 2 C), pelleted using centrifugation at 5000×g for 30 minutes, and resuspended six times in 50 mL cold sterile distilled water. After the final wash, spores were suspended in 50 ml of de-ionized sterile water and the master stock with a spore titer of 1.5×10^{10} Colony Forming Units (CFU)/mL was enumerated, by serial dilution plating. The stock was stored at 4°C until used within 30 days. Spore stocks were periodically checked by performing the Schaeffer-Fulton spore staining procedure to confirm the spore integrity and spore:vegetative cells ratio. The spore:vegetative cells percent ratio was 80:20. The stock was heat-shocked at 65°C for 30 minutes before use to render all vegetative cells non-viable. Working stocks with a titer of heat-resistant 1.5×10^9 CFU/mL were prepared by appropriately diluting the master stock with sterile water and confirmed by serial dilution and enumeration on tryptic soy agar (TSA) plates.

Panel Construction

Panels of stainless steel, PT (pressure-treated) lumber, and brick veneer were assembled in 1.2 m×1.2 m size. All panels were constructed with 1.1 cm thick 1.2 m×1.2 m oriented strand board (OSB) as a backing material (Home Depot, Cat. No. 386-081). Stainless steel (Durrett Sheppart Steel, Baltimore, Maryland) panels were composed of eight individual sheets of 30.5 cm×30.5 cm in size. The sheets of T-304 No. 2B finish 20 gauge stainless steel sheets were glued to the OSB backing board with construction adhesive to form a single four 1.2 m by 1.2 m panel. PT lumber (Home Depot, Cat. No. 155-400) panels were constructed by assembling 8 boards measuring 1.2 m in length, 14 cm in width, and 1.1 cm in thickness as well as one board measuring 1.2 m in length, 10.2 cm in width, and 1.1 cm in thickness to achieve a PT lumber panel of desired size. The PT lumber was secured to the OSB with a single 3.2 cm exterior screw (Home Depot Cat. No. 131-537) at each end of the board. Brick panels were constructed per manufacturer instruction by securing a metal grid (Brickit.com, Bohemia, NY, Cat. No. MGMOD48x8) to OSB panels with construction adhesive and 1.3 cm exterior screws. 1.3 cm think brick veneer (Brickit.com, Bohemia, NY, Cat. No. TSMODKINGW) was then secured to the metal grid using construction adhesive.

Panel Inoculation

Panels constructed for decontamination testing were inoculated with 1,280 individual 10-µL drops of previously described working stock of BG spores evenly distributed throughout the panel with the use of an electric micro-pipettor to achieve a final calculated total spore load of $\sim 2.1 \times 10^9$ spores per panel based on previous

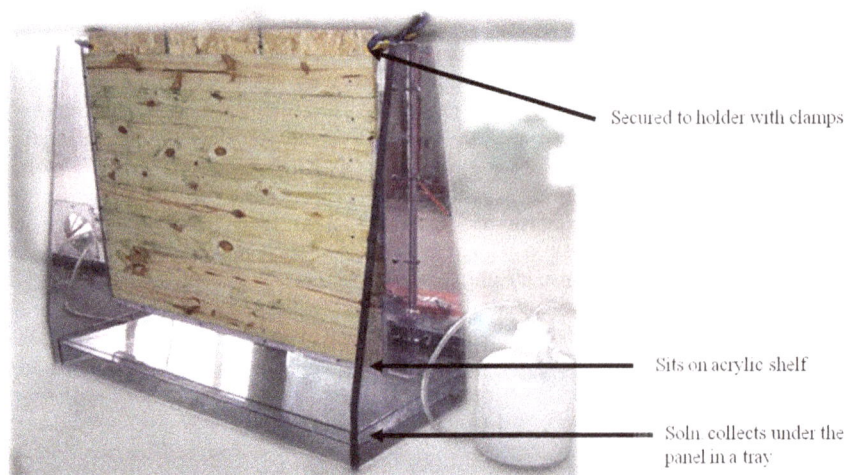

Secured to holder with clamps

Sits on acrylic shelf

Soln collects under the panel in a tray

Figure 1. Sampling flow chart.

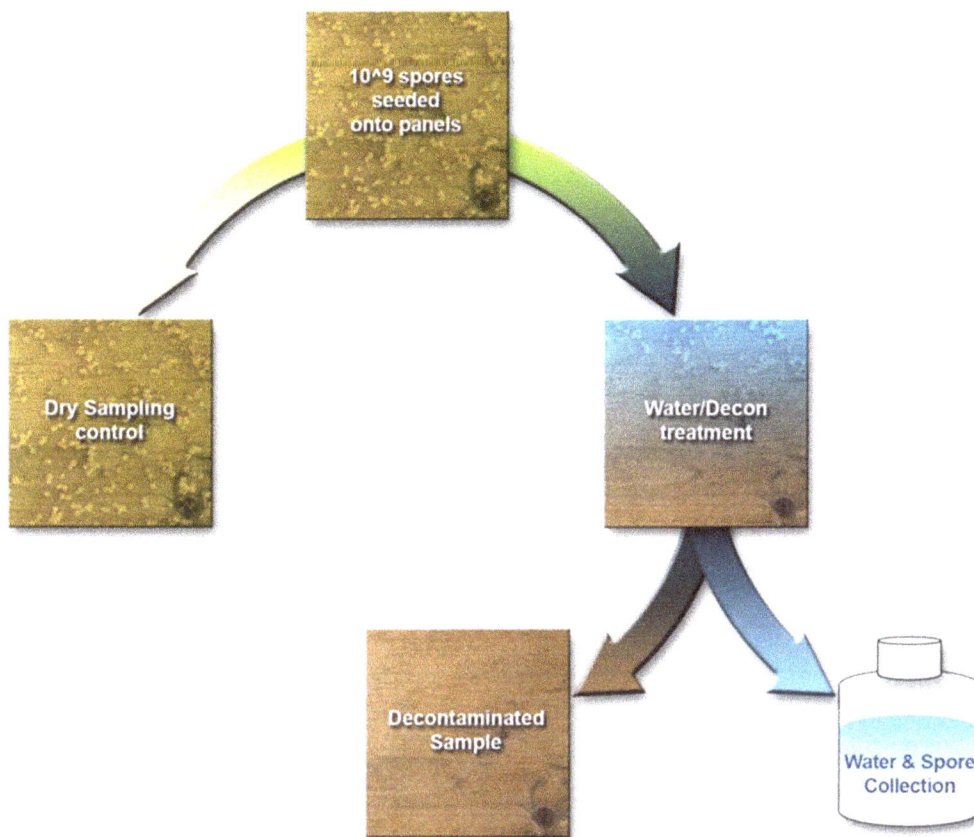

Figure 2. Panel holder for decontamination application.

suspension enumeration. Inoculation was done in a preparation area outside of the 81 m^3 ambient decontamination chamber and after inoculation, moved into the chamber and placed onto panel holders as described below. Because temperature and RH within the chamber are subject to the surrounding conditions, the panels were set aside to allow the spore suspension to dry for 24 h. Only the required number of panels to be tested the following day were inoculated with spores at any given time, and used within 48 h of inoculation. Panels were designated as control panels or sample panels and underwent treatment as detailed in Figure 1. Due to the large size of the panels, we were unable to sterilize each panel prior to testing. However, controls were taken periodically during testing to confirm that panels were not being contaminated with spores from natural flora nor testing procedures (data not shown).

Application of Decontamination Technologies

Inoculated panels were attached vertically to a specially constructed panel holder (designed by Dr. Rastogi and fabricated by the Advanced Design & Manufacturing Team at the Aberdeen Proving Grounds, Edgewood Area, APGEA) with a wood clamp in both the upper right and upper left corners. The run-off from the subsequent application of the decontamination solution was collected at the bottom (Figure 2). Contaminated panels were treated with one of three disinfectants, Peridox, a peroxide technology, (Clean Earth Technologies, Earth City MO, Cat. No. Per-1), 1:10 pH amended Ultra Clorox Germicidal bleach (pH 7+ 0.1, as described in Tomasino et al., 12), and CASCAD, a combination of peroxide and hypochlorite (Allen Vanguard Technologies, Ottawa, ON, Cat. No. GP2100-730, GCE2000-950, and GPX-4000). Peridox and CASCAD were diluted per manufacturer's directions. In addition, the control panels were sprayed with water to assess recovery of the inoculated spores. Each disinfectant was filled into a low pressure 15.1 L backpack sprayer (Agri Supply Co., Garner, NC, Cat. No. 59540) and the respective decontaminant (or water in the case of control panels) was sprayed on to the appropriate panels from a distance of

Table 1. Relative spore recovery from untreated panels.

Panel Type	% Recovery	Log Spores Recovered (SD)
Steel	76	9.2+/−(0.3)
Brick	1	7.3+/−(1.5)
Lumber	1	7.5+/−(1.8)

n = 3.

Table 2. Spores recovered in the runoff.

Panel Type	% Recovery	Log Spores Recovered (SD)
Steel	24	8.7+/−(0.4)
Brick	8	8.2+/−(1.1)
Lumber	16	8.6+/−(0.3)

n = 3.

~46 cm. Panels were visually monitored to ensure that they remain wet for a contact time of 30 min with the decontamination solutions. Panels were allowed to become visually dry by standing in the vertical position for 2 h prior to sampling.

After an initial decon application, the panels were set aside for two weeks, during which time no panels were sampled. Panels were then subsequently treated with a reapplication of the respective decontaminant. The only exception was with CASCAD

Decontamination Efficacy of Different Treatment on Spores of *Bacillus globigii*

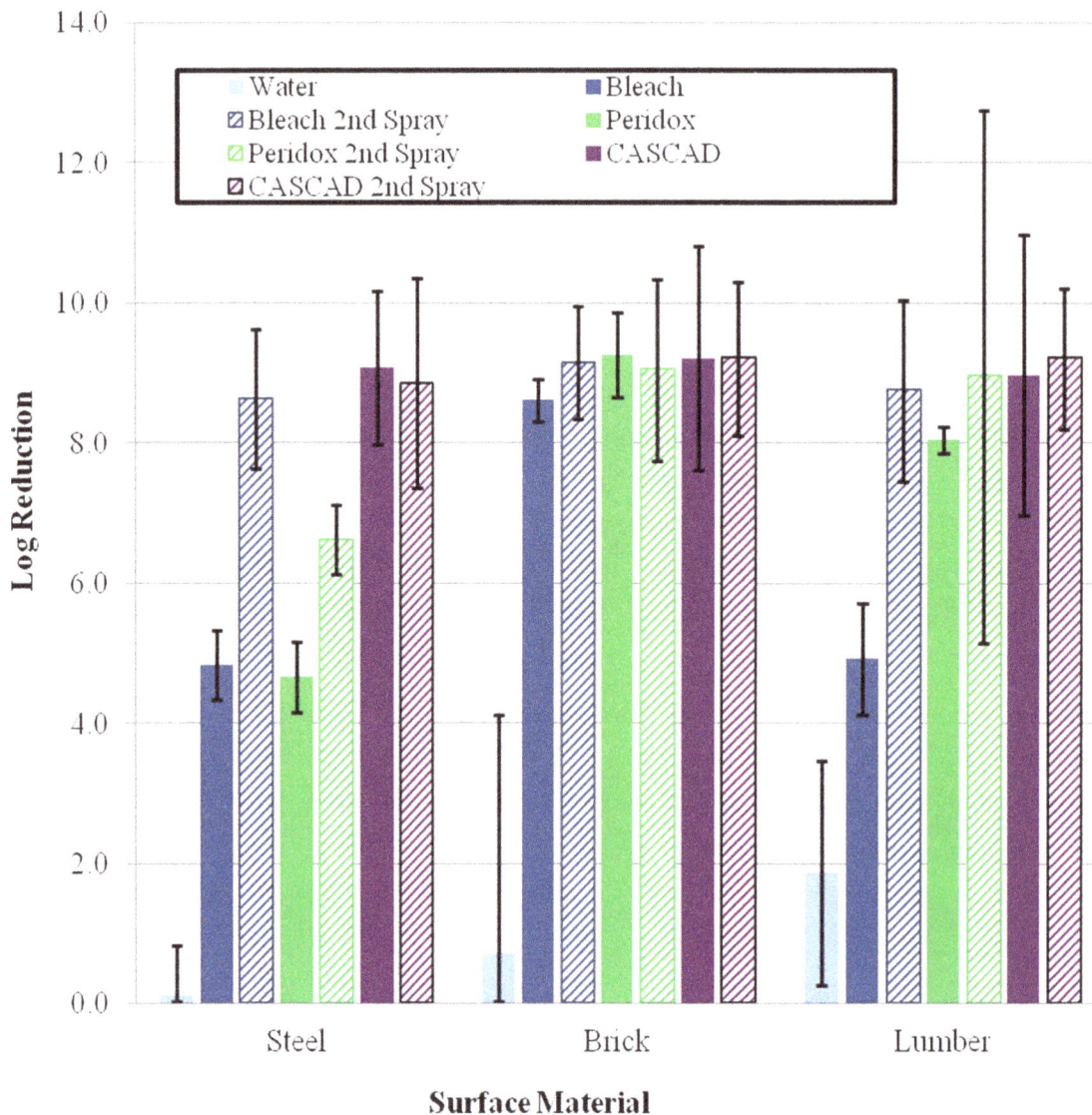

Figure 3. Sporicidal efficacy of three COTS disinfectants.

where half of the panels received a reapplication of CASCAD solution while the other half were rewetted with distilled water. Post-reapplication sampling of the panels was performed identically as described for the initial application.

Sampling Stainless Steel Panels

Stainless steel panels were sampled in 30.5 cm by 30.5 cm sections with each section sampled using 1/8 polyurethane wipe (VWR, Inc., Bridgeport, NJ, SterileWipe* LP Wiper, ITW Texwipe*, Cat. No. TWTX3211). Sixteen pre-wetted wipes were used per panel. Each wipe section was folded into fourths providing four wiping surfaces, with each surface used on the same 30.5-cm^2 section of stainless steel panel. Each panel was wiped ten times along the width, refolded to expose a clean wipe surface, and then wiped 20 times along the length. This procedure was repeated an additional time for a total of four individual wiping events per panel. After wiping, the wipes were placed into individual 50-ml conical tubes each containing 20-ml of recovery solution, PBST (phosphate saline buffer, pH 7.4 containing 0.04% Tween-80). The 16 wipes were processed as individual samples and subsequently the data was pooled. The tubes containing polyurethane wipes, were vortexed for ten minutes using a large capacity mixer (Glas-Col; catalog no. 099A-LC1012; Terra Haute, IN). After vortexing, the tubes were sonicated in a sonic bath (Branson 5510; Branson Ultrasonics Corporation, Danbury, CT) for an additional ten minutes. After spore extraction, the samples were plated in triplicate, using a spiral-plater (Spiral BioTech Autoplate 4000; Advanced Instruments, Norwood, MA). Plates were incubated over-night at 37°C and colonies were counted the next day using a Q-Count instrument (Advanced Instruments, Norwood, MA). The recovered CFUs were recorded and Coefficient of Variations (C.V.s) calculated. The total CFUs estimated per panel were compared to the spore number inoculated and percent recovery was calculated as follows: number of spores recovered/spore number inoculated ×100.

Sampling Brick and Lumber Panels

The two porous surface materials, PT lumber and brick, were sampled with a vacuum sock technology (Midwest Filtration co, Cincinnati, OH, vacuum filter sock collection kit, and Omega HEPA Vacuum, Cat No. FAB-20-01-001A and 950-A1-00-120). Each panel was sampled with a single vacuum sock. The nozzle of the collection tube was placed approximately 1.3 cm above the surface. The nozzle was slowly moved back and forth across the surface using left-to-right horizontal strokes to collect spores. This procedure was repeated two more times touching the nozzle to the surface of panel and using top-to-bottom vertical strokes and left-to-right horizontal strokes. The nozzle was removed from the vacuum hose and vacuum sock removed from the filtration nozzle. The vacuum sock filter was first cut and then placed into 50 mL conical tubes containing 35 mL PBST and pushed down to submerge the filter in the fluid. The spores were extracted from the sock filter after vortexing and sonication as described above.

Sampling Liquid Run-off

Run-off samples were collected from the collection tray under the panel stand. Both test and control panels were maintained wetted for 30 min by repeated sprays. The runoff sample volumes were collected, measured, recorded, and aliquots were serially diluted and plated as described above. Because of the large runoff volumes collected, the samples from test panels were expected to be dilute. Therefore, in order to keep the detection limit low, a 25 mL aliquot of the run-off sample from the collection tray was filtered through a 0.2 μm syringe filter. The filter was then rinsed twice by passing 25 mL of sterile distilled water through the filter. Filters were then placed into conical tubes with extraction buffer and processed as described above. In a separate study, the spore release efficiency from the filter was determined to be 60–80% (data not shown).

Sampling Analysis, Data Handling, and Statistical Treatment

Samples from each of the 16 wipes used to sample stainless steel were enumerated and the data was pooled for each panel. The vacuum filters used to sample lumber and brick, were serially-diluted and plated in triplicates, and the mean of each triplicate recorded. Mean colony-forming units (CFU) counts for each data set were calculated by averaging respective run-off and surface sampled spores. The total CFU numbers were transformed into log10 values. Since three experimental runs, each with three panels were performed, the log(CFU) values were averaged (mean of the logs) and SD values calculated. The log reduction (LR) values were computed by subtracting log(CFU) values from treated panels from that from control panels. For percent recovery (%RE) values were calculated by dividing the mean recoverable CFUs from the sampling material from control panels by total number of spores inoculated onto the panels. Control panels were treated just like test panels, with the exception that water was used in place of disinfectant, and the spore recovery was performed in triplicates.

Results

Sampling Recovery Efficiency

Polyester wipes were used to sample spores off steel panels and vacuum socks were used to sample spores from the other two porous panel types. Recovery efficiencies of sampling technologies were estimated by the number of viable spores recovered from control panels which had not been sprayed with any solution. Greater than 9.2 logs of spores were recovered from the stainless steel panels, which represented approximately 76% of the spores inoculated onto the panels (Table 1). Recovery efficiencies from brick and lumber were significantly lower, approximately 7.3 and

Table 3. Log10 reduction after initial decon application of decontamination solution (SD*).

Panel Type	Disinfectant Type Applied			
	Water	Bleach	Peridox	CASCAD
Steel	0.1+/−(0.7)	4.8+/−(0.5)	4.7+/−(0.5)	9.1+/−(1.1)
Brick	0.7+/−(3.4)	8.6+/−(0.3)	9.3+/−(0.6)	9.2+/−(1.6)
Lumber	1.9+/−(1.6)	4.9+/−(0.8)	8.0+/−(0.2)	9.0+/−(2.0)

* = Standard deviation, **n = 9**.

Table 4. Pair-wise Efficacy Analysis of Decontamination Technologies.

| | First application of Decontamination Technology | | |
	Stainless Steel	**Brick**	**Lumber**
Bleach			
Peridox	NSD	NSD	p<0.005
CASCAD	p<0.05	NSD	NSD
Peridox			
CASCAD	p<0.1	NSD	NSD
	Second application of Decontamination Technology		
	Stainless Steel	**Brick**	**Lumber**
Bleach			
Peridox	p<0.1	NSD	NSD
CASCAD	NSD	NSD	NSD
Peridox			
CASCAD	p<0.2	NSD	NSD
	Reapplication of Decontamination Technology		
	Stainless Steel	**Brick**	**Lumber**
Bleach	p<0.05	NSD	p<0.05
Peridox	NSD	NSD	NSD
CASCAD	NSD	NSD	NSD

All decontamination technologies, when compared to water, resulted in p<0.0001; *NSD = No statistical difference.*

7.5 logs, respectively, which accounted for <1% of the spores inoculated onto these panels.

Control Spores Collected in Runoff

Inoculated panels were sprayed with water to estimate mechanical spore removal from each panel type. Approximately 8.7 logs were recovered in the water runoff from stainless steel panels, which represented 24% of the spores inoculated onto the panels. (Table 2). From the other two panel types, brick and lumber, approximately 8.2 logs or 8%, and 8.6 logs or 16% of the spores inoculated onto the panels were respectively recovered in the water runoff (Table 2).

Efficacy of Decontamination Technologies

All panel types were treated with two applications of decontamination solution. The first application of decontamination solution on stainless steel panels resulted in 4.8, 4.7, and 9.1 log reduction (LR) in number of viable spores when treated with Ultra Clorox Germicidal bleach, Peridox, and CASCAD respectively (Table 3, Table 4, Figure 3). After the second application, the LR values increased to 8.6 (with Bleach), and 6.6 (with Peridox), and remained constant at 8.9 with CASCAD with no statistical difference between the three technologies (Table 4, Table 5, Fig. 3). With the exception of Peridox, two decontaminant applications result in comparable LR values on stainless steel panels (Table 4).

The sporicidal efficacy of all three decontaminants on two porous surfaces, PT lumber and brick, was comparable (LR values of 8.7, 8.9, and 9.2) after second application (Table 4), even though the LR value was significantly lower for bleach on PT lumber after the first application. (Table 3, Table 4, Table 5, and Figure 3).

Table 5. Cumulative log10 reduction after reapplication of decontamination solution (SD*).

| Panel Type | Disinfectant Type Applied | | |
	Bleach	**Peridox**	**CASCAD**
Steel	8.6+/−(1.0)	6.6+/−(0.5)	8.9+/−(1.5)
Brick	9.1+/−(0.8)	9.0+/−(1.3)	9.2+/−(1.1)
Lumber	8.7+/−(1.3)	8.9+/−(3.8)	9.2+/−(1.0)

* = Standard deviation, **n = 9**.

Discussion

Biological sampling and recovery from environmental surfaces is a complex issue, and is typically in the range of 5–60%, especially when inoculated as liquid suspension (7, 12). The spores inoculated on surfaces get partitioned three ways. One, a fraction of spores is irretrievable due to spore lodging into the pores and spore adhesion to the surface matrix. Second, a fraction of surface remains on the sampling tool surfaces, i.e. wipes or vacuum socks. Finally, the third fraction, which is retrieved from the surface by the sampling tools and those released from such sampling matrices. Sporicidal efficacy is determined from the fraction of spores that are accountable in the third fraction, resulting in not accounting for those in the other two fractions. These sampling limitations suggest that additional studies are needed to improve spore recovery by sampling tools.

A methodological approach with the goal of providing operational testing in the context of consequence management following a wide-area release and to assess the sporicidal efficacy of three COTS disinfectants, Ultra Clorox Germicidal bleach, Peridox, and CASCAD, are summarized in this paper. Even though the manufacturer's recommended contact times are ≥ 30 min, it is unreasonable and unrealistic to expect that the vertical surfaces be kept wetted for this long of a period of time in a large area environment. Approximately, 7.6 L of decontaminant was sprayed to ensure a contact time of 30 min for bleach and Peridox. Respraying was performed, every 2–5 min depending on the temperature and RH on a given test day, even though it may be unrealistic in 'real-life' scenario. With CASCAD, approximately 5.7 L per panel was used, and required only a single reapplication. This was due to the foaming/sticking properties of this decontaminant. Typical temperature and RH at the time of testing in June through August were >80% RH and >27°C. Both these physical parameters affected the total volume and the number of times, a given decontamination technology was applied.

With one 30 min application, CASCAD outperformed the other two decontaminants on stainless steel panels and significantly outperformed bleach on lumber panels. This result was not surprising as decontamination attempts using bleach on pinewood has previously been reported as ineffective [12]. The chemical composition of PT lumber can neutralize the active OCl^- species in bleach. On brick, however, all three decontamination technologies performed similarly, which varies with the previously reported performance of Peridox on brick [13]. The Peridox performance discrepancy could in part be due to the type of brick (and the components) used in the two studies. On steel panel, quick run-off of Peridox solution from the smooth vertical surface could have resulted in poor contact times leading to poor performance.

Although both brick and PT lumber are porous materials, and the stainless steel panels maintained their integrity after application of decontamination technology and did not corrode, the effectiveness of bleach to decontaminate these materials greatly differed while that was not true for Peridox and CASCAD suggesting that porosity alone is not responsible for decontamination efficacy (Table 4). It is likely that CASCAD outperformed the other two decontamination technologies due to the foaming, greater adhesion to the surfaces, and/or a higher chlorine content (10 fold higher concentration when following manufacturer's recommendation) compared to the Ultra Clorox Germicidal bleach solution.

While the efficiency and efficacy of the different decontamination technologies varied based on the technology used and the surface treated, sample to sample variation was a common underlying observed phenomenon. For each of the initial technology and surface combination, 10 out of 12 combinations produced CVs>50%, and with the re-sprays, only one combination resulted in a CV<49%. The high CV value is an indication of the potential difficulty in assessing and achieving consistent and efficient large-scale decontaminations. The variation in recovery of viable spores could result from a combination of factors, such as of the size of the panel (1400x larger than typical coupon size of 10-sq-cm), method of inoculation (suspension inoculated as small droplets), and errors associated in consistent application of decontamination technology, and most importantly variations in sampling of large panels.

One notable source of error in this study was the use of the vacuum socks technology for spore recovery from porous materials. A 99% reduction in the number of spores recovered from the porous materials without use of decontaminants documents a significant reduction in spore recovery with the use of current technology. The inconsistent results obtained from this sampling technology suggests limitations for environmental sampling applications. A study by Brown, et al. (2007) evaluated the vacuum filter sock and has identified several characteristics, including pore diameters of over a micron, contributing to the inefficiency of this particular sampling device [14].

In addition to a number of factors affecting sampling efficiencies, inherent characteristics of the surface materials, including porosities, chemistry, and the effects of spore surface composition on adhesion forces to a given surface type are not well understood [15]. A large gap exists with respect to our understanding in how varying porosities of surface material might protect spores from decontaminants. Additionally, if the biological agent is applied to porous materials as a wet aerosol, or applied to a wet surface, or comes into contact with rain prior to decontamination applications, the number of spore in the water run-off from the matrix is unknown and would likely over-estimate LR values [16]. In control experiment in which water was sprayed onto the panels, only 9% of the spores deposited onto the brick and 17% of the spores deposited onto the lumber panels were accounted for in runoff or by vacuum sampling as contrasted to near 100% mass balance accountability from the stainless steel panels. Current capability to estimate the penetration of agent into porous surfaces such as brick and lumber is lacking. The authors acknowledge that the recovery data presented here is a reflection of a number of factors which influence the ability to sample, recover, and culture spores. While the number of spores recovered using the current sampling technology for porous materials and the number of spores collected in the run-off have been quantified, those embedded within the matrix of the brick and lumber after sampling trials is unknown but does contribute to the number of available recoverable spores reported in this study.

Additionally, even though the panels were sampled while visually dry, the impact of retained moisture within the porous matrix of brick and lumber on spore recovery of is unknown. An improvement in vacuum-based or other porous material sampling devices as well as a fundamental understanding of effects of adhesive forces on physical interaction of bio-agent with surface materials is critical to improving the recovery efficiency and decontamination efficacy assessment of wide-area response and recovery efforts.

Author Contributions

Conceived and designed the experiments: JME VKR. Performed the experiments: JME VKR JPS. Analyzed the data: JME VKR JPS. Contributed reagents/materials/analysis tools: JME VKR. Wrote the paper: JME VKR.

References

1. Justice TUSDo (2010) Amerithrax Investigative Summary. 92 p.
2. Ritter S (2001) Anthrax cleanup - Hart Senate office building decontaminated. Chemical & Engineering News 79: 13–13.
3. (2001) EPA alters plan to rid Senate building of anthrax spores. Chemical & Engineering News 79; 18–18.
4. Canter DA, Gunning D, Rodgers P, O'Connor L, Traunero C, et al. (2005) Remediation of Bacillus anthracis contamination in the U.S. Department of Justice mail facility. Biosecur Bioterror 3: 119–127.
5. Han Y, Applegate B, Linton RH, Nelson PE (2003) Decontamination of Bacillus thuringiensis spores on selected surfaces by chlorine dioxide gas. Journal of Environmental Health 66: 16–20.
6. Beuchat LR, Pettigrew CA, Tremblay ME, Roselle BJ, Scouten AJ (2005) Lethality of chlorine, chlorine dioxide, and a commercial fruit and vegetable sanitizer to vegetative cells and spores of Bacillus cereus and spores of Bacillus thuringiensis. Journal of Industrial Microbiology & Biotechnology 32: 301–308.
7. Rastogi VK, Wallace L, Smith LS, Ryan SP, Martin B (2009) Quantitative method to determine sporicidal decontamination of building surfaces by gaseous fumigants, and issues related to laboratory-scale studies. Appl Environ Microbiol 75: 3688–3694.
8. Davies A, Pottage T, Bennett A, Walker J (2011) Gaseous and air decontamination technologies for Clostridium difficile in the healthcare environment. J Hosp Infect 77: 199–203.
9. Otter JA, Puchowicz M, Ryan D, Salkeld JA, Cooper TA, et al. (2009) Feasibility of routinely using hydrogen peroxide vapor to decontaminate rooms in a busy United States hospital. Infect Control Hosp Epidemiol 30: 574–577.
10. de Boer HE, van Elzelingen-Dekker CM, van Rheenen-Verberg CM, Spanjaard L (2006) Use of gaseous ozone for eradication of methicillin-resistant Staphylococcus aureus from the home environment of a colonized hospital employee. Infect Control Hosp Epidemiol 27: 1120–1122.
11. Wilson SC, Wu C, Andriychuk LA, Martin JM, Brasel TL, et al. (2005) Effect of chlorine dioxide gas on fungi and mycotoxins associated with sick building syndrome. Appl Environ Microbiol 71: 5399–5403.
12. Tomasino SF, Rastogi VK, Wallace L, Smith LS, Hamilton MA, et al. (2010) Use of Alternative Carrier Materials in AOAC Official Method(SM) 2008.05, Efficacy of Liquid Sporicides Against Spores of Bacillus subtilis on a Hard, Nonporous Surface, Quantitative Three-Step Method. Journal of Aoac International 93: 259–276.
13. Calfee MW (2010) Biological Agent Decontamination Technology Testing. RESEARCH Triangle Park, NC 27711: U.S. Environmental Protection Agency, Office of Research and development.
14. Brown GS, Betty RG, Brockmann JE, Lucero DA, Souza CA, et al. (2007) Evaluation of vacuum filter sock surface sample collection method for Bacillus spores from porous and non-porous surfaces. J Environ Monit 9: 666–671.
15. Edmonds JM (2009) Efficient methods for large-area surface sampling of sites contaminated with pathogenic microorganisms and other hazardous agents: current state, needs, and perspectives. Appl Microbiol Biotechnol 84: 811–816.
16. Edmonds JM, Collett PJ, Valdes ER, Skowronski EW, Pellar GJ, et al. (2009) Surface sampling of spores in dry-deposition aerosols. Appl Environ Microbiol 75: 39–44.

Purification and Characterization of an Extracellular, Thermo-Alkali-Stable, Metal Tolerant Laccase from *Bacillus tequilensis* SN4

Sonica Sondhi[1], Prince Sharma[1], Shilpa Saini[1], Neena Puri[2], Naveen Gupta[1]*

1 Department of Microbiology, BMS Block, Panjab University, Chandigarh, India, **2** Department of Industrial Microbiology, Guru Nanak Khalsa College, Yamunanagar, Haryana, India

Abstract

A novel extracellular thermo-alkali-stable laccase from *Bacillus tequilensis* SN4 (SN4LAC) was purified to homogeneity. The laccase was a monomeric protein of molecular weight 32 KDa. UV-visible spectrum and peptide mass fingerprinting results showed that SN4LAC is a multicopper oxidase. Laccase was active in broad range of phenolic and non-phenolic substrates. Catalytic efficiency (k_{cat}/K_m) showed that 2, 6-dimethoxyphenol was most efficiently oxidized by the enzyme. The enzyme was inhibited by conventional inhibitors of laccase like sodium azide, cysteine, dithiothreitol and β-mercaptoethanol. SN4LAC was found to be highly thermostable, having temperature optimum at 85°C and could retain more than 80% activity at 70°C for 24 h. The optimum pH of activity for 2, 6-dimethoxyphenol, 2, 2'-azino bis[3-ethylbenzthiazoline-6-sulfonate], syringaldazine and guaiacol was 8.0, 5.5, 6.5 and 8.0 respectively. Enzyme was alkali-stable as it retained more than 75% activity at pH 9.0 for 24 h. Activity of the enzyme was significantly enhanced by Cu^{2+}, Co^{2+}, SDS and CTAB, while it was stable in the presence of halides, most of the other metal ions and surfactants. The extracellular nature and stability of SN4LAC in extreme conditions such as high temperature, pH, heavy metals, halides and detergents makes it a highly suitable candidate for biotechnological and industrial applications.

Editor: Rafael Vazquez Duhalt, Center for Nanosciences and Nanotechnology, Mexico

Funding: These authors have no support or funding to report.

Competing Interests: The authors have declared that no competing interests exist.

* E-mail: gupta27_naveen@yahoo.com

Introduction

Laccases (benzenediol: oxygen oxidoreductases; EC 1.10.3.2) are multicopper oxidases (MCOs) which catalyze the oxidation of a wide variety of organic and inorganic compounds with concomitant four electron reduction of molecular oxygen to water. They catalyze the oxidation of both phenolic and non-phenolic substrates. In general, laccases oxidize phenols and aromatic amines such as methoxyphenols, phenols, polyphenols, anilines, aryl diamines, hydroxyindols, benzenethiols and some cyanide complexes of metals. Laccases are very useful enzymes with respect to their applications in industry. They have found use in industrial and biotechnological applications such as in biobleaching, xenobiotics bioremediation, textile dyes decolorization, biosensors, food industry etc [1].

Laccases are widely distributed in nature. They have been found in almost all spheres of life but have been most extensively studied in fungi including *Ascomycetes*, *Basidiomycetes* and *Deuteromycetes* [2]. Fungal laccases are not stable in extreme conditions like temperature, pH, salt etc. which exists in industry. Moreover, the production of fungal laccases in large quantity is problematic due to the accumulation of large amount of fungal biomass. Bacterial laccases have several significant properties which are not characteristics of fungal laccases like stability at high temperature and pH [3], salt tolerance [4] etc. Only a few bacterial laccases have been characterized till date but they could not be exploited

on an industrial scale as most of them are intracellular or spore bound [5]. There are some reports of extracellular laccase from *Sterptomyces* [6,7,8,9] which are difficult to produce in large quantity because of the problems associated with their filamentous growth [10], slow growth rate and expensive downstream processing [11], thus limiting their applications. Therefore, there is a need to study laccases which are thermo-alkali-stable and produced extracellularly by bacteria. In this regard, previously we have isolated *Bacillus tequilensis* SN4 from the activated sludge of paper mill effluent treatment plant which produces a laccase extracellularly in the culture supernatant. Moreover, on preliminary analysis, this laccase was found to be highly thermostable (optima at 80–90°C) and alkali-stable which makes this laccase a potential candidate for application in industry [12]. In this study we have purified and characterized the enzyme with respect to properties which are important for its industrial applications.

Materials and Methods

Chemicals

Guaiacol, 2, 2'-azino bis[3-ethylbenzthiazoline-6-sulfonate] (ABTS), syringaldazine (SGZ), sephadex G-150 and DEAE-cellulose were purchased from Sigma (USA). Other substrates *viz.* pyrogallol, α-naphthol, p-phenylenediamine (PPD), resorcinol, 2, 6-dimethoxyphenol (DMP), catechol, 3,4-dihydroxy-phenylalanine (L-DOPA) and tyrosine were purchased from Hi-media. Other chemicals used were of analytical grade.

Microorganism and growth conditions

The bacterial strain used in this study, *Bacillus tequilensis* SN4 MTCC no. 11828 (GenBank accession no. *KF150708*) producing extracellular thermo-alkali-stable laccase was previously isolated in our laboratory from activated sludge of paper mill effluent treatment plant [12].

Laccase production

Laccase was produced in M162 medium containing [13], 0.6% yeast extract, 0.2% tryptone, 100 μM CuSO$_4$ and 300 μM MnSO$_4$. The medium was inoculated with 0.3% of 24 h old culture of *B. tequilensis* SN4. Flasks were kept at 30°C; 150 rpm agitation rate for 96 h. After incubation, the culture was centrifuged at 7826 x g for 15 min. The supernatant was used as extracellular enzyme.

Laccase Assay

The enzyme assay was performed at 85°C using 2 mM DMP as substrate in 0.1 M Tris-HCl buffer (pH 8.0) for 5 min. The change in absorbance due to oxidation of DMP was monitored at 470 nm ($\varepsilon = 14800$ M^{-1} cm^{-1}). One unit of laccase was defined as the amount of the enzyme required to transform 1 μmol substrate per min under standard assay conditions. Specific activity was calculated as U mg^{-1} of protein.

Protein estimation

Protein concentration was determined by the method of Lowry et al. [14]. Bovine serum albumin was used as the standard. The proteins eluted from column chromatography were monitored by taking absorbance at 280 nm.

Purification of SN4 laccase

The laccase from *B. tequilensis* SN4 was purified to homogeneity by using a combination of purification techniques. In the first step, proteins were precipitated with acetone [15]. Chilled acetone (60%) was added to the crude enzyme, kept for 2–3 h at −20°C and then centrifuged at 7826 x g for 10 min at 4°C. The pellet was kept in open to evaporate residual acetone and dissolved in 0.1 M Tris-HCl buffer (pH 8.0). The acetone precipitated proteins were applied to Sephadex G-150 (40×1.5 cm^2) column pre-equilibrated with 0.1 M Tris-HCl buffer (pH 8.0). The protein was eluted with the same buffer at a flow rate of 1.0 ml min^{-1}. The active fractions were pooled and concentrated by polyethylene glycol (PEG) [16] and then applied onto DEAE-cellulose column (15×1.2 cm^2) pre-equilibrated with 0.1 M Tris-HCl buffer (pH 8.0). The bound proteins were eluted with a linear gradient of NaCl (0–1.0 M) at a flow rate of 0.8 ml min^{-1}. The active fractions were pooled and concentrated through PEG and dialyzed against same buffer to remove NaCl. The purity of the enzyme was determined by running SDS-PAGE gel electrophoresis.

SDS-PAGE and Activity Staining of Laccase Enzyme

To determine the purity of the protein and the molecular weight of laccase from *B. tequilensis* SN4, SDS- PAGE was performed under complete denaturing conditions [17]. Protein samples were heated for 5 min in the presence of SDS and β-mercaptoethanol. Electrophoresis was done with 5% stacking and 14% separating gel, stained with Coomassie Brilliant Blue R-250 dye and destained with methanol: acetic acid: distilled water (4:1:5). However, for activity staining, samples were heated for 5 min in

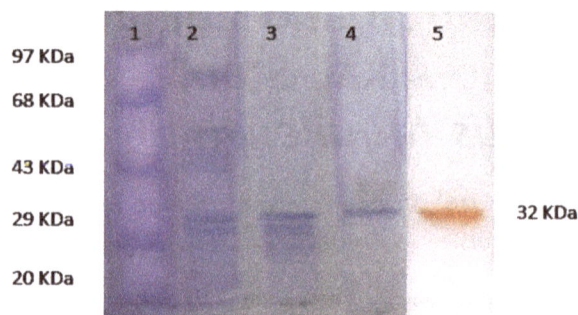

Figure 1. SDS-PAGE analysis of SN4LAC from *B. tequilensis* SN4. (Protein samples were denatured by heating for 5 min in the presence of SDS and β-mercaptoethanol): Lane 1: Protein markers, Lane 2: Acetone precipitated proteins, Lane 3: Sephadex-150 Column purified enzyme, lane 4: DEAE-Cellulose anion exchange Column purified enzyme, Lane 5: Activity staining; purified laccase stained with guaiacol (samples were heated for 5 min in the presence of SDS without β-mercaptoethanol).

the presence of SDS without β-mercaptoethanol (because β-mercaptoethanol inhibited the SN4 laccase activity).The staining of the gel was done with 2 mM guaiacol in 0.1 M Tris-HCl buffer (pH 8.0) at 60°C for half an hour. Laccase activity band was indicated by the development of reddish-brown color. Standard molecular weight protein markers (14–97 KDa) were used to calculate the molecular weight of laccase.

Absorbance spectrum and MALDI-TOF analysis of SN4 Laccase

The UV-visible spectrum of the purified laccase was determined in 0.1 M Tris-HCl buffer (pH 8.0) from 200–800 nm. Identification of protein by peptide mass fingerprinting was carried out by MALDI-TOF/TOF analysis of purified protein. The band corresponding to laccase activity was excised from coomassie stained SDS-PAGE gel, digested with trypsin and peptides were extracted by the method of Shevchenko et al. [18]. For subsequent peptide spectra acquisition and analysis, the matrix-assisted laser desorption/ionization-time of flight (MALDI-TOF) was performed using AB SCIEX MALDI-TOF/TOF 5800. Mass spectrometry data were compared with database in the NCBI and Swiss Prot databases using the Mascot search algorithm.

Substrate specificity and Kinetic Characteristics of SN4 laccase

The ability of SN4 laccase to oxidize several phenolic and non-phenolic substrates *viz.* ABTS, SGZ, guaiacol, pyrogallol, α-naphthol, PPD, resorcinol, DMP, catechol, L-DOPA and tyrosine was determined, at different pH values ranging from 1.0–10.0 [1.0–2.0 (0.1 M KCl-HCl buffer), 2.5–3.5 (0.1 M glycine-HCl buffer), 4.0–5.5 (0.1 M acetate buffer), 6.0–7.5 (0.1 M phosphate buffer), 8.0–9.0 (0.1 M tris-HCl buffer) and 9.5–10.0 (0.1 M carbonate-bicarbonate buffer)]. The relative rate of oxidation for each substrate (at their optimum pH) was compared using the enzyme activity with DMP as 100%.

For analyzing the kinetic properties of laccase, three conventional substrates oxidized by SN4 laccase *viz.* guaiacol, DMP and ABTS were taken at concentration of 100 to 5000 μM. Michaelis-Menton coefficient (K_m) were determined by plotting Line-Weaver Burk plot for each substrate. K_{cat} and V_{max} were also calculated for each substrate.

Table 1. Summary of purification procedures for *B. tequilensis* SN4 laccase.

Purification steps	Total activity (U)	Total protein (mg)	Specific activity (U/mg)	Yield (%)	Purification fold
Crude (Culture supernatant)	5678.94	540	10.52	100	1
Acetone precipitation (60%)	3599.28	195	18.46	63.38	1.75
Sephadex G-150	1507.52	15.79	95.47	26.55	9.07
DEAE-cellulose	757.49	2.53	299.40	13.34	28.46

Effect of inhibitors on laccase activity

The effect of known inhibitors of laccase *viz.* sodium azide (NaN_3), ethylenediaminetetraacetic acid (EDTA), diethylenetriaminepentaacetic acid (DTPA), cysteine monohydrate, dithiothreitol (DTT) and β-mercaptoethanol were studied at 1, 5 and 10 mM concentrations. The enzyme was incubated with the inhibitors for half an hour at 37°C with constant shaking at 150 rpm. The residual activity was then analyzed as per standard assay conditions.

Effect of temperature on laccase activity and stability

Effect of temperature on purified laccase was determined by oxidation of DMP at temperature ranging from 55–100°C at an interval of 5°C. The maximum enzyme activity was taken as 100% and relative activities were plotted.

Thermostability of the enzyme was measured over the temperature range of 65–85°C by incubating the enzyme in thin-wall test tubes for a time period of 0–24 h. At different time intervals, aliquots were withdrawn and residual activity was determined under standard assay conditions.

Effect of pH on laccase activity and stability

The pH dependence of laccase activity was determined for guaiacol (2 mM), ABTS (2 mM), DMP (2 mM) and SGZ (50 μM) as substrate by performing the enzyme assay at different pH values ranging from 1.0, 1.5 and 2.0 (0.1 M KCl-HCl buffer), 2.5, 3.0 and 3.5 (0.1 M Glycine-HCl buffer), 4.0, 4.5, 5.0 and 5.5 (0.1 M Acetate buffer), 6.0, 6.5, 7.0 and 7.5 (0.1 M phosphate buffer), 8.0, 8.5 and 9.0 (0.1 M Tris-HCl buffer) and 9.5 and 10.0 (0.1 M carbonate-bicarbonate buffer). The reaction with guaiacol, ABTS and DMP was carried out for 5 min and SGZ for 1 min. The oxidation of substrates was monitored at 465 nm for guaiacol ($\varepsilon = 12000$ M^{-1} cm^{-1}), 420 nm for ABTS ($\varepsilon = 36000$ M^{-1} cm^{-1}), 470 nm for DMP ($\varepsilon = 14800$ M^{-1} cm^{-1}), and 525 nm for SGZ ($\varepsilon = 64000$ M^{-1} cm^{-1}). The maximum enzyme activity was taken as 100% and relative activities were plotted.

Stability of the enzyme was measured at pH 8.0, 8.5 and 9.0 by incubating the enzyme in buffers of various pH values for 0–24 h at 65°C. At various time intervals, aliquots of enzyme were withdrawn and the residual activity was determined as per standard assay conditions.

Effect of metal ions, halides and surfactants on laccase activity

The effect of halides, metal ions and surfactants on laccase activity was studied by preincubating the enzyme for 30 min at 37°C, 150 rpm with 0.5–1.0 mM conc. of various metal ions including $FeSO_4.7H_2O$, $CuSO_4.5H_2O$, $NiSO_4.6H_2O$, $LiSO_4.H_2O$, $CaSO_4.2H_2O$, $CoSO_4.7H_2O$, $MnSO_4.H_2O$, $HgSO_4$, $ZnSO_4.7H_2O$, $MgSO_4.7H_2O$ and $Al_2(SO_4)_3.16H_2O$; 100–500 mM conc. of various halides NaF, NaCl, NaBr, NaI and 0.1–1.0 mM concentration of surfactants including non-ionic (triton X-100, tween-20, tween-80), anionic (SDS), cationic (CTAB). Enzyme without any agent was taken as control. The residual activity was then analyzed as per standard assay conditions.

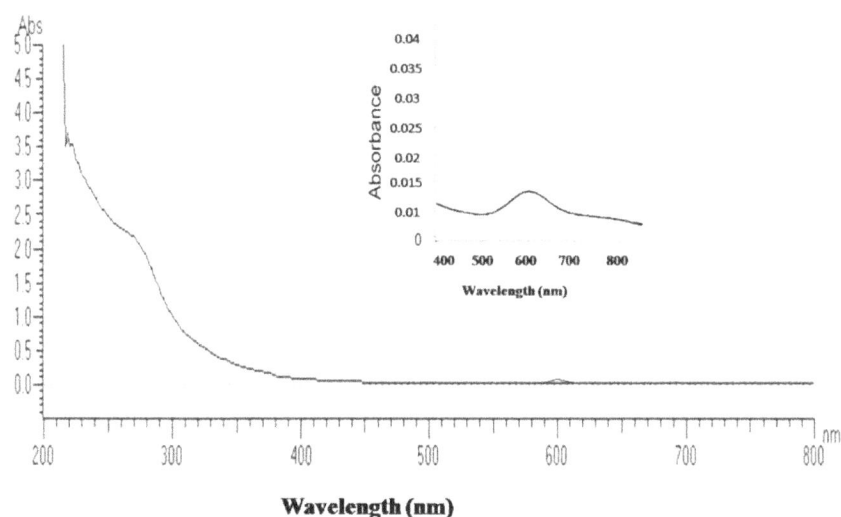

Figure 2. UV-visible spectrum (200–700 nm) of purified SN4LAC. The inset figure shows the zoomed image of the spectrum of SN4LAC in the range of 400–800 nm.

gi|321314354 spore copper-dependent laccase [*Bacillus subtilis* BSn5]

Score: 147; Mass: 58674; Matches:12(7); Sequences: 12(7); emPAI: 0.80

1 MTLEKFVDAL	PIPDTLKFVQ	QSKEKTYYEV	TMEECTHQLH	RDLPPTRLWG
51 YNGLFPGPTI	EVKRNENVYV	KWMNNLPSTH	FLPIDHTIHH	SDSQHEEPEV
101 KTVVHLHGGV	TPDDSDGYPE	AWFSKDFEQT	GFYFKREVYH	YPNQQRGAIL
151 WYHDHAMALT	RLNVYAGLVG	AYIIHDPKEK	RLKLPSDEYD	VPLLITDRTI
201 NEDGSLFYPS	APENPSPSLP	NPSIVPAFCG	ETILVNGKVW	FYLEVEPRKY
251 RFRVINASNT	RTYNLSLDNG	GEFIQIGSDG	GLLPRSVKLN	SFSLAPAERY
301 DIIIDFTAYE	GESIILANSA	GCGGDVNPET	DANIMQFRVT	KPLAQKDESR
351 KPKYLASYPS	VQHERIQNIR	TLKLAGTQDE	YGRFVLLLNN	KRWHDPVTEA
401 PKVGTTEIWS	IINPTRGTHP	IHLHLVSFKV	LDRRPFDIAR	YQESGELSYT
451 GPAVPPPPSE	KGWKDTIQAH	AGEVLRIAAT	FGPYSGRYVW	HCHILEHEDY
501 DMMRPMDITD	PHK			

Figure 3. Peptide sequence showing identity with spore bound copper dependent laccase of *Bacillus subtilis* **BSn5.** Peptide sequences detected by tryptic digestion of SN4 laccase was shown in bold red.

Statistical Analysis

All the experiments were carried out in triplicates and the mean ± standard deviation has been plotted. Data was analyzed using analysis of variance (ANOVA) by Sigma Stat version 2.03 and values which were statistically significant (p value <0.05) were taken.

Results

Protein Purification and Molecular weight determination

Extracellular secretion of SN4 laccase made its purification easier in comparison to all other known bacterial laccases which are either intracellular or spore bound. The purification was carried out using a combination of routine chromatography procedures. After precipitation with 60% acetone; the protein was applied on Sephadex G-150 column; laccase active fractions were pooled and subjected to DEAE-cellulose column and eluted with NaCl (Figure S1 in File S1). After final step, the enzyme was purified to 28.46 fold with a yield of 13.34% (Table 1). The SDS-PAGE analysis of purified protein showed a band of 32 KDa which corresponded to the activity staining band of laccase (Fig. 1). This purified laccase was designated as SN4LAC.

UV-visible spectrum and MALDI-TOF analysis of SN4LAC

UV-visible spectrum of purified SN4LAC showed an absorption peak at 600 nm (corresponding to the T1 copper centre) and a

Table 2. Substrate profile of Laccase from *B. tequilensis* SN4.

Substrate	ε (M^{-1} cm^{-1})	λ_{max} (nm)	Optimum pH	Relative activity (%)
DMP	14800	468	8.0	100
ABTS	36000	420	5.5	93.90±1.56
SGZ	64000	525	6.5	95.82±0.78
Guaiacol	12000	465	8.0	88.36±0.54
α-Naphthol	2200	330	6.0	93.95±1.25
Catechol	2211	450	7.0	19.21±1.38
Pyrogallol	35000	450	6.5	35.79±2.78
Tyrosine	12000	278	-	ND
PPD	14685	450	7.0	51.67±1.45
Resorcinol	6220	340	-	ND
L-DOPA	3600	475	9.0	69.17±0.24

ND- not detected.
Values represent mean ± SD (n = 3).

Table 3. Kinetic properties of purified SN4LAC in comparison to other laccases.

Substrates	Bacillus tequilensis SN4			Bacillus sp. HR03 [23]			Streptomyces ipomea [8]		
	K_m (mM)	k_{cat} (s^{-1})	k_{cat}/K_m (mM^{-1} s^{-1})	K_m (mM)	k_{cat} (s^{-1})	k_{cat}/K_m (mM^{-1} s^{-1})	K_m (mM)	k_{cat} (s^{-1})	k_{cat}/K_m (mM^{-1} s^{-1})
ABTS	1.404±0.08	67.04	47.82	0.535	127	237	0.4	9.99	24.975
DMP	0.840±0.012	73.15	87.05	0.053	3	56.60	4.2	4.20	1
Guaiacol	3.258±0.096	62.96	19.34	-	-	-	-	-	-

Values represent mean ± SD (n = 3).

slight shoulder at 330 nm (corresponding to the T3 binuclear copper centre) (Fig. 2). On MS/MS analysis, the protein was identified with significant protein scores (p>0.05) from Mascot searches of peptide mass fingerprints (Figure S2 in File S2). On MALDI-TOF analysis of 32 KDa protein band showed homology with laccases of other strains of *B. subtilis* being maximum with spore bound copper dependent laccase from *B. subtilis* BSn5 [19] with a score of 147 and query coverage of 34% (Fig. 3).

Substrate specificity and Kinetic properties of SN4LAC

SN4LAC was able to oxidize phenolic as well as non phenolic substrates. The optimum pH for each substrate was calculated and the rate of oxidation for different substrates was compared at their respective pH optima. SN4LAC oxidized o-phenols (in order of DMP>guaiacol>L-DOPA>pyrogallol>catechol), p-phenols (PPD) and other substrates of laccase such as ABTS, α-naphthol and SGZ, but poorly oxidized m-phenols (resorcinol) and no activity towards tyrosine was observed (Table 2).

The reaction rate of SN4LAC was dependent on substrate concentration and followed Michaelis-Menton kinetics. The K_m and k_{cat} for ABTS was 80±4 μM and 291±2.7 s^{-1}, for DMP was 680±27 μM and 11±0.1 s^{-1} and for guaiacol was 3.289±0.06 and 63±0.1 (Table 3).

Effect of inhibitors on laccase activity

SN4LAC was inhibited by the common inhibitors of laccase (Table 4). When reducing agents *viz.* cysteine, DTT and β-mercaptoethanol were added, no enzyme activity could be observed. 10 mM concentration of NaN$_3$, EDTA and DTPA decreased the enzyme activity to 32%, 22.8% and 23.62% respectively.

Temperature and pH optima and stability

The purified SN4LAC showed maximum activity in the temperature range of 80–90°C having optima at 85°C (Fig. 4a). The enzyme could retain 50% activity even at 100°C. The SN4LAC retained more than 80% activity at 70°C and was completely stable at 65°C for 24 h. The half life of SN4LAC was 4 h, 3 h and 1 h at 75°C, 80°C and 85°C respectively (Fig. 4b).

The pH optima of purified SN4LAC for four different substrates *viz.* ABTS, DMP, SGZ and guaiacol was found to be 5.5, 8.0, 6.5 and 8.0 respectively (Fig. 5a). Enzyme was found to be highly stable in the alkaline pH range. The enzyme could retain 75% activity even after 24 h incubation at pH 9.0 (Fig. 5b).

Effect of halides and metal ions on SN4LAC activity

SN4LAC retained 75–80% activity at 500 mM concentration of halides. Cu^{2+} and Co^{2+} increased the SN4LAC activity to 126% and 150% respectively at 5.0 mM concentration and in the presence of other metal ions, enzyme retained 70–80% activity. Hg^{2+} and Fe^{2+} inhibited the laccase activity to 27% and 40% respectively (Table 5).

Effect of surfactants on SN4LAC activity

The effect of various surfactants at 0.1, 0.5 and 1.0 mM concentration was studied on SN4LAC activity. Enzyme was quite stable in the presence of cationic and anionic detergents. An increase in enzyme activity in the presence of CTAB (38%) and SDS (20%) was observed at 0.1 mM concentration. In the presence of non-ionic detergents, the SN4LAC was stable at lower concentrations, however, at higher concentrations [above Critical Micelle Concentration (CMC)] the SN4LAC activity decreased by 15–30% (Table 6).

Table 4. Effect of inhibitors on SN4LAC activity.

Inhibitors	Conc. (mM)	Relative activity (%)
Control	----	100
Sodium azide	1	76.95±3.82
	5	53.86±1.40
	10	32.57±1.22
DTPA	1	33.99±.66
	5	31.82±0.72
	10	23.81±1.06
EDTA	1	42.45±1.03
	5	27.83±0.18
	10	21.53±1.18
Cysteine	1	63.75±1.65
	5	1.84±0.05
	10	1.45±0.05
Dithiothreitol	1	20.56±1.28
	5	0
	10	0
B-mercaptoethanol	1	0
	5	0
	10	0

Values represent mean ± SD (n = 3) relative to untreated control sample.

Discussion

Apart from *Streptomyces* [6,7,8,9] most of the reported bacterial laccases to date are either intracellular or spore bound [5] making their industrial application unfeasible. SN4LAC is an extracellular and highly thermo-alkali-stable laccase and is thus an attractive candidate for industrial applications [12]. In this study, laccase from *B. tequilensis* SN4 (SN4LAC) has been purified and characterized. SN4LAC was purified to homogeneity with a purification fold of 28.46 and yield of 13.34%. SN4LAC is a monomeric protein with a molecular weight of 32 KDa. In contrast, the molecular weight of all other fungal [2] and bacterial laccases [5] is in the range of 50–100 KDa. This difference in

molecular mass makes SN4LAC an interesting protein for studying structure-function relationship of laccases.

The UV-visible spectrum (200–800 nm) of purified SN4LAC showed characteristic peak at 600 nm corresponding to the presence of Type 1 copper center and a shoulder at 330 nm corresponding to the presence of Type 3 copper center, a characteristic of blue laccases [20] confirming it to be belonging to multicopper oxidase family. True laccase nature of SN4LAC was further supported by MALDI-TOF analysis of purified protein showed homology with laccases of other strains of *B. subtilis* being maximum with spore bound copper dependent laccase from *B.subtilis* BSn5 [19]. The extracellular laccase like

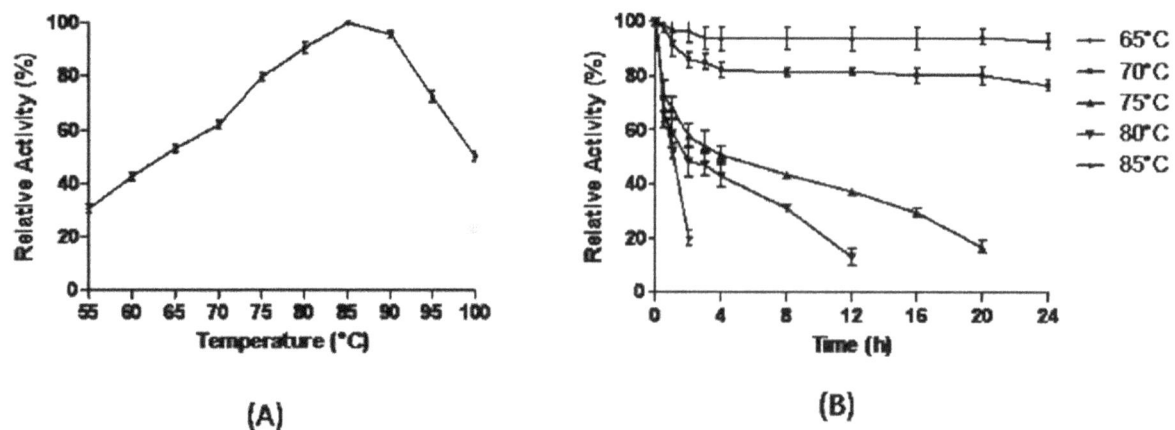

Figure 4. Effect of temperature on SN4LAC activity. (a) Optimum temperature of laccase activity (b) Stability of enzyme at various temperatures.

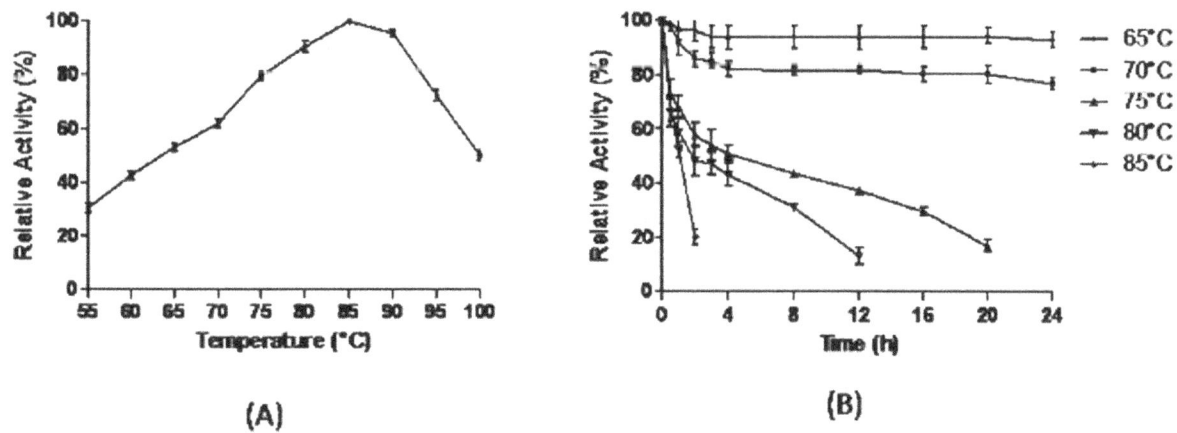

Figure 5. Effect of pH on SN4LAC activity. (a) Optimum pH for different substrates (b) Stability of laccase at different pH values.

Table 5. Effect of metal ions and halides on SN4LAC activity.

Metal Ions	Conc. (mM)	Relative Activity (%)
Control	-	100
$CuSO_4.5H_2O$	1.0	106.56±1.36
	5.0	126.09±1.54
$CaSO_4.2H_2O$	1.0	83.66±1.58
	5.0	75.47±0.92
$NiSO_4.6H_2O$	1.0	86.25±1.85
	5.0	76.84±1.58
$HgSO_4$	1.0	51.99±1.47
	5.0	26.60±1.22
$MgSO_4.7H_2O$	1.0	99.18±0.70
	5.0	76.99±1.22
$MnSO_4.H_2O$	1.0	83.59±1.10
	5.0	79.48±0.66
$LiSO_4.H_2O$	1.0	82.38±1.51
	5.0	77.81±0.84
$CoSO_4.7H_2O$	1.0	117.98±2.50
	5.0	150.69±1.81
$FeSO_4.7H_2O$	1.0	76.43±0.96
	5.0	41.47±1.37
$ZnSO_4.7H_2O$	1.0	77.34±1.33
	5.0	67.47±1.61
$Al_2(SO_4)_3.16H_2O$	1.0	82.94±2.63
	5.0	71.07±2.21
NaF	100	85.77±1.45
	500	75.91±2.58
NaCl	100	85.61±3.23
	500	75.23±1.36
NaBr	100	76.88±4.56
	500	75.33±3.25
NaI	100	84.69±3.45
	500	79.02±2.25

Values represent mean ± SD (n = 3) relative to untreated control sample.

Table 6. Effect of surfactants on SN4LAC activity.

Surfactants	Conc. (mM)	Relative activity (%)
Control	----	100
Tween 20	0.1	89.70±1.58
	0.5	78.60±2.56
	1.0	72.85+2.48
Tween 80	0.1	88.90±2.17
	0.5	76.85±1.12
	1.0	70.04±1.58
Triton X-100	0.1	100±0.58
	0.5	95.32±2.48
	1.0	84.54±2.45
SDS	0.1	120.58±1.45
	0.5	112.15±1.85
	1.0	105.80±1.78
CTAB	0.1	138.90±1.27
	0.5	133.50±0.45
	1.0	132.45±2.54

Values represent mean ± SD (n = 3) relative to untreated control sample.

protein from *Bacillus* sp. ADR did not show peak at 600 nm and was also unable to oxidize laccase specific substrates SGZ and ABTS, thus is not a true laccase [21].

SN4LAC showed wide substrate specificity. It was able to oxidize non-phenolic as well as phenolic substrates. However, the rate of oxidation of phenolic substrates varied with the nature and substitution on the phenolic ring. This difference in oxidation due to substitution on phenolic ring has been reported in case of other fungal as well as bacterial laccases [22]. It is known that laccase and tyrosinase have an overlapping range of substrates; the ability of an enzyme to oxidize SGZ and ABTS, with an inability to oxidize tyrosine, is an indicator of true laccase activity [2]. As SN4LAC was able to oxidize SGZ and ABTS but not tyrosine, it is a true laccase. The K_m of SN4LAC towards the conventional substrates showed that binding affinity of SN4LAC was in the order of DMP>ABTS>guaiacol indicating that DMP is the most suitable substrate for SN4LAC with lowest K_m and maximum V_{max}. Similar results have been observed for Cot A laccase from *Bacillus* sp. HR03 [23]. Further, K_m of SN4LAC for DMP is much less [8] or comparable [9] than extracellular laccases from *Streptomyces* (Table 3).

SN4LAC was also inhibited by the known inhibitors of laccase. The inhibition by sodium azide can be explained by the binding of N_3^- to the trinuclear copper center, that affect internal electron transfer, which ultimately affect the overall oxidation process catalyzed by laccase [24]. EDTA and DTPA deprive the Cu^{2+} ions present at type 1 copper centre and inhibit the enzyme activity, revealing the role of Cu^{2+} ion in laccase function [25]. The enzyme activity was not observed when reducing agents such as DTT, cysteine and β-mercaptoethanol were added to the reaction mixture. This can be due to reduction of the oxidized substrate by the sulfhydryl groups of the redox reagents [26]; this has also been observed with other laccases [3].

The purified SN4LAC has temperature optima of 85°C, slightly less than the crude enzyme (90°C). The SN4LAC is completely stable for 24 h at 65°C and retained more than 80% activity at 70°C. The half life of SN4LAC was 4 h, 3 h and 1 h at 75°C,

80°C and 85°C respectively. SN4Lac is more stable than extracellular laccase from *Streptomyces*. The highest stable laccase from *S. cyaneus* CECT 3335 is having temperature optima of 70°C and retained 75% activity for 24 h only at 50°C [6]. Moreover, SN4 laccase has been found to be more thermostable in comparison to other bacterial laccases [23,25].

Due to the difference in the redox potential of the Type 1 Copper of laccase and the substrate, the optimum pH of enzyme activity varies with the type of substrate used [1]. The optimum pH of SN4LAC for SGZ, DMP and guaiacol was observed to be 6.5, 8.0 and 8.0 respectively. The pH optima of the phenolic substrates towards alkaline range can be explained by the redox potential difference between the phenol and the T1 copper of laccase, (the driving force for electron transfer) which increases with increase in pH [4]. With ABTS, the optimum pH of SN4LAC is 5.5 which can be explained by the non-phenolic nature of ABTS [23]. The optimum pH of SN4LAC activity for various substrates is higher than that reported for other *Bacillus* spp. [22,27]. This shift in pH to alkaline range indicates that SN4LAC is alkali-stable. This fact is supported by the pH stability experiments in which SN4LAC retained 75% activity even after 24 h incubation at pH 9.0. This stability makes it a suitable candidate to be used in industries where high pH is required.

Halides as well as metal ions are known to bind to enzymes and alter their stability [28]. In the presence of halides, SN4LAC retained 75–80% activity whereas other bacterial laccases have been reported to retain only 20–50% activity [3,23]. Cu^{2+} and Co^{2+} increased the SN4LAC activity; similar to the results reported by Murugesan et al. [29] for laccase from *Ganoderma lucidum*. The positive effect of Cu^{2+} on laccase activity can be explained due to the filling of type 1 Copper binding site by Cu ions [1]. Fe^{2+} and Hg^{2+} inhibited the activity of SN4LAC, similar results have been reported for laccase from *Ganoderma lucidum* [29]. It has been reported that Hg^{2+} ions have strong affinity for sulfhydral (-SH) groups, binding to which causes distortion of enzyme structure [30]. Inhibitory effect of Fe^{2+} may be due to its interaction with electron transport system of laccase [29]. The enzyme was almost stable in the presence of other metal ions including Ca^{2+}, Mg^{2+}, Mn^{2+}, Li^{2+}, Zn^{2+}, Ni^{2+} and Al^{3+}. Stability of SN4LAC in presence of most of the metal ions makes it suitable for applications where they are present in high concentrations e.g. pulp and paper industry, wastewater containing heavy metals [1].

Ionic surfactants have been reported to inhibit laccase activity in most of the cases [3,31]. However, the stimulation of activity by ionic surfactants has been reported for laccase from *Azospirillum lipoferum* [32]. SN4LAC activity was also stimulated by ionic surfactants i.e. CTAB and SDS. This stimulation can be explained on the basis of hypothesis given by Moore and Flurkey [33] that binding of these surfactants (generally, below the CMC) to the enzyme may cause the alterations in its enzymatic and physical characteristics.

In the presence of non-ionic detergents, the SN4LAC was stable at lower concentration, however, in contrast to ionic detergents SN4LAC activity decreased at higher concentrations of these detergents. Decrease in enzyme activity at high concentration of non-ionic detergents can be due to the reason that concentration of these detergents is higher than their CMC in the assay mixture. Difference in stability in the presence of non-ionic detergents as compared to ionic detergents at same concentration can be due to the fact that the CMC of non-ionic detergents is known to be lower by 1 order than that of ionic detergents. Thus, SN4LAC is also stable in the presence of ionic and non-ionic surfactants below their CMC. Stability of SN4LAC in the presence of surfactants makes it further useful for application in the treatment of industrial waste like dye degradation in effluent from textile industry.

Conclusion

A novel thermo-alkali-stable extracellular laccase from *Bacillus tequilensis* SN4 has been purified to homogeneity. Purified enzyme is smaller in size than other known laccases. This makes it an interesting protein for structure-function studies. Detailed characterization showed that this laccase is highly stable at high temperature and pH. Moreover, it can work in the presence of various halides, metal ions, surfactants etc. These characteristics make it an ideal candidate for industrial applications where such extreme conditions exist.

Author Contributions

Conceived and designed the experiments: S. Sondhi S. Saini. Performed the experiments: S. Sondhi. Analyzed the data: PS. Contributed reagents/materials/analysis tools: NG. Wrote the paper: S. Sondhi NG. Contributed in the revision of manuscript: NP.

References

1. Shraddha, Shekhar R, Sehgal S, Kamthania M, Kumar A (2011) Laccase: Microbial Sources, Production, Purification, and Potential Biotechnological Applications. Enzyme Res 2011:217861.

2. Brijwani K, Rigdon A, Vadlani PV (2010) Fungal Laccases: Production, Function, and Applications in Food Processing. Enzyme Res 2010:149748.

3. Zhang C, Zhang S, Diao H, Zhao H, Zhu X, et al. (2013) Purification and characterization of a temperature and pH-stable laccase from the spores of *Bacillus Vallismortis* fmb-103 and its application in the degradation of malachite green. J Agric Food Chem 61: 5468–5473.

4. Ruijssenaars HJ, Hartmans S (2004) A cloned *Bacillus halodurans* multicopper oxidase exhibiting alkaline laccase activity. Appl Microbiol Biotechnol 65: 177–182.

5. Sharma P, Goel R, Capalash N (2007) Bacterial laccases. World J Microbiol Biotechnol 23: 823–832.

6. Arias ME, Arenas M, Rodriguez J, Soliveri J, Ball AS, et al. (2003) Kraft pulp biobleaching and mediated oxidation of a nonphenolic substrate by laccase from *Streptomyces cyaneus* CECT 3335. Appl Environ Microbiol 69(4): 1953–1958.

7. Niladevi KN, Prema P (2007) Effect of inducers and process parameters on laccase production by *Streptomyces psammoticus* and its application in dye decolorization. Bioresour Techol 99: 4583–4589.

8. Molina-Guijarro JM, Perez J, Munoz-Dorado J, Guillen F, Moya R, et al. (2009) Detoxification of azo dyes by a novel pH-versatile, salt-resistant laccase from *Streptomyces ipomoea*. Inter Microbiol 12:13–21.

9. Gunne M, Urlacher VB (2012) Characterization of the alkaline laccase Ssl1 from *Streptomyces viceus* with unusual properties discovered by genome mining. PLoS ONE 7(12): e52360. doi:10.1371/journal.pone.0052360.

10. Van Wezel GP, Krabben P, Traag BA, Keijser BJF, Kerste R, et al. (2006) Unlocking Streptomyces spp. for use as sustainable industrial production platforms by morphological engineering. Appl Environ Microbiol 72(10): 6863–6863.

11. Gomes J, Menawat AS (1998) Fed-batch bioproduction of spectinomycin. Adv Biochem Eng Biotechnol 59:1–46.

12. Sondhi S, Sharma P, George N, Chauhan PS, Puri N, et al. (2014) An extracellular thermo-alkali-stable laccase from *Bacillus tequilensis* SN4, with a potential to biobleach softwood pulp. 3 Biotech DOI 10.1007/s13205-014-0207-z.

13. Degryse E, Glansdorff N, Pierard A (1978) A comparative analysis of extreme thermophilic bacteria belonging to the genus Thermus. Arch Microbiol 117:189–196.

14. Lowry OH, Rosebrough NJ, Farr AL, Randall RJ (1951) Protein measurement with folin phenol reagent. J Biol Chem 193: 265–275.

15. Chauhan PS, George N, Sondhi S, Puri N, Gupta N (2014) An overview of purification strategies for microbial mannanases. Int J Pharm Bio Sci 5: 176–192.

16. Degerli N, Akpinar MA (2001) A novel concentration method for concentrating solutions of protein extracts based on dialysis techniques. Anal Biochem 297: 192–194.

17. Laemmli UK (1970) Cleavage of structural proteins during assembly of the head of bacteriophage T4. Nature 277: 680–682.

18. Shevchenko A, Tomas H, Havli J, Olsen JV, Mann M (2006) In-gel digestion for mass spectrometric characterization of proteins and proteomes. Nat Protoc 1: 2856–2860.

19. Deng Y, Zhu Y, Wang P, Zhu L, Zheng J, et al. (2011) Complete genome sequence of *Bacillus subtilis* BSn5, an endophytic bacterium of amorphophallus konjac with antimicrobial activity for the plant pathogen *Erwinia carotovora* subsp. *Carotovora*. J Bacteriol 193(8): 2070–2071.

20. Martins LO, Soares CM, Pereira MM, Teixeira M, Costa T, et al. (2002) Molecular and biochemical characterization of a highly stable bacterial laccase that occurs as structural component of the *Bacillus subtilis* Endospore Coat. J Biol Chem 277(21):18849–18859.

21. Telke AA, Ghodake GS, Kalyani DC, Dhanve RS, Govindwar SP (2011) Biochemical characteristics of a textile dye degrading extracellular laccase from a *Bacillus* sp. ADR. Bioresour Technol 102:1752–1756.

22. Koschorreck K, Richter SM, Ene AB, Roduner E, Schmid RD, et al. (2008) Cloning and characterization of a new laccase from *Bacillus licheniformis* catalyzing dimerization of phenolic acids. Appl Microbiol Biotechnol 79: 217–224.

23. Mohammadian M, Fathi-Roudsari M, Mollania N, Badoei-Dalfard A, Khajeh K (2010) Enhanced expression of a recombinant bacterial laccase at low temperature and microaerobic conditions: Purification and biochemical characterization. J Ind Microbiol Biotechnol 37: 863–869.

24. Ryan S, Schnitzhufer W, Tzanov T, Cavaco-Paulo A, Gubitz GM (2003) An acid-stable laccase from *Sclerotium rolfsii* with potential for wool dye decolourization. Enzyme Microb Technol 33: 766–774.

25. Kaushik G, Thakur IS (2013) Purification, characterization and usage of thermotolerant laccase from *Bacillus* sp. for biodegradation of synthetic dyes. Appl Biochem Microbiol 49(4): 352–359.

26. Johannes C, Majcherczyk A (2000) Laccase activity tests and laccase inhibitors. J Biotechnol 78: 193–199.

27. Reiss R, Ihssen J, Meyer LT (2011) *Bacillus pumilus* laccase: a heat stable enzyme with a wide substrate spectrum. BMC Biotechnol 11:9.

28. Admafio NA, Sarpong NS, Mensah CS, Obodai M (2012) Extracellular laccase from *Pleurotus ostreatus* strain EM-1: Thermal stability and response to metal ions. Asian J Biochem 7(3): 143–150.

29. Murugesan K, Kim Y, Jeon J, Chang Y (2009) Effect of metal ions on reactive dye decolorization by laccase from *Ganoderma lucidum*. J Hazard Mater 168: 523–529.

30. Baldrian P, Gabriel J (2002) Copper and cadmium increase laccase activity in *Pleurotus ostreatus*. FEMS Microbiol Lett 206: 69–74.

31. Robles A, Lucas R, Martinez-canamero M, Omar NB, Peres R, et al. (2002) Characterization of laccase activity produced by the hyphomycete *Chalara* (syn. Thielaviopsis) *paradoxa* CH32. Enzyme Microb Technol 31: 516–522.

32. Diamantidis G, Effosse A, Potier P, Bally R (2000) Purification and characterization of the first bacterial laccase in the rhizospheric bacterium *Azospirillum lipoferum*. Soil Biol Biochem 32 919–927.

33. Moore BM, Flurkey WH (1990) SDS activation of a plant poly-phenoloxidase. J Biological Chem 265: 4982–4988.

Spatial Segregation and Aggregation of Ectomycorrhizal and Root-Endophytic Fungi in the Seedlings of Two *Quercus* Species

Satoshi Yamamoto[1]*, **Hirotoshi Sato**[1], **Akifumi S. Tanabe**[2], **Amane Hidaka**[3], **Kohmei Kadowaki**[1], **Hirokazu Toju**[1]

1 Graduate School of Human and Environmental Studies, Kyoto University, Kyoto, Japan, 2 National Research Institute of Fisheries Science, Fisheries Research Agency, Yokohama, Japan, 3 Network center of Forest and Grassland Survey in Monitoring Sites 1000 Project, Japan Wildlife Research Center, c/o Filed Science Center for Northern Biosphere, Hokkaido University, Tomakomai, Japan

Abstract

Diverse clades of mycorrhizal and endophytic fungi are potentially involved in competitive or facilitative interactions within host-plant roots. We investigated the potential consequences of these ecological interactions on the assembly process of root-associated fungi by examining the co-occurrence of pairs of fungi in host-plant individuals. Based on massively-parallel pyrosequencing, we analyzed the root-associated fungal community composition for each of the 249 *Quercus serrata* and 188 *Quercus glauca* seedlings sampled in a warm-temperate secondary forest in Japan. Pairs of fungi that co-occurred more or less often than expected by chance were identified based on randomization tests. The pyrosequencing analysis revealed that not only ectomycorrhizal fungi but also endophytic fungi were common in the root-associated fungal community. Intriguingly, specific pairs of these ectomycorrhizal and endophytic fungi showed spatially aggregated patterns, suggesting the existence of facilitative interactions between fungi in different functional groups. Due to the large number of fungal pairs examined, many of the observed aggregated/segregated patterns with very low *P* values (e.g., < 0.005) turned non-significant after the application of a multiple comparison method. However, our overall results imply that the community structures of ectomycorrhizal and endophytic fungi could influence each other through interspecific competitive/facilitative interactions in root. To test the potential of host-plants' control of fungus–fungus ecological interactions in roots, we further examined whether the aggregated/segregated patterns could vary depending on the identity of host plant species. Potentially due to the physiological properties shared between the congeneric host plant species, the sign of hosts' control was not detected in the present study. The pyrosequencing-based randomization analyses shown in this study provide a platform of the high-throughput investigation of fungus–fungus interactions in plant root systems.

Editor: Minna-Maarit Kytöviita, Jyväskylä University, Finland

Funding: This work was financially supported by the Funding Program for Next Generation World-Leading Researchers of Cabinet Office, the Japanese government (to H.T.; GS014). The funders had no role in study design, data collection and analysis, decision to publish, or preparation of the manuscript.

Competing Interests: The authors have declared that no competing interests exist.

* E-mail: s_yamamoto@terra.zool.kyoto-u.ac.jp

Introduction

In terrestrial ecosystems, various functional groups of root-associated fungi interact with plants [1]. Ectomycorrhizal and arbuscular mycorrhizal fungi are mutualistic partners of more than 80% of terrestrial plant species, enhancing host plant growth and survival by transporting soil nutrients [2-5], protecting host plants from pathogens and herbivores [6–8], and reducing competition among co-occurring plant individuals/species [9,10]. Diverse clades of root endophytic fungi, which do not form mycorrhizae, also interact with diverse phylogenetic groups of plants [11,12]. Some clades of those endophytic fungi are known to enhance the nutritional conditions of host plants [13,14], whereas many others inhabit plant roots as commensalistic symbionts or parasites [15]. Those different functional groups of root-associated fungi often co-occur in plant roots [16,17], and understanding the assembly processes of those ecologically and phylogenetically diverse fungi in root systems is one of the major challenges in fungal ecology.

Competitive and facilitative interactions between fungal species are considered to be the major factors responsible for the community organization of fungi in roots [18–20]. Fungal species in roots can compete with each other by impeding the colonization of others [21–24] or by expelling other species from host roots [25]. These competitive interactions between fungal species in roots, importantly, are expected to result in fine-scale segregated distributions of competing species [26,27]. Interactions between fungal species can be competitive even between fungi in different functional or phylogenetic groups. In the roots of a *Eucalyptus* plant, for example, ectomycorrhizal fungi prevent arbuscular mycorrhizal fungi from infecting and proliferating in the host [28,29]. On the other hand, the presence of a fungal species in roots does not always negatively affect the colonization of others. A morphological observation of fungal hyphae in roots revealed that ectomycorrhizal and root-endophytic fungi coexisted in a single root system of a *Pinus* tree, presumably because the two functional groups of fungi occupied different habitats within the roots [30].

Likewise, recent molecular studies have demonstrated the coexistence of ectomycorrhizal and endophytic fungi within roots [31–33]. In these cases, interactions between fungal species may be neutral or even facilitative.

Although competitive or facilitative interactions between fungal species can be assessed by inoculation experiments [22,24,34], many of the mycorrhizal and endophytic fungi that dominate root-associated fungal communities in natural forests are unculturable. Therefore, such inoculation-based experimental studies are applicable to only a part of fungus–fungus interactions occurring in the wild. An alternative research approach for elucidating the nature of fungal interspecific interactions is to examine the pattern of the presence/absence of fungal species in host individuals and thereby examine the spatial segregation and aggregation (co-occurrence) patterns. Spatial segregation could result from competitive interactions or differences in niches. On the other hand, spatial aggregation could arise from either species sorting [35], in which a pair of fungi with shared habitat requirements come to exist in a particular root, or from facilitative interactions. Therefore, community-wide analyses of the spatial segregation and aggregation of root-associated fungi are expected to reveal the species pairs that have a role in shaping fungal community structures. Indeed, some recent studies on root-associated fungi have examined such segregation/aggregation patterns and inferred the patterns of possible fungus–fungus ecological interactions [27,36]

While previous studies analyzed the segregation/aggregation of pairs of root-associated fungi regardless of the effects of host plant species [26,27,36], such spatial patterns representing fungus–fungus competitive/facilitative interactions are expected to vary depending on host plant species. For example, if plant species with lower photosynthetic rates provide less carbohydrate to their root symbiont communities, it may promote competition for limited carbon resource and cause competitive exclusion between root-associated fungi [37]. Host plants' preference for a particular microenvironment (e.g., soil moisture) may also indirectly affect the relative competitive ability of root-associated fungi that interact with each other in root systems. By comparing segregation/aggregation patterns of root-associated fungi among different host plant species, we can examine such hypothetical plant-mediated processes of fungus–fungus interactions.

By identifying segregated and aggregated distributions of pairs of root-associated fungi on two oak species, we determined the patterns of fungal interspecific interactions and examined the dependence of such fungus–fungus interactions on background plant species. In a temperate forest in Japan, we first analyzed the community composition of root-associated fungi for the seedlings of co-occurring deciduous and evergreen oak species based on the pyrosequencing of fungal internal transcribed spacer (ITS) sequences. We then conducted a randomization test to detect the spatial segregation and aggregation of pairs of root-associated fungi. Furthermore, we examined whether or not the identity of host-plant species was associated with the spatial segregation or aggregation patterns of root-associated fungi.

Materials and Methods

Study area and sampling

The seedling samples were collected in a warm-temperate secondary forest on Mt. Yoshida located in Kyoto City, Japan (35.026°N, 135.786°E). No specific permissions were required for the location/activity. We confirmed that the field study did not involve endangered or protected species. The climate of Kyoto City is characterized by humid summers and dry winters: mean temperature and precipitation over the recent 30 years are 26.0 °C and 566.5 mm in summer (from June to August), and 5.6 °C and 164.9 mm in winter (from December to February) [38]. Mt. Yoshida (alt 121 m), which is a small hill with an area of 14.3 ha, is covered mainly by the two oak species *Quercus serrata* and *Q. glauca* (Fagaceae), while *Pinus densiflora* (Pinaceae) and *Ilex pedunculosa* (Aquifoliaceae) co-occur in the canopy layer. The two dominant oak species belong to different subgenus and have different ecological properties: *Q. serrata* is a deciduous species that occurs in the early stages of the secondary succession of temperate forests [39], while *Q. glauca* is an evergreen species whose seedlings occur both sunny and shaded understory of warm-temperate secondary forests [40]. Note that our field research site is not privately owned and sampling seedling in the research area is not banned.

From 20 to 31 May 2011, seedlings of each *Quercus* species were sampled at a minimum interval of 1 m: the number of sampled seedlings was 261 for *Q. serrata* and 199 for *Q. glauca*. The 460 sampling positions were recorded with a GPS device (Germin, GPSMAP 62S; Fig. 1). The size of sampled seedlings was 20–30 cm in height. To sample the seedlings, we dug to a depth of ca. 25 cm, taking great care not to damage the root tips of the seedlings. The amount of the dug soil of each root system was approximately 3,000 cm^3. The sampled seedlings were individually stored in sealed plastic bags in an ice chest. On the same day of the fieldwork, we randomly collected ten 2-cm fragments of terminal roots per seedling in the laboratory: note that there were seedlings with less than ten 2-cm fragments of terminal roots and hence the number of root fragments collected per seedling ranged from five to ten. The terminal roots were stored in 1.5-ml tubes with 70% ethanol at –20°C until DNA extraction.

DNA extraction, PCR, and pyrosequencing

To remove soil adhering to roots, 1-mm zirconium balls were introduced into the sample tubes and then the tubes were shaken at 18 Hz for 3 min using TissueLyser II (Qiagen, Hilden, Germany) as detailed elsewhere [33]. After the cleaning, five terminal roots per sample were subjected to fungal DNA extraction. Terminal roots were transferred to a new tube and were pulverized with 4-mm zirconium balls by shaking at 20 Hz for 3 min. DNA extraction was conducted with the cetyltrimethylammonium bromide (CTAB) method [41].

To selectively amplify the fungal ITS2 region from the extracted DNA of the plant roots, a nested polymerase chain reaction (PCR) method was applied [33]. In the first PCR step, we amplified the

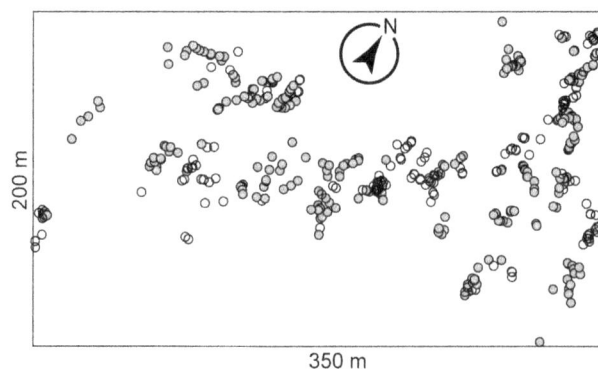

Figure 1. Distribution of the *Quercus* seedlings analyzed on Mt. Yoshida. Circles indicate the sampling locations of the seedlings. In total, 249 *Q. serrata* (gray) and 188 *Q. glauca* (white) seedlings were subjected to the randomization analyses of *C* and *T* scores.

entire ITS region using a fungus-specific primer (ITS1-F_KYO2 [42]) and a universal primer (ITS4 [43]) with the Ampdirect Plus (Shimadzu, Kyoto, Japan) buffer system. The ITS2 region was then amplified using a fusion primer of ITS3_KYO2 [42], which included the 454-adapter-A sequence and a 8-mer multiplex identifier (MID) tag sequence designed by Hamady *et al.* [44] to identify the source seedling (5'-CCA TCT CAT CCC TGC GTG TCT CCG ACT CAG [adapter A]–NNNNNNNN [MID]–GAT GAA GAA CGY AGY RAA [ITS3_KYO2] -3'), and a reverse fusion primer of ITS4, which included the 454-adapter-B sequence (5'-CCT ATC CCC TGT GTG CCT TGG CAG TCT CAG [adapter B]–TCC TCC GCT TAT TGA TAT GC [ITS4]-3'). The PCR products of all the 460 seedling samples were pooled in a new tube. We then purified the pooled PCR amplicons using ExoSAP-IT (GE Healthcare, Little Chalfont, Buckinghamshire, UK) and a PCR purification kit (Qiagen, Venlo, The Netherlands) before pyrosequencing using 454 GS Junior (Roche Diagnostics, Indianapolis, IN, USA). Because we did not obtain enough data in the first pyrosequencing run, we conducted an additional emulsion PCR and a pyrosequencing run. The pyrosequencing data were deposited to a public repository (DDBJ DRA: DRA000926).

Constructing operational taxonomic units and taxonomic identification

The pyrosequencing data were processed following the method of Toju *et al.* [33]. The low-quality 3'-tails of the pyrosequencing reads obtained were trimmed based on a threshold sequence quality value of 27. The reads were then filtered by a minimum sequence length of 150 bp excluding forward primer sequences and MID tags. To obtain the molecular operational taxonomic units (OTUs), the remaining reads were assembled as follows. The reads were sorted by seedling samples using the sample-specific MID tags and assembled into contigs for each sample with a minimum sequence similarity cutoff of 97% using Assams v0.1.2012.05.24 [45] (see also [46] for detailed assembling process), which is a parallelized pipeline for implementing the assembler program Minimus [47]. This within-sample assembling helped to avoid overestimates of OTU richness [48]. Possible chimeric sequences were detected and removed using UCHIME v4.2.40 [49] with a minimum score to report a chimera of 0.1. The within-sample contigs that passed the chimera removal process were subjected to further assembling of among-sample contigs with a sequence similarity cutoff of 97% using Assams, and the among-sample contigs were analyzed in the following statistical analyses as fungal OTUs. Note that the downstream statistical results did not qualitatively differ from those obtained based on 93% and 95% cutoff similarity settings in the among-sample clustering (see Results).

We conducted a BLAST-search of the fugal OTUs using the NCBI nt database on 22 February 2013. We also attempted identification based on the lowest common ancestor (LCA) algorithm [50], of which the results were much more conservative than the BLAST top-hit matches. Specifically, the query-centric auto-k-nearest-neighbor (QCauto) method implemented in the program Claident v0.1.2012.05.21 [51,52] was applied using the reference-sequence information of the "all_genus" and "all_underclass" sequence databases and the NCBI-Taxonomy information of the "all_genus" and "all_underclass" taxonomy databases (see [51] for details of those databases). The query OTU sequences are shown in Data S1.

Binary data matrices

Based on the pyrosequencing dataset, we obtained a binary matrix that depicted the presence (1) or absence (0) of fungal OTUs in each of the 261 *Q. serrata* and 199 *Q. glauca* seedling samples (Data S2). Before obtaining the binary matrix, seedling samples with less than 20 pyrosequencing reads were excluded: 10 and nine seedling samples were excluded for *Q. serrata* and *Q. glauca*, respectively. In addition, OTUs representing less than 5% of the sample-total reads were excluded from each sample to reduce among-sample variance in α-diversity that resulted from variance in sequencing effort (i.e., variance in the number of sequencing reads among samples [mean = 129.2, SD = 69.7]). In this process, singletons and rare OTUs, which were expected to contain high proportions of pyrosequencing errors in their sequences [53], were eliminated. The eleven seedling samples that included the sequences of plants other than Fagaceae or *Quercus* spp. were also excluded from the data set: these samples were contaminated by DNAs of *Ilex*, *Prunus*, and Ericaceae plants, which commonly occurred in the study forest. Consequently, 249 *Q. serrata* and 188 *Q. glauca* seedlings were subjected to the following analyses (Data S2).

Fungal diversity and spatial autocorrelation

Based on the presence/absence data matrix of fungal OTUs (Data S2), the diversity and spatial structure of the fungal communities on the two *Quercus* species were evaluated. To assess the species richness of fungi, species accumulation curves (Mao Tau curves) were drawn for each of the two host species using the function 'specaccum' in the Vegan package [54] of R (version 3.0.2 [55]). To evaluate spatial autocorrelation in fungal OTU composition within the study site, a Mantel correlogram analysis was applied to each host plant species. In the analysis, we calculated Mantel's correlation (r) between dissimilarity in fungal OTU composition (i.e., β-diversity) and Euclidean distance spanning sampling positions (999 permutations). For the calculation of β-diviersity, we used Raup-Crick metric [56], which could minimize statistical artifacts resulting from difference in α-diversity among samples (see [56]). In addition, to test the spatial autocorrelation of the occurrence of each fungal OTU within the study site, we conducted a Moran's I analysis [57] for each fungal OTU that occurred on 10 or more seedling samples using the R package Ape 3.0-11 [58,59].

Comparison of the root-associated fungal community structure between *Q. serrata* and *Q. glauca*

Prior to the statistical analysis of spatial segregation and aggregation of root-associated fungi, we examined differences in fungal community structure between *Q. serrata* and *Q. glauca* by PERMANOVA [60]. In this analysis, we measured dissimilarity in fungal OTU composition between seedling samples based on Raup–Crick β-diversity, and then tested for differences in the centroid of the fungal community structure of each *Quercus* species in multivariate space (9,999 permutations). The difference in the homogeneity of multivariate dispersion (the variance of β-diversity) between the fungal communities of the two hosts was also examined by PERMDISP [61] as implemented in Vegan.

We also examined the presence of host-specific fungal OTUs in the data set. A test using the multinomial species classification method (CLAM [62]) was performed to classify fungal OTUs preferentially associated with either host species and OTUs commonly found on both host species. A multinomial model was used to examine the statistical significance of respective fungal OTUs' preferences for host plants with a specialization threshold value of 2/3 ("supermajority" rule [62]). Because Bonferroni correction generally returns too stringent results in CLAM analysis [62], an α value of 0.001 was used as the threshold of statistical significance. The Vegan package of R was used in this analysis.

Segregation and aggregation of pairs of fungi in roots

We used the Checkerboard and Togetherness scores (C score and T score [63,64]) as indices of spatial segregation and aggregation of fungal OTUs in the seedling samples, respectively. The C score is defined as $(R_i - S) \times (R_j - S)$, where R_i and R_j represent the total number of occurrences of species i and j, respectively, and S is the number of co-occurrences [63]. The T score is defined as $S(N + S - R_i - R_j)$, where N is the number of seedlings analyzed [64]. In the presence of antagonistic interspecific interactions or the differentiation of niches, the observed C score of each pair of species is expected to be greater than that obtained by randomization under a null model. However, in the presence of facilitative interspecific interactions or shared habitat requirements, the observed T score is expected to be greater than that obtained by randomization under a null model. In contrast, a lack of significance suggests that species co-occur randomly.

For the dataset of each host plant, we tested the significance of C and T scores with 100,000 randomizations using the Bipartite package 2.0-1 [65] of R (Test 1). Observed and randomized C and T scores were standardized to range from 0 (the possible lowest level of segregation in terms of C scores and the possible lowest level of aggregation in terms of T scores) to 1 (the possible highest level of segregation in terms of C scores and the possible highest level of aggregation in terms of T scores) [66]. In each test for Q. serrata or Q. glauca dataset, OTUs observed in five or more seedlings and OTU pairs whose sum of seedling-sample counts were 25 or more were used because the statistical significance of C and T scores was difficult to examine for fungal OTU pairs with fewer sample counts. In addition to the examination for each host plant, the randomization analysis was applied to the whole dataset including both Q. serrata and Q. glauca seedling samples: OTUs observed in 10 or more Q. serrata and Q. glauca seedlings and OTU pairs whose sum of seedling-sample counts were 50 or more were used (Test 2). False discovery rate (FDR; Benjamini-Hochberg method) control [67] was applied to each randomization analysis.

We further assessed whether or not each pair of fungal OTUs was associated with each other in different ways on different host plant species. To this end, for each pair of fungal OTUs, we calculated the difference of C scores between the two host plant species (i.e., $C_{serrata} - C_{glauca}$ where $C_{serrata}$ and C_{glauca} were standardized C scores on Q. serrata and Q. glauca, respectively). Likewise, the difference of T scores between the two host plant species ($T_{serrata} - T_{glauca}$, where $T_{serrata}$ and T_{glauca} were standardized T scores on Q. serrata and Q. glauca, respectively) was calculated for each pair of fungal OTUs. The significance of the difference of C or T scores on different host plants was tested based on 100,000 randomizations (Test 3). In this analysis, we used the fungal OTU pairs that were included in the dataset of both host plants in the Test 1. The fungal OTU pairs analyzed in the Test 1 of both host plants were used. FDR control was also applied to the analysis.

Finally, to assess whether whole community of root-associated fungi in the study site show segregated/aggregated patterns, we tested the significance of C and T scores with 100,000 randomizations. We also applied the analysis to each sub-dataset including a taxonomic or functional group of fungi (Ascomycota, Basidiomycota, and ectomycorrhizal fungal sub-datasets).

Results

Assembling and identification of molecular OTUs

In total, 65,150 reads were obtained by pyrosequencing (18,667 and 46,483 reads in the first and second GS Junior runs, respectively). Mean length of those reads were 348 (SD = 69.8) bp for the first run and 362 (SD = 55.29) bp for the second run. Only 0.175% of the reads were those of plants. The total numbers of OTUs were 1869, 1079 and 785 based on sequence cutoff similarities of 97, 95 and 93%, respectively; the numbers of singletons were 940, 414 and 270, respectively. After removing seedling samples with less than 20 pyrosequencing reads, those with ITS reads of plants other than Quercus, and OTUs representing less than 5% of the sample-total reads, the binary data matrix of the fungal community included 319, 274 and 242 OTUs with 97, 95 and 93%-cutoff similarities, respectively (Data S2).

Of the 319 OTUs detected with a cutoff sequence similarity of 97%, 89.7% were identified at the phylum level, 58.9% at the order level, 50.4% at the family level, and 40.7% at the genus level (Fig. 2). From Q. serrata and Q. glauca seedlings, 94 and 86 basidiomycete, 96 and 98 ascomycete, and three and one glomeromycete OTUs were detected, respectively. Of the 319 OTUs, 34 occurred in 10 or more seedling samples. Among these 34 most common OTUs, 14 were assigned to ectomycorrhizal genera. Of the remaining 23 OTUs, three and 19 were respectively assigned to Basidiomycota and Ascomycota at the phylum level but their genera remained unidentified; the remaining one OTU could not be identified even at the phylum level by the QCauto method (Table 1). BLAST searches against the NCBI nr/nt database (Table 1) indicated that the commonly observed ascomycete OTUs were allied to genera or species that had been generally detected from living plant tissue in previous studies (i.e., possibly endophytic ascomycetes; e.g., Catenulifera [68], Pezicula [teleomorph of Cryptosporiopsis [69]], Lophodermium [70], and Cladophialophora [teleomorph of Capronia [71]], except for an OTU (OTU 121) related to soil fungi in the genus Archaeorhizomyces [72]. The list of commonly observed fungi included Mycena, Oidiodendron, and Glomeromycota, which are known as saprobes or ericoid/arbuscular mycorrhizal fungi (Table S1). We found no inconsistency between the QCauto-based and BLAST-based identification results, although results by the QCauto method were more conservative than those of BLAST (Table 1). Taxonomic diversity of OTUs were qualitatively similar among identification results with different sequence-similarity cutoffs (see Tables S1 and S2 for results at 93% and 95% cutoffs).

Fungal diversity and spatial autocorrelation

Of the 319 OTUs detected with a cutoff sequence similarity of 97%, 103 occurred on both Q. serrata and Q. glauca, while 115 and 101 OTUs occurred only on either Q. serrata or Q. glauca, respectively (Fig. 1a). Species accumulation curves did not reach a plateau for either host-plant species (Fig. 1b). The average number of OTUs per seedling was almost similar between the two host plant species but it was statistically higher on Q. serrata than on Q. glauca (3.51 and 3.20 on Q. serrata and Q. glauca, respectively; $t = 2.12$, df = 414.377, $P < 0.05$).

A Mantel correlogram analysis indicated that the fungal OTU composition of the examined seedling samples was spatially autocorrelated at a very small spatial scale within the study site: i.e., the scale of the autocorrelation was < ca. 8 m for Q. serrata seedlings and < ca. 2 m for Q. glauca seedlings (Fig. 3c, d). In a Moran's I analysis, significant spatial autocorrelation within the study site

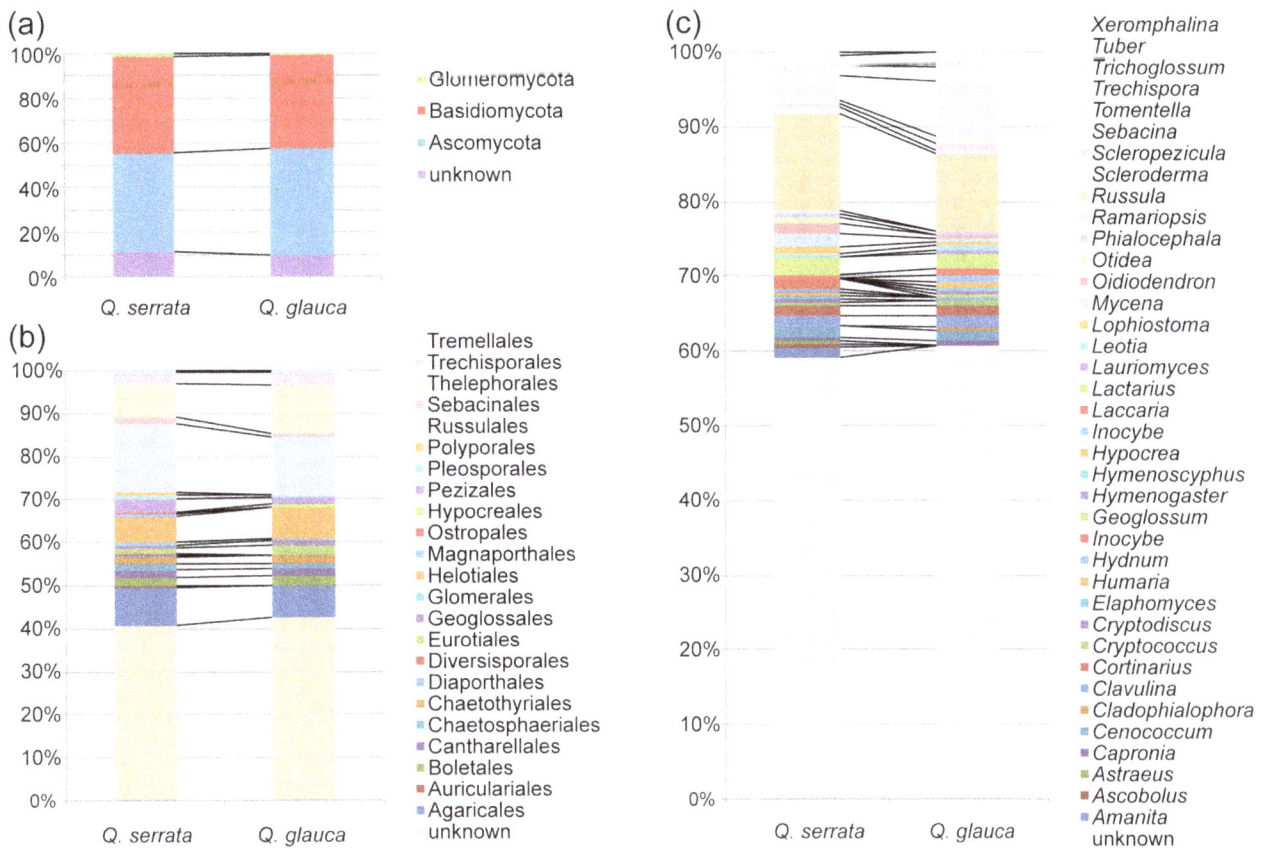

Figure 2. Taxonomic compositions of fungal OTUs on *Quercus serrata* and *Q. glauca*. (a) Phylum-level compositions of fungal OTUs. (b) Order-level compositions of fungal OTUs. (c) Genus-level compositions of fungal OTUs.

was observed for only seven of the 34 common fungal OTUs examined (Table 1).

Comparison of root-associated fungal community structures between Q. serrata and Q. glauca

Although PERMANOVA analysis indicated a significant difference in community structures between the two host species ($F = 4.79$, $P = 0.0002$), the R^2 value was very low ($R^2 = 0.0109$), suggesting that the effects of host plant identity on fungal OTU compositions were small in the present dataset. In fact, the community compositions of root-associated fungi on *Q. serrata* and *Q. glauca* were largely similar to each other (Fig. 2). The PERMDISP analysis showed significant differences in the among-sample β-diversity (Raup-Crick) of fungal communities between the two host-plant species. The β-diversity of the fungal communities was greater on *Q. glauca* than on *Q. serrata* (average distance between each fungal community and the centroid of fungal community were 0.5765 for *Q. serrata* and 0.6283 for *Q. glauca*, $F = 12.61$, $P = 0.0004$). Consistent results were obtained in the analyses based on 95% and 93% cutoff sequence similarities: for simplicity, results at 97% cutoff sequence similarity are shown in the following statistical analyses.

The CLAM test revealed the presence of 18 OTUs commonly associated with both *Quercus* species and an OTU exclusively associated with *Q. serrata*, although many OTUs were too rare to be assigned host preference (Table 1). The OTU that exclusively occurred on *Q. serrata* (OTU 1089) showed 99% sequence similarity to *Lactarius quietus* (JF273529). The list of fungi commonly

associated with both plant species included fungal OTUs related to various ectomycorrhizal fungi, e.g., *Cenococcum geophilum* (OTU 167), *Russula cerolens* (OTU 193), and *Thelephora terrestris* (OTU 211), and those related to possibly endophytic fungi in the orders Chaetothyriales and Helotiales, e.g., *Catenulifera luxurians* (OTU 757) and *Pezicula* sp. (OTU 329).

Segregation and aggregation of pairs of fungi in roots

When the seedling samples of the two host plants were analyzed separately (Test 1), no pair of fungal OTUs showed significant C or T scores after adjusting P values based on FDR control. However, many pairs of fungal OTUs displayed low (< 0.005) P values without FDR control (Table 2). In the C score analysis, an ectomycorrhizal ascomycete in the genus *Cenococcum* (OTU 167) and a possibly endophytic ascomycete (OTU 757) displayed segregated distribution on *Q. serrata* (Table 2). In the T score analysis, eight pairs of fungal OTUs displayed aggregated patterns with low (< 0.005) P values on either of the two host plants. Among the eight pairs, six were pairs of an ectomycorrhizal basidiomycete (*Russula* [OTU 185, 193, 509, or 1135] or *Lactarius* [OTU 205]) and a possibly endophytic ascomycete fungus (OTU 115 or 331; Table 2). The remaining two pairs were those of possibly endophytic ascomycetes (Table 2).

In the analysis in which the datasets of the two host plants were combined (Test 2), three pairs of fungal OTUs displayed statistically significant aggregated patterns even after FDR control (Table 3). Among the three pairs, two were those of an ectomycorrhizal basidiomycete (*Russula* [OTU 193] or *Lactarius*

Table 1. List of molecular OTUs occurring in 10 or more seedlings.

ID	CLAM test Preffered host	Moran's I	Number of occurrences Q. serrata	Q. glauca	BLAST top-hit result Description	TS	QC	E value	Identity	Accession	Taxonomic assignment using the QCauto method	Type
757	Both	0.0215	106	56	Catenulifera luxurians	462	91%	6E-127	293/314	GU727560	phylum: Ascomycota; class: Leotiomycetes	
167	Both	0.0053	40	31	Cenococcum geophilum	555	91%	1E-154	309/313	JQ711949	class: Dothideomycetes; genus: Cenococcum	EcM
115	Both	0.0109	36	20	Catenulifera brevicollaris	483	91%	5E-133	299/317	GU727561	subkindom: Dikarya; phylum: Ascomycota	
329	Both	0.0192	31	16	Pezicula sp.	499	91%	5E-138	299/313	AB731133	order: Helotiales; family: Dermateaceae	
1845	Both	0.0144	27	20	Cryptosporiopsis sp.	375	91%	8E-101	275/310	JN601680	class: Leotiomycetes; order: Helotiales	
387	Both	-0.0127	29	16	Leptodontidium sp.	486	80%	4E-134	270/273	DQ069033	class: Leotiomycetes; order: Helotiales	
193	Both	0.0149	26	16	Russula cerolens	675	93%	0E+00	384/393	JN681168	family: Russulaceae; genus: Russula	EcM
199	Both	0.0527*	18	20	Cryptosporiopsis sp.	379	91%	6E-102	277/312	JN601680	subkindom: Dikarya; phylum: Ascomycota	
203	Both	-0.0169	19	18	Lophodermium jiangnanense	267	92%	5E-68	265/321	GU138714	phylum: Ascomycota; class: Leotiomycetes	
205	Both	0.0201	12	18	Arcangeliella camphorata	678	93%	0E+00	408/426	EU644700	family: Russulaceae; genus: Lactarius	EcM
331	Both	0.0073	19	9	Cladophialophora carrionii	497	93%	2E-137	326/352	HM803232	order: Chaetothyriales; family: Herpotrichiellaceae	
121	Both	0.0635*	17	9	Archaeorhizomyces finlayi	159	90%	9E-36	255/291	JQ912673	subkindom: Dikarya; phylum: Ascomycota	
1089	Q. serrata	0.0310	25	0	Lactarius quietus	767	89%	0E+00	419/421	JF273529	species: Lactarius quietus	EcM
211	Both	0.0903*	11	12	Thelephora terrestris	647	92%	0E+00	374/386	JX030236	family: Thelephoraceae; genus: Thelephora	EcM
169	Both	0.0730*	14	5	Tomentella sp.	689	90%	0E+00	375/376	JF2735461	family: Thelephoraceae; genus: Tomentella	
823	Both	0.0174	14	5	Tomentella sp.	593	80%	2E-166	327/330	HE814132	order: Thelephorales; family: Thelephoraceae	
867	Both	0.0056	17	2	Rhizoscyphus ericae	473	93%	3E-130	294/312	JQ711893	subkindom: Dikarya; phylum: Ascomycota	
425	Both	0.0567*	7	11	Clavulina sp.	710	90%	0E+00	384/384	JF273519	family: Clavulinaceae; genus: Clavulina	EcM
185		0.0137	15	1	Russula japonica	577	87%	2E-161	344/358	AB509603	family: Russulaceae; genus: Russula	EcM
375	Both	-0.0035	9	7	Absconditella lignicola	241	89%	3E-60	219/260	FJ904669	kingdom: Fungi; subkingdom: Dikarya	
411		-0.0016	7	7	Tomentella sp.	612	93%	6E-172	369/387	FM955848	order: Thelephorales; family: Thelephoraceae	
527		0.0057	4	10	Graddonia coracina	353	91%	4E-94	277/317	JQ256423	subkindom: Dikarya; phylum: Ascomycota	
207		-0.0006	11	1	Cryptosporiopsis sp.	374	93%	3E-100	276/312	JN601680	subkindom: Dikarya; phylum: Ascomycota	
393		0.0482*	5	7	Penicillium sp.	272	92%	1E-69	286/348	FJ379804	order: Eurotiales; family: Trichocomaceae	
517		0.0120	10	2	Cenococcum geophilum	540	91%	3E-150	305/311	HM189732	class: Dothideomycetes; genus: Cenococcum	EcM
1135		0.2397*	5	7	Russula sp.	712	85%	0E+00	388/389	HE814200	family: Russulaceae; genus: Russula	EcM
157		0.0058	10	1	Cortinarius sp.	492	91%	8E-136	295/309	JQ272415	subkindom: Dikarya; phylum: Ascomycota	
349		-0.0122	8	3	Russula vesca	549	92%	5E-153	365/394	AY606965	family: Russulaceae; genus: Russula	EcM
423		-0.0116	5	6	Brevicellicium olivascens	315	96%	2E-82	315/382	JN649327	class: Agaricomycetes; order: Trechisporales	
1157		0.0019	8	3	Russula sp.	676	88%	0E+00	366/366	AB531451	family: Russulaceae; genus: Russula	EcM
113		-0.0025	3	7	Cenococcum geophilum	520	91%	4E-144	306/317	EU427331	class: Dothideomycetes; genus: Cenococcum	EcM
153		0.0181	6	4	Elaphomyces decipiens	505	100%	1E-139	327/352	EU837229	family: Elaphomycetaceae; genus: Elaphomyces	EcM
195		0.0289	8	2	Parmelia sp.	436	88%	4E-119	274/293	HQ671309	phylum: Ascomycota; class: Dothideomycetes	EcM

Table 1. Cont.

ID	CLAM test	Moran's I	Number of occurrences		BLAST top-hit result						Taxonomic assignment using the QCauto method	Type
	Preffered host		Q. serrata	Q. glauca	Description	TS	QC	E value	Identity	Accession		
623		0.0126	3	7	Sphaerosporella sp.	313	94%	7E-82	248/285	JQ711781	subkindom: Dikarya; phylum: Ascomycota	

Legend: Columns indicate the results of CLAM tests and Moran's I analyses, the number of occurrences in each Q. serrata (n = 249) and Q. glauca (n = 188), taxonomic identification results based on BLAST searches and the QCauto method, and the putative fungal functional type. In the column of BLAST top-hit result, total Blast score (TS), query coverage (QC), and identity (number of identical sites/number of the sites aligned to those of the BLAST top-hit sequence) are shown.
* P < 0.05 after FDR control.

[OTU 205]) and a possibly endophytic ascomycete (OTUs 115 or 331). The remaining pair was that of possibly endophytic ascomycetes (OTUs 199 and 1845). Note that such aggregation of pairs of ectomycorrhizal and endophytic fungi or those of endophytic fungi was also observed in the datasets based on 93% or 95% cutoff similarities (Table S2). In addition to the above-mentioned pairs, two pairs of an ectomycorrhizal basidiomycete (*Russula* [OTU 1135] or *Lactarius* [OTU 1089]) and a possibly edophytic ascomycete (OTU 115 or 331) and a pair of possibly root-endophytic ascomycete fungi displayed aggregated patterns with low (< 0.005) P values (Table 3). Likewise, segregated patterns with low P values were observed for a pair of an ectomycorrhizal ascomycete (*Cenococcum* [OTU 167]) and a possibly endophytic ascomycete (OTU 757), although the patterns were non-significant after FDR control (Table 3).

When the effects of host plants on the segregation/aggregation of fungal OTUs were examined (Test 3), no pair of fungal OTUs showed significant difference of C or T scores between *Q. serrata* and *Q. glauca* after FDR control; even without FDR control, no fungal pair showed low (< 0.005) P values (Table S3). These results indicate that the identity of host plant species did not affect the segregation/aggregation patters of root-associated fungi in the present dataset.

In the community-scale analysis of C scores (Table 4), significantly segregated patterns were observed within the entire community of the observed fungi and within the basidiomycete and ectomycorrhizal fugal sub-communities, while no sign of segregation was observed within the ascomycete sub-community. As expected by the C score analysis, observed values of togetherness (T) were very low (Table 5).

Discussion

In the present study, the sign of segregated or aggregated patterns was detected for a small number of fungal OTU pairs. Although the results may reflect the rarity of competitive or facilitative interactions between root-associated fungi in the study forest, there are potential statistical issues that may have hampered the detection of significant patterns. To make clear the potential pitfalls in pyrosequencing-based high-throughput analyses of segregated/aggregated distributions of fungi, we start with discussing problems related to multiple comparisons and the power of randomization tests.

When the checkerboard (C) and togetherness (T) scores of pairs of fungal OTUs were examined on each of the two host plant species (Test 1), no fungal pair showed statistically significant segregation nor aggregation patterns after FDR control (Table 2). Given that more than 179 pairs of fungal OTUs were examined in our pyrosequencing-based analysis, the application of the multiple comparison method might make the results prone to type II errors [73]. In the studies of segregation/aggregation patterns of root-associated fungi, such a statistical issue has been generally underappreciated. For example, a previous study discussed spatial segregation or aggregation of fungal species without any multiple-comparison adjustment of P-values [26], thereby making their results prone to type I errors (false positives) rather than type II errors (false negatives). Therefore, by applying multiple comparison methods to the datasets of the previous studies, one may be able to screen for pairs of fungi with strong sign of segregated/aggregated patterns.

In addition to the possible effects of multiple comparison methods, problems related to sample size could affect the results of randomization test. Intriguingly, when the samples of the two host species were pooled in the C and T score analysis (Test 2), three

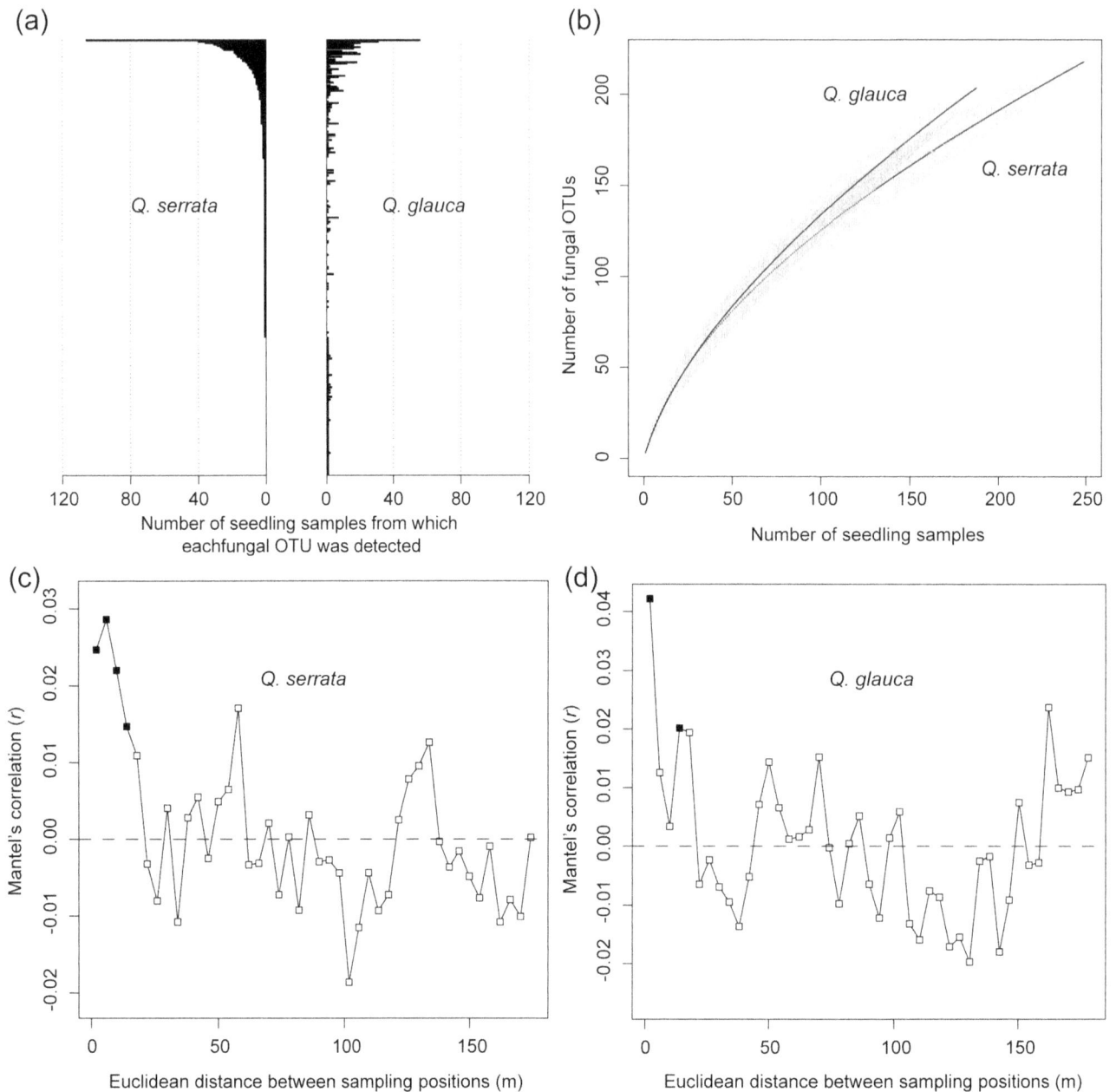

Figure 3. Root-associated fungal diversity on *Quercus serrata* and *Q. glauca*. (a) Number of seedling samples from which each fungal OTU was detected. (b) Accumulation curves of fungal OTUs against the number of *Q. serrata* or *Q. glauca* seedlings. The solid line and the gray area denote the expected mean OTU richness and its standard deviation, respectively. (c, d) Spatial autocorrelation analysis of fungal OTU compositions. A Mantel correlogram analysis was conducted to evaluate the extent of the spatial autocorrelation of root-associated fungal OTU compositions. Mantel's r statistic representing the correlation between dissimilarity in fungal OTU compositions (Raup-Crick β-diversity) and Euclidean distance spanning sampling positions is shown for each distance class. Filled symbols represent significantly positive spatial autocorrelation. (c) Analysis on *Q. serrata* with intervals of ca. 2 m. (d) Analysis on *Q. glauca* with intervals of ca. 2 m.

pairs of fungal OTUs showed statistically significant aggregation even after FDR control (Table 3). Given that the identity of host plants was not associated with the segregation/aggregation patterns of root-associated fungi in the current dataset (Test 3; Table S3), the fact that significant segregation/aggregation was observed in the simultaneous analysis of *Q. serrata* and *Q. glauca* seedlings (Test 2) but not in the independent analysis of them (Test 1) might be attributed to the larger number of samples in Test 2 than in Test 1. Importantly, when the sample counts of examined fungi (i.e., the number of seedling samples in which respective

fungi occurred) are small, C or T scores of randomized data matrices frequently take the minimum (0) or maximum (1) values and hence deviate from normal distribution (Fig. S1), possibly reducing the power of the randomization tests. Therefore, in the present study, sample counts of fungi would have been insufficient to detect segregation/aggregation patterns independently on each host plant (Test 1), despite the intensive sampling of seedlings in the forest (249 *Q. serrata* and 188 *Q. glauca* seeding samples). Overall, the comparison of the results between Test 1 and Test 2 suggests that fungal pairs with strong segregation/aggregation

Table 2. List of fungal OTU pairs that displayed segregated or aggregated patterns when each host plant species was analyzed independently (Test 1).

Pair of OTUs				C or T score	P	FDR§
IDs (Taxonomic information)†						
Q. serrata						
Segregation (C score analysis)						
757	(class: Leotiomycetes)	167	(genus: *Cenococcum*)*	0.679	0.0042	1.000
Aggregation (T score analysis)						
509	(genus: *Russula*)*	115	(phylum: Ascomycota)	0.072	0.0010	0.300
193	(genus: *Russula*)*	115	(phylum: Ascomycota)	0.133	0.0016	0.300
331	(family: Herpotrichiellaceae)	185	(genus: *Russula*)*	0.075	0.0034	0.409
Q. glauca						
Aggregation (T score analysis)						
331	(family: Herpotrichiellaceae)	205	(genus: *Lactarius*)*	0.115	0.0007	0.124
375	(subkingdom: Dikarya)	203	(class: Leotiomycetes)	0.093	0.0024	0.154
199	(phylum: Ascomycota)	1845	(order: Helotiales)	0.150	0.0030	0.154
115	(phylum: Ascomycota)	1135	(genus: *Russula*)*	0.092	0.0038	0.154
193	(genus: *Russula*)*	115	(phylum: Ascomycota)	0.131	0.0043	0.154

Legend: The significance of C or T scores was examined based on a randomization test for each pair of fungal OTUs (100,000 permutations).
†For each fungal OTU, taxonomic information based on the QCauto method is shown. Asterisks indicate possibly ectomycorrhizal OTUs.
§Adjusted P values (FDR control).

signs become detectable by increasing sample size (and thereby sample counts; Fig. S1).

Although the abovementioned statistical issues should be treated with caution, the present results have important implications for competitive or facilitative interactions between root-associated fungi. Albeit non-significant when FDR control was applied, a segregated pattern with a low P value (< 0.005) was observed between an ectomycorrhizal fungus in the ascomycete genus *Cenococcum* and a possibly endophytic ascomycete fungus (Tables 2 and 3). This result is intriguing, given that most previous studies investigated the potential competitive interactions *within* ectomycorrhizal or root-endophytic fungal communities [19,20,26,27] (see also [30], Tables 4 and 5) and did not assume that

ectomycorrhizal fungi could affect the spatial distribution of endophytic (or other types of non-ectomycorrhizal) fungi and vice versa. However, as suggested by our present results, competition for space or resources within root systems could occur not only between fungi with similar ecological or physiological properties but also between fungi in different functional groups. Similar phenomena were reported in a study of root-associated fungi on an ericaceous plant [36]. Combined with the findings of the previous study, our present results suggest that simultaneous analysis of multiple functional groups would give novel insights into the community dynamics of root-associated fungi.

In addition to spatial segregation, aggregated patterns of pairs of fungi were observed in this study (Tables 2 and 3). These pairs

Table 3. List of fungal OTU pairs that displayed segregated or aggregated patterns when the seedling samples of the two *Quercus* species were analyzed simultaneously (Test 2).

Pair of OTUs				C or T score	P	FDR§
IDs (Taxonomic information)†						
Segregation (C score analysis)						
757	(class: Leotiomycetes)	167	(genus: *Cenococcum*)*	0.698	0.0014	0.359
Aggregation (T score analysis)						
193	(genus: *Russula*)*	115	(phylum: Ascomycota)	0.131	0.0000	0.003
199	(phylum: Ascomycota)	1845	(order: Helotiales)	0.101	0.0004	0.032
331	(family: Herpotrichiellaceae)	205	(genus: *Lactarius*)*	0.072	0.0004	0.032
375	(subkingdom: Dikarya)	203	(class: Leotiomycetes)	0.054	0.0014	0.084
115	(phylum: Ascomycota)	1135	(genus: *Russula*)*	0.052	0.0021	0.104
331	(family: Herpotrichiellaceae)	1089	(species: *Lactarius quietus*)*	0.054	0.0043	0.180

Legend: The significance of C or T scores was examined based on a randomization test for each pair of fungal OTUs (100,000 permutations).
†For each fungal OTU, taxonomic information based on the QCauto method is shown. Asterisks indicate possibly ectomycorrhizal OTUs.
§Adjusted P values (FDR control).

Table 4. Analysis of segregated patterns (i.e., C score analysis) within each taxonomic or functional group of fungi.

Subset of community	Q. serrata			Q. glauca			Both host species		
	C score	P (FDR)	SES[§]	C score	P (FDR)	SES[§]	C score	P (FDR)	SES[§]
All	0.9741	**0.0091**	2.37	0.9757	0.0544	1.69	0.9823	**0.0127**	2.27
Ascomycota	0.9609	0.2553	0.69	0.9622	0.9157	−1.42	0.9718	0.6499	−0.37
Basidiomycota	0.9849	**0.0004**	3.25	0.9859	**0.0080**	2.67	0.9895	**0.0008**	3.54
EcM	0.9814	**0.0018**	2.84	0.9791	0.0544	1.66	0.9860	**0.0046**	2.68

Legend: The significance of C scores was examined based on a randomization test (100,000 permutations). FDR control was applied to each (sub-)dataset: significant P values are indicated in bold.
[§]SES indicates standard effect size.

Table 5. Analysis of aggregated patterns (i.e., T score analysis) within each taxonomic or functional group of fungi.

Subset of community	Q. serrata			Q. glauca			Both host species		
	T score	P	SES[§]	T score	P	SES[§]	T score	P	SES[§]
All	0.0004	0.9962	−2.51	0.0004	0.9975	−2.45	0.0004	1.0000	−3.43
Ascomycota	0.0007	0.4216	0.17	0.0008	0.1083	1.26	0.0008	0.1720	0.96
Basidiomycota	0.0005	1.0000	−4.15	0.0004	1.0000	−3.64	0.0005	1.0000	−5.54
EcM fungi	0.0005	1.0000	−4.06	0.0004	0.9979	−2.56	0.0005	1.0000	−4.95

Legend: The significance of T scores was examined based on a randomization test (100,000 permutations). FDR control was applied to each (sub-)dataset.
[§]SES indicates standard effect size.

included those of an ectomycorrhizal fungus in the genus *Russula* or *Lactarius* and a possibly-endophytic ascomycete fungus (Tables 2 and 3). Such co-occurrence of ectomycorrhizal and possibly endophytic fungi in single root system has been inferred from other lines of evidence [30-33], suggesting that specific pairs of fungi in different functional groups can interact with each other, potentially in facilitative or mutualistic ways. However, aggregated patterns can be observed between fungal species with similar physiological requirements for root environments, and the observed coexistence of possibly endophytic ascomycete fungi (Tables 2 and 3) may be attributed to shared preferences for habitats. Given that we focused on the segregated/aggregated patterns of fungi at the scale of host seedling individuals, potential vertical heterogeneity of microenvironments (e.g., soil nutrient availability) and the resultant partitioning of niches within root systems might have produced the observed aggregated patterns. Therefore, further studies are required to confirm whether the observed aggregated patterns reflect actual fungus–fungus ecological interactions in root systems. For example, experimental-inoculation methods will help to examine what kinds of mechanisms are responsible for the co-occurrence of those specific pairs of fungi.

Despite the proposition that fungus–fungus competitive or facilitative interactions could vary depending on host plant species, no significant difference in the levels of spatial segregation or aggregation was observed between *Q. serrata* and *Q. glauca* samples. The lack of host effects on the segregated or aggregated patterns may be partly attributed to the phylogenetic closeness of the two host species. Given that the root-associated fungal community composition was highly similar between the congeneric plant species (Fig. 2; cf. [74]), the environmental conditions experienced by root-associated fungi might be almost identical between the two host species. Hence, the use of congeneric host plant species in the present study may have precluded the detection of hosts' effects on fungus–fungus interactions. Therefore, the present results do not necessarily mean that host plants generally have no impact on the nature or strength of fungus–fungus interactions in roots: comparative analyses of C or T scores on phylogenetically-distant host plant species are awaited to further discuss the potential effects of host plants on the assembly processes of root-associated fungi.

As shown in this study, pyrosequencing technologies allow high-throughput profiling of root-associated fungal communities in hundreds (or more) of host individuals and will offer a breakthrough in the investigation of fungus–fungus interactions in roots. In addition, sequence-based taxonomic assignment (identification) potentially allows us to assess intraspecific genotypic diversity [75], providing further opportunities to infer fungus–fungus ecological/evolutionary interactions. Meanwhile, this pyrosequencing approach is based entirely on observational data on the co-occurrence of fungal OTUs in roots, and hence, it only provides insights into the potential consequences of competitive or facilitative interactions between root-associated fungal species/

taxa. Therefore, complementary experimental studies are important to reveal the nature of fungus–fungus ecological interactions. For example, some experimental approaches have separated the mechanisms and consequences of interspecific interactions between fungal species [23,24,30] and others have quantitatively evaluated the effect of fungus–fungus interactions on host's performance [75,76]. In combination with such experimental studies, the pyrosequencing-based high-throughput profiling of fungal communities will contribute to our understanding of the assembly processes of ecologically diverse root-associated fungi.

Supporting Information

Figure S1 Histograms of C and T scores that were obtained from the randomization of simulated data. In each combination of sample size (N) and sample counts of OTUs (R_i and R_j), the histogram of C or T scores of randomized data matrix were obtained. C and T scores tend to take the maximum (1) and minimum (0) values, respectively, when sample sizes (N) or sample counts of fungal OTUs (R_i and R_j) are small. Likewise, when R_i and R_j are much smaller than N, C and T scores tend to take the extreme values. Thus, all of sample size, sample counts of fungal OTUs, and the balance between them should be carefully inspected when screening pairs of fungal OTUs prior to randomization tests of C or T scores

Table S1 Results of taxonomic assignment using QCauto method.

Table S2 Tables with a list of fungal OTU (defined with cutoff similarities of 93 and 95%) pairs that displayed segregated or aggregated patterns.

Table S3 Tables with a list of fungal OTU (defined with cutoff similarity of 97%) pairs that displayed segregated or aggregated patterns.

Acknowledgments

We thank Nana Adachi for inputting data.

Author Contributions

Conceived and designed the experiments: HT HS SY. Performed the experiments: SY HT HS AH. Analyzed the data: SY AT KK. Wrote the paper: SY.

References

1. Smith SE, Read DJ (2008) Mycorrhizal Symbiosis. Third Edition. Cambridge: Academic Press. 787 p.

2. Hattingh MJ, Gray LE, Gerdemann JW (1973) Uptake and translocation of 32P-labeled phosphate to onion roots by endomycorrhizal fungi. Soil Sci 116: 383–387. doi:10.1097/00010694-197311000-00007.

3. Simard SW, Perry DA, Jones MD, Myrold DD, Durall DM, et al. (1997) Net transfer of carbon between ectomycorrhizal tree species in the field. Nature 388: 579–582.

4. Lindahl B, Stenlid J, Olsson S, Finlay R (1999) Translocation of 32P between interacting mycelia of a wood-decomposing fungus and ectomycorrhizal fungi in

microcosm systems. New Phytol 144: 183–193. doi:10.1046/j.1469-8137.1999.00502.x.

5. Grelet G-A, Johnson D, Paterson E, Anderson IC, Alexander IJ (2009) Reciprocal carbon and nitrogen transfer between an ericaceous dwarf shrub and fungi isolated from *Piceirhiza bicolorata* ectomycorrhizas. New Phytol 182: 359–366. doi:10.1111/j.1469-8137.2009.02813.x.

6. Borowicz VA (2001) Do arbuscular mycorrhizal fungi alter plant-pathogen relations? Ecology 82: 3057–3068. doi:10.2307/2679834.

7. Hartley SE, Gange AC (2009) Impacts of plant symbiotic fungi on insect herbivores: mutualism in a multitrophic context. Annu Rev Entomol 54: 323–342. doi:10.1146/annurev.ento.54.110807.090614.

8. Koricheva J, Gange AC, Jones T (2009) Effects of mycorrhizal fungi on insect herbivores: a meta-analysis. Ecology 90: 2088–2097. doi:10.1890/08-1555.1.

9. Booth MG (2004) Mycorrhizal networks mediate overstorey-understorey competition in a temperate forest. Ecol Lett 7: 538–546. doi:10.1111/j.1461-0248.2004.00605.x.

10. van der Heijden MGA, Bardgett RD, van Straalen NM (2008) The unseen majority: soil microbes as drivers of plant diversity and productivity in terrestrial ecosystems. Ecol Lett 11: 296–310. doi:10.1111/j.1461-0248.2007.01139.x.

11. Jumpponen A, Trappe JM (1998) Dark septate endophytes: a review of facultative biotrophic root-colonizing fungi. New Phytol 140: 295–310.

12. Jumpponen A (2001) Dark septate endophytes - are they mycorrhizal? Mycorrhiza 11: 207–211. doi:10.1007/s005720100112.

13. Usuki F, Narisawa K (2007) A mutualistic symbiosis between a dark septate endophytic fungus, Heteroconium chaetospira, and a nonmycorrhizal plant, Chinese cabbage. Mycologia 99: 175–184. doi:10.3852/mycologia.99.2.175.

14. Newsham KK (2011) A meta-analysis of plant responses to dark septate root endophytes. New Phytol 190: 783–793. doi:10.1111/j.1469-8137.2010.03611.x.

15. Reininger V, Sieber TN (2012) Mycorrhiza reduces adverse effects of dark septate endophytes (DSE) on growth of conifers. PLoS ONE 7: e42865. doi:10.1371/journal.pone.0042865.s004.

16. Toju H, Sato H, Yamamoto S, Kadowaki K, Tanabe AS, et al. (2013) How are plant and fungal communities linked to each other in belowground ecosystems? A massively parallel pyrosequencing analysis of the association specificity of root-associated fungi and their host plants. Ecol Evol 3: 3112–3124. doi:10.1002/ece3.706.

17. Toju H, Yamamoto S, Sato H, Tanabe AS (2013) Sharing of Diverse Mycorrhizal and Root-Endophytic Fungi among Plant Species in an Oak-Dominated Cool–Temperate Forest. PLoS ONE 8: e78248. doi:10.1371/journal.pone.0078248.

18. Kennedy P (2010) Ectomycorrhizal fungi and interspecific competition: species interactions, community structure, coexistence mechanisms, and future research directions. New Phytol 187: 895–910. doi:10.1111/j.1469-8137.2010.03399.x.

19. Saunders M, Glenn AE, Kohn LM (2010) Exploring the evolutionary ecology of fungal endophytes in agricultural systems: using functional traits to reveal mechanisms in community processes. Evol Appl 3: 525–537. doi:10.1111/j.1752-4571.2010.00141.x.

20. Koide RT, Fernandez C, Petprakob K (2011) General principles in the community ecology of ectomycorrhizal fungi. Ann For Sci 68: 45–55. doi:10.1007/s13595-010-0006-6.

21. Mamoun M, Olivier JM (1993) Competition between Tuber melanosporum and other ectomycorrhizal fungi under two irrigation regimes. Plant Soil 149: 211–218. doi:10.1007/BF00016611.

22. Jansa J, Smith FA, Smith SE (2008) Are there benefits of simultaneous root colonization by different arbuscular mycorrhizal fungi? New Phytol 177: 779–789. doi:10.1111/j.1469-8137.2007.02294.x.

23. Hortal S, Pera J, Parladé J (2008) Tracking mycorrhizas and extraradical mycelium of the edible fungus Lactarius deliciosus under field competition with Rhizopogon spp. Mycorrhiza 18: 69–77. doi:10.1007/s00572-007-0160-3.

24. Kennedy P, Peay K, Bruns T (2009) Root tip competition among ectomycorrhizal fungi: are priority effects a rule or an exception? Ecology 90: 2098–2107. doi:10.1890/08-1291.1.

25. Wu B, Nara K, Hogetsu T (1999) Competition between ectomycorrhizal fungi colonizing Pinus densiflora. Mycorrhiza 9: 151–159. doi:10.1007/s005720050300.

26. Koide RT, Xu B, Sharda J, Lekberg Y, Ostiguy N (2005) Evidence of species interactions within an ectomycorrhizal fungal community. New Phytol 165: 305–316. doi:10.1111/j.1469-8137.2004.01216.x.

27. Pickles BJ, Genney DR, Anderson IC, Alexander IJ (2012) Spatial analysis of ectomycorrhizal fungi reveals that root tip communities are structured by competitive interactions. Mol Ecol 21: 5110–5123. doi:10.1111/j.1365-294X.2012.05739.x.

28. Chilvers GA, Lapeyrie FF, Horan DP (1987) Ectomycorrhizal vs endomycorrhizal fungi within the same root system. New Phytol 107: 441–448. doi:10.1111/j.1469-8137.1987.tb00195.x.

29. Chen YL, Brundrett MC, Dell B (2000) Effects of ectomycorrhizas and vesicular–arbuscular mycorrhizas, alone or in competition, on root colonization and growth of Eucalyptus globulus and E. urophylla. New Phytol 146: 545–555. doi:10.1046/j.1469-8137.2000.00663.x.

30. Wagg C, Pautler M, Massicotte HB, Peterson RL (2008) The co-occurrence of ectomycorrhizal, arbuscular mycorrhizal, and dark septate fungi in seedlings of four members of the Pinaceae. Mycorrhiza 18: 103–110. doi:10.1007/s00572-007-0157-y.

31. Tedersoo L, Pärtel K, Jairus T, Gates G, Põldmaa K, et al. (2009) Ascomycetes associated with ectomycorrhizas: molecular diversity and ecology with particular reference to the Helotiales. Environ Microbiol 11: 3166–3178. doi:10.1111/j.1462-2920.2009.02020.x.

32. Kernaghan G, Patriquin G (2011) Host associations between fungal root endophytes and boreal trees. Microb Ecol 62: 460–473. doi:10.1007/s00248-011-9851-6.

33. Toju H, Yamamoto S, Sato H, Tanabe AS, Gilbert GS, et al. (2013) Community composition of root-associated fungi in a Quercus-dominated temperate forest: "codominance" of mycorrhizal and root-endophytic fungi. Ecol Evol 3: 1281–1293. doi:10.1002/ece3.546.

34. Wilson JM, Tommerup IC (1992) Interactions between fungal symbionts: VA mycorrhizae. In: Allen M, editor. New York: Chapman and Hall. pp. 199–248.

35. Leibold MA, Holyoak M, Mouquet N, Amarasekare P, Chase JM, et al. (2004) The metacommunity concept: a framework for multi-scale community ecology. Ecol Lett 7: 601–613. doi:10.1111/j.1461-0248.2004.00608.x.

36. Gorzelak MA, Hambleton S, Massicotte HB (2012) Community structure of ericoid mycorrhizas and root-associated fungi of Vaccinium membranaceum across an elevation gradient in the Canadian Rocky Mountains. Fungal Ecol 5: 36–45. doi:10.1016/j.funeco.2011.08.008.

37. Izzo A, Nguyen DT, Bruns TD (2006) Spatial structure and richness of ectomycorrhizal fungi colonizing bioassay seedlings from resistant propagules in a Sierra Nevada forest: comparisons using two hosts that exhibit different seedling establishment patterns. Mycologia 98: 374–383. doi:10.3852/mycologia.98.3.374.

38. Japan Meteorological Agency (2010) Climate statistics. Japan Meteological Agency. Available: http://www.jma.go.jp/jma/indexe.html. Accessed 1 November 2013.

39. Fujihara M (1996) Development of secondary pine forests after pine wilt disease in western Japan. J Veg Sci 7: 729–738. doi:10.2307/3236384.

40. Cho M, Kawamura K, Takeda H (2005) Sapling architecture and growth in the co-occurring species Castanopsis cuspidata and Quercus glauca in a secondary forest in western Japan. J For Res 10: 143–150. doi:10.1007/s10310-004-0124-9.

41. Sato H, Yumoto T, Murakami N (2007) Cryptic species and host specificity in the ectomycorrhizal genus Strobilomyces (Strobilomycetaceae). Am J Bot 94: 1630–1641. doi:10.3732/ajb.94.10.1630.

42. Toju H, Tanabe AS, Yamamoto S, Sato H (2012) High-coverage ITS primers for the DNA-based identification of ascomycetes and basidiomycetes in environmental samples. PLoS ONE 7: e40863. doi:10.1371/journal.pone.0040863.t004.

43. White TJ, Bruns T, Lee S, Taylor J (1990) Amplification and direct sequencing of fungal ribosomal RNA genes for phylogenetics. In: Innis MA, Gelfand DH, Sninsky JJ, White TJ, editors. PCR protocols: a guide to methods and applications. San Diego: Academic Press. pp. 315–322.

44. Hamady M, Walker JJ, Harris JK, Gold NJ, Knight R (2008) Error-correcting barcoded primers for pyrosequencing hundreds of samples in multiplex. Nat Meth 5: 235–237. doi:10.1038/nmeth.1184.

45. Tanabe AS (2012) Assams. A software distributed by the author. Available: http://www.fifthdimension.jp/products/assams/. Accessed 20 December 2013.

46. Toju H, Sato H, Tanabe AS (2014) Diversity and spatial structure of belowground plant–fungal symbiosis in a mixed subtropical forest of ectomycorrhizal and arbuscular mycorrhizal plants. PLoS ONE 9: e86566. doi: 10.1371/journal.pone.0086566.

47. Sommer DD, Delcher AL, Salzberg SL, Pop M (2007) Minimus: a fast, lightweight genome assembler. BMC Bioinformatics 8: 64. doi:10.1186/1471-2105-8-64.

48. Kunin V, Engelbrektson A, Ochman H, Hugenholtz P (2010) Wrinkles in the rare biosphere: pyrosequencing errors can lead to artificial inflation of diversity estimates. Environ Microbiol 12: 118–123. doi:10.1111/j.1462-2920.2009.02051.x.

49. Edgar RC, Haas BJ, Clemente JC, Quince C, Knight R (2011) UCHIME improves sensitivity and speed of chimera detection. Bioinformatics 27: 2194–2200. doi:10.1093/bioinformatics/btr381.

50. Huson DH, Auch AF, Qi J, Schuster SC (2007) MEGAN analysis of metagenomic data. Genome Res 17: 377–386. doi:10.1101/gr.5969107.

51. Tanabe AS (2012) Claident. A software distributed by the author. Available: http://www.fifthdimension.jp/products/claident/. Accessed 20 December 2013.

52. Tanabe AS, Toju H (2013) Two new computational methods for universal DNA barcoding: A benchmark using barcode sequences of bacteria, archaea, animals, fungi, and land plants. PLoS ONE 8: e76910. doi:10.1371/journal.pone.0076910.

53. Tedersoo L, Nilsson RH, Abarenkov K, Jairus T, Sadam A, et al. (2010) 454 Pyrosequencing and Sanger sequencing of tropical mycorrhizal fungi provide similar results but reveal substantial methodological biases. New Phytol 188: 291–301. doi:10.1111/j.1469-8137.2010.03373.x.

54. Oksanen J, Blanchet FG, Kindt R, Legendre P, Minchin PR, et al. (n.d.) Vegan: Community ecology package. R package version 2.0-5. Available: http://CRAN.R-project.org/package = vegan. Accessed 20 December 2012.

55. R Development Core Team (2013) R: A Language and Environment for Statistical Computing. Vienna: R Foundation for Statistical Computing.

56. Chase JM, Kraft NJB, Smith KG, Vellend M, Inouye BD (2011) Using null models to disentangle variation in community dissimilarity from variation in α-diversity. Ecosphere 2: art24. doi:10.1890/ES10-00117.1.

57. Gittleman JL, Kot M (1990) Adaptation: Statistics and a null model for estimating phylogenetic effects. Syst Zool 39: 227. doi:10.2307/2992183.

58. Paradis E, Claude J, Strimmer K (2004) APE: Analyses of Phylogenetics and Evolution in R language. Bioinformatics 20: 289–290. doi:10.1093/bioinformatics/btg412.

59. Paradis E (2012) Analysis of Phylogenetics and Evolution with R. Second Edition. New York: Springer. 386 p.

60. Anderson MJ (2001) A new method for non-parametric multivariate analysis of variance. Austral Ecol 26: 32–46. doi:10.1111/j.1442-9993.2001.01070.pp.x.

61. Anderson MJ (2006) Distance-based tests for homogeneity of multivariate dispersions. Biometrics 62: 245–253. doi:10.1111/j.1541-0420.2005.00440.x.

62. Chazdon RL, Chao A, Colwell RK, Lin S-Y, Norden N, et al. (2011) A novel statistical method for classifying habitat generalists and specialists. Ecology 92: 1332–1343. doi:10.1890/10-1345.1.

63. Stone L, Roberts A (1990) The checkerboard score and species distributions. Oecologia 85: 74–79. doi:10.1007/BF00317345.

64. Stone L, Roberts A (1992) Competitive exclusion, or species aggregation? Oecologia 91: 419–424. doi:10.1007/BF00317632.

65. Dormann CF, Gruber B, Fründ J (2008) Introducing the bipartite package: analysing ecological networks. R news 8/2: 8–11.

66. Dormann CF, Fründ J, Blüthgen N, Gruber B (2009) Indices, graphs and null models: analyzing bipartite ecological networks. Open Ecol J 2: 7–24. doi:10.2174/1874213000902010007.

67. Benjamini Y, Hochberg Y (1995) Controlling the false discovery rate: a practical and powerful approach to multiple testing. J R Stat Soc Series B (Methodological) 57: 289–300.

68. Bogale M, Orr M-J, O'hara MJ, Untereiner WA (2010) Systematics of *Catenulifera* (anamorphic Hyaloscyphaceae) with an assessment of the phylogenetic position of *Phialophora hyalina*. Fungal Biol 114: 396–409. doi:10.1016/j.funbio.2010.02.006.

69. Wang W, Tsuneda A, Gibas CF, Currah RS (2007) *Cryptosporiopsis* species isolated from the roots of aspen in central Alberta: identification, morphology, and interactions with the host, in vitro. Can J Bot 85: 1214–1226. doi:10.1139/B07-086.

70. Lantz H, Johnston PR, Park D, Minter DW (2011) Molecular phylogeny reveals a core clade of Rhytismatales. Mycologia 103: 57–74. doi:10.3852/10-060.

71. Davey ML, Heegaard E, Halvorsen R, Kauserud H, Ohlson M (2013) Amplicon-pyrosequencing-based detection of compositional shifts in bryophyte-associated fungal communities along an elevation gradient. Mol Ecol 22: 368–383. doi:10.1111/mec.12122.

72. Rosling A, Cox F, Cruz-Martinez K, Ihrmark K, Grelet G-A, et al. (2011) Archaeorhizomycetes: Unearthing an ancient class of ubiquitous soil fungi. Science 333: 876–879. doi:10.1126/science.1206958.

73. Narum SR (2006) Beyond Bonferroni: Less conservative analyses for conservation genetics. Conserv Genet 7: 783–787. doi:10.1007/s10592-005-9056-y.

74. Tedersoo L, Mett M, Ishida TA, Bahram M (2013) Phylogenetic relationships among host plants explain differences in fungal species richness and community composition in ectomycorrhizal symbiosis. New Phytol 199: 822–831. doi:10.1111/nph.12328.

75. Kennedy PG, Hortal S, Bergemann SE, Bruns TD (2007) Competitive interactions among three ectomycorrhizal fungi and their relation to host plant performance. J Ecology 95: 1338–1345. doi:10.1111/j.1365-2745.2007.01306.x.

76. Mack KML, Rudgers JA (2008) Balancing multiple mutualists: asymmetric interactions among plants, arbuscular mycorrhizal fungi, and fungal endophytes. Oikos 117: 310–320. doi:10.1111/j.2007.0030-1299.15973.x.

Exo-Metabolome of *Pseudovibrio* sp. FO-BEG1 Analyzed by Ultra-High Resolution Mass Spectrometry and the Effect of Phosphate Limitation

Stefano Romano[1]*, Thorsten Dittmar[2], Vladimir Bondarev[1], Ralf J. M. Weber[3], Mark R. Viant[3], Heide N. Schulz-Vogt[4]

1 Max Planck Institute for Marine Microbiology, Bremen, Germany, **2** Research Group for Marine Geochemistry, Institute for Chemistry and Biology of the Marine Environment (ICBM), University of Oldenburg, Oldenburg, Germany, **3** School of Biosciences, University of Birmingham, Birmingham, United Kingdom, **4** Department of Biological Oceanography, Leibniz-Institute for Baltic Sea Research Warnemuende (IOW), Rostock, Germany

Abstract

Oceanic dissolved organic matter (DOM) is an assemblage of reduced carbon compounds, which results from biotic and abiotic processes. The biotic processes consist in either release or uptake of specific molecules by marine organisms. Heterotrophic bacteria have been mostly considered to influence the DOM composition by preferential uptake of certain compounds. However, they also secrete a variety of molecules depending on physiological state, environmental and growth conditions, but so far the full set of compounds secreted by these bacteria has never been investigated. In this study, we analyzed the exo-metabolome, metabolites secreted into the environment, of the heterotrophic marine bacterium *Pseudovibrio* sp. FO-BEG1 via ultra-high resolution mass spectrometry, comparing phosphate limited with phosphate surplus growth conditions. Bacteria belonging to the *Pseudovibrio* genus have been isolated worldwide, mainly from marine invertebrates and were described as metabolically versatile *Alphaproteobacteria*. We show that the exo-metabolome is unexpectedly large and diverse, consisting of hundreds of compounds that differ by their molecular formulae. It is characterized by a dynamic recycling of molecules, and it is drastically affected by the physiological state of the strain. Moreover, we show that phosphate limitation greatly influences both the amount and the composition of the secreted molecules. By assigning the detected masses to general chemical categories, we observed that under phosphate surplus conditions the secreted molecules were mainly peptides and highly unsaturated compounds. In contrast, under phosphate limitation the composition of the exo-metabolome changed during bacterial growth, showing an increase in highly unsaturated, phenolic, and polyphenolic compounds. Finally, we annotated the detected masses using multiple metabolite databases. These analyses suggested the presence of several masses analogue to masses of known bioactive compounds. However, the annotation was successful only for a minor part of the detected molecules, underlining the current gap in knowledge concerning the biosynthetic ability of marine heterotrophic bacteria.

Editor: Tilmann Harder, University of New South Wales, Australia

Funding: This study was funded by the European Research Council, Grant No. 203364 and the Max Planck Society. The funders had no role in study design, data collection and analysis, decision to publish, or preparation of the manuscript.

Competing Interests: The authors have declared that no competing interests exist.

* E-mail: sromano@mpi-bremen.de

Introduction

Microorganisms dynamically interact with their environment, they are influenced by its composition and, in turn, they influence its composition. This reciprocity has an effect on bacterial gene expression, protein synthesis, and metabolite uptake and production. In the ocean the dissolved organic matter (DOM), which consists of a collection of reduced carbon compounds often containing heteroatoms (e.g. N, P, S), is the result of these interconnected processes. Photosynthetic and non-photosynthetic bacteria can release metabolites into the environment according to their physiological state [1]. Examples are compounds secreted for nutrient acquisition (e.g. siderophores), for communication (e.g. homoserine lactones), and for interspecies competition (e.g. antibiotics). Several studies have investigated the effect of the activity of photosynthetic bacteria on DOM composition (reviewed in [1] and [2]), whereas the composition of the DOM produced by heterotrophic bacteria is almost unknown. Special attention has been paid to metabolites of biotechnological interest, but little is known about the full suite of compounds produced by bacteria under different nutrient regimes and growth phases, resulting in a general lack of information on the influence that the metabolism of marine heterotrophic bacteria has on oceanic DOM composition [2].

Metabolomics is the field of science that aims to characterize and quantify metabolites, or low molecular weight molecules, originating from cellular activity under a given set of physiological conditions. This collection of metabolites is termed the metabolome [3], which can be partitioned into the so called endo-metabolome (all intracellular metabolites) and the exo-metabolome (all extracellular metabolites) [4–6]. Metabolomics is a ''downstream'' approach and reflects the final response of cells to specific environmental conditions and it completes and integrates the associated techniques of proteomics and transcriptomics [3]. Microbial metabolomic studies have already been performed for different purposes, e.g. to elucidate metabolic pathways, to investigate the response of bacterial metabolism to environmental stresses, to support bacterial identification, and to diagnose bacterial infections [7–13]. Such studies have the potential to provide new insights into the composition of the metabolites secreted by marine heterotrophic bacteria and into their influence on the oceanic DOM composition.

Among the different analytical techniques, high resolution accurate mass (HRAM) mass spectrometry has acquired a predominant position in metabolomic studies [14]. Among others, Fourier transform ion cyclotron resonance mass spectrometry (FT-ICR-MS) is emerging as the most promising technology since it provides accurate mass measurement with ppm or sub-ppm error. It allows to obtain ultra-high resolved profiles with thousands of accurate masses, which in principle can be transformed into real elementary composition [15–18]. Therefore, it permits high-throughput screening of intracellular and extracellular metabolites providing overall information on bacterial metabolism.

This technique was successfully employed to analyze the variation in the endo-metabolome during bacterial growth, in studies of metabolic diversity among different ecotypes and in analyzing bacterial response to stress conditions [19–22]. However, studies that analyze the bacterial exo-metabolome during growth and in response to nutrient limitation are missing. In the present manuscript we report a detailed analysis of the exo-metabolome of strain FO-BEG1, which belongs to the genus *Pseudovibrio*. These are heterotrophic *Alphaproteobacteria* distributed worldwide and they have been detected especially in association with marine invertebrates [23,24]. Bacteria belonging to this genus have often been shown to produce bioactive secondary metabolites, and they are considered a potential source of new molecules of medical interest [23,25–27].

We investigated the composition of the secreted metabolites during bacterial growth, and we analyzed the effect of phosphate limitation. Phosphate limitation was chosen because it is a common environmental condition encountered in many marine systems [28–30], and it has been described to have a significant effect on primary and secondary metabolism [31,32]. We report here the astonishing diversity of the exo-metabolome of strain FO-BEG1 and the drastic effect that phosphate limitation has on its composition. These data shed new light onto the complexity of the metabolites secreted by heterotrophic marine bacteria and onto the effect that their metabolic state can have on the composition of DOM in the ocean.

Materials and Methods

Growth conditions

Strain FO-BEG1 was cultivated in the carbohydrate/mineral medium (CM) as described by Shieh et al. [33] and modified by Bondarev et al., [23]. For the phosphate surplus condition ($+P_i$) phosphate was added to a final concentration of 1.4 mmol L^{-1}, whereas no phosphate was added to the phosphate limited ($-P_i$) medium. Under $-P_i$ conditions the final phosphate concentration was 0.1 mmol L^{-1}, and derived from the buffer used for preparing the vitamin solutions. Erlenmeyer flasks of 250 mL were filled with 100 mL of medium and inoculated with 100 µL of a pre-culture grown under $+P_i$ conditions. Cultures were incubated at 28°C in the dark and shaken at 120 rpm. We monitored bacterial growth by means of Optical Density (OD) measured at 600 nm using an Eppendorf BioPhotometer (Eppendorf AG, Hamburg, Germany). The OD_{600} was then correlated with the cell number, determined

using a Thoma chamber (Brand GmbH, Wertheim, Germany; data not shown). All experiments were performed and sampled in independent experimental triplicates.

Solid phase extraction of dissolved organic matter (SPE-DOM), dissolved organic carbon (DOC) measurements, and Fourier transform ion cyclotron resonance mass spectrometry (FT-ICR-MS) of DOM

For both $-P_i$ and $+P_i$ cultures, samples were collected immediately after the inoculation (T0) and in the exponential growth phase (T1). Additionally, samples at the end of the logarithmic phase (T2) and during the stationary phase (T3) were collected for the $-P_i$ cultures. One more set of samples was collected also in the stationary phase (T2) of $+P_i$ cultures. Cells were removed via centrifugation at 10,000 × *g* for 10 min at 5°C, the supernatant was then filtered into 150 mL combusted glass serum-bottles using Acrodisc 25 mm syringe filters with a 0.2 µm pore size GHP membrane (Pall LifeSciences, Ann Arbor, MI, USA), acidified to pH 2 with 2 mol L^{-1} HCl, and stored at 4 °C until further analyses. We collected the samples from all biological triplicates in both $+P_i$ and $-P_i$ conditions, with the exception of T0.

DOM of the cell-free supernatants was extracted according to the solid phase extraction of dissolved organic matter (SPE-DOM) method described by Dittmar et al. [34]. The extraction was performed using Bond Elute PPL (Agilent Technologies, Wildbronn, Germany) cartridges with a styrene-divinylbenzene (SDVB) polymer modified with a property surface able to retain also the most polar classes of analytes. DOC content of each extract was analyzed using a Shimadzu TOC-VCPH total organic carbon analyzer (Shimadzu, Kyoto, Japan). The extracted DOM samples were then diluted with a mixture of methanol (MS grade) and ultra-pure water (50:50 v/v) to yield a DOC concentration of 20 mg L^{-1} carbon, filtered using a 0.2 µm pore size PTFE filter (Rotilabo, Carl Roth GmbH, Karlsruhe, Germany), and analyzed with a solariX FT-ICR-MS (Bruker Daltonik GmbH, Bremen, Germany) with a 15.0 Tesla magnet and equipped with an electrospray ionization (ESI) source. To maximize our analytical window, all samples were analyzed on the ESI-FT-ICR-MS in positive and negative ionization mode. We minimize the formation of adducts (and dimers of analyte compounds) by applying a gentle in-source collision-induced dissociation (CID) energy. This breaks apart larger adducts (including dimers), but no covalent bonds. All data were acquired with a time domain size of 4 megawords and with a detection range of *m/z* (mass to charge ratio) 150 to 2,000. For each run, 500 broadband scans were accumulated. All the mass spectra acquired under both positive and negative mode were analyzed with the Data Analysis version 4.0 SP4 software package (Bruker Daltonik GmbH).

Calibration of the mass spectra was performed as follows: one replicate of $-P_i$ T3 was spiked with 0.05 ppm L-arginine (Sigma-Aldrich, Steinheim, Germany), used for the ESI-negative analyses, or with 0.05 ppm Tuning-mix (Agilent Technologies, Palo Alto, CA, USA), used for the ESI-positive analyses. The resulting mass spectra were calibrated internally with reference mass lists, and molecular formulae were assigned for the remaining peaks in the spectra using the Data Analysis software. For the ESI-negative mode, the molecular formulae were assigned to an elemental composition in the following ranges: C 1–∞, H 1–∞, O 1–∞, N 0–4, S 0–2, P 0–1, and allowing errors lower than 0.2 ppm. A mass list with more than 300 masses in the range 150–800 *m/z* was obtained and used to calibrate all other acquired mass spectra. Due to the diversity of the samples, the calibration list was adjusted manually to cover always the full detected mass range with at least 40 calibration points. For the ESI-positive mode, the molecular formulae were assigned to an elemental composition in the following ranges: C 1–∞, H 1–∞, O 1–∞, N 0–4, S 0–2, P 0–2, Na 0–1, allowing errors lower than 0.2 ppm. A mass list containing 25 masses covering the range 295–850 m/z was used to calibrate the other acquired mass spectra, using at least 5 calibration points. All linear calibrations resulted in an average mass error of below 0.05 ppm. Additionally, the instrument was externally calibrated with an in-house marine deep sea DOM reference sample (mass accuracy of less than 0.1 ppm). Before each sample set, blank checks with methanol/ultrapure water 1:1 were measured.

Sample comparison, molecular formulae assignment, filtration of the datasets, and statistical analysis

Comparison of the mass spectra and isotope (^{13}C) identification were performed for the data obtained from both ionization modes, whereas the formulae assignment was done only for the ESI-negative mode. Sodium adducts frequently occur in ESI-positive mode and are considered in our molecular formulae assignment routine. However, other adducts, such as NH_4^+, are not easily identifiable, because those cannot be distinguished from other compounds where the same combination of elements is covalently bound. Such a distinction would require extensive additional analyses, such as fragmentation experiments (MS/MS) in the FT-ICR-MS. This was beyond the scope of this study. Because of the inherent uncertainty, we restricted our main analysis to ESI-negative data. With our ESI settings, the ionization in negative mode is highly reproducible and due to loss of H$^+$. Other possible adducts (e.g. Cl$^-$) can be identified by their unique isotope patterns but were not present in our mass spectra. The computational procedures were performed using an in-house Matlab routine developed by the Max Planck Research Group for Marine Geochemistry. The molecular formulae were assigned in the elemental composition in the following ranges: C 1–40, H 1–∞, O 1–∞, N 0–4, S 0–2, P 0–1, no Na, Fe, Cl and allowing a mass error of maximum 500 ppb. Only peaks with signal to noise ratio > 4 were considered and only formulae with a minimum H/C ration of 0.3 and a maximum O/C ratio of 1 were accepted. All detected ions were singly charged, as indicated by the mass difference between isotopologues (of ^{12}C versus ^{13}C). Therefore, all detected m/z values were equivalent to molecular masses.

The ion intensities of the m/z detected in both ionization modes were normalized by dividing the intensity of each mass by the sum of the 500 highest intensities measured in the respective mass spectra. This normalization procedure was performed independently for each measurement. The normalization was performed after removing all singlets, i.e. masses detected only in one sample out of the seventeen analyzed. In order to have an overview of the similarity among the samples, we performed a non-metric multidimensional scaling (NMDS) for the datasets obtained in both ionization modes, using the Bray-Curtis similarity index for the calculation of the distance matrices. Minimum-spanning trees between all samples were constructed to visualize pairwise sample similarities. Nearest neighbors, i.e. the most similar samples, were identified and graphically connected.

In order to reduce contingent noise and to consider only the molecules produced by the bacteria, we further filtered both datasets using the following criteria: we removed all masses detected in the samples $-P_i$ and $+P_i$ T0 that did not at least double their normalized ion intensity during the experiment; we removed all masses that were not present in at least all triplicates of one condition at a specific time point; we removed all masses that could contain the isotope ^{13}C. The filtered datasets were newly analyzed by means of NMDS, but the samples collected at T0 for both growth conditions were not considered, due to the significant

alteration in their m/z composition derived by the filtration of the datasets. A minimum-spanning tree between all samples was newly constructed. In order to verify the statistical reliability of the clustering observed in all NMDS plots, bootstrap analyses with 1000 reiterations were performed on dendrograms constructed for the similarity matrices obtained using the Bray-Curtis index. The paired group algorithm was used for the construction of the dendrograms. NMDSs, the relative stress values, which are a measure that reflects the degree of deviation of NMDS distances from original matrix distances, the minimum-spanning trees, the construction of the dendrograms, the calculation of the cophenetic correlation coefficients, which are a measure of how faithfully the dendrograms preserve the pairwise distances between the data points, and the bootstrap analyses were carried out by means of the PAST program [35]. Subsequently, in order to identify the unique masses present per time point under both conditions and the masses shared among the growth stages, we created Venn diagrams considering only masses present in all triplicates at the respective time points.

The elemental composition and the modified aromaticity index (AI$_{mod}$; [36]) of each molecular formula assigned to the m/z detected in ESI-negative mode were used to divide them into molecular categories according to criteria modified after Šantl-Temkiv et al. [37]. For this analysis, we excluded all masses for which multiple molecular formulae were obtained. We divided the molecular formulae into the following categories: peptides (if the molecular formula has an H/C ratio between 1.5 and 2, an O/C ratio lower than 0.9 and includes N), sugars (if the molecular formula has an O/C ratio equal or higher than 0.9 and an AI$_{mod}$ lower than 0.5), saturated fatty acids (if the molecular formula has an H/C ratio equal or higher than 2 and an O/C ratio lower or equal to 0.9), unsaturated aliphatic compounds (if the molecular formula has an H/C ratio between 1.5 and 2, an O/C ratio lower than 0.9 and does not contain N), highly unsaturated compounds (if the molecular formula has an AI$_{mod}$ lower than 0.5, an H/C ratio lower than 1.5, and an O/C ratio lower than 0.9), phenols (if the molecular formula has an AI$_{mod}$ equal or higher than 0.5 and less than 12 C atoms), and polyphenols (if the molecular formula has an AI$_{mod}$ equal or higher than 0.5 and 12 or more C atoms). We emphasize that this categorization is not unambiguous, and alternative structures may exist for a given molecular formula. However, this subdivision provides a helpful overview of likely structures behind the identified molecular formulae.

Metabolite and pathway annotation

The masses detected in all three biological replicates at each time point in ESI-negative mode were putatively annotated (i.e. level 2 of metabolite identification as defined by the Metabolomics Standards Initiative [38]) using the "transformation mapping" approach [39], after correcting the mass values for the H$^+$ loss. This method is based on mapping an experimentally-derived empirical formula difference for a pair of peaks to a known empirical formula difference between substrate-product pairs derived from the KEGG database (Kyoto Encyclopedia of Genes and Genomes [40]). To reduce the number of false positive assignments only metabolites that occurred in one of the *Pseudovibrio* sp. FO-BEG1 pathways (KEGG identifier: psf) were selected for annotation (as listed in KEGG on July 2013). Furthermore, we annotated the obtained masses considering the molecules reported in the Human Metabolome Database [41] and Drug Bank [42]. An additional annotation was performed using a sub-set of compounds reported in the Dictionary of Natural Products Online [43] obtained after performing a search based on the word "bacteria" typed in the property field "Type of organism word". Since in these three databases the pathways for the compounds are not indicated we could not apply the "transformation mapping" approach; therefore, the annotation was based on one to one matches between the detected masses and the masses of the known compounds, allowing always an error of $\leq$ 1 ppm.

Results

Measurement of the DOC released during bacterial growth and FT-ICR-MS analysis

Phosphate limitation repressed the growth of *Pseudovibrio* sp. FO-BEG1, leading to a final cell density 2.5–3.5 times lower than the one observed under phosphate surplus conditions (Fig. 1A). Under $-P_i$ conditions, a slightly higher amount of solid phase extractable dissolved organic carbon (SPE-DOC) was produced during the first half of the exponential phase (T1; Fig. 1B). As observed in T0, the SPE extraction did not retain the provided glucose, which alone would correspond to 60 mmol L^{-1} DOC. Therefore, the measured DOC represented the organic compounds produced and secreted by *Pseudovibrio* sp. FO-BEG1. At T1 under both conditions only around 2 mmol L^{-1} of glucose was taken up by the cells (Romano et al., unpublished data), resulting in a conversion of the initial carbon source in SPE-DOC of 0.4% for $-P_i$ cultures and 0.3% for $+P_i$ cultures. Despite the lower growth, under $-P_i$ conditions the SPE-DOC concentration increased to 267 ± 58 µmol L^{-1} and 511 ± 192 µmol L^{-1} at the end of the logarithmic (T2) and in the middle stationary (T3) phase, respectively. At both growth stages the glucose consumed was around 5 mmol L^{-1} (Romano et al., unpublished data). Consequently, in both cases the SPE-DOC represented 0.9% of the used glucose. Compared to this, the SPE-DOC concentration under $+P_i$ conditions during the stationary phase was more than three times lower (144 ± 18 µmol L^{-1}; Fig. 1B), representing 0.2% of the consumed glucose.

The raw data obtained from the ESI-negative FT-ICR-MS analysis consisted of 23,892 masses ranging from 154 m/z to 1,930 m/z. After normalization of the ion intensities, we performed a non-metrical multidimensional scaling (NMDS) in order to evaluate the similarities among the samples (Fig. 2A). As the stress value of the NMDS plot was 0.06, it could be considered a good representation of the calculated distance matrix and thus of the similarity among the samples. The samples collected at T1 for each biological triplicate under both $-P_i$ and $+P_i$ conditions clustered together and were clearly separated from the samples collected during the rest of the growth period (Fig. 2A). All biological triplicates of the $-P_i$ conditions collected at the end of the logarithmic phase and in the stationary phase (T2 and T3) were completely divergent from the samples collected under $+P_i$ stationary phase (T2). Moreover, the samples T2 and T3 for the $-P_i$ conditions also clustered separately in the plot (Fig. 2A). The bootstrap analysis of the dendrogram constructed for the Bray-Curtis similarity matrix revealed that the divergence among the samples described above was statistically highly significant, since during the 1000 reiterations always the same clustering occurred (Fig. S1A). In ESI-positive mode 17,859 masses were detected, ranging from 153 m/z to 1,999 m/z. The NMDS plot obtained for this dataset was characterized by a stress value of 0.07, therefore, it could be considered a good representation of the distance matrix as well (Fig. S2A). All samples had a similar clustering as the one observed in the ESI-negative NMDS plot. One of the main difference was the higher divergence between one $-P_i$ replicate ($-P_i$ III T1) and the other replicates collected at the same time point. However, the minimum spanning tree showed that this

sample shared the highest degree of similarity with the other samples collected under the same growth stage. Additionally, the samples collected at T2 and T3 under $-P_i$ conditions showed a higher degree of similarity (Fig. S2A). The bootstrap analysis performed on the respective dendrogram revealed that in > 75% of the cases the samples clustered consistently with the NMDS groups, indicating that the divergences described above were statistically significant (Fig. S3A)

In order to consider only those metabolites that were produced by the strain under the respective conditions, we removed from the datasets all compounds that were already present at T0 and did not at least double their ion intensities during the investigated growth period. Moreover, only compounds present in all biological triplicates at a certain time point and growth condition were further considered. This filtration reduced the ESI-negative dataset to 8,381 masses ranging from an m/z value of 154 to 998. The NMDS plot (Fig. 2B) performed for this new dataset showed the same clustering pattern as the one constructed for the unreduced dataset (Fig. 2A). 7,499 masses ranging from 163 to 1,234 m/z were obtained after the filtration of the ESI-positive dataset and, as for the negative mode, the new NMDS plot constructed using these masses showed a clustering consistent with the one of the unreduced dataset (Fig. S2B). The only exception was the higher similarity among the samples $+P_i$ T2 and the ones collected at T1 under both phosphate regimes. For both ionization modes, the bootstrap analyses performed on the dendrograms suggested that the clustering of the samples observed in the NMDS plots was statistically significant (Fig. S1B, S3B). Only for the filtered data obtained in ESI-positive mode the divergence among the samples collected at T2 and T3 under phosphate limited condition was not statistically significant, since their divergence occurred in < 50% of the reiterations (Fig. S3B).

In the Venn diagram constructed considering the masses obtained in ESI-negative mode (Fig. 3), it was evident that the samples collected during the logarithmic growth phase under both $+P_i$ and $-P_i$ conditions presented 23 and 100 unique masses, respectively. These samples shared 202 masses never detected in the stationary phase. Independent of the condition and the growth phase, we detected 573 masses shared among all samples. The samples collected at the end of the logarithmic and in the stationary phase under $-P_i$ conditions (T2 and T3) overall showed 1,088 unique masses never detected in the other time points, whereas in the samples collected in the stationary phase under $+P_i$ conditions we detected 832 unique masses (Fig. 3). A highly similar distribution was observed in the Venn diagram obtained for the ESI-positive mode (Fig. S4). The samples collected at the end of the logarithmic phase and in stationary phase under $-P_i$ conditions showed a higher number of masses (total of 2,220) than the samples collected in the stationary phase under $+P_i$ conditions. In contrast to the results obtained in ESI-negative mode, a higher number of masses (108) was shared among all phosphate limited samples, and a lower number of masses (122) was shared among all samples independent of the condition or growth stage.

The higher variability among replicates observed in ESI-positive mode was likely due to multiple ionization mechanisms, which can result, for example, in ammonium or sodium adduction, both ions present in our culturing medium.

Conversion of masses obtained in ESI-negative mode into molecular formulae and annotation of metabolites

Of the 8,381 masses detected in ESI-negative mode after the filtration of the dataset described above, we were able to assign molecular formulae to 4,914. Isotopologues were not included in

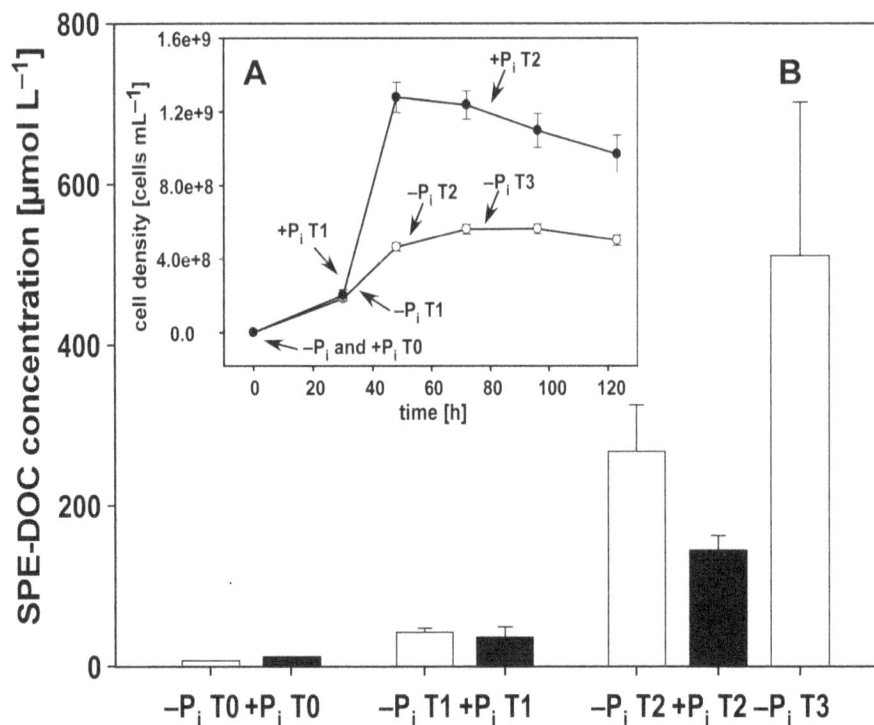

Figure 1. Bacterial growth (A) and concentrations of solid phase extractable dissolved organic carbon (SPE-DOC) (B). The bars (**B**) represent the average concentrations of SPE-DOC measured in the solid phase extracts of the biological triplicates collected during growth under both $+P_i$ (black) and $-P_i$ conditions (white). The inner panel (**A**) shows the cell growth, measured as cell density over time, for the two tested conditions. Filled circles represent the cultures growing under $+P_i$ conditions and empty circles represent the cultures growing under $-P_i$ conditions. Error bars indicate the standard deviation of biological triplicates

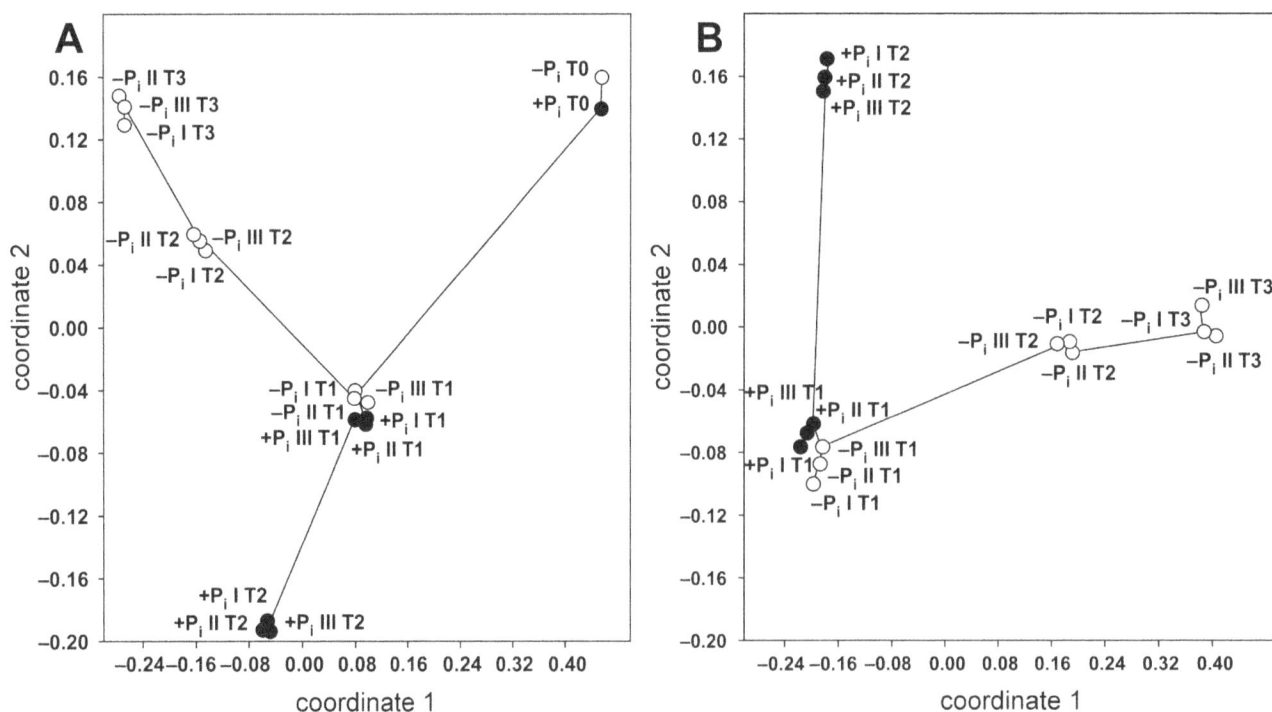

Figure 2. Similarity among the FT-ICR-MS samples analyzed in ESI-negative mode during bacterial growth under $+P_i$ and $-P_i$ conditions. Non metrical multidimensional scaling (NMDS) was performed by employing the Bray-Curtis similarity index and using the data of the unfiltered (**A**) and filtered (**B**) datasets. All biological triplicates of $+P_i$ (filled circles) and $-P_i$ (empty circles) conditions are shown. Nearest neighbor samples (i.e. most similar) are connected to visualize pairwise sample similarities. The stress value for both plots is 0.06.

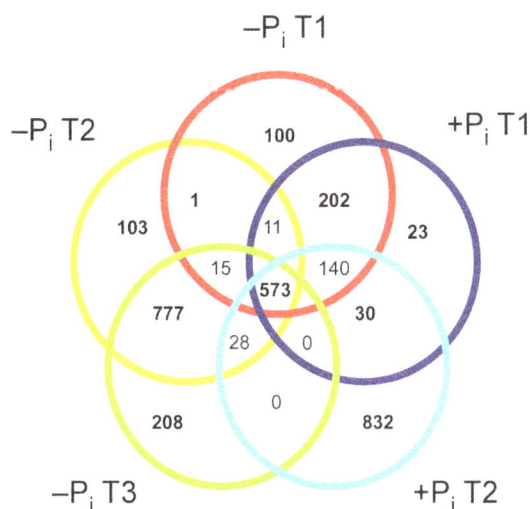

Figure 3. Venn diagram showing unique and shared masses detected in ESI-negative mode in all biological triplicates of the different samples. Only masses detected in all biological triplicates for each time point were considered.

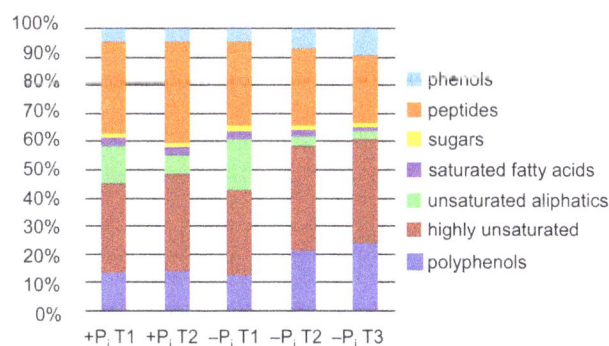

Figure 4. Percentages of molecular formulae attributed to molecular categories. Only masses detected in all biological triplicates for each time point were considered.

the number of assigned molecular formulae. Of these, 4,122 were unique molecular formulae, i.e. only one molecular formula could be assigned to the respective m/z value, corresponding to 49% of the m/z values present in the filtered dataset. A greater percentage of molecular formulae could be assigned to the masses obtained from samples collected at T1 under both $+P_i$ and $-P_i$ conditions (Table 1). Under $+P_i$ conditions an increase in the relative number of formulae containing nitrogen was observed from logarithmic to stationary phase, whereas the percentage of these compounds decreased under $-P_i$ conditions (Table 1). Interestingly, during bacterial growth under $-P_i$ conditions the relative amount of molecular formulae containing sulfur increased strongly from 45% to 65% of the total assigned formulae (Table 1).

After calculating the modified aromaticity index (AI_{mod}) we assigned the obtained molecular formulae to specific molecular categories and calculated their relative abundances at different time points (Fig. 4). In agreement with the similarity observed in the NMDS plots, at T1 the composition of the secreted metabolites was similar in both treatments. The major components of the exo-metabolome were compounds with molecular formulae assigned to peptides and highly unsaturated molecules. Only under $-P_i$ conditions, a pronounced increase of highly unsaturated, phenolic and polyphenolic compounds and a

decrease in peptides and unsaturated aliphatic compounds could be observed during stationary phase.

The ultra-high resolution of the FT-ICR-MS results in precise masses that can be compared and assigned to known compounds present in pathways described for the considered organism and collected in target databases such as KEGG (Kyoto Encyclopedia of Genes and Genomes). The metabolite names reported in the pathways of strain *Pseudovibrio* sp. FO-BEG1 in this database were used to annotate the masses obtained from the FT-ICR-MS analysis. The annotation strategy was based on mapping an experimentally-derived empirical formula difference for a pair of m/z to an empirical formula difference calculated for substrate-product pairs retrieved from KEGG [39]. It was previously shown that this approach can reduce the false positive rate of putative metabolite annotation by more than fourfold in comparison to searching a compound database using a one to one match approach (peak by peak search), while maintaining a minimal false negative rate [39]. A molecular name could be assigned only to a minor proportion of compounds detected in ESI-negative mode (less than 3%; Dataset S1). For the masses detected in ESI-positive mode, the percentage was even lower. For the reasons outlined above, we did not further consider the ESI-positive results for the annotation of metabolites. We could annotate 85 masses for the sample $-P_i$ T1 and of them 55 were assigned to unique metabolites (1.8% of the detected masses in all triplicates). The number of masses assigned to unique metabolites decreased to 46 (65 total annotated masses) for the samples $-P_i$ T2 and to 30 for $-P_i$ T3 (37 total annotated masses), representing 1.3% and 1.2% of the detected masses in all triplicates, respectively. 49 and 64 masses

Table 1. Overview of the data obtained from the ESI-negative FT-ICR-MS analysis.

	$+P_i$ T1	$+P_i$ T2	$-P_i$ T1	$-P_i$ T2	$-P_i$ T3
detected masses	2596	4206	3112	3566	2479
unique molecular formulae	1578 (60.8%)	2426 (57.7%)	1931 (62.1%)	1876 (52.6%)	1241 (50.1%)
formulae containing nitrogen	1193 (75.6%)	2085 (85.9%)	1362 (70.5%)	1387 (73.9%)	813 (65.5%)
formulae containing sulfur	648 (41.0%)	1087 (44.8%)	859 (44.5%)	1138 (60.7%)	802 (64.6%)
formulae containing phosphorus	221 (14.0%)	374 (15.4%)	247 (12.8%)	301 (16.0%)	176 (14.2%)

The number of masses detected in all biological triplicates of each time point is shown. The data refer to the dataset obtained after applying the filtration criteria described in the Materials and Methods section. Values in brackets represent the percentages of masses to which a unique molecular formula could be assigned and the percentages of unique msolecular formula containing heteroatoms. Isotopologues of assigned molecular formulae are not counted as assigned. Overall, a unique molecular formula could be assigned to 4,122 masses, corresponding to 49% of the obtained m/z.

could be assigned to unique metabolites (73 and 97 total annotated masses) in the samples $+P_i$ T1 and $+P_i$ T2 (1.8% and 1.5% of the detected masses in all triplicates), respectively. Most of the annotated compounds were intermediates in the metabolism of the amino acids lysine, tyrosine, tryptophan and phenylalanine (Dataset S2 and Fig S5). In all samples, except $-P_i$ T3, several metabolites were also annotated in the pathways of the purine metabolism (Dataset S2 and Fig. S5).

In order to identify possible molecules of biotechnological relevance, we performed three additional annotations using the masses obtained in ESI-negative mode and targeting the Drug Bank (Dataset S3), the Human Metabolome Database (HMDB; Dataset S4), and the Dictionary of Natural Products (DNP; Dataset S5). When the Drug Bank was chosen as target, the maximum number of annotated masses was 327 and was obtained for the sample $+P_i$ T2. The same sample presented the highest number of annotate masses (211) also in the annotation performed targeting the DNP. Whereas, using the HMDB, 186 was the maximum number of masses annotated, and it was obtained for the data of the sample $-P_i$ T1. In these databases the bio-synthetic pathways are not reported; therefore, the "transformation mapping" approach could not be applied. Consequently, the data obtained have to be considered with caution since a high number of false positive annotations can occur. The annotation performed using the HMDB was in line with the results obtained using the KEGG database, showing mostly intermediate metabolites of amino acid and nucleotide metabolisms (Dataset S4). Although the annotation performed using the Drug Bank database resulted in a higher number of assignments, in every sample several m/z were annotated as plant metabolites (e.g. epigallocatechin, commonly found in tea leaves; ginkgolide-A, produced by *Ginkgo biloba*) or compounds of synthetic origin (e.g. ibuprofen; Dataset S3). This suggests a high number of false positive annotations; therefore, these data will not be further discussed except when consistent with the other annotation approaches we applied. Finally, the annotation performed using the DNP resulted in the assignment of several masses to compounds having antibacterial, signaling, and enzymatic inhibiting activities (Dataset S5). Interestingly, only in the samples $-P_i$ T2 and $-P_i$ T3 the m/z 210.952904 was annotated as tropodithietic acid, which is a potent antibiotic produced by bacteria belonging to the *Roseobacter* clade and the *Pseudovibrio* genus [23,44].

Discussion

In order to quantify and characterize the metabolites released by strain *Pseudovibrio* sp. FO-BEG1 into the medium during growth and to evaluate the effect of phosphate limitation on them, we performed an ultra-high resolution mass spectrometry analysis of the bacterial exo-metabolome. Mass spectrometry is the most widely used approach in metabolomic studies [14]. In particular high resolution accurate mass (HRAM) mass spectrometry instruments are receiving progressively more attention, owing to their ability to resolve highly complex samples and to yield accurate mass measurements, which allow precise calculations of the elemental composition [15,16,18].

When cells growing under $-P_i$ conditions entered stationary phase, they released three times more solid phase extractable dissolved organic carbon (SPE-DOC) than cells growing under $+P_i$ conditions (Fig. 1). We are confident that the SPE-DOC concentrations and the number of metabolites obtained are not biased by the presence of compounds derived from the cultivation medium because, as shown by the amount of SPE-DOC at T0, the SPE method did not retain significant quantities of organic compounds present in the medium (Fig. 1B). Moreover, during the filtration of the datasets we removed all m/z (mass to charge ratio) that were detected at T0 and did not at least double their ion intensities during the experiment. Therefore, all compounds originally present in the medium and not used by the cells during bacterial growth were excluded from the analyses.

It has been known for several years that low phosphate concentrations can induce the production of secondary metabolites ([31] and references therein), which would suggest that under $-P_i$ conditions a higher fraction of the carbon source provided was used by *Pseudovibrio* sp. FO-BEG1 for the production of such compounds. In addition, it is known that phosphate limitation can trigger membrane lipid rearrangement, with the substitution of phosphorous-containing with phosphorous-free lipids [45,46], a phenomenon that we also observed for *Pseudovibrio* sp. FO-BEG1 (Romano et al., unpublished data). Therefore, it is reasonable to hypothesize that due to the membrane rearrangement more cytosolic metabolites could leak out from the cells, explaining the higher production of SPE-DOC under $-P_i$ conditions. Consistently, nutrient leakage was also described in a marine yeast strain growing under phosphate limited conditions [47]. Other studies showed that bacteria can convert from 5 to 15% of the provided carbon into DOC [48–50], which is one order of magnitude higher than observed in our experiments. However, a precise comparison is difficult because in all mentioned examples different medium composition, growth parameters, and analytic procedures were used.

Rosselló-Móra et al. [19] and Antón et al. [22] analyzed the endo- and the exo-metabolome of different *Salinibacter ruber* isolates during the classification of different ecotypes, and reported that the isolates can be distinguished by their metabolic profiles. Moreover, Brito-Echeverria et al. [20] analyzed the endo- and exo-metabolome of different *Salinibacter* strains in response to different stress conditions, and reported that the exo-metabolome was affected to a greater extent than the endo-metabolome. In all studies, the analyses were performed via FT-ICR-MS, and they are the first reports that provide information about the complexity of the bacterial exo-metabolome. In line with these observations, our analysis revealed that *Pseudovibrio* sp. FO-BEG1 produced and released at least many hundreds of compounds into the medium, and that the composition of this DOC was greatly affected by phosphate limitation. In this respect, FT-ICR-MS represents an ideal and powerful technique to unravel this complexity. We could clearly show that the exo-metabolome composition differs during different growth phases and between the two tested conditions (Fig. 2, 3, S1, S2, S3, S4). These data are consistent with previous studies, which applying low resolution techniques reported that the metabolites secreted by bacteria can change during different growth phases and in response to environmental stresses [8,11,21,51]. One interesting difference between the two phosphate regimes was the higher amount of compounds containing sulfur detected under $-P_i$ conditions (Table 1). This suggests that phosphate limitation also influences the sulfur metabolism of *Pseudovibrio* sp. FO-BEG1, increasing the amount of sulfur released into the environment in the form of DOM.

The presence of unique masses detected only at specific time points under both conditions shows a dynamic cycling of organic compounds. Molecules produced during the beginning of the logarithmic growth phase were then taken up again when cells entered stationary phase. A similar phenomenon was observed in a study that investigated the effect of grazing on the DOC production in a pure culture of *Pseudomonas chlororaphis* [50]. Interestingly, even though for each sampling point and each condition we identified hundreds of unique masses, we also detected 573 and 122 masses in ESI-negative and positive mode, respectively, which were always present in our samples independent of the growth stage or the growth condition (Fig. 3, S4). It would be interesting to verify whether this "core" exo-metabolome is affected by other environmental changes or it represents a distinctive "metabolic signature" of the strain.

It has been suggested that the trophic status of the environment affects DOM composition via shaping the ecological processes that are responsible for its production [2]. Productive, nutrient rich regions have significant DOM production directly from photosynthesis, whereas oligotrophic, nutrient poor regions have significant DOM production from grazing processes [52,53]. This difference was attributed to the complexity of the microbial food web in different environments, with the oligotrophic regions having a more effective microbial loop compared to the classical food web described in the productive regions [54]. Our data suggest that in order to understand DOM composition the effect of the environmental nutrient regimes on bacterial physiology should not be underestimated. As we show, it can greatly affect both the amount and the composition of the produced organic compounds.

Comparing the variation of the metabolome of *Escherichia coli* and *Saccharomyces cerevisiae* in response to carbon and nitrogen limitation, Brauer et al. [10] unexpectedly showed global metabolic trends remarkably conserved among these two distantly related microorganisms. Therefore, in order to verify the presence of shared metabolic responses, which could indicate the presence of highly conserved regulatory schemes, it would be of great interest to compare the variations of the exo-metabolome in response to nutrient limitation among different bacteria. Here we show that the nutrient regime greatly influenced the DOM secreted by *Pseudovibrio* sp. FO-BEG1 into the environment. Therefore, it is reasonable to expect that by extending these kinds of studies to different marine bacteria the influences of microbes on DOM composition in natural environments characterized by particular trophic conditions could be better understood.

The molecular formula assignment allowed us to classify the detected masses in molecular categories, giving a broad overview of the types of compounds released during growth. Under phosphate limitation, we observed a higher production of phenolic and polyphenolic compounds when cells entered stationary phase (Fig. 4). Production of phenol was described for the strains *Pseudovibrio* sp. D323 and L4-8 [55,56]. The crude extract of the spent medium of the latter strain showed a strong antioxidant activity, which is consistent with our finding that strain FO-BEG1 produces different types of phenols and polyphenols, known for their antioxidant properties [57]. Higher production of these compounds under $-P_i$ conditions could be related to the increased oxidative stress that cells growing under phosphate limitation might experience [58–60], and which we also inferred for strain FO-BEG1 from the comparison of the protein expression between $+P_i$ and $-P_i$ conditions (Romano et al., unpublished data).

Some of the detected phenolic and polyphenolic compounds could be, for example, tropone derivates. These molecules are commonly produced by bacteria of the *Roseobacter* clade and can have algaecide and antibacterial activity, as, for example, the potent antibiotic tropodithietic acid (TDA; [61,62]). Previous experiments using high performance liquid chromatography suggested that a compound with the same retention time and UV-visible spectra as the TDA standard was produced by *Pseudovibrio* sp. FO-BEG1 under $-P_i$ conditions when cells entered stationary phase (Romano et al., unpublished data). During the FT-ICR-MS analyses, we identified the m/z 210.952904 with the molecular formula assigned $C_8H_4O_3S_2$, which was, considering also its peculiar isotopic patterns due to the presence of two sulfur

atoms per molecule, consistent with being TDA. This compound was detected only under phosphate limitation and its ion intensity increased from T2 to T3. Consistently, when the Dictionary of Natural Products (DNP) was used as target for annotating the detected masses, the previously mentioned m/z was assigned to TDA, and to thiotropocin and troposulfenin (Dataset S5) which are tautomers of TDA [63] and are known to be produced by *Pseudomonas* spp. [64]. In addition, in the same samples, the m/z 226,947816 was annotated as hydroxytropodithietic acid, which was suggested to derive from the hydroxylation of TDA [65]. Members of the *Roseobacter* clade produce TDA together with an uncharacterized yellow pigment [44] and consistently also the *Pseudovibrio* cultures growing under $-P_i$ conditions developed an intense yellow coloration when entered stationary phase (Romano et al., unpublished data). Altogether, this data support our interpretation that TDA was produced during the stationary phase under $-P_i$ conditions.

The annotation using the DNP resulted in the attribution of several masses to molecules previously described in marine bacteria, including members of the *Roseobacter* clade (Datasets S5; e.g. 3-(4-Hydroxy-3-nitrophenyl)propanoic acid, cyclo(glutamylglycylprolyl), cyclo(glutamylglycylserylprolyl), homo-ξ-rhodomycinone). Among those, several masses were annotated as cyclic dipeptides produced by *Roseobacter* strains isolated from marine sponges. Cyclic dipeptides are molecules with antibacterial properties and biological and pharmacological effects on cells of higher organisms. It was suggested that they could play a role in bacterial and prokaryote-eukaryote communication [66–69]. In the last years several cyclic dipeptides have been isolated from marine organisms such as sponges and algae and from many marine prokaryotes, suggesting that the ability to produce these compounds is widespread among marine bacteria [68]. Considering the phylogenetic and physiological similarity between *Roseobacter*- and *Pseudovibrio*-related bacteria, and the recurrent association between both cyclic dipeptides and *Pseudovibrio* with marine sponges [23,68], it is reasonable to speculate that *Pseudovibrio* sp. FO-BEG1 released such compounds into the medium. This information offers a solid base for further chemical characterization of these compounds, which could represent new molecules of biotechnological interest.

When the KEGG database was used as target, most of the metabolites assigned to the detected masses were compounds involved in the synthesis of mainly aromatic amino acids (e.g. tyrosine, tryptophan, phenylalanine; Dataset S1, S2, Fig. S5) and nucleotides. Consistently, the annotation performed using the HMDB and the Drug Bank also suggested the presence of intermediates of these metabolisms as, for example, the shikimate pathway (e.g. erythrose-4-phosphate, shikimate-3-phosphate; Datasets S3, S4), which is responsible for the biosynthesis of aromatic amino acids. Release of these compounds was also observed in the analysis of the exo-metabolome of other bacterial and yeast strains [70,71]. In conditions of "overflow metabolism", i.e. conditions with an excess of carbon or energy source or in the presence of nutrient limitation, intermediates of different metabolic pathways can be released [72]. Recent evidence suggests that this is a common phenomenon in different microorganisms when they are cultivated under conditions of non-inhibited carbon uptake [70]. Aromatic amino acids are key intermediates in the production of aromatic secondary metabolites [73] suggesting that strain FO-BEG1 is potentially producing such compounds, which, however, are of unknown structure. Unlike observed for other microorganisms [70], no masses were annotated as metabolic intermediates of central metabolic pathways such as the tricarboxylic acid cycle. The main reason for this is that most of these metabolites have a low molecular mass, e.g. fumarate 116.07 Da, which fells outside the m/z range chosen for our analysis (150-2,000 m/z).

Among the identified compounds, a smaller number of metabolites could be annotated for the samples collected at T3 under $-P_i$ conditions. However, the majority of the annotated compounds belonged to the same pathways identified in KEGG for the other samples. Under phosphate limitation, the number of formulae annotated in the pathway "tyrosine metabolism" and "tryptophan metabolism" using KEGG, and the intermediates of the shikimate pathway detected in HMDB and Drug Bank, decreased strongly from T1 to T3, indicating that these metabolites were taken up again by the cells when they entered stationary phase. This uptake of previously released metabolites was likely done to satisfy specific anabolic needs under this growth conditions.

Production and release of amino acids by bacterial communities was also reported by Kawasaki and Benner [49], and these compounds were shown to be important constituents of DOC in some coastal areas [74,75], environments where also *Pseudovibrio* strains were often isolated [33,76]. It is worth pointing out that comparing the list of molecular formulae retrieved from our exo-metabolome study with a list of formulae detected in DOM of the deep North Pacific Ocean [77], we found only 83 shared compounds. However, comparing our data with a list of molecular formulae detected in DOM during and after a phytoplankton bloom in the North Sea (Dittmar et al., unpublished data), we detected 729 matches (18% of the masses with unique molecular formulae assigned) and 91% of them were always present in the natural samples, irrespective of the occurrence of the phytoplankton bloom (Dataset S6). This indicates that, at least on a molecular formula level, a large fraction of the detected compounds are indeed part of natural DOM, and their presence does not seem to be directly related to the immediate activity of primary producers. Consistently, also Kujawinski et al. [78] showed that some molecules detected in a pure culture of "*Candidatus* Pelagibacter ubique" were present in open-ocean DOM.

Our approach represents a high-throughput way of performing metabolomic studies, and it was adequate to capture the diversity of the metabolites released by the bacterium into the environment. However, the translation of the analytical information into existent biological knowledge by using the available tools showed two major drawbacks. The first one regards the reliability of the annotation performed using the HMDB, the Drug Bank and the DNP. By using a one to one match approach (peak by peak search), we were able to assign metabolite names to up to 8% of the detected masses. Especially using the first two databases a high number of double assignments and of non-bacterial metabolites were obtained, suggesting a high rate of false positive identifications. These results confirm previous reports that underlined the limit of single match annotation even when a mass error below 1 ppm is adopted [39,79]. As pointed out previously, these approaches require the application of orthogonal filtration processes, such as isotopic abundance patterns or "transformation mapping", which can significantly decrease the number of false positive assignments, generating more reliable information which can be integrated into a biological contest [39,79].

To increase the confidence in the annotation, we applied the "transformation mapping" approach, using the metabolic pathways of *Pseudovibrio* sp. FO-BEG1 reported in KEGG. Applying this method we were able to annotate less than 3% of the detected masses. The main drawback of this strategy is the incompleteness of the databases used, which can reduce the annotation efficiency by overlooking metabolites that are known, but not yet integrated into the database. For instance, even though we have strong evidence that TDA was produced during the stationary phase under $-P_i$ conditions, and the annotation performed using the DNP identified one mass consistent with being TDA, we could not identify it during the annotation processes using KEGG. The reason is the absence of the biosynthetic pathway for TDA among the annotated ones in *Pseudovibrio*. Databases such as KEGG are mostly restricted to genome-reconstruction pathways. Wrongly annotated genes and absence of compounds for which the biosynthetic routs have not been completely elucidated yet can decrease the number of identified molecules in metabolomic studies, and limit the capabilities of techniques such as FT-ICR-MS. This underlines the lack of knowledge we have about the biosynthetic ability of marine bacteria and also the necessity to create more comprehensive databases, containing information about both primary and secondary metabolites.

Conclusions

In this study we investigated in detail the exo-metabolome of a marine heterotrophic bacterium using ultra-high resolution mass spectrometry. Our work shows that HRAM instruments represent promising tools to unravel the complexity of the metabolites secreted from microorganisms. We show that the exo-metabolome is unexpectedly large and diverse, it is characterized by a dynamic recycling of compounds, and it is drastically affected by the physiological state of the strain. Our data clearly illustrate that phosphate limitation triggered a pronounced increase in the secretion of DOC and at the same time greatly affected its composition, leading to an increased production of functionalized phenols and polyphenols. A Part of the molecular formulae discovered in the exo-metabolome was also detected in natural marine DOM. Therefore, future studies on the exo-metabolomes of different strains and DOM from different locations might help to understand to what extent the compounds secreted by heterotrophic bacteria influence the oceanic DOM composition. The discrepancy between the number of measured masses and the number of annotated molecules obtained using different databases underlines the gap in our knowledge concerning the biosynthetic ability of marine bacteria, indicating the necessity of further work directed to chemically characterize the secreted metabolites. However, the integrated metabolic annotation we performed using multiple databases gave us a first glimpse of the composition of the secreted compounds, suggesting that the large bacterial exo-metabolome can represent a "chemical reservoir" for the discovery of new molecules of biotechnological interest. Our data underline the great biosynthetic ability of heterotrophic bacteria and suggest that, using the words of Traxler and Kolter [80], "*the chemical landscape inhabited and manipulated by bacteria is vastly more complex and sophisticated than previously thought*".

Supporting Information

Figure S1 Bootstrap analyses performed on the dendrograms obtained using the paired group algorithm and the Bray-Curtis similarity index calculated for the FT-ICR-MS samples analyzed in ESI-negative mode. Since the cophenetic correlation coefficients were $> 90\%$, the dendrograms can be considered a reliable representation of the similarity matrices. 1000 reiterations were allowed for the bootstrap analyses. Dendrograms were constructed using the data of the unfiltered (**A**) and filtered (**B**) datasets. All biological triplicates of $+P_i$ and $-P_i$ conditions are shown.

Figure S2   Similarity among the FT-ICR-MS samples analyzed in ESI-positive mode during bacterial growth under +P$_i$ and −P$_i$ conditions. Non metrical multidimensional scaling (NMDS) was performed by employing the Bray-Curtis similarity index and using the data of the unfiltered (**A**) and filtered (**B**) datasets. All biological triplicates of +P$_i$ (filled circles) and −P$_i$ (empty circles) conditions are shown. Nearest neighbor samples (i.e. most similar) are connected to visualize pairwise sample similarities. The stress value for **A** is 0.07 and for **B** is 0.08.

Figure S3   Bootstrap analyses performed on the dendrograms obtained using the paired group algorithm and the Bray-Curtis similarity index calculated for the FT-ICR-MS samples analyzed in ESI-positive mode. Since the cophenetic correlation coefficients were > 95%, the dendrograms can be considered a reliable representation of the similarity matrices. 1000 reiterations were allowed for the bootstrap analyses. Dendrograms were constructed using the data of the unfiltered (**A**) and filtered (**B**) datasets. All biological triplicates of +P$_i$ and −P$_i$ conditions are shown.

Figure S4   Venn diagram showing unique and shared masses detected in ESI-positive mode in all biological triplicates of the different samples. Only masses detected in all biological triplicates for each time point were considered.

Figure S5   Number of metabolites annotated in the metabolic pathways of Pseudovibrio sp. FO-BEG1 collected in the KEGG database. The masses obtained from the ESI-negative FT-ICR-MS analysis were annotated using the MI-Pack package. The bars of each color, representing the different time points, indicate the absolute number of metabolites annotated in the respective pathways reported in the KEGG database.

Dataset S1   Metabolites annotated using as target the metabolic pathways of Pseudovibrio sp. FO-BEG1 reported in the KEGG database.

Dataset S2   Metabolic pathways present in the KEGG database, which contain the annotated metabolites.

Dataset S3   Metabolites annotated using the Drug Bank as target.

Dataset S4   Metabolites annotated using the Human Metabolome Database as target.

Dataset S5   Metabolites annotated using the Dictionary of Natural Products as target.

Dataset S6   List of molecular formulae shared between the exo-metabolome of Pseudovibrio sp. FO-BEG1 and the North Sea DOM.

Acknowledgments

We are very grateful to K. Klaproth for technical support with the FT-ICR-MS analyses and I. Ulber and M. Friebe for help in the DOM extraction and DOC measurements. We are indebted to Dr. J. Niggemann for sharing DOM data of the North Sea. We acknowledge Dr. A. Ramette and G. Jessen for the constructive discussions concerning the statistical analysis.

Author Contributions

Conceived and designed the experiments: SR HNSV TD. Performed the experiments: SR VB. Analyzed the data: SR TD. Contributed reagents/materials/analysis tools: SR TD RJMW MRV HNSV. Wrote the paper: SR.

References

1. Carlson CA (2002) Production and removal processes. In: Hansell DA, Carlson CA, editors. Biogeochemistry of marinine Dissolved Organic Matter. San Diego: CA: Elsevier. pp. 91–152.

2. Kujawinski EB (2011) The impact of microbial metabolism on marine dissolved organic matter. Ann Rev Mar Sci 3: 567–599.

3. Oliver SG, Winson MK, Kell DB, Baganz F (1998) Systematic functional analysis of the yeast genome. Trends Biotechnol 16: 373–378.

4. Allen J, Davey HM, Broadhurst D, Heald JK, Rowland JJ, et al. (2003) High-throughput classification of yeast mutants for functional genomics using metabolic footprinting. Nat Biotechnol 21: 692–696.

5. Fiehn O (2001) Combining genomics, metabolite analysis, and biochemical modelling to understand metabolic networks. Comp Funct Genomics 2: 155–168.

6. Mapelli V, Olsson L, Nielsen J (2008) Metabolic footprinting in microbiology: methods and applications in functional genomics and biotechnology. Trends Biotechnol 26: 490–497.

7. Cundy KV, Willard KE, Valeri LJ, Shanholtzer CJ, Singh J, et al. (1991) Comparison of traditional Gas-Chromatography (Gc), headspace Gc, and the microbial identification library Gc system for the identification of Clostridium difficile. J Clin Microbiol 29: 260–263.

8. Shnayderman M, Mansfield B, Yip P, Clark HA, Krebs MD, et al. (2005) Species-specific bacteria identification using differential mobility spectrometry and bioinformatics pattern recognition. Anal Chem 77: 5930–5937.

9. Carlier JP (1990) Identification by quantitative gas-chromatography of common Fusobacterium species isolated from clinical specimens. Lett Appl Microbiol 11: 133–136.

10. Brauer MJ, Yuan J, Bennett BD, Lu WY, Kimball E, et al. (2006) Conservation of the metabolomic response to starvation across two divergent microbes. Proc Natl Acad Sci U S A 103: 19302–19307.

11. Coucheney E, Daniell TJ, Chenu C, Nunan N (2008) Gas chromatographic metabolic profiling: a sensitive tool for functional microbial ecology. J Microbiol Methods 75: 491–500.

12. Boersma MG, Solyanikova IP, Van Berkel WJH, Vervoort J, Golovleva LA, et al. (2001) F-19 NMR metabolomics for the elucidation of microbial degradation pathways of fluorophenols. J Ind Microbiol Biotechnol 26: 22–34.

13. Rinas U, Hellmuth K, Kang RJ, Seeger A, Schlieker H (1995) Entry of Escherichia coli into stationary phase is indicated by endogenous and exogenous accumulation of nucleobases. Appl Environ Microbiol 61: 4147–4151.

14. Want EJ, Nordstrom A, Morita H, Siuzdak G (2007) From exogenous to endogenous: The inevitable imprint of mass spectrometry in metabolomics. J Proteome Res 6: 459–468.

15. Junot C, Madalinski G, Tabet JC, Ezan E (2010) Fourier transform mass spectrometry for metabolome analysis. Analyst 135: 2203–2219.

16. Aharoni A, Ric de Vos CH, Verhoeven HA, Maliepaard CA, Kruppa G, et al. (2002) Nontargeted metabolome analysis by use of Fourier transform ion cyclotron mass spectrometry. Omics 6: 217–234.

17. Brown SC, Kruppa G, Dasseux JL (2005) Metabolomics applications of FT-ICR mass spectrometry. Mass Spectrom Rev 24: 223–231.

18. Marshall AG (2004) Accurate mass measurement: taking full advantage of nature's isotopic complexity. Physica B-Condensed Matter 346: 503–508.

19. Rosselló-Móra R, Lucio M, Peña A, Brito-Echeverria J, López-López A, et al. (2008) Metabolic evidence for biogeographic isolation of the extremophilic bacterium Salinibacter ruber. ISME J 2: 242–253.

20. Brito-Echeverria J, Lucio M, López-López A, Antón J, Schmitt-Kopplin P, et al. (2011) Response to adverse conditions in two strains of the extremely halophilic species Salinibacter ruber. Extremophiles 15: 379–389.

21. Takahashi H, Kai K, Shinbo Y, Tanaka K, Ohta D, et al. (2008) Metabolomics approach for determining growth-specific metabolites based on Fourier transform ion cyclotron resonance mass spectrometry. Anal Bioanal Chem 391: 2769–2782.

22. Antón J, Lucio M, Peña A, Cifuentes A, Brito-Echeverria J, et al. (2013) High metabolomic microdiversity within co-occurring Isolates of the extremely halophilic bacterium Salinibacter ruber. PLoS ONE 8(5): e64701. doi:64710.61371/journal.pone.0064701.

23. Bondarev V, Richter M, Romano S, Piel J, Schwedt A, et al. (2013) The genus *Pseudovibrio* contains metabolically versatile bacteria adapted for symbiosis. Environ Microbiol 15: 2095–2113.

24. Enticknap JJ, Kelly M, Peraud O, Hill RT (2006) Characterization of a culturable alphaproteobacterial symbiont common to many marine sponges and evidence for vertical transmission via sponge larvae. Appl Environ Microbiol 72: 3724–3732.

25. O' Halloran JA, Barbosa TM, Morrissey JP, Kennedy J, O' Gara F, et al. (2011) Diversity and antimicrobial activity of *Pseudovibrio* spp. from Irish marine sponges. J Appl Microbiol 110: 1495–1508.

26. Flemer B, Kennedy J, Margassery LM, Morrissey JP, O'Gara F, et al. (2012) Diversity and antimicrobial activities of microbes from two Irish marine sponges, *Suberites carnosus* and *Leucosolenia* sp. J Appl Microbiol 112: 289–301.

27. Kennedy J, Baker P, Piper C, Cotter PD, Walsh M, et al. (2009) Isolation and analysis of bacteria with antimicrobial activities from the marine sponge *Haliclona simulans* collected from Irish waters. Mar Biotechnol (NY) 11: 384–396.

28. Thingstad TF, Krom MD, Mantoura RFC, Flaten GAF, Groom S, et al. (2005) Nature of phosphorus limitation in the ultraoligotrophic eastern Mediterranean. Science 309: 1068–1071.

29. Cotner JB, Ammerman JW, Peele ER, Bentzen E (1997) Phosphorus-limited bacterioplankton growth in the Sargasso Sea. Aquat Microb Ecol 13: 141–149.

30. Wu JF, Sunda W, Boyle EA, Karl DM (2000) Phosphate depletion in the western North Atlantic Ocean. Science 289: 759–762.

31. Martín JF (2004) Phosphate control of the biosynthesis of antibiotics and other secondary metabolites is mediated by the PhoR-PhoP system: an unfinished story. J Bacteriol 186: 5197–5201.

32. Martín JF, Sola-Landa A, Santos-Beneit F, Fernández-Martínez LT, Prieto C, et al. (2011) Cross-talk of global nutritional regulators in the control of primary and secondary metabolism in *Streptomyces*. Microb Biotechnol 4: 165–174.

33. Shieh WY, Lin YT, Jean WD (2004) *Pseudovibrio denitrificans* gen. nov., sp. nov., a marine, facultatively anaerobic, fermentative bacterium capable of denitrification. Int J Syst Evol Microbiol 54: 2307–2312.

34. Dittmar T, Koch B, Hertkorn N, Kattner G (2008) A simple and efficient method for the solid-phase extraction of dissolved organic matter (SPE-DOM) from seawater. Limnol Oceanogr Methods 6: 230–235.

35. Hammer O, Harper DAT, Ryan PD (2001) PAST: Paleontological statistics software package for education and data analysis. Palaeontologia Electronica 4: 4–9. http://palaeo-electronica.org/2001_1/past/issue1_01.htm

36. Koch BP, Dittmar T (2006) From mass to structure: an aromaticity index for high-resolution mass data of natural organic matter. Rapid Commun Mass Spectrom 20: 926–932.

37. Šantl-Temkiv T, Finster K, Dittmar T, Hansen BM, Thyrhaug R, et al. (2013) Hailstones: a window into the microbial and chemical inventory of a storm cloud. PLoS ONE 8(1): e53550. doi:53510.51371/journal.pone.0053550.

38. Sumner LW, Amberg A, Barrett D, Beale MH, Beger R, et al. (2007) Proposed minimum reporting standards for chemical analysis Chemical Analysis Working Group (CAWG) Metabolomics Standards Initiative (MSI). Metabolomics 3: 211–221.

39. Weber RJM, Viant MR (2010) MI-Pack: increased confidence of metabolite identification in mass spectra by integrating accurate masses and metabolic pathways. Chemometr Intell Lab Syst 104: 75–82.

40. Kanehisa M, Goto S (2000) KEGG: kyoto encyclopedia of genes and genomes. Nucleic Acids Res 28: 27–30.

41. Wishart DS, Knox C, Guo AC, Eisner R, Young N, et al. (2009) HMDB: a knowledgebase for the human metabolome. Nucleic Acids Res 37: D603–610.

42. Wishart DS, Knox C, Guo AC, Cheng D, Shrivastava S, et al. (2008) DrugBank: a knowledgebase for drugs, drug actions and drug targets. Nucleic Acids Res 36: D901–906.

43. Dictionary of Natural Products (2014) Chapman & Hall/CRC Informa, London Version 22.2.

44. Bruhn JB, Nielsen KF, Hjelm M, Hansen M, Bresciani J, et al. (2005) Ecology, inhibitory activity, and morphogenesis of a marine antagonistic bacterium belonging to the *Roseobacter* clade. Appl Environ Microbiol 71: 7263–7270.

45. Benning C, Huang ZH, Gage DA (1995) Accumulation of a novel glycolipid and a betaine lipid in cells of *Rhodobacter sphaeroides* grown under phosphate limitation. Arch Biochem Biophys 317: 103–111.

46. Minnikin DE, Baddiley J, Abdolrah.H (1972) Variation of polar lipid composition of *Bacillus subtilis* (Marburg) with different growth conditions. FEBS Lett 27: 16–18.

47. Robertson BR, Button DK (1979) Phosphate-limited continuous culture of *Rhodotorula rubra* - Kinetics of transport, leakage, and growth. J Bacteriol 138: 884–895.

48. Ogawa H, Amagai Y, Koike I, Kaiser K, Benner R (2001) Production of refractory dissolved organic matter by bacteria. Science 292: 917–920.

49. Kawasaki N, Benner R (2006) Bacterial release of dissolved organic matter during cell growth and decline: Molecular origin and composition. Limnol Oceanogr 51: 2170–2180.

50. Gruber DF, Simjouw JP, Seitzinger SP, Taghon GL (2006) Dynamics and characterization of refractory dissolved organic matter produced by a pure bacterial culture in an experimental predator-prey system. Appl Environ Microbiol 72: 4184–4191.

51. Barreto MC, Frisvad JC, Larsen TO, Mogensen J, San-Romão MV (2011) Exometabolome of some fungal isolates growing on cork-based medium. Eur Food Res Technol 232: 575–582.

52. Maranón E, Cermeno P, Fernández E, Rodríguez J, Zabala L (2004) Significance and mechanisms of photosynthetic production of dissolved organic carbon in a coastal eutrophic ecosystem. Limnol Oceanogr 49: 1652–1666.

53. Nagata T (2000) Production mechanisms of dissolved organic matter; Kirchman DL, editor New York.Wiley-Liss 121–152.

54. Teira E, Pazo MJ, Serret P, Fernandez E (2001) Dissolved organic carbon production by microbial populations in the Atlantic Ocean. Limnol Oceanogr 46: 1370–1377.

55. Penesyan A, Tebben J, Lee M, Thomas T, Kjelleberg S, et al. (2011) Identification of the antibacterial compound produced by the marine epiphytic bacterium *Pseudovibrio* sp. D323 and related sponge-associated bacteria. Mar Drugs 9: 1391–1402.

56. Roué M, Quévrain E, Domart-Coulon I, Bourguet-Kondracki ML (2012) Assessing calcareous sponges and their associated bacteria for the discovery of new bioactive natural products. Nat Prod Rep 29: 739–751.

57. Scalbert A, Johnson IT, Saltmarsh M (2005) Polyphenols: antioxidants and beyond. Am J Clin Nutr 81: 215s–217s.

58. Yuan ZC, Zaheer R, Finan TM (2005) Phosphate limitation induces catalase expression in *Sinorhizobium meliloti*, *Pseudomonas aeruginosa* and *Agrobacterium tumefaciens*. Mol Microbiol 58: 877–894.

59. Gerard F, Dri AM, Moreau PL (1999) Role of *Escherichia coli* RpoS, LexA and H-NS global regulators in metabolism and survival under aerobic, phosphate-starvation conditions. Microbiology 145: 1547–1562.

60. Moreau PL (2004) Diversion of the metabolic flux from pyruvate dehydrogenase to pyruvate oxidase decreases oxidative stress during glucose metabolism in nongrowing *Escherichia coli* cells incubated under aerobic, phosphate starvation conditions. J Bacteriol 186: 7364–7368.

61. Seyedsayamdost MR, Carr G, Kolter R, Clardy J (2011) Roseobacticides: small molecule modulators of an algal-bacterial symbiosis. J Am Chem Soc 133: 18343–18349.

62. Thiel V, Brinkhoff T, Dickschat JS, Wickel S, Grunenberg J, et al. (2010) Identification and biosynthesis of tropone derivatives and sulfur volatiles produced by bacteria of the marine *Roseobacter* clade. Org Biomol Chem 8: 234–246.

63. Greer EM, Aebisher D, Greer A, Bentley R (2008) Computational studies of the tropone natural products, thiotropocin, tropodithietic acid, and troposulfenin. Significance of thiocarbonyl-enol tautomerism. J Org Chem 73: 280–283.

64. Kintaka K, Ono H, Tsubotani S, Harada S, Okazaki H (1984) Thiotropocin, a new sulfur-containing 7-membered-ring antibiotic produced by a *Pseudomonas* sp. J Antibiot (Tokyo) 37: 1294–1300.

65. Liang L (2003) Investigation of secondary metabolites of North Sea bacteria: fermentation, isolation, structure elucidation and bioactivity. University of Göttingen.136

66. Degrassi G, Aguilar C, Bosco M, Zahariev S, Pongor S, et al. (2002) Plant growth-promoting *Pseudomonas putida* WCS358 produces and secretes four cyclic dipeptides: cross-talk with quorum sensing bacterial sensors. Curr Microbiol 45: 250–254.

67. Holden MT, Ram Chhabra S, de Nys R, Stead P, Bainton NJ, et al. (1999) Quorum-sensing cross talk: isolation and chemical characterization of cyclic dipeptides from *Pseudomonas aeruginosa* and other gram-negative bacteria. Mol Microbiol 33: 1254–1266.

68. Huang R, Zhou X, Xu T, Yang X, Liu Y (2010) Diketopiperazines from marine organisms. Chem Biodivers 7: 2809–2829.

69. Prasad C (1995) Bioactive cyclic dipeptides. Peptides 16: 151–164.

70. Paczia N, Nilgen A, Lehmann T, Gätgens J, Wiechert W, et al. (2012) Extensive exometabolome analysis reveals underestimated overflow metabolism in various microorganisms. Microb Cell Fact 11: 122.

71. Behrends V, Ebbels TMD, Williams HD, Bundy JG (2009) Time-resolved metabolic footprinting for nonlinear modeling of bacterial substrate utilization. Appl Environ Microbiol 75: 2453–2463.

72. Kramer R (1994) Secretion of amino-acids by bacteria - Physiology and mechanism. FEMS Microbiol Rev 13: 75–93.

73. Herrmann KM (1995) The shikimate pathway as an entry to aromatic secondary metabolism. Plant Physiol 107: 7–12.

74. Coble PG (1996) Characterization of marine and terrestrial DOM in seawater using excitation emission matrix spectroscopy. Mar Chem 51: 325–346.

75. Yamashita Y, Tanoue E (2003) Distribution and alteration of amino acids in bulk DOM along a transect from bay to oceanic waters. Mar Chem 82: 145–160.

76. Hosoya S, Yokota A (2007) *Pseudovibrio japonicus* sp nov., isolated from coastal seawater in Japan. Int J Syst Evol Microbiol 57: 1952–1955.

77. Rossel PE, Vähätalo AV, Witt M, Dittmar T (2013) Molecular composition of dissolved organic matter from a wetland plant (*Juncus effusus*) after photochemical and microbial decomposition (1.25 yr): Common features with deep sea dissolved organic matter. Org Geochem 60: 62–71.

78. Kujawinski EB, Longnecker K, Blough NV, Del Vecchio R, Finlay L, et al. (2009) Identification of possible source markers in marine dissolved organic matter using ultrahigh resolution mass spectrometry. Geochim Cosmochim Acta 73: 4384–4399.

79. Kind T, Fiehn O (2006) Metabolomic database annotations via query of elemental compositions: mass accuracy is insufficient even at less than 1 ppm. BMC bioinformatics 7: 234.

80. Traxler MF, Kolter R (2012) A massively spectacular view of the chemical lives of microbes. Proc Natl Acad Sci U S A 109: 10128–10129.

Levels of Germination Proteins in *Bacillus subtilis* Dormant, Superdormant, and Germinating Spores

Yan Chen[1], W. Keith Ray[2], Richard F. Helm[2], Stephen B. Melville[1], David L. Popham[1]*

1 Department of Biological Sciences, Virginia Tech, Blacksburg, Virginia, United States of America, 2 Department of Biochemistry, Virginia Tech, Blacksburg, Virginia, United States of America

Abstract

Bacterial endospores exhibit extreme resistance to most conditions that rapidly kill other life forms, remaining viable in this dormant state for centuries or longer. While the majority of *Bacillus subtilis* dormant spores germinate rapidly in response to nutrient germinants, a small subpopulation termed superdormant spores are resistant to germination, potentially evading antibiotic and/or decontamination strategies. In an effort to better understand the underlying mechanisms of superdormancy, membrane-associated proteins were isolated from populations of *B. subtilis* dormant, superdormant, and germinated spores, and the relative abundance of 11 germination-related proteins was determined using multiple-reaction-monitoring liquid chromatography-mass spectrometry assays. GerAC, GerKC, and GerD were significantly less abundant in the membrane fractions obtained from superdormant spores than those derived from dormant spores. The amounts of YpeB, GerD, PrkC, GerAC, and GerKC recovered in membrane fractions decreased significantly during germination. Lipoproteins, as a protein class, decreased during spore germination, while YpeB appeared to be specifically degraded. Some protein abundance differences between membrane fractions of dormant and superdormant spores resemble protein changes that take place during germination, suggesting that the superdormant spore isolation procedure may have resulted in early, non-committal germination-associated changes. In addition to low levels of germinant receptor proteins, a deficiency in the GerD lipoprotein may contribute to heterogeneity of spore germination rates. Understanding the reasons for superdormancy may allow for better spore decontamination procedures.

Editor: Daniel Paredes-Sabja, Universidad Andres Bello, Chile

Funding: Research reported in this publication was supported by the National Institute of Allergy and Infectious Disease of the National Institutes of Health under award number R21AI088298. The content is solely the responsibility of the authors and does not necessarily represent the official views of the National Institutes of Health. The mass spectrometry resources used in this work are maintained in part through funding by the Fralin Life Science Institute at Virginia Tech and the Agricultural Experiment Station Hatch Program at Virginia Tech (CRIS Project Number: VA-135981). The funders had no role in study design, data collection and analysis, decision to publish, or preparation of the manuscript.

Competing Interests: The authors have declared that no competing interests exist.

* E-mail: dpopham@vt.edu

Introduction

Bacterial endospores are metabolically dormant and resistant to a variety of anti-microbial treatments due to their protective structures and dehydrated spore core [1,2]. These spores can survive for decades in the absence of nutrients. However, they are able to return to a metabolically active state through a series of events termed spore germination. Once spores lose many of their resistance properties during germination, they can then be easily eliminated by routine decontamination methods [3]. Since the spores of *Bacillus* and *Clostridium* species cause food spoilage and are infectious agents in several human diseases [4], the development of methods or reagents that stimulate highly efficient germination across a spore population could greatly simplify decontamination efforts and reduce morbidity and mortality.

Procedures used for triggering spore germination do not achieve 100% efficiency due to heterogeneity in germination rate within spore populations. Studies of single germinating *B. cereus* and *Clostridium* spores indicated that the spore germination heterogeneity results from the variation in time of initiation of rapid Ca^{2+}-dipicolinic acid (DPA) release (T_{lag}) [5,6]. Subpopulations of *B. subtilis* spores termed superdormant spores can be isolated following multiple rounds of germination with saturating nutrient germinant levels [7]. These superdormant spores exhibit extremely poor germination response to the germinant used for isolation, but will germinate to varying degrees when triggered with germinants that utilize other germinant receptors [7,8]. Individual germinating superdormant spores exhibit longer times for initiation of rapid Ca^{2+}-DPA release relative to initial dormant spore populations [8]. However, once rapid Ca^{2+}-DPA release is initiated, the rate of release is similar for all spores. Thus one may hypothesize that the state of superdormancy is related to processes occurring prior to Ca^{2+}-DPA release.

Four groups of proteins have been implicated to be involved in the early steps of germination: 1) germinant receptors; 2) DPA channel proteins; 3) germination-specific lytic enzymes (GSLEs) and their partner proteins; and 4) lipoproteins potentially involved in transducing germinant-binding signals [3,9]. In *B. subtilis*, three major germinant receptors (GRs) have been characterized: GerA, GerB, and GerK [10–12]. Each GR is comprised of at least A, B, and C subunits (some receptors have D subunits encoded within or associated with the receptor operon [13]) and is localized to the spore inner membrane. The A and B subunits are believed to be integral membrane proteins with multiple transmembrane domains. The C subunits are putative lipoproteins based on their

N-terminal signal peptides and on the effect of a *gerF* mutation, which eliminates the only protein diacylglycerol transferase in this species, on their function [14].

Previous genetic studies of these GRs illustrated their germinant specificity. GerA alone responds to L-alanine or L-valine, while GerB and GerK are required for germination with a mixture of L-asparagine, D-glucose, D-fructose, and potassium ions (AGFK) [12,15]. The binding of nutrient germinants to their cognate GR or GRs initiates irreversible germination activation [3], and via an unclear pathway, results in the opening of DPA channels and the rapid release of this abundant spore solute. Proteins encoded by the *spoVA* operon are involved in DPA uptake during sporulation as well as release during spore germination. Since SpoVA proteins are transcribed exclusively in the developing forespore and some appear to be integral membrane proteins, they are most likely localized to the inner spore membrane [16–18]. The PrkC protein has been identified as an alternate class of germinant receptor that recognized the presence of peptidoglycan fragments in the medium [19].

Complete germination requires that the thick layer of spore cortex peptidoglycan be degraded by GSLEs [20–22]. SleB is a key GSLE [21], and some evidence indicates that it and a co-expressed protein involved in SleB stabilization, YpeB, are localized to the inner spore membrane in the dormant spore [23]. Spores with a *gerD* deletion mutation had a dramatically slower response to nutrient germinants utilizing any of the Ger receptors [24]. Lipoproteins involved in germination, including GerAC, GerBC, GerKC and GerD, are believed to be anchored in the spore inner membrane by a covalently attached lipid [14]. Spores of a *B. subtilis gerF* null mutant also lacked both the GerAC and GerD proteins [25]. This mutant exhibited a significant defect in germination with a greater effect on germination triggered through the GerA receptor relative to responses via the GerB and GerK receptors [14,25].

Several studies have indicated that the abundance of germination-associated proteins can impact the rate of spore germination. Overexpression of the GerA receptor significantly increased the germination rate triggered by its corresponded germinants but did not affect GerB and GerK abundance or germination function [26]. In contrast, overexpression of SpoVA proteins increased germination rates triggered through any germinant receptor [27]. It is hypothesized, based upon quantitative Western blot analyses, that a significant reduction in the amount of a Ger receptor could be the reason for spore superdormancy [28].

In an effort to provide additional insight into the mechanisms of germination, we developed a multiple-reaction monitoring (MRM) mass spectrometry assay [29] to quantify 11 germination proteins believed to be associated with the spore inner membrane. MRM assays are based upon the analyses of peptides specific to the target protein (proteotypic peptides), which become surrogates for protein abundance. The method has high specificity and sensitivity for target protein quantification, and permits reproducible analyses of multiple samples. MRM analyses were performed on membrane preparations obtained from dormant, rapidly germinating, and superdormant spore samples. The results of these analyses indicate that the GerD lipoprotein level can contribute to the heterogeneity of spore germination rate and superdormancy.

Materials and Methods

Spore Sample Preparation

The *B. subtilis* strain used was PS832, a prototrophic laboratory derivative of strain 168. Spores were prepared on 2xSG [30] agar plates without antibiotics. Spores were harvested after 72 h incubation at 37°C and purified by water washing and centrifugation through a 50% sodium diatrizoate (Sigma) layer as described [31]. All spores used in this work were 99% free of vegetative cells and were stored in deionized water at 4°C until analysis.

A 10-ml suspension of dormant spores at an optical density at 600 nm (OD_{600}) of 20 in water were heat-activated at 75°C for 30 min and cooled on ice for at least 10 min. The spores were then germinated at 37°C and at an OD_{600} of 2 with 10 mM L-valine in 25 mM Tris-HCl buffer (pH 7.4). The germination of spores was terminated after the OD_{600} dropped to 50% of the initial value. Germinated spores were collected by centrifugation at 12,000xg for 5 min at 4°C, quickly washed with cold deionized water, centrifuged again, and frozen at −80°C. Examination by phase-contrast microscopy indicated that >95% of the spores in these preparations had germinated.

Superdormant spores were isolated and characterized as described previously [7]. Briefly, dormant spores at OD_{600} of 1 were germinated as described above for 2 h and collected by centrifugation. The pellet was washed with deionized water, suspended in 20% w/v sodium diatrizoate, and centrifuged through a 50% w/v sodium diatrizoate solution (13,000×g for 45 min) to separate dormant spores from germinated spores. The dormant spore pellets were collected and washed thoroughly with deionized water. These dormant spores were subjected to another 2 h round of germination and were separated by density gradient centrifugation again. The final superdormant spore pellet was washed thoroughly with deionized water and stored at 4°C.

Superdormant Spore Characterization

For phenotypic studies, isolated superdormant spores as well as initial dormant spores were germinated with nutrient germinants: 10 mM L-Valine or AGFK (13 mM L-asparagine, 13 mM D-glucose, 13 mM D-fructose, 13 mM KPO_4 [pH 7.4]); or the non-nutrient germinant 60 mM Ca^{2+}-DPA [pH 7.4]. Prior to nutrient-triggered germination, spores were heat-activated in water at 75°C for 30 min and then briefly cooled on ice. Germination was initiated by diluting spores to an OD_{600} of 0.2 in germination solutions and incubating at 37°C. Germination was monitored as the change in OD_{600} over time. Spores used for Ca^{2+}-DPA germination were not heat-activated and the germination was at 30°C. To assess Ca^{2+}-DPA germination, 100 spores were examined by phase-contrast microscopy at several incubation time points.

Preparation of Spore Membrane Fractions

Spore membrane fractions were prepared by a modification of previously described methods [32–34]. Dormant, germinated, and superdormant spores prepared as described above were lyophilized. The dry spores (~19 mg for germinated spores and ~24 mg for dormant and superdormant spores) were pulverized with 100 mg of glass beads in a dental amalgamator (Wig-L-Bug) at 4,600 rpm for pulses of 30 s each, with 30 s pauses on ice between pulses. Spore disruption was monitored by suspending a small sample of spore material in H_2O and observing under phase-contrast microscopy. Once >80% of spores were disrupted, the dry powder was suspended in 0.5 ml of 4°C extraction buffer (10 mM Tris-HCl [pH 7.4], 1 mM EDTA, 2 mg/ml RNase A, 2 mg/ml DNase I, 1 mM phenylmethylsulfonyl fluoride (PMSF)). The suspension was centrifuged (6,000×g, 10 min, 4°C) and the resultant supernatant was centrifuged again (13,000×g, 10 min, 4°C) to remove insoluble material. The remaining supernatant was subjected to ultracentrifugation (100,000×g, 60 min, 4°C). The resulting supernatant was considered the spore core soluble fraction and was stored at −80°C. The resulting pellet, designated the crude spore membrane fraction, was homogenized in 1 ml high salt buffer (20 mM Tris-HCl [pH 7.5], 10 mM EDTA, 1 M NaCl, 1 mM PMSF) and was gently shaken for 30 min at 4°C. The homogenate was subjected to ultracentrifugation again as described above. The remaining pellet was homogenized in 1 ml alkaline buffer (100 mM Na$_2$CO$_3$-HCl [pH 11], 10 mM EDTA, 100 mM NaCl, 1 mM PMSF) and was again subjected to ultracentrifugation. After a final wash with 1 ml TE buffer (10 mM Tris-HCl [pH 7.4], 1 mM EDTA, 1 mM PMSF), the resulting pellet was homogenized in 200 μl TE buffer, flash frozen, and stored at −80°C until analysis. The protein concentration was determined by acid hydrolysis and amino acid analysis [35] with comparison to a standard set of amino acids (Sigma).

Protein Digestion

Proteins in spore membrane fractions (70 μg) were precipitated with 1 mL of acetone −20°C overnight and collected by centrifugation for 20 min at 12,000 g. Protein was resuspended in 250 μl of freshly-prepared 8 M urea, 20 mM Tris-HCl, pH 8.0 to give a final protein concentration of 1 mg/ml. Proteins were denatured by the addition of 27.8 μl of freshly-prepared 45 mM dithiothreitol, 20 mM Tris-HCl, pH 8.0, and incubation for 1 h at 37°C. Free cysteines were alkylated by the addition of 30.9 μl of freshly-prepared 100 mM iodoacetamide, 20 mM Tris-HCl, pH 8.0, incubation at room temperature in the dark for 30 min. Unreacted iodoacetamide was inactivated by the addition of 102.9 μl of freshly-prepared 45 mM dithiothreitol, 20 mM Tris-HCl, pH 8.0. Proteins were digested by the addition of 1.03 ml of 20 mM Tris-HCl, pH 8.0, and 5 μg trypsin in 10 μl 50 mM acetic acid followed by incubation overnight at 37°C with shaking. Trifluoroacetic acid was added to a final concentration of 0.25% and formic acid was added to a final concentration of 1%. The pH was measured and additional formic acid was added until the pH was at or below 3. Conditioning of 0.1 ml OMIX C18 solid phase extraction cartridges used 0.2 ml methanol, followed by 0.2 ml 50% acetonitrile, 0.1% TFA and finally 0.2 ml 2% acetonitrile, 0.1% TFA. A protein sample was applied to the cartridge, which was then washed three times with 0.2 ml 2% acetonitrile, 0.1% TFA. Peptides were eluted with 0.2 ml 75% acetonitrile, dried, and resuspended in 0.02 ml solvent A (2:98 acetonitrile:water containing 0.1% formic acid).

Liquid Chromatography and Mass Spectrometry

Thirteen germination-related membrane proteins (Table 1) were initially targeted for MRM method development, with a list of potential proteotypic tryptic peptides generated using the Enhanced Signature Peptide Prediction tool [36] using a cutoff value of 0.6. Peptides were synthesized by JPT Peptide Technologies GmbH Inc., and were directly infused into the mass spectrometer for determination of target fragment ions and ionization conditions. For each synthesized peptide, elution times were identified, the dominant precursor ion of predicted m/z (Q1 ion) was identified and fragmented, and dominant fragment ions of expected m/z (Q3 ions) were identified and quantified. Limits of quantification (LOQ) (Table 1) were determined using the established MRM methods and dilution series from 10–1500 fmol of each synthetic peptide (Fig. 1).

Proteins in spore membrane fractions were solubilized with 20 mM Tris-HCl [pH 8.0], 8 M urea, 45 mM dithiothreitol at a final protein concentration of 1 mg/ml, followed by a 37°C overnight trypsin digestion at 20:1 (w/w) protein:Trypsin ratio. The tryptic peptides were desalted and concentrated using OMIX C18 microextraction pipette tips (Varian) following the manufac-

turer's protocol. Peptides was separated using an Eksigent Nano 2-D liquid chromatography system connected to a 100×0.075 mm Magic C18AQ (200 Å, 3 μm, Bruker) column packed in-house using an eFRIT fused silica capillary (Phoenix S&T). Ten microliters of each sample was first loaded onto a C18 trap cartridge at 10 μl/min for 15 minutes using solvent A (2:98 acetonitrile:water containing 0.1% formic acid). The trap cartridge was switched in-line with the analytical column and the trap and column were flushed with 95% solvent A, 5% solvent B (98:2 acetonitrile:water containing 0.1% formic acid) for 5 minutes at 300 nl/min. This was followed by a linear gradient to 86% solvent A over 5 minutes then a linear gradient to 71% solvent A over 45 minutes and finally a linear gradient to 35% solvent A over 5 minutes. The column was flushed for 2 minutes with 35% solvent A and reequilibrated at the starting conditions for 13 minutes prior to the next sample injection. The eluent was introduced into an AB Sciex 4000 QTrap mass spectrometer controlled by Analyst 1.4.2 software (AB Sciex) via a nano-electrospray source (Phoenix S&T). The mass spectrometer was operated in positive ion mode utilizing an MRM method containing precursor/product ion transitions corresponding to peptides described below. Dwell time for each transition was 40 ms and the total cycle time was 6.6 seconds. The first quadrupole was operated at low resolution while the third quadrupole was set to unit resolution. Ion spray voltage was 2400 V, curtain and sheath gases were 12 (arbitrary units), interface heater temperature was 120°C and the entrance potential was 10 V for all transitions. CAD gas was set to medium corresponding to a vacuum of 3.1×10^{-5} Torr.

Data Collection and Refining

When determining which of the identified Q3 ion peak areas were suitable for quantitative comparisons across all samples, we applied the following raw data refining criteria. 1) The retention time of a Q3 ion in all samples should be the same as that determined for the corresponding synthetic peptide. Q3 ions that did not have consistent retention times were excluded from further analysis. 2) If a quantified peptide had less than two quantifiable Q3 ions, the peptide was excluded from further analysis. 3) If the peak area of a Q3 ion was below established limited of quantification, the Q3 ion was excluded from further analysis. 4) Among all nine samples, if the Q3 ion peaks in more than three samples had S/N ratio values less than 10, then the Q3 ion was excluded from further analysis. (The end section of each Q3 ion spectrum was considered as base line (noise) when collecting the S/N ratio for limit of quantification evaluation.).

Within each biological replicate set, there were three membrane fraction samples: dormant, germinated, and superdormant. Three biological replicates were derived from three independent spore preparations. For each quantified Q3 ion, peak area ratios between two membrane fractions were calculated only within a biological replicate set. Ratios were then compared across biological replicates. Theoretically, if a protein's abundance was the same in two different samples, the peak area ratios for the Q3 ions of its peptides would be 1. Among all Q3 ion peak area ratios calculated, those of proteins GerAA, GerBA, and GerKA were always close to 1.0. We took these proteins to represent unchanged proteins within the samples, and pooled their Q3 ion peak area ratios to represent the level of physiological variance. For each comparison group, we then evaluated the significance of a protein change by comparing peak area ratios of the protein to this unchanged protein peak area ratio pool using a two samples student t-test. In addition, for each protein, we evaluated the significance of two comparison groups using the Student's t-test.

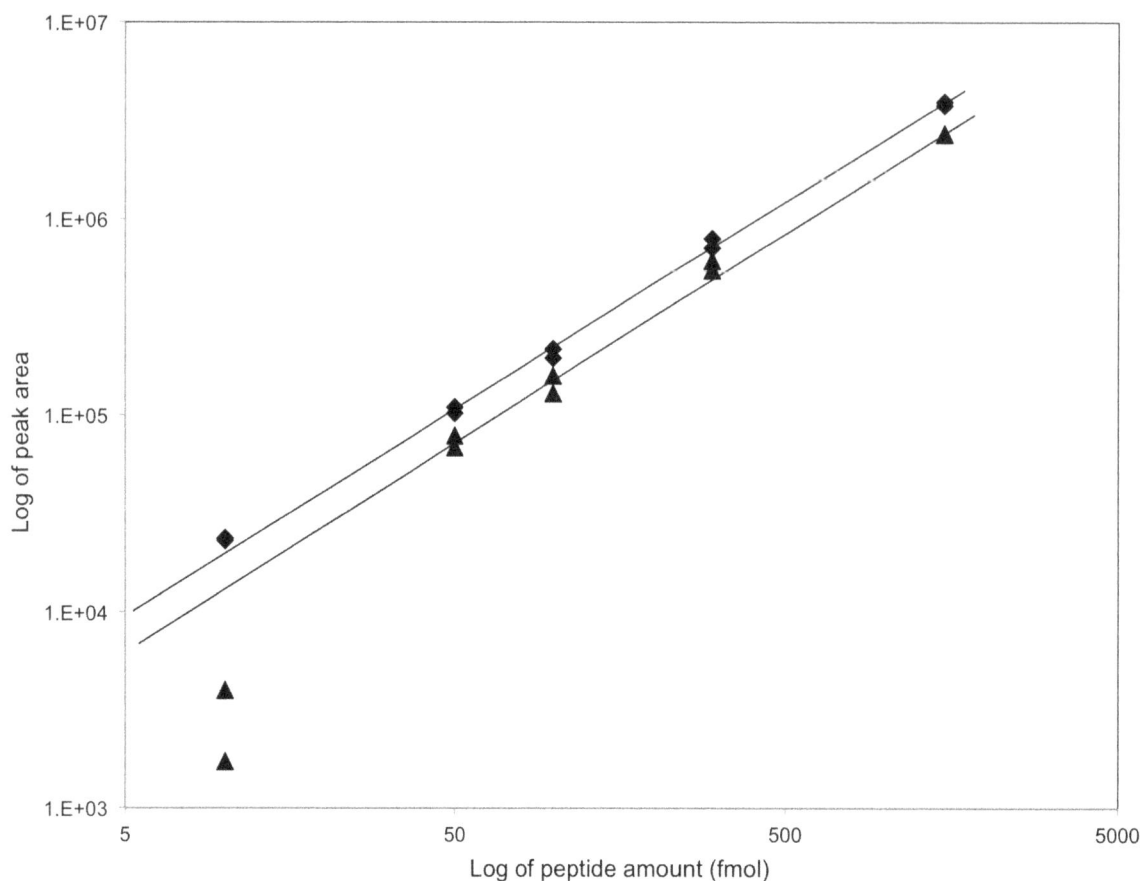

Figure 1. Example determinations of MRM Limit of Quantitation (LOQ) using synthetic peptides. Dilution series of peptides SLDEPSSEVVER (◆), which is proteotypic for the *B. subtilis* GerBA protein, and EAYSDDVPEGQVVK (▲), which is proteotypic for the *B. subtilis* PrkC protein, were subjected in duplicate to MRM analysis with detection of the parent ion 493.9 m/z and the fragment ion 798.3 m/z for GerBA or the parent ion 768.7 m/z and the fragment ion 1085.6 m/z for PrkC (Table 1). Fragment ion peak areas were plotted against peptide amount, and best-fit lines were applied. LOQ's were defined as the lowest concentration at which the response was still linear or the lowest concentration at which the contribution due to noise was such that reproducible results could be obtained (peak area $\geq 1 \times E^{+4}$).

Both tests used two-tailed, unequal variance p values, and statistical significance for both t-tests was set at p<0.05.

Results

Isolation and Characterization of Spore Populations

Three independent preparations of *B. subtilis* dormant spores were germinated using L-valine, with downstream processing producing rapidly germinating and superdormant spore populations. The yield of superdormant spores was $1.09 \pm 0.16\%$ (n = 3); somewhat less than the 3.8% yield in a previous publication [7]. Two reported characteristics of superdormant spore populations isolated using L-valine are that the superdormant spores germinate poorly with L-valine as well as with germinants that use a different germinant receptor, with the superdormant spores being as viable as the initial dormant spores when germinated with non-nutrient germinants [7]. Our superdormant spores also germinated slowly with L-valine, in comparison with the rapid germination of the initial dormant spores (Fig. 2A). However, when using AGFK as germinant, which acts through different germinant receptors than does L-valine [12], the superdormant spores germinated more rapidly than the initial dormant spores (Fig. 2B). In addition, the superdormant spores also reached a higher efficiency of germination based on a greater OD_{600} decrease than the initial dormant

spores. While our results are different from those of the original description of superdormant spores [7], similar observations were reported for superdormant spores isolated in a more recent study [8]. The effect of a non-nutrient germinant on the superdormant spores was tested using Ca^{2+}-DPA, which causes activation of the GSLE CwlJ [37], bypassing part of the germination apparatus that may be deficient in superdormant spores. The superdormant spores completed Ca^{2+}-DPA-triggered germination as efficiently as the initial dormant spores after an initial lag period (Fig. 2C), similar to a previous report [8]. In summary, the results of the phenotypic analyses support the claim that spores isolated after extensive L-valine germination can be classified as superdormant. To verify that these spores were not superdormant due to a genetic alteration, they were germinated and spread on plates, and 10 randomly selected colonies were selected, cultured, sporulated, and tested for germination rate. Similar to a previous report [7], spore populations produced by these strains germinated equivalently to those of the wild type strain.

Quantification of Spore Membrane Proteins by MRM Assays

Membrane samples were prepared from dormant, germinated, and superdormant spores and were used to quantify the targeted germination-related proteins relative to the total protein concen-

Table 1. Peptide Details for MRM Analysis of the *B. subtilis* germination proteins.

Protein	Proteotypic peptide sequence (position in protein)	Parent Ion m/z (charge state)	Fragment ion m/z (ion)	LOQ fmoles (peak area)[a]
GerAA	LDQLDARPVETAK (77–89)	486.2 (+3)	672.0 (y_{12}^{+2})	10 ($3 \times E^{+5}$)
		486.2 (+3)	614.5 (y_{11}^{+2})	10 ($3 \times E^{+5}$)
	DEETLTLDQVK (64–74)	646.1 (+2)	703.4 (y_6^{+1})	10 ($8 \times E^{+4}$)
	VSSALFNGR (233–241)	476.3 (+2)	765.3 (y_7^{+1})	10 ($1 \times E^{+4}$)
		476.3 (+2)	374.1 (y_7-NH_3^{+2})	10 ($1 \times E^{+4}$)
GerAC	ADVTGLGNEVR (321–331)	565.8 (+2)	744.3 (y_7^{+1})	50 ($1 \times E^{+4}$)
		565.8 (+2)	845.5 (y_8^{+1})	100 ($1 \times E^{+4}$)
		565.8 (+2)	687.4 (y_6^{+1})	50 ($1 \times E^{+4}$)
GerBA	TSDPNLVIK (156–164)	493.9 (+2)	683.2 (y_6^{+1})	10 ($2 \times E^{+4}$)
		493.9 (+2)	798.3 (y_7^{+1})	10 ($2 \times E^{+4}$)
	SLDEPSSEVVER (124–135)	674.1 (+2)	902.3 (y_8^{+1})	10 ($4 \times E^{+4}$)
		674.1 (+2)	316.2 (y_5^{+2})	10 ($3 \times E^{+4}$)
	VESSLLEGR (235–243)	495.4 (+2)	761.3 (y_7^{+1})	10 ($1 \times E^{+4}$)
GerBC	GILTEDQNPNENSFSK (279–294)	897.3 (+2)	922.4 (y_8^{+1})	50 ($1 \times E^{+4}$)
		897.3 (+2)	1036.6 (y_9^{+1})	50 ($1 \times E^{+4}$)
	GNAADVFTK (135–143)	461.8 (+2)	609.4 (y_5^{+1})	10 ($1 \times E^{+4}$)
		461.8 (+2)	680.3 (y_6^{+1})	50 ($1 \times E^{+4}$)
GerKA	ERPVLISPSLAK (31–42)	437.4 (+3)	595.3 (b_5^{+1})	10 ($2 \times E^{+4}$)
		437.4 (+3)	515.2 (y_5^{+1})	10 ($2 \times E^{+4}$)
		437.4 (+3)	602.1 (y_6^{+1})	50 ($1 \times E^{+4}$)
	SIQEPSTQVSFR (159–170)	690.2 (+2)	921.7 (y_8^{+1})	10 ($1 \times E^{+4}$)
		690.2 (+2)	329.0 (b_3^{+1})	50 ($1 \times E^{+4}$)
		690.2 (+2)	1050.5 (y_9^{+1})	50 ($1 \times E^{+4}$)
	EVGSSSDVIIR (50–60)	581.5 (+2)	789.5 (y_7^{+1})	10 ($1 \times E^{+4}$)
GerKC	TLDFTEAQYGR (166–176)	651.0 (+2)	330.2 (b_3^{+1})	10 ($1 \times E^{+4}$)
		651.0 (+2)	723.5 (y_6^{+1})	50 ($1 \times E^{+4}$)
GerD	NIFEDTDFAEGFAK (90–103)	802.5 (+2)	1376.6 (y_{12}^{+1})	100 ($1 \times E^{+4}$)
		802.5 (+2)	422.1 (y_4^{+1})	300 ($1 \times E^{+4}$)
		802.5 (+2)	1229.6 (y_{11}^{+1})	300 ($7 \times E^{+5}$)
SpoVAC	SEGLVLGVATNM(ox)FK (109–122)	741.5 (+2)	883.2 (y_8^{+1})	50 ($1 \times E^{+4}$)
		741.5 (+2)	996.3 (y_9^{+1})	50 ($1 \times E^{+4}$)
	SEGLVLGVATNMFK (109–122)	733.6 (+2)	867.3 (y_8^{+1})	300 ($1 \times E^{+4}$)
		733.6 (+2)	980.7 (y_9^{+1})	300 ($1 \times E^{+4}$)
SpoVAD	ETIPTIAHGVVFER (320–333)	523.9 (+3)	409.3 (y_{11}^{+3})	50 ($1 \times E^{+4}$)
		523.9 (+3)	613.6 (y_{11}^{+2})	50 ($1 \times E^{+4}$)
	QLMEDAVNVALQK (57–69)	730.2 (+2)	459.3 (y_4^{+1})	100 ($1 \times E^{+4}$)
		730.2 (+2)	672.3 (y_6^{+1})	100 ($1 \times E^{+4}$)
YpeB	IGVFSYVPVENK (326–337)	676.7 (+2)	586.1 (y_5^{+1})	50 ($1 \times E^{+4}$)
		676.7 (+2)	935.3 (y_8^{+1})	50 ($1 \times E^{+4}$)
	TIPKPAITEAEAK (372–384)	457.2 (+3)	577.9 (y_{11}^{+2})	10 ($2 \times E^{+5}$)
		457.2 (+3)	929.5 (y_9^{+1})	10 ($5 \times E^{+4}$)
	VALDDGEVVGFSAR (349–362)	718.2 (+2)	636.3 (y_6^{+1})	100 ($1 \times E^{+5}$)
		718.2 (+2)	537.4 (y_5^{+1})	100 ($3 \times E^{+5}$)
PrkC	EAASGYLEDNGLK (508–520)	684.3 (+2)	789.3 (y_7^{+1})	10 ($1 \times E^{+4}$)
		684.3 (+2)	1096.5 (y_{10}^{+1})	10 ($1 \times E^{+4}$)
	EAYSDDVPEGQVVK (525–538)	768.7 (+2)	756.3 (y_7^{+1})	50 ($1 \times E^{+4}$)
		768.7 (+2)	1085.6 (y_{10}^{+1})	50 ($1 \times E^{+4}$)
	TEIGDVTGQTVDQAK (429–443)	781.8 (+2)	947.6 (y_9^{+1})	50 ($1 \times E^{+4}$)
		781.8 (+2)	1218.8 (y_{12}^{+1})	10 ($1 \times E^{+4}$)

[a]LOQ is the limit of quantitation determined for this fragment ion, as described in Materials and Methods.

tration. The total protein in each sample was determined by amino acid analysis. SDS-PAGE analysis of total proteins was consistent with this quantification and revealed essentially identical protein band patterns across biological replicates (Fig. 3).

The MRM assays centered on the detection of 11 of the 13 proteins expected to be membrane associated and involved in spore germination (Table 1). As peptides of varying compositions exhibit different ionization efficiencies, we determined LOQs for each peptide. We were not able to identify any proteotypic peptides for GerAB that were even predicted to function well in an MRM assay, and we were not able to obtain quantifiable MRM data for GerBB and GerKB due to the fact that the signal for the proteotypic peptides designed for these integral membrane proteins were below the limit of detection. Nonetheless, we were able to quantify the A and C subunits of the germinant receptors. The ratios of GerAA, GerBA, and GerKA between dormant and superdormant spores were very close to 1.0. In contrast, the amounts of GerAC and GerKC in superdormant spores were 3.4 and 1.9-fold lower than the amounts in dormant spores (Fig. 4A). These decreases of GerAC (p = 0.002) and GerKC (p = 0.023) were statistically significant. GerBC, however, showed no significant difference in amount between superdormant and dormant spores (Fig. 4A).

GerD is a lipoprotein that is localized predominantly to the spore inner membrane [32] and functions in both GerA and GerB/K-mediated germination responses [24]. GerD was 1.8-fold less abundant in membranes isolated from superdormant spores in comparison to those from dormant spores (Fig. 4A). PrkC, SpoVAC, SpoVAD, and YpeB exhibited no significant difference in abundance between superdormant and dormant spore samples (Fig. 4A).

The relative amounts of GerAA, GerBA, and GerKA in germinated spore samples were similar to the amounts of these proteins in those from dormant spores. In contrast, the amounts of GerAC, GerBC, and GerKC in germinated spore membranes decreased 1.9, 1.6, and 2.4-fold respectively in comparison to dormant spores (Fig. 4B). Previous western blot results showed that membrane-associated GerD decreased during spore germination [32]. This was confirmed in the MRM assays, which showed that membrane-associated GerD decreased 3.5-fold during spore germination (Fig. 4B). The decreases of GerAC (p = 0.009), GerKC (p = 0.032), and GerD (p < 0.0001) were statistically significant.

To date, there is no report regarding PrkC function in a nutrient germinant receptor-mediated pathway. The MRM assays indicated that membrane-associated PrkC was significantly decreased 3.8-fold in amount after spore germination (Fig. 4B). Similarly, three proteotypic YpeB peptides decreased 6.8-fold in amount during spore germination (Fig. 4B). This result is consistent with a previous observation of YpeB degradation during germination [23]. A fourth YpeB peptide was clearly quantifiable in dormant spore samples, but was undetectable in germinated spore membrane fractions. This peptide is near the YpeB N-terminus (residues 57–67), whereas the three detectable YpeB peptides were closer to the C terminus.

SpoVAC and SpoVAD proteins were chosen as representatives of the spoVA-encoded proteins. SpoVAC is predicted to be an integral membrane protein [18,38], and SpoVAD is more likely to be a peripheral membrane protein based on its crystal structure [39]. While SpoVAD could be detected in both dormant and germinated membrane fraction samples by Western-blot (data not shown), the MRM assays indicated that SpoVAC and SpoVAD significantly increased 1.3 and 1.7-fold respectively in germinated

spore membrane samples in comparison to those of dormant spores (Fig. 4B).

The germinant receptor A subunit amounts detected were similar in superdormant and germinated spore samples. While membranes of both superdormant and germinated spores had significantly less GerAC and GerKC proteins relative to dormant spores, the decrease in GerAC was significantly greater than that of GerKC and was 1.6-fold lower in superdormant spores in comparison to germinated spores (Fig. 4C). Similarly, while GerD was also less abundant in membranes from both superdormant and germinated spores than in dormant spores, the difference was greater in the germinated spores, such that GerD was 2-fold more abundant in superdormant spores than in germinated spores (Fig. 4C). This difference was statistically significant (p < 0.0001). If germination-induced protein changes initiated but did not progress past GerD in superdormant spores, the difference in abundance for proteins involved in later germination events in comparison to germinated spores should be similar to a dormant/germinated spores comparison. Indeed, our results showed that YpeB and PrkC were 4.9 and 2.9-fold more abundant in samples from superdormant spores than in those from germinated spores, and SpoVAC and SpoVAD were 1.6 and 1.5-fold less abundant in superdormant spore samples than in germinated spore samples (Fig. 4C).

Discussion

Spore germination starts at the spore inner membrane with the interaction of germinant with Ger receptor proteins and progresses through core rehydration and cortex breakdown. Deficiencies in Ger receptors and associated proteins, Ca^{2+}-DPA channels, and lytic enzymes can potentially inhibit the germination process, leading to the production of superdormant spores. In this work, L-valine superdormant spores responded poorly to valine but germinated well with AGFK; a result that is different from the initial report of superdormant spores [7] but consistent with a later report [8]. In contrast to a high yield (12%) of superdormant spores isolated with AGFK in previous work [7], we were unable to isolate any superdormant spores using AGFK. The reasons for these differences are not clear, but raise the possibility that there may be multiple pathways to superdormancy, and slight differences in the method of preparation and isolation may result in significantly different superdormant spore populations.

Our quantitative MRM assays were performed on membrane samples derived from broken spores. Results published while this work was in progress indicate that these samples may represent a subset of the spore membrane fraction, as extensive chemical extraction is required to recover the full amount of several spore membrane proteins [40]. While the sample analyzed here may not represent the entire spore membrane fraction, two lines of evidence indicate that the samples recovered from each spore type are similar fractions. The relative abundance of the majority of the proteins analyzed was the same in dormant and superdormant spores, and the amounts of the integral membrane A subunits of the Ger receptors were not altered in germinated spores, indicating that the membrane fractions obtained from the different spore types were comparable in their protein complements.

We interpret the similarity in Ger receptor A protein abundance in dormant and superdormant spore fractions to imply that the isolated superdormant population is not delayed in germination due to a low level of Ger receptors. These results differ from a previous publication that reported GerAA and GerAC protein levels were 7-fold lower and other GR subunits 3-fold lower in

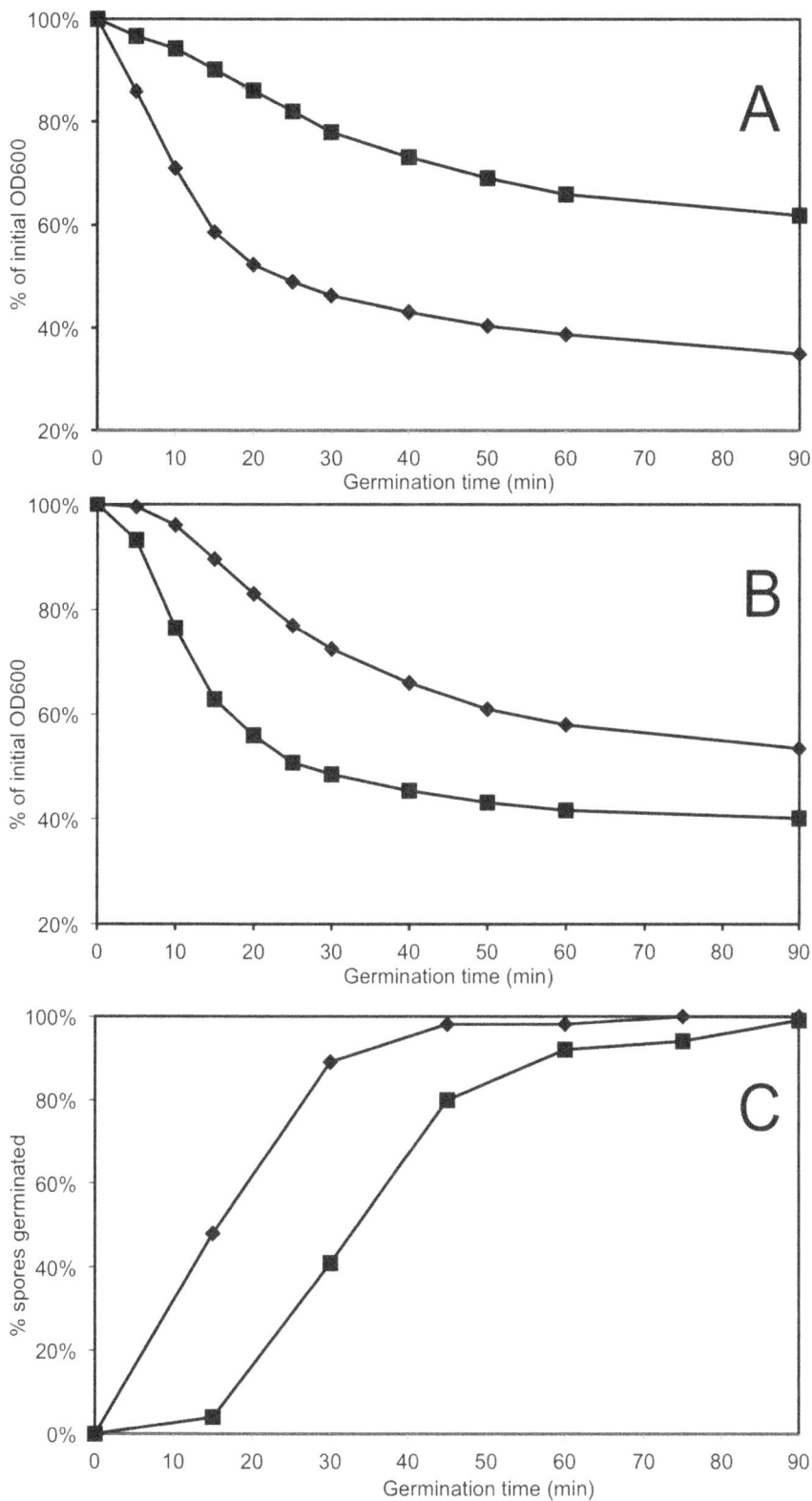

Figure 2. Germination of dormant and superdormant spores with nutrient and non-nutrient germinants. Superdormant spores of *B. subtilis* strain PS832 (wild type) were isolated following prolonged germination with 10 mM L-Valine as described in materials and methods. Squares (■) indicate isolated superdormant spores, and diamonds (◆) indicate the initial dormant spores used for isolation. A) Germination with 10 mM L- Valine; B) Germination with AGFK; C) Germination with Ca^{2+}-DPA. Data from one biological replicate of dormant and superdormant spores is shown. Analyses of the other two biological replicates produced very similar results.

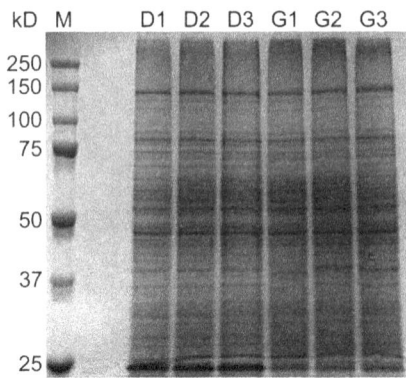

Figure 3. Gel electrophoresis of membrane-associated spore proteins. Membrane preparations were obtained from dormant (D) and germinated (G) spores produced from three independent spore preparations (1, 2, and 3). Protein concentrations in membrane preparations were determined by quantitative amino acid analyses, and identical protein amounts were loaded onto a 9% polyacrylamide gel. Sizes of protein standard markers (M) are indicated on the left. Proteins were stained using Coomassie blue.

superdormant spores [28]. We did find that the C subunits of the GerA and GerK receptors were decreased in samples from our superdormant spores. Germination-induced decreases in C subunit abundance, relative to dormant spores, were in one case less (GerAC) and in other cases slightly greater (GerBC and GerKC) than those observed in superdormant spores. As the C subunits are believed to be present in a stoichiometric association with the A subunits in the membranes of dormant spores [40], a decrease in the C subunits may indicate a change that takes place during early germination or may indicate a change in the maintenance of subunit association with the membrane during membrane fraction washing and isolation. Published data indicate that GerD, a lipoprotein like the C subunits, can be partially extracted by a high salt wash similar to that used in this study and is released from the membrane during germination [32]. The decreased abundance of C subunits observed in superdormant spore samples could be either a cause of superdormancy or a result of the superdormant spore isolation process. Protein associations could decrease during Ger receptor activation upon germinant binding, and the membrane-association may be less stable as the spore membrane regains fluidity during germination [41]. Prolonged exposure to germinants during the superdormant spore isolation may result in the loss of C subunits from the membrane, with a blockage in the germination pathway at a point downstream of the Ger receptors.

An absence of GerD was previously shown to result in a germination initiation defect [24,27]. Previous studies also indicated that the germination receptors and GerD co-localize to a discrete cluster on the membrane [42] and that SpoVA proteins can associate with Ger receptors [43], and thus we expect that the comparable membrane factions we derived from different spore types would contain similar amounts of GerD. Superdormant spore membrane samples had 2-fold less GerD than those from dormant spores, and germinated spore samples contained even less GerD. Similar to the case for the Ger receptor C subunits, the decreased level of GerD in superdormant spores could be either a cause of superdormancy or a result of a partial germination response that is blocked at a subsequent point.

Other germination-active proteins, PrkC, SpoVAC, SpoVAD, and YpeB were present in dormant and superdormant spore

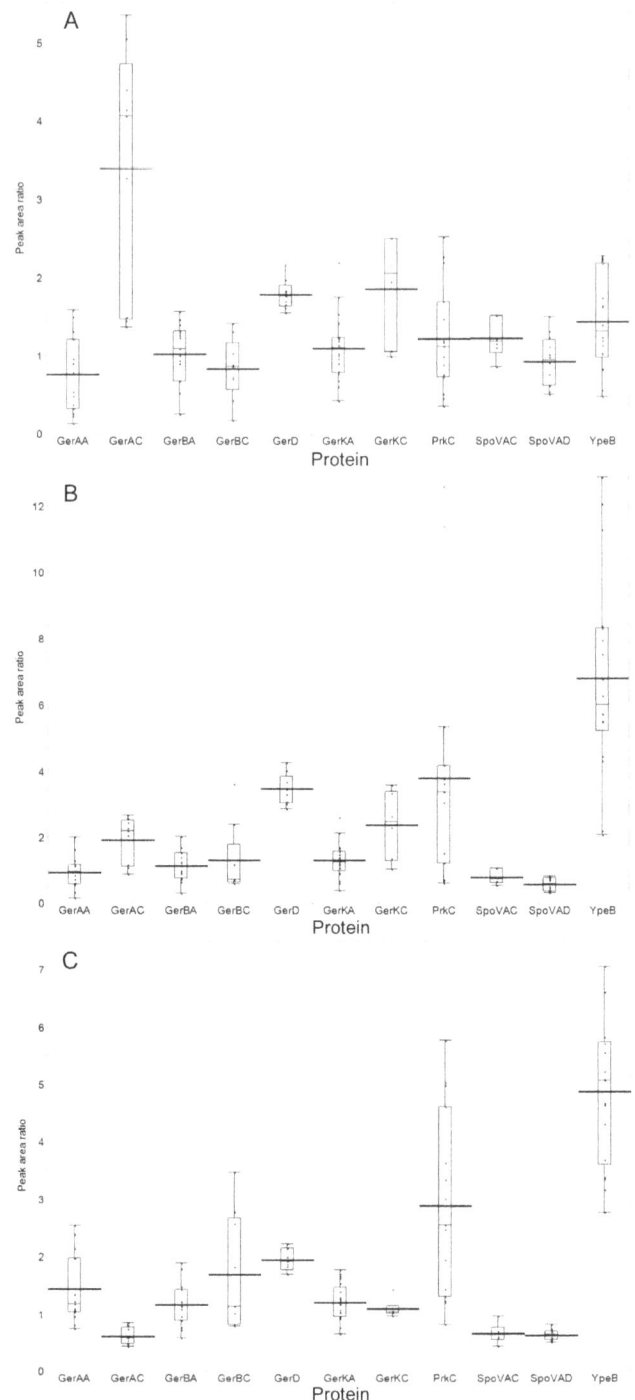

Figure 4. Relative quantities of germination proteins in dormant, superdormant, and germinated spores. Proteins were quantified in membrane samples by MRM analyses. The relative abundance of each protein is expressed as the ratio of each particular product ion peak area detected in dormant versus superdormant (A), dormant versus germinated (B), or superdormant versus germinated (C) spore samples derived from the same spore preparation. Ratios determined for samples from three independent spore preparations were then pooled. Each box and whisker plot indicates individual product ion ratio values (dots), 25–75 percentiles (boxes) and full ranges (whiskers), excluding statistically determined outliers. Means are indicated by the short lines traversing the boxes and medians are shown by the lines traversing the boxes.

membrane preparations in similar quantities, but did show changes in abundance during germination. This suggests that any early germination-related events that have taken place within the superdormant spores did not progress to later events such as SpoVA-assisted DPA release [17] or YpeB-related cortex degradation [23]. Interestingly, the abundance of membrane-associated SpoVAC and SpoVAD proteins increased during germination. As this increase is taking place well before any new protein synthesis, it must represent an increased association of the proteins with the membrane or with other membrane-bound proteins, most likely other products of the SpoVA DPA-transport complex. Several studies indicate that SpoVAD localizes to the spore inner membrane [44,45], and recent structural analyses of this protein show that it is likely to be a peripheral membrane protein [39,46]. In our western-blot analysis of SpoVAD, it was detected in both soluble and membrane fractions, suggesting a weak membrane association (data not shown). Another novel finding is that PrkC abundance decreased significantly during spore germination triggered by L-Val. Although this protein was previously demonstrated to be a germinant receptor that responds to muropeptides [19], there is no report of its activity in other germinant receptor-mediated pathways.

The most dramatic change in protein abundance during germination was that of YpeB, which is required for incorporation of SleB in spore and thus for normal cortex degradation during germination [23]. This result is consistent with a previous observation of YpeB degradation during germination. The 52 kDa YpeB is processed to a ~30 kDa product during germination [23]. The 6.8-fold decrease in YpeB abundance we observed was calculated using those peptides we could detect in both dormant and germinated spore samples, all of which were in the C-terminal half of the protein. One peptide nearer the N-terminus of YpeB was detected in dormant spore samples but was undetectable in germinated spore samples, indicating a decrease of >14-fold. This differential loss of peptides indicates that the more stable 30-kDa portion of YpeB represents a C-terminal portion. Ongoing studies in our lab are consistent with this (data not shown). It is not clear if the 6.8-fold decrease in the observed YpeB peptides is due to protein degradation, a decrease in membrane association, or both. Because the N-terminus of YpeB apparently contains an uncleaved signal peptide, proteolytic removal of this domain would be expected to decrease YpeB-membrane association.

Studies of spore germination heterogeneity have shown that the major variable in kinetics of nutrient-triggered spore germination is the germination initiation time, termed T_{lag}, with the range of superdormant spores' T_{lag} times being significantly greater than that of dormant spores [8]. Previous studies also showed that *gerD* spores had significantly longer T_{lag} times than wild-type spores [27]. The results from our work in relation to those previous efforts lead us to propose that decreased abundance of GerD can be a contributing factor in superdormancy.

Author Contributions

Conceived and designed the experiments: YC WKR RFH SBM DLP. Performed the experiments: YC WKR DLP. Analyzed the data: YC WKR RFH SBM DLP. Contributed reagents/materials/analysis tools: YC WKR RFH DLP. Wrote the paper: YC WKR RFH SBM DLP.

References

1. Leggett MJ, McDonnell G, Denyer SP, Setlow P, Maillard JY (2012) Bacterial spore structures and their protective role in biocide resistance. J Appl Microbiol 113: 485–498.
2. Riesenman PJ, Nicholson WL (2000) Role of the spore coat layers in *Bacillus subtilis* spore resistance to hydrogen peroxide, artificial UV-C, UV-B, and solar UV radiation. Appl Environ Microbiol 66: 620–626.
3. Setlow P (2003) Spore germination. Curr Opin Microbiol 6: 550–556.
4. Mallozzi M, Viswanathan VK, Vedantam G (2010) Spore-forming Bacilli and Clostridia in human disease. Future Microbiol 5: 1109–1123.
5. Kong L, Zhang P, Setlow P, Li YQ (2010) Characterization of bacterial spore germination using integrated phase contrast microscopy, Raman spectroscopy, and optical tweezers. Anal Chem 82: 3840–3847.
6. Wang G, Zhang P, Paredes-Sabja D, Green C, Setlow P, et al. (2011) Analysis of the germination of individual *Clostridium perfringens* spores and its heterogeneity. J Appl Microbiol 111: 1212–1223.
7. Ghosh S, Setlow P (2009) Isolation and characterization of superdormant spores of *Bacillus* species. J Bacteriol 191: 1787–1797.
8. Zhang P, Kong L, Wang G, Scotland M, Ghosh S, et al. (2012) Analysis of the slow germination of multiple individual superdormant *Bacillus subtilis* spores using multifocus Raman microspectroscopy and differential interference contrast microscopy. J Appl Microbiol 112: 526–536.
9. Moir A (2006) How do spores germinate? J Appl Microbiol 101: 526–530.
10. Corfe BM, Sammons RL, Smith DA, Mauel C (1994) The *gerB* region of the *Bacillus subtilis* 168 chromosome encodes a homologue of the *gerA* spore germination operon. Microbiology 140 (Pt 3): 471–478.
11. Zuberi AR, Feavers IM, Moir A (1985) Identification of three complementation units in the *gerA* spore germination locus of *Bacillus subtilis*. J Bacteriol 162: 756–762.
12. Moir A, Kemp EH, Robinson C, Corfe BM (1994) The genetic analysis of bacterial spore germination. J Appl Bacteriol 77: 9S–16S.
13. Ramirez-Peralta A, Gupta S, Butzin XY, Setlow B, Korza G, et al. (2013) Identification of new proteins that modulate the germination of spores of *Bacillus* species. J Bacteriol 195: 3009–3021.
14. Igarashi T, Setlow B, Paidhungat M, Setlow P (2004) Effects of a *gerF* (*lgt*) mutation on the germination of spores of *Bacillus subtilis*. J Bacteriol 186: 2984–2991.
15. Atluri S, Ragkousi K, Cortezzo DE, Setlow P (2006) Cooperativity between different nutrient receptors in germination of spores of *Bacillus subtilis* and reduction of this cooperativity by alterations in the GerB receptor. J Bacteriol 188: 28–36.
16. Tovar-Rojo F, Chander M, Setlow B, Setlow P (2002) The products of the *spoVA* operon are involved in dipicolinic acid uptake into developing spores of *Bacillus subtilis*. J Bacteriol 184: 584–587.
17. Vepachedu VR, Setlow P (2007) Role of SpoVA proteins in release of dipicolinic acid during germination of *Bacillus subtilis* spores triggered by dodecylamine or lysozyme. J Bacteriol 189: 1565–1572.
18. Fort P, Errington J (1985) Nucleotide sequence and complementation analysis of a polycistronic sporulation operon, *spoVA*, in *Bacillus subtilis*. J Gen Microbiol 131: 1091–1105.
19. Shah IM, Laaberki MH, Popham DL, Dworkin J (2008) A eukaryotic-like Ser/Thr kinase signals bacteria to exit dormancy in response to peptidoglycan fragments. Cell 135: 486–496.
20. Popham DL, Helin J, Costello CE, Setlow P (1996) Muramic lactam in peptidoglycan of *Bacillus subtilis* spores is required for spore outgrowth but not for spore dehydration or heat resistance. Proc Natl Acad Sci U S A 93: 15405–15410.
21. Boland FM, Atrih A, Chirakkal H, Foster SJ, Moir A (2000) Complete spore-cortex hydrolysis during germination of *Bacillus subtilis* 168 requires SleB and YpeB. Microbiology 146 (Pt 1): 57–64.
22. Ishikawa S, Yamane K, Sekiguchi J (1998) Regulation and characterization of a newly deduced cell wall hydrolase gene (*cwlJ*) which affects germination of *Bacillus subtilis* spores. J Bacteriol 180: 1375–1380.
23. Chirakkal H, O'Rourke M, Atrih A, Foster SJ, Moir A (2002) Analysis of spore cortex lytic enzymes and related proteins in *Bacillus subtilis* endospore germination. Microbiology 148: 2383–2392.
24. Pelczar PL, Igarashi T, Setlow B, Setlow P (2007) Role of GerD in germination of *Bacillus subtilis* spores. J Bacteriol 189: 1090–1098.
25. Stewart KA, Yi X, Ghosh S, Setlow P (2012) Germination Protein Levels and Rates of Germination of Spores of *Bacillus subtilis* with Overexpressed or Deleted Genes Encoding Germination Proteins. J Bacteriol 194: 3156–3164.
26. Cabrera-Martinez RM, Tovar-Rojo F, Vepachedu VR, Setlow P (2003) Effects of overexpression of nutrient receptors on germination of spores of *Bacillus subtilis*. J Bacteriol 185: 2457–2464.
27. Wang G, Yi X, Li YQ, Setlow P (2011) Germination of individual *Bacillus subtilis* spores with alterations in the GerD and SpoVA proteins, which are important in spore germination. J Bacteriol 193: 2301–2311.
28. Ghosh S, Scotland M, Setlow P (2012) Levels of germination proteins in dormant and superdormant spores of *Bacillus subtilis*. J Bacteriol 194: 2221–2227.
29. Picotti P, Rinner O, Stallmach R, Dautel F, Farrah T, et al. (2010) High-throughput generation of selected reaction-monitoring assays for proteins and proteomes. Nat Methods 7: 43–46.

30. Leighton TJ, Doi RH (1971) The stability of messenger ribonucleic acid during sporulation in *Bacillus subtilis*. J Biol Chem 254: 3189–3195.

31. Nicholson WL, and P Setlow. (1990) Sporulation, germination, and outgrowth. Molecular methods for bacillus. Chichester, England: John Wiley & Sons Ltd. P. 391–450.

32. Pelczar PL, Setlow P (2008) Localization of the germination protein GerD to the inner membrane in *Bacillus subtilis* spores. J Bacteriol 190: 5635–5641.

33. Hahne H, Wolff S, Hecker M, Becher D (2008) From complementarity to comprehensiveness-targeting the membrane proteome of growing *Bacillus subtilis* by divergent approaches. Proteomics 8: 4123–4136.

34. Paidhungat M, Setlow P (2001) Localization of a germinant receptor protein (GerBA) to the inner membrane of *Bacillus subtilis* spores. J Bacteriol 183: 3982–3990.

35. González-Castro MJ, López-Hernández J, Simal-Lozano J, Oruña-Concha MJ (1997) Determination of amino acids in green beans by derivitization with phenylisothiocyanate and high-performance liquid chromatography with ultraviolet detection. J Chrom Sci 35: 181–185.

36. Fusaro VA, Mani DR, Mesirov JP, Carr SA (2009) Prediction of high-responding peptides for targeted protein assays by mass spectrometry. Nat Biotechnol 27: 190–198.

37. Paidhungat M, Ragkousi K, Setlow P (2001) Genetic requirements for induction of germination of spores of *Bacillus subtilis* by Ca(2+)-dipicolinate. J Bacteriol 183: 4886–4893.

38. Paredes-Sabja D, Setlow B, Setlow P, Sarker MR (2008) Characterization of *Clostridium perfringens* spores that lack SpoVA proteins and dipicolinic acid. J Bacteriol 190: 4648–4659.

39. Forouhar F, Su M, Seetharaman J, Fang F, Xiao R, et al. (2010) Crystal structure of Stage V sporulation protein AD (SpoVAD) from *Bacillus subtilis*, Northeast Structural Genomics Consortium Target SR525. PDB id: 3LM6.

40. Stewart KA, Setlow P (2013) Numbers of individual nutrient germinant receptors and other germination proteins in spores of *Bacillus subtilis*. J Bacteriol 195: 3575–3582.

41. Cowan AE, Olivastro EM, Koppel DE, Loshon CA, Setlow B, et al. (2004) Lipids in the inner membrane of dormant spores of *Bacillus* species are largely immobile. Proc Natl Acad Sci USA 101: 7733–7738.

42. Griffiths KK, Zhang J, Cowan AE, Yu J, Setlow P (2011) Germination proteins in the inner membrane of dormant *Bacillus subtilis* spores colocalize in a discrete cluster. Mol Microbiol 81: 1061–1077.

43. Vepachedu VR, Setlow P (2007) Analysis of interactions between nutrient germinant receptors and SpoVA proteins of *Bacillus subtilis* spores. FEMS Microbiol Lett 274: 42–47.

44. Korza G, Setlow P (2013) Topology and accessibility of germination proteins in the *Bacillus subtilis* spore inner membrane. J Bacteriol 195: 1484–1491.

45. Vepachedu VR, Setlow P (2005) Localization of SpoVAD to the inner membrane of spores of *Bacillus subtilis*. J Bacteriol 187: 5677–5682.

46. Li Y, Davis A, Korza G, Zhang P, Li YQ, et al. (2012) Role of a SpoVA protein in dipicolinic acid uptake into developing spores of *Bacillus subtilis*. J Bacteriol 194: 1875–1884.

8

FLS2-BAK1 Extracellular Domain Interaction Sites Required for Defense Signaling Activation

Teresa Koller, Andrew F Bent*

Department of Plant Pathology, University of Wisconsin – Madison, Madison, Wisconsin, United States of America

Abstract

Signaling initiation by receptor-like kinases (RLKs) at the plasma membrane of plant cells often requires regulatory leucine-rich repeat (LRR) RLK proteins such as SERK or BIR proteins. The present work examined how the microbe-associated molecular pattern (MAMP) receptor FLS2 builds signaling complexes with BAK1 (SERK3). We first, using *in vivo* methods that validate separate findings by others, demonstrated that flg22 (flagellin epitope) ligand-initiated FLS2-BAK1 extracellular domain interactions can proceed independent of intracellular domain interactions. We then explored a candidate SERK protein interaction site in the extracellular domains (ectodomains; ECDs) of the significantly different receptors FLS2, EFR (MAMP receptors), PEPR1 (damage-associated molecular pattern (DAMP) receptor), and BRI1 (hormone receptor). Repeat conservation mapping revealed a cluster of conserved solvent-exposed residues near the C-terminus of models of the folded LRR domains. However, site-directed mutagenesis of this conserved site in FLS2 did not impair FLS2-BAK1 ECD interactions, and mutations in the analogous site of EFR caused receptor maturation defects. Hence this conserved LRR C-terminal region apparently has functions other than mediating interactions with BAK1. *In vivo* tests of the subsequently published FLS2-flg22-BAK1 ECD co-crystal structure were then performed to functionally evaluate some of the unexpected configurations predicted by that crystal structure. In support of the crystal structure data, FLS2-BAK1 ECD interactions were no longer detected in *in vivo* co-immunoprecipitation experiments after site-directed mutagenesis of the FLS2 BAK1-interaction residues S554, Q530, Q627 or N674. In contrast, *in vivo* FLS2-mediated signaling persisted and was only minimally reduced, suggesting residual FLS2-BAK1 interaction and the limited sensitivity of co-immunoprecipitation data relative to *in vivo* assays for signaling outputs. However, Arabidopsis plants expressing FLS2 with the Q530A+Q627A double mutation were impaired both in detectable interaction with BAK1 and in FLS2-mediated responses, lending overall support to current models of FLS2 structure and function.

Editor: Vladimir N. Uversky, University of South Florida College of Medicine, United States of America

Funding: This research was funded by U.S. Department of Energy Basic Energy Biosciences Grant DE-FG02-02ER15342 and University of Wisconsin Hatch Funding to AFB. The funders had no role in study design, data collection and analysis, decision to publish, or preparation of the manuscript.

Competing Interests: The authors have declared that no competing interests exist.

* Email: afbent@wisc.edu

Introduction

Plants use pattern-recognition receptors (PRRs) as a first layer of defense against pathogens [1,2]. In order to engineer plants with improved pathogen recognition abilities, it is important to understand the molecular details underlying the interaction of PRRs not only with their ligands but also with their co-receptors, immediate downstream targets and other partner proteins that facilitate appropriate signaling. Several PRRs have been identified in different plant species [reviewed in 1,2]. PRRs are localized at the plasma membrane where they monitor the apoplastic space for microbe-associated molecular patterns (MAMPs), damage-associated molecular patterns (DAMPs) and apoplastic effectors. Most known PRRs are receptor-like kinases (RLKs) or receptor-like proteins (RLPs). Both receptor types consist of an extracellular domain for ligand perception and a transmembrane domain, but only the RLKs have an intracellular kinase domain. Two of the best characterized PRRs, FLS2 and EFR [3,4], carry large extracellular domains (ECDs, ectodomains) that predominantly consist of a leucine-rich repeat (LRR) domain [5,6]. The genomes of Arabidopsis and other plants each encode hundreds of LRR receptor-like kinases (LRR-RLKs) with 4 to 28 repeat units of the LRR [7].

Receptors typically exhibit high specificity for ligands with which they interact, but cells also contain co-receptors and regulatory proteins that function together with receptors and do not necessarily exhibit specificity for only a single type of ligand [8,9]. These co-receptors and regulatory proteins can be important facilitators or suppressors of signaling activation. They also allow signaling crosstalk at the plasma membrane, helping to coordinate appropriate downstream signaling in the presence of diverse endogenous and exogenous extracellular ligands. Important examples of regulatory/co-receptor RLKs include the SERK family members [8,10], BIR family members [11,12] and SOBIR1 [13].

SERK proteins have been identified in many different plant species. In Arabidopsis the family consists of five members (SERK1, SERK2, SERK3/BAK1, SERK4 and SERK5). They all have five LRRs in their ectodomain, share high overall sequence similarity and have redundant functions to various degrees. SERK proteins (mainly SERK3, also known as BAK1) have been shown to be involved in plant immunity in Arabidopsis, tomato and rice, through interactions with the receptors FLS2, EFR, PEPR1, PEPR2, Xa21, Ve1 and Eix1 [14–19]. The BAK1 co-receptor also contributes to somatic embryogenesis [20,21] and to plant development through interaction with the brassinolide hormone receptor BRI1 [22,23]. Despite impressive progress, much remains unknown about how the SERK proteins participate in all these different cell signaling tasks, and about the spatial expression of SERK proteins [24]. Studies of the SERK proteins are impeded by the redundant functions among family members and by pleiotropic effects when multiple SERK proteins are knocked out. As an example, $bak1^-$ Arabidopsis plants only have partially disrupted FLS2 signaling outputs [14,15,25,26]. A possible means of circumventing this problem of SERK functional redundancy, adopted in the present study, is to identify the specific SERK interaction site of a partner receptor and then mutate that site. If all SERKs interact with a specific receptor at similar amino acids, this approach should impair the interaction of the receptor with all SERK family members.

Recent X-ray crystallography studies provided detailed insight into the interaction of the ectodomain of BAK1 with the ectodomains of FLS2 and BRI1 [27,28], and the interaction of the ectodomains of SERK1 and BRI1 [29]. In all three cases the respective ligand promotes interaction between the ectodomains of the main receptors (FLS2 and BRI1) and the SERK co-receptors (BAK1 and SERK1). The ligand binds to the LRR domain of the main receptor, but the LRR domain of the SERK co-receptor also has multiple direct contacts with the ligand. It is surprising to see these fine-tuned co-receptor/ligand interactions, considering how many different known and potential unknown receptors and ligands BAK1 and SERK1 are able to interact with. Similar residues of the BAK1 and SERK1 ectodomains are involved in their interactions with FLS2 and BRI1. However, the residues on FLS2 and BRI1 ectodomains predicted to be used for the interactions with their SERK co-receptors are very different, not only in sequence but also in their location within the receptor LRR domain [27–29]. In BRI1 the residues interacting with co-receptors are located at the island domain, the last LRR, and the juxtamembrane domain, all close to the transmembrane domain. However, in FLS2 the BAK1-interacting residues in the crystal structure are located 108–300 amino acids from the predicted transmembrane domain, at repeats #18 to 26 of the LRR domain. This predicts a relatively recumbent orientation for the FLS2 ectodomain, bent down toward the plasma membrane (see Figure S3A).

FLS2 mediates perception of bacterial flagellin protein, an abundant MAMP, and FLS2 recognizes in particular a ∼20 amino acid region that is relatively conserved across flagellins from diverse Gram-negative bacteria [1,30]. Many aspects of FLS2 structure and function have been characterized [reviewed in 31]. There is a third surprising feature of the FLS2-flg22-BAK1 ECD co-crystal structure [27]. Most research regarding FLS2 utilizes as ligand, in place of flagellin protein, a 22 amino acid "flg22" peptide whose sequence matches the recognized domain of *Pseudomonas aeruginosa* flagellin, or utilizes other small peptides based on similar sequences from various bacteria [30,32,33]. The FLS2-flg22-BAK1 ECD co-crystal structure predicts a tight pocket for the flg22 peptide, which may not be compatible with (allow sufficient space or sufficient ligand flexibility for) analogous binding of flg22 domains embedded within full-length flagellin proteins (discussed below).

In this study we first explored the possibility that a relatively universal SERK interaction site has evolved in the LRR domains

of different SERK-interacting LRR-RLKs. We also showed that flg22-dependent FLS2 interaction with BAK1 occurs via the FLS2 extracellular domains – a result subsequently shown by alternative methods by Sun *et al.* (2103). We then performed site-directed mutagenesis and functional testing of predicted LRR-RLK receptor/SERK co-receptor interaction residues, and obtained *in vivo* evidence that supports models suggested by the recently published receptor/co-receptor co-crystal structures of truncated FLS2 and BAK1. The overall goal of this study was to furnish a more clear understanding of the requirements for formation of a signaling-competent plant basal immune system MAMP receptor – an understanding that may be essential to allow future engineering of PRRs with broadened or otherwise improved performance.

Methods

Arabidopsis and *Nicotiana benthamiana* transformation

The floral dip method was used to stably transform Arabidopsis $fls2^-$ and efr^- plants. T1 seedlings were selected on 0.5x MS plates containing 25 mg/L kanamycin and 25 mg/L hygromycin. Leaves of 4-week-old *Nicotiana benthamiana* plants were infiltrated with *Agrobacterium tumefaciens* GV3101 containing the binary plasmids [34]. Proteins were harvested two days after *Agrobacterium tumefaciens* infiltration.

Co-immunoprecipitation

Transiently transformed leaf tissue from *Nicotiana benthamiana* was infiltrated with 1 μM flg22 or 1 μM elf18, or with water for mock infiltration. After 2 minutes the leaf tissues were blotted dry and frozen in liquid N_2. Then 200 mg of tissues were ground in 200 μl protein extraction buffer (50 mM Tris pH 7.5, 150 mM NaCl, 0.5% Triton X-100, 1x plant protease inhibitor cocktail (Sigma-Aldrich)). After centrifugation 300 μl supernatant was incubated with 3 μl 9E10 anti-myc antibody (Sigma-Aldrich or Covance) and rotated at 4°C for 1 h. 50 μl Protein A (Thermo Scientific) was added and the tubes were rotated at 4°C for an additional 2 h. After 3x washing with protein extraction buffer and 1x washing with ddH$_2$O the beads were resuspended in 60 μl loading buffer and boiled at 95°C for 5 min. After centrifugation the supernatant was separated on two 8% SDS-PAGE gels. For protein detection the antibodies anti-HA-HRP, anti-myc rabbit and goat-anti-rabbit-HRP (Sigma-Aldrich) were used.

Conservation mapping

Mapping of conserved regions of predicted LRR surfaces was performed using the Repeat Conservation Mapping (RCM) program at www.plantpath.wisc.edu/RCM [35], with heat map coloration range set to the minimal and maximal conservation scores of the data within each figure. The LRR domain sequences were obtained from The Arabidopsis Information Resource (TAIR) website at www.arabidopsis.org and from the National Center for Biotechnology Information (NCBI) website at www. ncbi.nlm.nih.gov. The following FLS2 non-Brassicaceae sequences were used: *Populus trichocarpa* (XP_002305701.1); *Vitis vinifera* (XP_002272319.2); *Glycine max* (XP_003532650.1); *Lotus japonicus* (AER60531.1); *Ricinus communis* (XP_002519723.1); *Sorghum bicolor* (XP_002448543.1); *Oryza sativa Japonica* (CAE02151.2); *Oryza sativa Indica* (CAH68341.1); *Hordeum vulgare* (BAJ89141.1); *Brachypodium distachyon* (XP_003581675.1).

Site-directed mutagenesis

Point mutations were generated according to the QuikChange mutagenesis kit (Agilent Technologies) on pENTR plasmids (Invitrogen) containing FLS2, FLS2-NoKinase or EFR with 35 S or native promoters [36]. Gateway LR Clonase II (Invitrogen) was used to transfer the construct into the binary plasmids pGWB13 or pGWB14 [37].

EndoH assay

Leaf tissues (60 mg) from Arabidopsis T1 plants or from transiently transformed *Nicotiana benthamiana* plants were ground in 2x SDS buffer and boiled for 5 min at 95°C. After centrifugation for 10 min at 14000 rpm at 4°C supernatants were digested with Endoglycosidase H (New England BioLabs) as per manufacturer's suggestion and separated on 8% SDS-PAGE gel. Proteins were detected using anti-HA-HRP antibody (Sigma-Aldrich).

Seedling growth inhibition

T1 Arabidopsis seedlings were grown for 6 days on 0.5x MS plates with 25 mg/L kanamycin and 25 mg/L hygromycin and 200 mg/L cefotaxime. 24 seedlings per genotype, representing 24 independent transformation events, were transferred to 24-well-plates containing 1 ml 0.5x MS liquid media per well. 12 seedlings per genotype were grown for 14 days in wells containing 1 μM flg22 and 12 seedlings per genotype were grown for 14 days in wells containing only 0.5 x MS. Seedlings were then blotted dry and weighed. The weight of each flg22-treated seedling was divided by the average weight of the mock treated seedlings of the same genotype from the same experiment, prior to determination of experiment means and standard errors.

Oxidative burst

Seven leaf discs were taken from six-week-old T1 Arabidopsis plants and incubated overnight in 1% DMSO solution. Peptide solution was added to the leaf discs and luminescence was measured by a plate reader for 0–30 min after addition of flg22 peptide. For measurement each leaf disc was in 100 μl peptide solution containing 0.5 μl 2 mg/ml horseradish peroxidase, 0.5 μl 2 mg/ml luminol in DMSO and 1 μM flg22.

Results and Discussion

Extracellular domain of FLS2 can mediate interaction with BAK1 in the presence of flg22

Full-length FLS2 and BAK1 do not detectably interact until exposure to flg22 or similar flagellin ligands, at which time interaction is immediately observed [14,15,38]. Flg22-elicited immune signaling then requires phosphorylation events among the respective kinase domains [26,38,39]. We hypothesized that the FLS2-BAK1 interaction is mediated not only intracellularly by the respective kinase domains, but also by interaction of the ectodomains. To test this we used a truncated FLS2 carrying the N-terminal ∼70% of the protein including the LRR and transmembrane domains but not the predicted intracellular domains (*FLS2-NoKinase-HA*; [36]). *FLS2-NoKinase-HA* was expressed in *Nicotiana benthamiana* together with a plasmid encoding a full-length, epitope tagged *BAK1-Myc*. The transiently transformed leaves were treated with flg22 and co-immunoprecipitation experiments were performed. BAK1 and FLS2-NoKinase interact in the presence of flg22, indicating that the kinase domain of FLS2 is not needed for interaction with BAK1 *in planta* (Figure 1).

Sun *et al.* 2013 also showed ECD mediation of FLS2-BAK1 interaction [27]. Their work utilized *in vitro* mixing experiments with purified recombinant proteins, or mutated BAK1 expressed in Arabidopsis protoplasts. Our results with mutated FLS2, tested in transgenic whole plants with *FLS2* expressed under control of *FLS2* promoter sequences, are complimentary and in agreement with the results of Sun *et al.* (2103), and reveal that intracellular/kinase domain interactions of these proteins are not required for flg22-stimulated FLS2-BAK1 interaction. It is also interesting to note the previously published finding that FLS2-FLS2 interaction occurs *in planta*, with either full-length FLS2 or FLS2-NoKinase constructs [36]. At least some FLS2 exists *in planta* in FLS2-FLS2 complexes, prior to and after flagellin or flg22 exposure. FLS2-BAK1 interaction after exposure to flg22 did not appreciably deplete the overall presence of co-immunoprecipitable FLS2-FLS2 complexes [36]. Hence findings that FLS2 and BAK1 interact via LRR domains suggest either that FLS2-FLS2 interactions utilize a different side or face of the FLS2 LRR than the region that interacts with BAK1, or that different sub-pools of FLS2 are at any given moment interacting with FLS2 or BAK1. The results of Albert *et al.* (2013) and Cao *et al.* (2013) are also relevant to these updated models of PRR receptor - co-receptor structure/function [39,40]. Those studies demonstrated that *in planta* responses to flg22 are retained when hybrid FLS2 and BAK1 proteins are expressed in which the kinase domains of FLS2 and BAK1 have been reciprocally swapped [40], and that flg22-mediated FLS2-BIK1 disassociation and FLS2-BAK1 association still occur when FLS2 kinase domain mutations are present that block defense signaling. Schulze *et al.* 2010 and Schwessinger *et al.* 2011 showed that kinase-dead BAK1 still interact with FLS2, but impair FLS2 signaling [26,38]. The evidence increasingly indicates that interactions of the FLS2 and BAK1 extracellular domains are a first step in flg22 perception that can proceed relatively independent of intracellular domain structural or functional interactions.

Identification of a conserved region in the C-terminal LRRs of BAK1-interacting receptors

The SERK family members have been shown to interact with several different transmembrane LRR-RLKs involved in plant immunity and development. It is not known if the SERK interaction sites of these receptors evolved independently or originate from a common and potentially conserved SERK interaction site. We hypothesized the latter and also hypothesized that, to facilitate spatial proximity of potentially interacting extracellular domains, the relatively small ectodomains of SERK proteins would interact near the C-terminal end of the large LRR ectodomains of those partner receptors. Using Repeat Conservation Mapping [35] we searched the last seven repeats of the LRRs of the known Arabidopsis BAK1-interacting proteins FLS2 (28 total repeats in the LRR domain), EFR (21 LRRs), BRI1 (25 LRRs) and PEPR1 (26 LRRs), looking for the patch of solvent-exposed amino acids in this region that is most conserved across the four proteins. A conserved region of interest was identified (Figure 2A). Separately, we compared the solvent exposed amino acids of the whole LRR domains of eleven non-Brassicaceae FLS2s (Figure 2B). Both conservation maps revealed a conserved region at a similar location in the C-terminal LRRs. We hypothesized that this may be a somewhat universal site for interaction with SERK proteins.

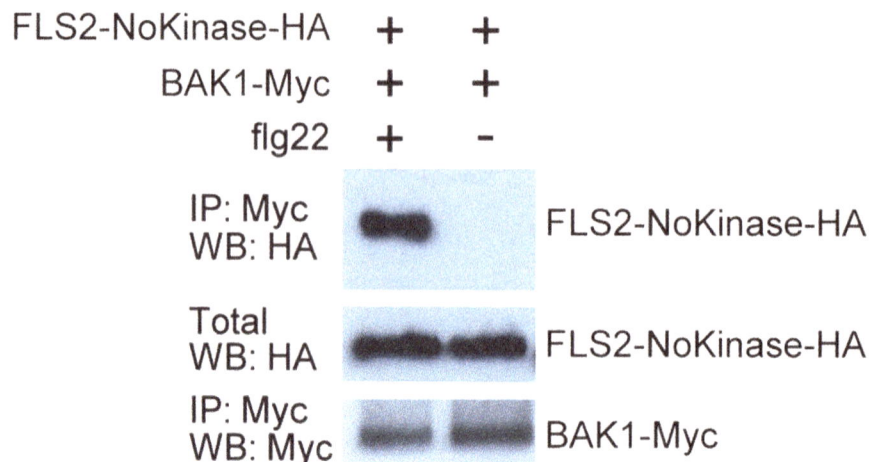

Figure 1. Extracellular domain of FLS2 can mediate interaction with BAK1. Co-immunoprecipitation experiments performed using *35S–FLS2-NoKinase-HA* (construct lacking the FLS2 intracellular domain) and *35S–BAK1-Myc* transiently expressed in *Nicotiana benthamiana* leaves by agroinfiltration. Samples were prepared for SDS-PAGE two days after agroinfiltration, two minutes after flg22 or water (mock) was infiltrated into leaves. IP: antibody used for immunoprecipitation prior to SDS-PAGE; WB: antibody used for immunodetection on protein blot; crude: SDS-PAGE and blotting of total (crude extract) protein samples. The experiment was repeated three times with similar results.

No disruption of FLS2-BAK1 interaction by mutations in the FLS2 LRR domain C-terminal region conserved among EFR, PEPR1, BRI1 and multiple FLS2s

Site-directed mutagenesis was carried out to alter residues in the identified conserved LRR C-terminal region of FLS2 and EFR (Figure 2; Figure S1A–E). D557 and S559 mutations in FLS2 were included as control mutations located in LRR sites analogous to N704/S706 and D728/S730, but outside of the conserved LRR C-terminus. The amino acids were replaced with similar yet bulkier residues in order to impair interactions. The resulting full-length receptors were expressed in *N. benthamiana* and co-immunoprecipitation experiments were then carried out, using BAK1-Myc for pull-down in the presence and absence of the corresponding ligands flg22 and elf18 in the case of FLS2 or EFR, respectively. The mutations in FLS2 did not abolish the interaction with BAK1 in the presence of flg22 (Figure 3A). As is common for agroinfiltration experiments, variable levels of expression were observed for any single transgene-encoded FLS2 protein across replicates within or between experiments, but none of the mutant proteins was reproducibly present at levels different from transgene-encoded wild-type FLS2. To ensure that interaction of the kinase domains of FLS2 and BAK1 was not masking non-interaction of mutated FLS2 and BAK1 ectodomains, the same mutations were also placed into FLS2-NoKinase constructs. In these FLS2-NoKinase variants the mutations again did not prevent interaction with BAK1 (Figure 3B).

Mutations in the conserved C-terminal LRR region of EFR cause EFR glycosylation/maturation defects

Mutations analogous to those of the preceding section were also engineered into *EFR*. These LRR domain C-terminal region mutations (Figure 2; Figure S1C, D) did cause disruption of interaction with BAK1 in the presence of elf18 (Figure 4A). This *in vivo* result could be attributable to direct impacts of the mutations on EFR-BAK1 interaction, or to defects in maturation and delivery of newly synthesized EFR out of the endoplasmic reticulum (ER) and golgi. Endoglycosidase H (EndoH) analyses were therefore conducted. EndoH cleaves incomplete glycosylation modifications present on proteins that have not successfully

passed through the ER and related endomembrane systems [41,42]. On the other hand, mature glycosylated proteins that are delivered to their functional location typically carry EndoH-resistant glycosylation [41,42]. Treatment of the EFR protein extracts with EndoH revealed defects in the mutated EFR proteins, both in *N. benthamiana* and in stable transgenic Arabidopsis *efr-* plants expressing transgene *EFR* constructs driven by native *EFR* promoter sequences (Figure 4B, C). The mutations we generated in FLS2 full-length and FLS2-NoKinase did not result in glycosylation defects (Figure S2A, B). Häweker *et al.* 2010 [42] and Sun at al. 2012 [36] showed that single amino acid changes in glycosylation sites in the EFR ectodomain result in protein degradation and several studies reported the importance of intact glycosylation enzymes for successful processing and function of EFR [43–47]. FLS2 is less sensitive to mutations in glycosylation sites [36,42]. The N590Q+S592T mutations that we placed in EFR are indeed in a Nx(S/T) predicted glycosylation site [48]. However, the EFR mutations D566E+S568T and D566F are not, yet they still disrupted correct EFR processing.

Taken together, the above results suggest that functional roles of the LRR C-terminal conserved domain of BAK1-interacting proteins (Figure 2) do not serve as a universal SERK protein interaction site. However, in EFR the integrity of this site is important for correct protein processing.

FLS2 mutations in proposed FLS2-BAK1 ECD interaction residues disrupt FLS2-BAK1 interaction in the presence of flg22

While the above work was in progress the crystal structure of FLS2-flg22-BAK1 ECD became available [27]. That important work identified in detail the interaction sites of the FLS2 and BAK1 ectodomains. Because the data are for *in vitro* crystallized protein complexes of isolated LRR domains, they may or may not capture the most functionally prominent *in vivo* configurations. Sun *et al.* [27] therefore functionally tested BAK1 mutations, and also tested FLS2 mutations in repeats #9, 11, 14 and 15 of the LRR that are predicted to mediate interaction with flg22. The sites on FLS2 predicted to mediate interaction with BAK1 did not receive mutational testing. In order to test *in vivo* the significance

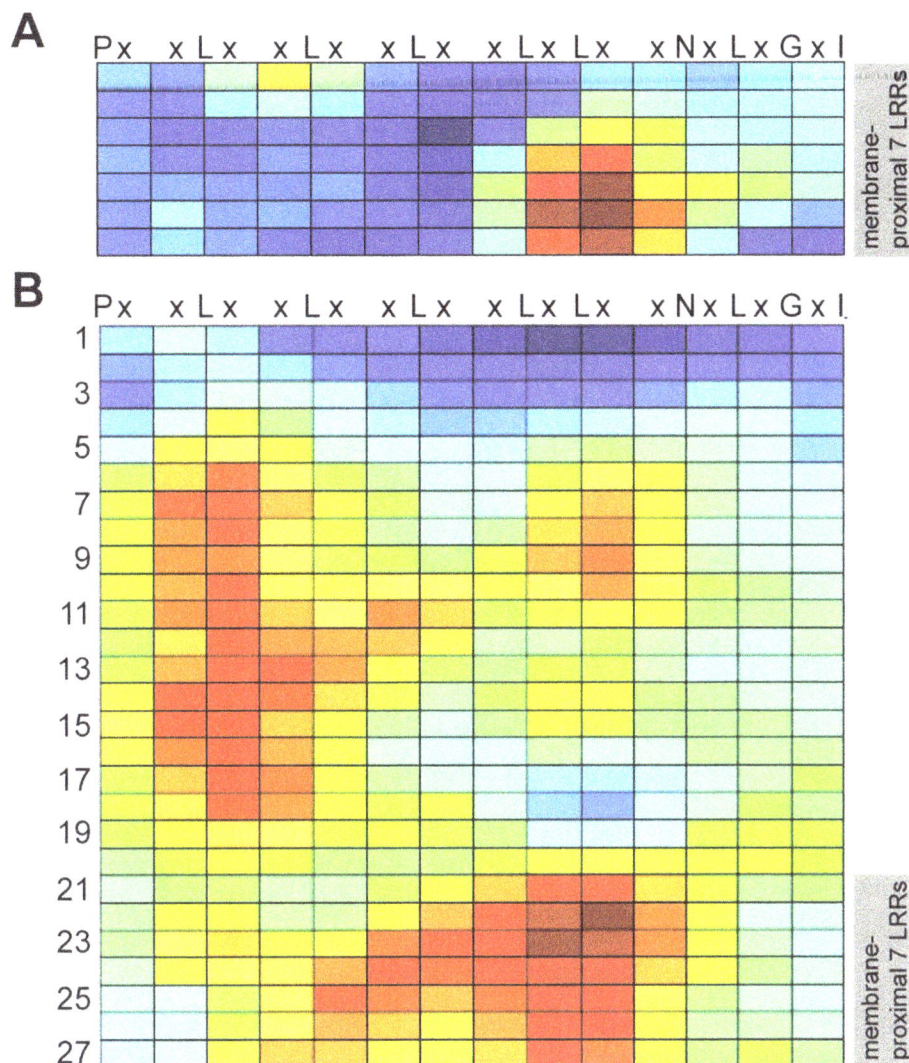

Figure 2. Repeat Conservation Mapping reveals conserved region near C-terminus of LRR domains of FLS2, EFR, BRI1 and PEPR1.
Each row represents one leucine-rich repeat (LRR) and each square represents one solvent-exposed "x" amino acid position (as per LRR consensus sequence shown at the top). Conservation score at each amino acid position is center-weighted score for the cluster of 15, 20 or 25 predicted solvent-exposed LRR amino acids surrounding that site; blue: least conserved, red: most conserved. For FLS2, the seven rows of (A) are the same repeats (same residues) as rows 21–27 of (B). (**A**) Conservation map generated by comparing the most C-terminal seven repeats of the LRR sequences of the BAK1 interacting proteins FLS2, EFR, BRI1 and PEPR1. (**B**) Conservation map generated by comparing the entire FLS2 LRR domain sequences from eleven non-Brassicaceae plant species.

of these FLS2 BAK1-interaction residues, which are likely to also mediate interaction of FLS2 with other SERK proteins, we performed site-directed mutagenesis on *FLS2-NoKinase* and full-length *FLS2*.

For FLS2 amino acids predicted in the crystal structure to form FLS2-BAK1 interaction sites [27], we changed single residues to alanine (small and relatively inactive) or to tryptophan (bulky). In addition to the single mutations we made two *FLS2-NoKinase* constructs with double mutations and one full-length *FLS2* construct with a double mutation. We had previously shown that the FLS2-NoKinase used in this work performed similarly to FLS2-full-length in flg22-dependent BAK1 co-immunoprecipitation experiments (Figure 3A, B). In *in vivo* tests of the newly predicted FLS2-BAK1 interaction sites, mutation of FLS2 residues Q530, S554, Q627 or N674 to tryptophan disrupted the flg22-stimulated interaction of FLS2-NoKinase with BAK1 (Figure 5A, B). The interaction was disrupted as well when FLS2 Q530 and

N674 were changed to alanine (Figure 5A, B). However, the FLS2 S554 and Q627 single mutations to alanine had much less impact on flg22-dependent interaction with BAK1 (Figure 5A, B), suggesting a stronger role for Q530 and N674 than S554 or Q627 in mediating FLS2-BAK1 interaction. The double alanine mutation Q530A+Q627A and the double tryptophan mutation S554W+Q627W disrupted BAK1 interaction as well (Figure 5C). The presence of abundant EndoH-insensitive bands suggested that FLS2 maturation had proceeded successfully for each of the representative FLS2 mutants S554A, Q627W and Q530A+Q627A (Figure S2C).

Arabidopsis *fls2⁻* plants carrying FLS2-Q530A+Q627A have impaired FLS2-mediated signaling outputs

To investigate if mutations in predicted FLS2 BAK1-interaction residues not only disrupt FLS2-BAK1 interactions in co-immu-

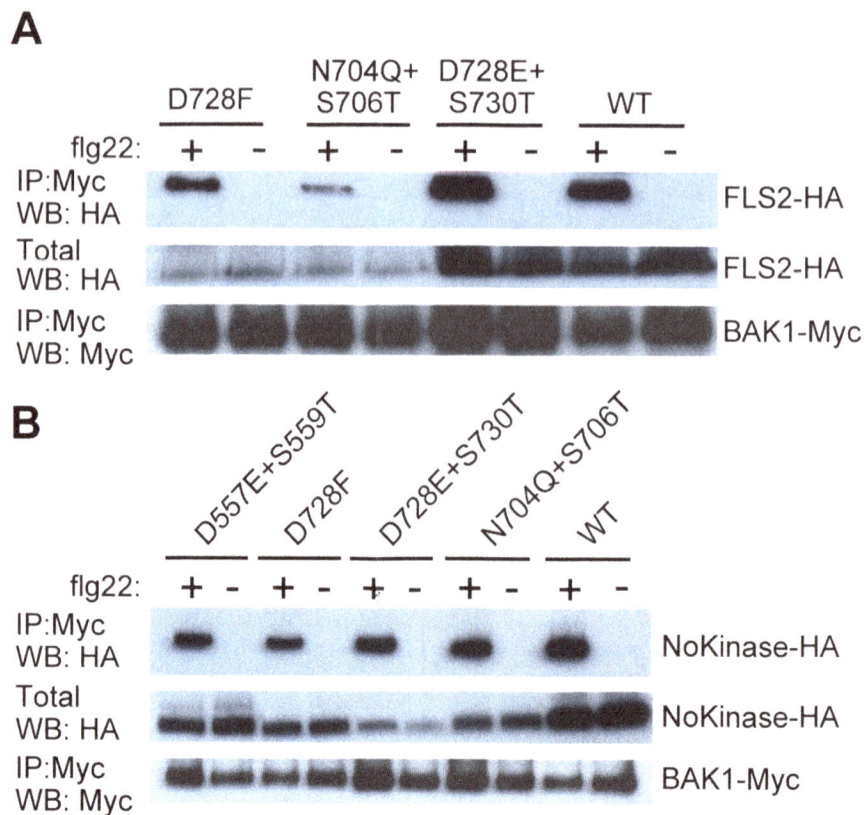

Figure 3. Mutations in the conserved C-terminal region of the FLS2 LRR domain did not have an impact on BAK1-FLS2 or BAK1-FLS2-NoKinase interaction. (**A**) Co-immunoprecipitation experiments performed using full-length P_{FLS2}-FLS2-HA, with mutations as indicated or WT (no mutations), and 35S–BAK1-Myc. (**B**) Co-immunoprecipitation experiments performed using 35S–FLS2-NoKinase-HA, with mutations as indicated or WT (no mutations), and 35S–BAK1-Myc. All samples in (A) and (B) are from *Nicotiana benthamiana*. Labeling as in Figure 1.

noprecipitation experiments but also have an impact on FLS2 signaling, we made the analogous single mutations and one of the double mutations in full-length *FLS2s*. We then tested FLS2 signaling in stably transformed *fls2⁻* Arabidopsis plants containing the mutated and HA-tagged full-length *FLS2s* under control of native *FLS2* promoter sequences. The two most widely used assays for FLS2 signaling were utilized: ROS burst assays and seedling growth inhibition assays [1]. Surprisingly, *in vivo* FLS2-mediated signaling persisted and was only minimally reduced in plants expressing most single-mutant forms of FLS2 (Figure 6A, B), including mutants that exhibited no detectable flg22-induced co-immunoprecipitation with BAK1 (Figure 5A, B). As a general trend across the multiple independent transgenic lines tested for each *FLS2* construct, mutations to alanine allowed stronger FLS2 signaling than mutations to tryptophan (Figure 6A, B). The results suggest that reduced-affinity or more transient interactions of FLS2 and BAK1 occur with many of the FLS2 mutants described in Figures 5 and 6, and that those interactions are sufficient for flg22-stimulated FLS2 signaling even if the stability of FLS2-BAK1 interactions is reduced below levels detectable in standard co-immunoprecipitation experiments. Although some FLS2 signaling capacity was still conferred by FLS2 constructs mutated at single predicted FLS2 BAK1-interaction sites, with the double mutation Q530A+Q627A FLS2-mediated signaling was significantly impaired (Figure 6C, D), supporting current models of FLS2 structure and function.

Alternative hypotheses, other than reduced-affinity or more transient interactions of FLS2 and BAK1, can be formulated

regarding the continued signaling by the FLS2 single mutants of Figure 5 and 6. For example, small sub-populations of FLS2 receptors (sufficient to initiate the levels of defense signaling observed in Figure 6) may exist in the cell that, because of different localization or post-translational modifications, continue to exhibit robust flg22-dependent interaction with BAK1 despite presence of mutations that disrupt interaction between most of the cellular FLS2 and BAK1. As another possibility, the single mutations that transition FLS2 away from high-affinity flg22-dependent binding with BAK1 may have allowed or even enhanced interaction with other SERK proteins, to an extent that allows defense signaling.

LRRs are a protein structure evolved to display widely varying surface amino acid combinations on a relatively invariant scaffold [5,6]. A previous study of over 1200 FLS2 LRR mutations of predicted LRR solvent-exposed residues at and adjacent to flg22 binding sites, carrying changes to all possible amino acids (i.e., not just to alanine), found that the vast majority of LRR surface mutations do not disrupt FLS2 function [49]. Thus the structural alterations caused by the FLS2 mutations of the present study are likely to be highly local. Their disruption of FLS2-BAK1 interactions detected via co-immunoprecipitation supports the relevance of the FLS2-flg22-BAK1 configuration in the published co-crystal structure. Mutation of FLS2 residues D557 and S559, which reside close to but outside of the BAK1-interaction residues in the solved crystal structure ([27], Figure S1), did not disrupt flg22-stimulated FLS2-BAK1 co-immunoprecipitation (Figure 3). Hence the functional disruption of signaling caused by the presumably additive effect of two alanine substitutions in FLS2

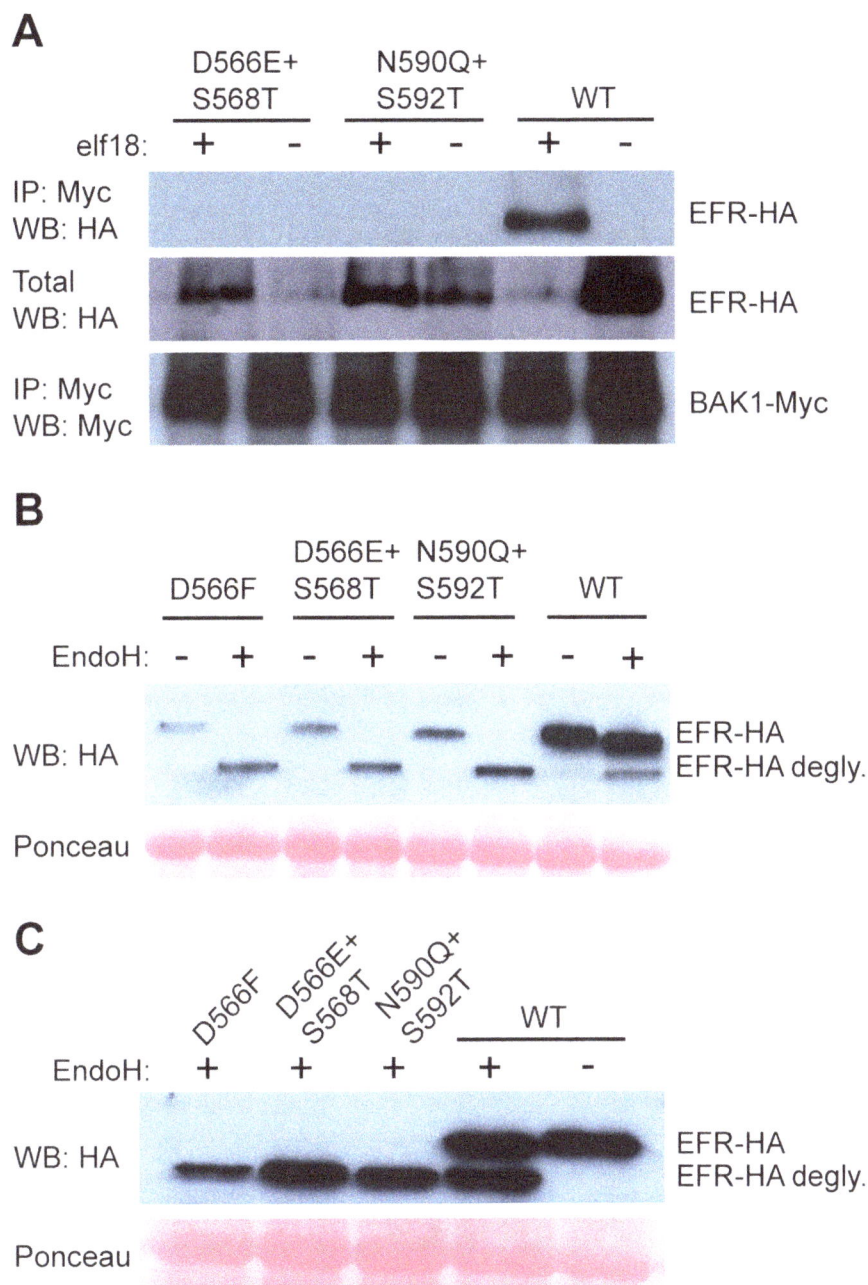

Figure 4. Mutations in the conserved C-terminal region of the EFR LRR domain disrupt EFR glycosylation and interaction with BAK1 in the presence of elf18. (**A**) Co-immunoprecipitation experiments performed using P_{EFR}-EFR-HA with mutations as indicated or WT (no mutations), and 35S–BAK1-Myc, in *Nicotiana benthamiana*. (**B, C**) Protein extracts from plants expressing P_{EFR}-EFR-HA with mutations as indicated, or WT (no mutations), not digested or digested with endoglycosidase H (EndoH). Samples in (B) are from *Nicotiana benthamiana*, samples in (C) are from stably transformed *efr* $^-$ *Arabidopsis* leaves. EndoH-resistant (mature) EFR is present in the EndoH-treated EFR wild type (WT) samples but is not detected for EFRs carrying the indicated mutations. Degly.: EFR pool deglycosylated by EndoH. Labeling as in Figure 1. Ponceau: blots treated with Ponceau stain to confirm even loading of total protein.

Q530A+Q627A provides further *in vivo* functional evidence indicating the requirement for this site both for FLS2-BAK1 interaction and for flg22 induction of FLS2-dependent immune signaling. Our results also indicate that, if SERK proteins other than BAK1 make residual contributions to FLS2 activation (as is suggested above and in the literature [14,15,25,26]), the FLS2 Q530A+Q627A mutations are sufficient to disrupt functional signaling mediated by those interactions as well.

Closing Observations

In this study we explored the idea of a universal SERK protein interaction site in the C-terminal repeats of the LRR ectodomains of receptors known to interact with SERK proteins. However, mutagenesis of a possible BAK1 interaction site in the ectodomains of FLS2 and EFR did not confirm this hypothesis. The subsequently available FLS2-flg22-BAK1 and BRI1-brassinolide-SERK1 extracellular domain crystal structures [27] [29], and the mutational studies in the present work, instead suggest a fine-tuned

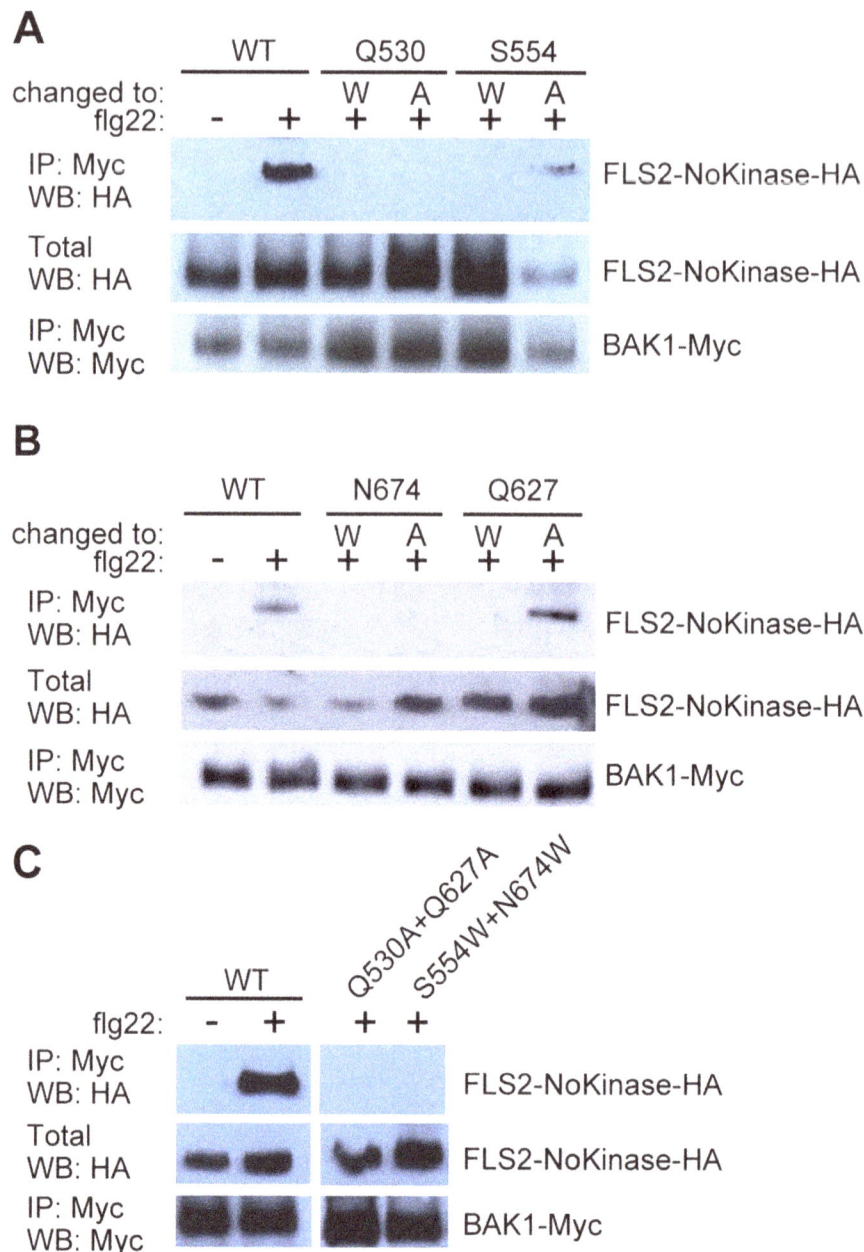

Figure 5. FLS2 residues Q530, S554, Q627 and N674 are important for FLS2-BAK1 ectodomain interaction in the presence of flg22.
Co-immunoprecipitation experiments performed in *N. benthamiana* with *35S–FLS2-NoKinase-HA* with mutations as indicated or WT (no mutations), and with *35S–BAK1-Myc*. Flg22-dependent interaction between FLS2-NoKinase and BAK1 not detected for (**A**) FLS2 carrying Q530A, Q530W or S554W mutations, (**B**) FLS2 carrying N674A, N674W or Q627W mutations, or (**C**) FLS2 carrying Q530A+Q627A or S554W+N674W double mutations. Labeling as in Figure 1.

interaction unique for each receptor/ligand/co-receptor complex. SERK1 and BAK1 use similar residues to interact with the BRI1 and FLS2 ectodomains, respectively. However, the SERK-interacting residues in the ectodomains of BRI1 and FLS2 are very different in terms of both the amino acid identities and their location along the large LRR macromolecule, and thus may have evolved separately.

The LRR surface region exhibiting conservation between FLS2, EFR, PEPR1 and BRI1 (Figure 2A) spans four repeats of the LRR, but overlaps with the larger region highlighted in Figure 2B that is conserved across diverse FLS2 proteins and spans the final seven repeats of the LRR. Within the larger conserved region, the

residues that are further to the left as shown in Figure 2B (or Figure S1E) encompass the BAK1 interaction site, but the residues on the right do not. The present study detected no impact of mutations in FLS2 in the Figure 2A conserved region, which in FLS2 is the same as the bottom right of the larger conserved region of Figure 2B. A previous study from our group [35] reported little or no functional impact of mutations in the upper-right area of the conserved region (the darkest red/most conserved area of Figure 2B). In that study, libraries of changes to all possible amino acids were made at the four FLS2 residues D605, S607, F633 and S634, directly above the N704/S706 and D728/S730 residues targeted in this study but in repeats #22 and 23 [35].

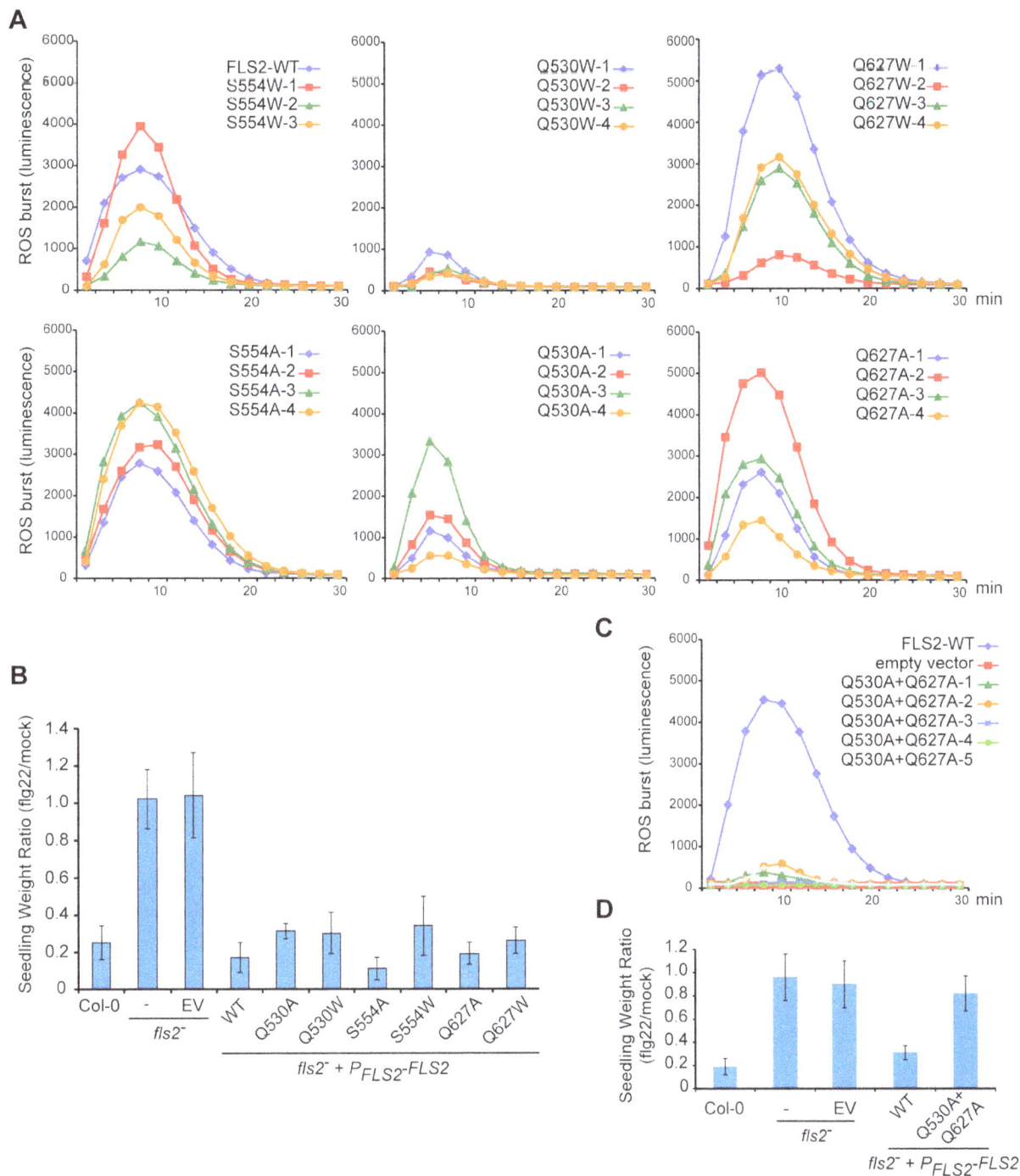

Figure 6. FLS2 signaling output impaired to various degrees in Arabidopsis _fls2⁻_ plants expressing FLS2 mutations that impact FLS2-BAK1 interaction. (A) Reactive oxygen species (ROS) production in response to flg22 in Arabidopsis Col-0 _fls2⁻_ plants stably transformed to express full-length FLS2 proteins carrying single mutations as noted, under control of _FLS2_ promoter sequences. For each mutation, ROS production was recorded for 30 min. and the average for seven separately monitored leaf discs is shown for each of four independent transgenic lines (or three lines for S554W). WT: Average ROS response for six independent _fls2⁻_ transformants expressing wild-type FLS2 (42 total leaf discs for WT), from same experiment. **(B)** FLS2-mediated seedling growth inhibition (SGI) in response to flg22, for plant lines as in (A). Mean and std. error of mean shown for six to eight independent transformants for each _FLS2_ construct. **(C)** ROS experiment as in (A), except with five independent lines expressing FLS2 Q530A+Q627A double mutations. **(D)** Seedling growth experiment as in (B), except with twelve independent lines expressing Q530A+Q627A double mutations.

Hence it is intriguing that this right side of the region highlighted in Figure 2B, which lies along the concave β-strand surface of repeats #21–27, is highly tolerant of mutations despite being relatively conserved across FLS2 proteins from diverse plant species. It remains of interest to discover the function of this portion of the FLS2 LRR.

As a separate but related matter, it is intriguing that the set of BAK1-interacting residues of FLS2 lie not only within regions highly conserved across FLS2 proteins from diverse plant species (e.g., Q627 and N674, Figure S1E), as might be expected, but also outside of conserved regions (e.g., Q530 and S554, Figure S1E). Figure 2B and Figure S1E show regions of LRR surface residue conservation in a comparison among FLS2s from non-Brassicaceae plant species (see Methods). But even in maps of conservation among FLS2s only from Brassicaceae species (see for example [35]), the region around Arabidopsis FLS2 residues Q627 and N674 is strongly conserved while Q530, S554 and adjacent BAK1-interacting residues [27] are in an LRR surface region that is less conserved. This raises the hypothesis that there is a functionally relevant diversification of SERKs and/or this upper portion of the SERK-interaction site of FLS2, even across Brassicaceae species.

The relevance of the FLS2-flg22-BAK1 co-crystal structure to actual configurations of the protein complex within plant cells would gain stronger support if more features of the crystal structure were reconciled with other findings regarding plant FLS2s and flagellin detection. We noted in the Introduction the concern that the co-crystal, made with flg22 peptide, may not allow enough space for docking of a full-length flagellin protein at the appropriate location. Figure S3 shows hypothetical alignments of the FLS2-flg22-BAK1 ECD structure (PDB ID: 4MN8) with the structure of one Salmonella flagellin protein (PDB ID: 3A5X), placing the flg22 region of 3A5X near the apparent flg22 binding sites of FLS2 and BAK1 while attempting to minimize co-occupancy of the same space by two different molecules. The FLS2 LRR, which is notably lacking in 'loop-out' or non-LRR-consensus regions, is likely to be relatively inflexible. Flagellin monomers in solution (not polymerized with other flagellins to form flagella-like structures) are likely to be more flexible than shown, particularly in the region of the flg22 residues that form a less ordered linker between two alpha-helical regions [50,51] (see Figure S3F). Nevertheless, space-filling models (e.g., Figure S3D) demonstrate the difficulty of docking a large flagellin onto the requisite FLS2 LRR sites while also allowing space for BAK1 and not allowing co-occupancy of identical space. Importantly, even in hypothesized configurations (not shown) that might allow space for a more flexible full-length flagellin to interact with FLS2 and BAK1, the flg22 residues within a flagellin protein are apparently constrained in ways that would restrict simultaneous interaction with the majority of the FLS2 LRR surface residues that interact with the elongated flg22 in the published FLS2-flg22-BAK1 co-crystal structure (e.g., Figure S3E). FLS2, flagellins and BAK1 may associate in vivo in configurations that depart significantly from the co-crystal structure. However, numerous aspects of the published FLS2-flg22-BAK1 co-crystal structure are substantiated by experimental evidence ([27]; references therein; present study). Hence we consider it equally likely that the published FLS2-flg22-BAK1 co-crystal is essentially correct in representing in vivo configurations, and predict that flagellin proteins within plants must be fragmented rather than intact in order to form the FLS2-flagellin-BAK1 complexes that elicit plant innate immune system activation.

In the future, it also will be interesting to compare more receptor/ligand/co-receptor signaling complexes in order to learn more about the functional plasticity of co-receptors. As one example, a ligand-mediated EFR-BAK1 ectodomain complex is likely to initiate EFR signaling. Interestingly, when EFR from Arabidopsis was transferred to Nicotiana benthamiana or tomato (which lack an endogenous EFR) it triggered an elf18-activated immune response, indicating functional interaction of AtEFR with SERK proteins from Nicotiana benthamiana and tomato [52]. Thus one or more SERK proteins apparently carry sufficient

structure-function plasticity to interact with different receptors even from diverse plant species, while complying with the fine-tuned sequence constraints of the resulting receptor/ligand/co-receptor complexes. For future engineering of PRR receptors with novel ligand specificities it will be important to ensure presence of an intact SERK protein interaction site in the ectodomain of the PRR, close to or overlapping with the ligand binding site, and ensure that the co-receptors also can form PRR/ligand/co-receptor complexes with the novel ligands for which new recognition specificity is sought.

Supporting Information

Figure S1 Mutation sites in FLS2 and EFR ECDs. (A) Sites subjected to site-directed mutagenesis in the FLS2 LRR domain. Only repeats 17–28 are shown (FLS2: total 28 repeats). Green: mutation sites in the conserved LRR domain C-terminus (see also (B, D, E)). Blue: "control" mutations; sites similar to N704/D728 and S706/S730 but outside of the conserved region; blue "control" sites also adjacent to but outside of FLS2 BAK1-interaction site. Orange: mutation sites based on FLS2 BAK1-interaction sites in the FLS2-flg22-BAK1 ECD co-crystal structure. **(B)** Mutation sites as described in (A), using same color scheme as in (A). Structure is PDB ID: 4MN8 with FLS2 backbone as black ribbon, BAK1 backbone as light blue ribbon, and flg22 backbone as red ribbon. Space-filling spheres show side-chains only for mutagenized sites. **(C)** Mutation sites in the EFR LRR domain. Only repeats 17–21 are shown (EFR: total 21 repeats). Green: mutation sites in the conserved LRR domain C-terminus (see also (D)). **(D, E)** Regional LRR surface conservation maps from Arabidopsis FLS2, EFR, PEPR1 and BRI1 (D) or eleven non-Brassicaceae FLS2s (E), as shown and described in Figure 2, with x's at the FLS2 and EFR LRR domain amino acid positions described above that were subjected to site-directed mutagenesis in the present study.

Figure S2 EndoH assay reveals no glycosylation defects in mutated FLS2 and FLS2-NoKinase. (A) Protein extracts from Arabidopsis $fls2^-$ leaves carrying P_{FLS2}-FLS2-HA (with mutations as indicated, or WT = no mutations), not digested (−) or digested (+) with endoglycosidaseH (EndoH). An EndoH-resistant protein pool (characteristic of mature glycosylated proteins) is visible in all EndoH-treated samples. **(B)** Protein extracts from Nicotiana benthamiana carrying 35S–FLS2-NoKinase-HA (with mutations as indicated), digested with EndoH. An EndoH-resistant protein pool is visible in all EndoH-treated samples. Mutations D557E+S559T were included as control mutations located in sites of a single LRR repeat analogous to D728E+S730T, but outside of the conserved LRR C-terminus. **(C)** Protein extracts from Arabidopsis $fls2^-$ seedlings carrying P_{FLS2}-FLS2-HA (with mutations as indicated, or WT = no mutations), digested with EndoH. An EndoH-resistant protein pool is visible in all EndoH-treated samples except for the empty vector (EV) control. Ponceau stained blot shows similar loading of total protein in all lanes including EV negative control. Degly.: FLS2 pools deglycosylated by EndoH.

Figure S3 Hypothetical docking of full-length flagellin structure (PDB ID: 3A5X) to FLS2-flg22-BAK1 structure (PDB ID: 4MN8) illustrates minimal space for flagellin inside FLS2 LRR, and constraints to flg22 contact with FLS2 LRRs #3–15 if flg22 region is held within full-length flagellin. (A), (B), (C) Flagellin, hypothetically positioned so that flg22 residues within full-length flagellin are near the flg22 binding sites of FLS2 and BAK1. PDB structures 3A5X and 4MN8 superimposed at same scale; (B) and (C) are 90° rotated views of (A). Light blue: flagellin; red: flg22 residues within flagellin (3A5X). Dark blue: FLS2 LRR; green: BAK1 LRR; yellow: flg22 co-crystallized with FLS2 and BAK1 LRRs (4MN8). (**D**) Same view as (C), with space-filling representation of flagellin to more clearly illustrate impossible overlap of flagellin and FLS2 residues in same spatial locations in this arrangement (and other arrangements) of 3A5X and 4MN8. FLS2 and BAK1 side-chains omitted for clarity. (**E**) PyMol alignment of flg22 (yellow, in structure 4MN8) and flg22 region within flagellin (red, in structure 3A5X). Lower portions of flg22 in (E) (the yellow residues that are not proximal to red residues) are the N-terminal 7 residues of flg22 that associate with FLS2 LRRs #3–7 (FLS2 and BAK1 not

shown, for clarity). (**F**) Full length flagellin (PDB structure 3A5X) colored as in (A) but shown by itself, showing that flg22 region forms a less-ordered hinge region between flanking pairs of alpha-helical bundles.

Acknowledgments

We thank Stephen Mosher and Adam Bayless for critical reading of the manuscript.

Author Contributions

Conceived and designed the experiments: TK AFB. Performed the experiments: TK. Analyzed the data: TK AFB. Contributed to the writing of the manuscript: TK AFB.

References

1. Boller T, Felix G (2009) A renaissance of elicitors: perception of microbe-associated molecular patterns and danger signals by pattern-recognition receptors. Annu Rev Plant Biol 60: 379–406. Available: http://www.ncbi.nlm.nih.gov/entrez/query.fcgi?cmd=Retrieve&db=PubMed&dopt=Citation&list_uids=19400727.

2. Wu Y, Zhou J-M (2013) Receptor-like kinases in plant innate immunity. J Integr Plant Biol 55: 1271–1286. Available: http://www.ncbi.nlm.nih.gov/pubmed/24308571.

3. Gómez-Gómez L, Boller T (2000) FLS2: an LRR receptor-like kinase involved in the perception of the bacterial elicitor flagellin in Arabidopsis. Mol Cell 5: 1003–1011. doi:S1097-2765(00)80265-8 [pii].

4. Zipfel C, Kunze G, Chinchilla D, Caniard A, Jones JDG, et al. (2006) Perception of the Bacterial PAMP EF-Tu by the Receptor EFR Restricts Agrobacterium-Mediated Transformation. Cell 125: 749–760. doi:10.1016/j.cell.2006.03.037.

5. Kobe B, Kajava A V (2001) The leucine-rich repeat as a protein recognition motif. Curr Opin Struct Biol 11: 725–732. Available: http://www.ncbi.nlm.nih.gov/pubmed/11751054.

6. Bella J, Hindle KL, McEwan PA, Lovell SC (2008) The leucine-rich repeat structure. Cell Mol Life Sci 65: 2307–2333. Available: http://www.ncbi.nlm.nih.gov/pubmed/18408889.

7. Lehti-Shiu MD, Zou C, Hanada K, Shiu S-H (2009) Evolutionary history and stress regulation of plant receptor-like kinase/pelle genes. Plant Physiol 150: 12–26. Available: http://www.pubmedcentral.nih.gov/articlerender.fcgi?artid=2675737&tool=pmcentrez&rendertype=abstract.

8. Chinchilla D, Shan L, He P, de Vries S, Kemmerling B (2009) One for all: the receptor-associated kinase BAK1. Trends Plant Sci 14: 535–541. Available: http://www.ncbi.nlm.nih.gov/pubmed/19748302.

9. Liebrand TW, van den Burg HA, Joosten MH (2014) Two for all: receptor-associated kinases SOBIR1 and BAK1. Trends Plant Sci 19: 123–132. Available: http://www.ncbi.nlm.nih.gov/pubmed/24238702.

10. Kim BH, Kim SY, Nam KH (2013) Assessing the diverse functions of BAK1 and its homologs in arabidopsis, beyond BR signaling and PTI responses. Mol Cells 35: 7–16. Available: http://www.ncbi.nlm.nih.gov/pubmed/23269431.

11. Gao M, Wang X, Wang D, Xu F, Ding X, et al. (2009) Regulation of cell death and innate immunity by two receptor-like kinases in Arabidopsis. Cell Host Microbe 6: 34–44. Available: http://www.ncbi.nlm.nih.gov/pubmed/19616764.

12. Halter T, Imkampe J, Mazzotta S, Wierzba M, Postel S, et al. (2013) The Leucine-Rich Repeat Receptor Kinase BIR2 Is a Negative Regulator of BAK1 in Plant Immunity. Curr Biol. Available: http://www.ncbi.nlm.nih.gov/pubmed/24388849.

13. Liebrand TWH, van den Berg GCM, Zhang Z, Smit P, Cordewener JHG, et al. (2013) Receptor-like kinase SOBIR1/EVR interacts with receptor-like proteins in plant immunity against fungal infection. Proc Natl Acad Sci U S A 110: 10010–10015. Available: http://www.pubmedcentral.nih.gov/articlerender.fcgi?artid=3683720&tool=pmcentrez&rendertype=abstract.

14. Chinchilla D, Zipfel C, Robatzek S, Kemmerling B, Nürnberger T, et al. (2007) A flagellin-induced complex of the receptor FLS2 and BAK1 initiates plant defence. Nature 448: 497–500. Available: http://www.ncbi.nlm.nih.gov/pubmed/17625569.

15. Heese A, Hann DR, Gimenez-Ibanez S, Jones AME, He K, et al. (2007) The receptor-like kinase SERK3/BAK1 is a central regulator of innate immunity in plants. Proc Natl Acad Sci U S A 104: 12217–12222. Available: http://www.pubmedcentral.nih.gov/articlerender.fcgi?artid=1924592&tool=pmcentrez&rendertype=abstract.

16. Bar M, Sharfman M, Ron M, Avni A (2010) BAK1 is required for the attenuation of ethylene-inducing xylanase (Eix)-induced defense responses by the decoy receptor LeEix1. Plant J 63: 791–800. Available: http://www.ncbi.nlm.nih.gov/pubmed/20561260.

17. Fradin EF, Zhang Z, Juarez Ayala JC, Castroverde CDM, Nazar RN, et al. (2009) Genetic dissection of Verticillium wilt resistance mediated by tomato Ve1. Plant Physiol 150: 320–332. Available: http://www.pubmedcentral.nih.gov/articlerender.fcgi?artid=2675724&tool=pmcentrez&rendertype=abstract.

18. Chen X, Zuo S, Schwessinger B, Chern M, Canlas PE, et al. (2014) An XA21-Associated Kinase (OsSERK2) regulates immunity mediated by the XA21 and XA3 immune receptors. Mol Plant. Available: http://www.ncbi.nlm.nih.gov/pubmed/24482436.

19. Postel S, Küfner I, Beuter C, Mazzotta S, Schwedt A, et al. (2010) The multifunctional leucine-rich repeat receptor kinase BAK1 is implicated in Arabidopsis development and immunity. Eur J Cell Biol 89: 169–174. Available: http://www.sciencedirect.com/science/article/pii/S0171933509003306.

20. Schmidt ED, Guzzo F, Toonen MA, de Vries SC (1997) A leucine-rich repeat containing receptor-like kinase marks somatic plant cells competent to form embryos. Development 124: 2049–2062. Available: http://www.ncbi.nlm.nih.gov/pubmed/9169851.

21. Hecht V, Vielle-Calzada JP, Hartog MV, Schmidt ED, Boutilier K, et al. (2001) The Arabidopsis SOMATIC EMBRYOGENESIS RECEPTOR KINASE 1 gene is expressed in developing ovules and embryos and enhances embryogenic competence in culture. Plant Physiol 127: 803–816. Available: http://www.pubmedcentral.nih.gov/articlerender.fcgi?artid=129253&tool=pmcentrez&rendertype=abstract.

22. Li J, Wen J, Lease KA, Doke JT, Tax FE, et al. (2002) BAK1, an Arabidopsis LRR receptor-like protein kinase, interacts with BRI1 and modulates brassinosteroid signaling. Cell 110: 213–222. Available: http://www.ncbi.nlm.nih.gov/pubmed/12150929.

23. Nam KH, Li J (2002) BRI1/BAK1, a receptor kinase pair mediating brassinosteroid signaling. Cell 110: 203–212. Available: http://www.ncbi.nlm.nih.gov/pubmed/12150928.

24. Belkhadir Y, Yang L, Hetzel J, Dangl JL, Chory J (2014) The growth-defense pivot: crisis management in plants mediated by LRR-RK surface receptors. Trends Biochem Sci. Available: http://www.ncbi.nlm.nih.gov/pubmed/25089011.

25. Roux M, Schwessinger B, Albrecht C, Chinchilla D, Jones A, et al. (2011) The Arabidopsis leucine-rich repeat receptor-like kinases BAK1/SERK3 and BKK1/SERK4 are required for innate immunity to hemibiotrophic and biotrophic pathogens. Plant Cell 23: 2440–2455. Available: http://www.pubmedcentral.nih.gov/articlerender.fcgi?artid=3160018&tool=pmcentrez&rendertype=abstract.

26. Schwessinger B, Roux M, Kadota Y, Ntoukakis V, Sklenar J, et al. (2011) Phosphorylation-dependent differential regulation of plant growth, cell death, and innate immunity by the regulatory receptor-like kinase BAK1. PLoS Genet 7: e1002046. Available: http://www.pubmedcentral.nih.gov/articlerender.fcgi?artid=3085482&tool=pmcentrez&rendertype=abstract.

27. Sun Y, Li L, Macho AP, Han Z, Hu Z, et al. (2013) Structural basis for flg22-induced activation of the Arabidopsis FLS2-BAK1 immune complex. Science 342: 624–628. Available: http://www.ncbi.nlm.nih.gov/pubmed/24114786.

28. Sun Y, Han Z, Tang J, Hu Z, Chai C, et al. (2013) Structure reveals that BAK1 as a co-receptor recognizes the BRI1-bound brassinolide. Cell Res 23: 1326–1329. Available: http://www.pubmedcentral.nih.gov/articlerender.fcgi?artid=3817550&tool=pmcentrez&rendertype=abstract.

29. Santiago J, Henzler C, Hothorn M (2013) Molecular mechanism for plant steroid receptor activation by somatic embryogenesis co-receptor kinases. Science 341: 889–892. Available: http://www.ncbi.nlm.nih.gov/pubmed/23929946.

30. Felix G, Duran JD, Volko S, Boller T (1999) Plants have a sensitive perception system for the most conserved domain of bacterial flagellin. Plant J 18: 265–276. Available: http://www.ncbi.nlm.nih.gov/pubmed/10377992.

31. Robatzek S, Wirthmueller L (2012) Mapping FLS2 function to structure: LRRs, kinase and its working bits. Protoplasma 21. Available: http://www.ncbi.nlm.nih.gov/pubmed/23053766.

32. Sun W, Dunning FM, Pfund C, Weingarten R, Bent AF (2006) Within-species flagellin polymorphism in *Xanthomonas campestris pv campestris* and its impact on elicitation of Arabidopsis FLAGELLIN SENSING2-dependent defenses. Plant Cell 18: 764–779. Available: http://www.pubmedcentral.nih.gov/articlerender.fcgi?artid=1383648&tool=pmcentrez&rendertype=abstract.

33. Mueller K, Bittel P, Chinchilla D, Jehle AK, Albert M, et al. (2012) Chimeric FLS2 Receptors Reveal the Basis for Differential Flagellin Perception in Arabidopsis and Tomato. Plant Cell 24: 2213–2224. Available: http://www.ncbi.nlm.nih.gov/pubmed/22634763.

34. Tai TH, Dahlbeck D, Clark ET, Gajiwala P, Pasion R, et al. (1999) Expression of the Bs2 pepper gene confers resistance to bacterial spot disease in tomato. Proc Natl Acad Sci U S A 96: 14153–14158. Available: http://www.pubmedcentral.nih.gov/articlerender.fcgi?artid=24206&tool=pmcentrez&rendertype=abstract.

35. Helft L, Reddy V, Chen X, Koller T, Federici L, et al. (2011) LRR conservation mapping to predict functional sites within protein leucine-rich repeat domains. PLoS One 6: e21614. Available: http://www.pubmedcentral.nih.gov/articlerender.fcgi?artid=3138743&tool=pmcentrez&rendertype=abstract.

36. Sun W, Cao Y, Jansen Labby K, Bittel P, Boller T, et al. (2012) Probing the Arabidopsis flagellin receptor: FLS2-FLS2 association and the contributions of specific domains to signaling function. Plant Cell 24: 1096–1113. Available: http://www.pubmedcentral.nih.gov/articlerender.fcgi?artid=3336135&tool=pmcentrez&rendertype=abstract.

37. Nakagawa T, Kurose T, Hino T, Tanaka K, Kawamukai M, et al. (2007) Development of series of gateway binary vectors, pGWBs, for realizing efficient construction of fusion genes for plant transformation. J Biosci Bioeng 104: 34–41. Available: http://www.ncbi.nlm.nih.gov/pubmed/17697981.

38. Schulze B, Mentzel T, Jehle AK, Mueller K, Beeler S, et al. (2010) Rapid heteromerization and phosphorylation of ligand-activated plant transmembrane receptors and their associated kinase BAK1. J Biol Chem 285: 9444–9451. Available: http://www.ncbi.nlm.nih.gov/entrez/query.fcgi?cmd=Retrieve&db=PubMed&dopt=Citation&list_uids=20103591.

39. Cao Y, Aceti DJ, Sabat G, Song J, Makino S-I, et al. (2013) Mutations in FLS2 Ser-938 dissect signaling activation in FLS2-mediated Arabidopsis immunity. PLoS Pathog 9: e1003313. Available: http://www.pubmedcentral.nih.gov/articlerender.fcgi?artid=3630090&tool=pmcentrez&rendertype=abstract.

40. Albert M, Jehle AK, Fürst U, Chinchilla D, Boller T, et al. (2013) A two-hybrid-receptor assay demonstrates heteromer formation as switch-on for plant immune receptors. Plant Physiol 163: 1504–1509. Available: http://www.ncbi.nlm.nih.gov/pubmed/24130196.

41. Maley F, Trimble RB, Tarentino AL, Plummer TH (1989) Characterization of glycoproteins and their associated oligosaccharides through the use of endoglycosidases. Anal Biochem 180: 195–204. Available: http://www.ncbi.nlm.nih.gov/pubmed/2510544.

42. Häweker H, Rips S, Koiwa H, Salomon S, Saijo Y, et al. (2010) Pattern recognition receptors require N-glycosylation to mediate plant immunity. J Biol Chem 285: 4629–4636. Available: http://www.pubmedcentral.nih.gov/articlerender.fcgi?artid=2836068&tool=pmcentrez&rendertype=abstract.

43. Farid A, Malinovsky FG, Veit C, Schoberer J, Zipfel C, et al. (2013) Specialized roles of the conserved subunit OST3/6 of the oligosaccharyltransferase complex in innate immunity and tolerance to abiotic stresses. Plant Physiol 162: 24–38. Available: http://www.pubmedcentral.nih.gov/articlerender.fcgi?artid=3641206&tool=pmcentrez&rendertype=abstract.

44. Li J, Zhao-Hui C, Batoux M, Nekrasov V, Roux M, et al. (2009) Specific ER quality control components required for biogenesis of the plant innate immune receptor EFR. Proc Natl Acad Sci U S A 106: 15973–15978. Available: http://www.ncbi.nlm.nih.gov/pubmed/19717464.

45. Liu Y, Li J (2013) A conserved basic residue cluster is essential for the protein quality control function of the Arabidopsis calreticulin 3. Plant Signal Behav 8: e23864. Available: http://www.ncbi.nlm.nih.gov/pubmed/23425854.

46. Su W, Liu Y, Xia Y, Hong Z, Li J (2012) The Arabidopsis homolog of the mammalian OS-9 protein plays a key role in the endoplasmic reticulum-associated degradation of misfolded receptor-like kinases. Mol Plant 5: 929–940. Available: http://www.pubmedcentral.nih.gov/articlerender.fcgi?artid=3399701&tool=pmcentrez&rendertype=abstract.

47. Von Numers N, Survila M, Aalto M, Batoux M, Heino P, et al. (2010) Requirement of a homolog of glucosidase II beta-subunit for EFR-mediated defense signaling in Arabidopsis thaliana. Mol Plant 3: 740–750. Available: http://www.ncbi.nlm.nih.gov/pubmed/20457640.

48. Kornfeld R, Kornfeld S (1985) Assembly of asparagine-linked oligosaccharides. Annu Rev Biochem 54: 631–664. Available: http://www.ncbi.nlm.nih.gov/pubmed/3896128.

49. Dunning FM, Sun W, Jansen KL, Helft L, Bent AF (2007) Identification and mutational analysis of Arabidopsis FLS2 leucine-rich repeat domain residues that contribute to flagellin perception. Plant Cell 19: 3297–3313. Available: http://www.pubmedcentral.nih.gov/articlerender.fcgi?artid=2174712&tool=pmcentrez&rendertype=abstract.

50. Yonekura K, Maki-Yonekura S, Namba K (2003) Complete atomic model of the bacterial flagellar filament by electron cryomicroscopy. Nature 424: 643–650. Available: http://www.ncbi.nlm.nih.gov/pubmed/12904785.

51. Maki-Yonekura S, Yonekura K, Namba K (2010) Conformational change of flagellin for polymorphic supercoiling of the flagellar filament. Nat Struct Mol Biol 17: 417–422. Available: http://www.ncbi.nlm.nih.gov/pubmed/20228803.

52. Lacombe S, Rougon-Cardoso A, Sherwood E, Peeters N, Dahlbeck D, et al. (2010) Interfamily transfer of a plant pattern-recognition receptor confers broad-spectrum bacterial resistance. Nat Biotechnol 28: 365–369. Available: http://www.ncbi.nlm.nih.gov/pubmed/20231819.

9

Genome Features of the Endophytic Actinobacterium *Micromonospora lupini* Strain Lupac 08: On the Process of Adaptation to an Endophytic Life Style?

Martha E. Trujillo[1]*, **Rodrigo Bacigalupe**[1], **Petar Pujic**[2], **Yasuhiro Igarashi**[3], **Patricia Benito**[1], **Raúl Riesco**[1], **Claudine Médigue**[4], **Philippe Normand**[2]

1 Departamento de Microbiología y Genética, Edificio Departamental, Campus Miguel de Unamuno, Universidad de Salamanca, Salamanca, Spain, **2** Université Lyon 1, Université de Lyon, CNRS-UMR5557 Ecologie Microbienne, Villeurbanne, France, **3** Biotechnology Research Center, Toyama Prefectural University, Kurokawa, Imizu, Toyama, Japan, **4** Genoscope, CNRS-UMR 8030, Atelier de Génomique Comparative, Evry, France

Abstract

Endophytic microorganisms live inside plants for at least part of their life cycle. According to their life strategies, bacterial endophytes can be classified as "obligate" or "facultative". Reports that members of the genus *Micromonospora*, Gram-positive Actinobacteria, are normal occupants of nitrogen-fixing nodules has opened up a question as to what is the ecological role of these bacteria in interactions with nitrogen-fixing plants and whether it is in a process of adaptation from a terrestrial to a facultative endophytic life. The aim of this work was to analyse the genome sequence of *Micromonospora lupini* Lupac 08 isolated from a nitrogen fixing nodule of the legume *Lupinus angustifolius* and to identify genomic traits that provide information on this new plant-microbe interaction. The genome of *M. lupini* contains a diverse array of genes that may help its survival in soil or in plant tissues, while the high number of putative plant degrading enzyme genes identified is quite surprising since this bacterium is not considered a plant-pathogen. Functionality of several of these genes was demonstrated *in vitro*, showing that Lupac 08 degraded carboxymethylcellulose, starch and xylan. In addition, the production of chitinases detected *in vitro*, indicates that strain Lupac 08 may also confer protection to the plant. *Micromonospora* species appears as new candidates in plant-microbe interactions with an important potential in agriculture and biotechnology. The current data strongly suggests that a beneficial effect is produced on the host-plant.

Editor: Holger Brüggemann, Aarhus University, Denmark

Funding: MET received financial support from the Spanish Ministerio de Economía y Competitividad under project CGL2009-07287. PN acknowledges financial support from the ANR (Sesam). The funders had no role in study design, data collection and analysis, decisión to publish, or preparation of the manuscript.

Competing Interests: The authors have declared that no competing interests exist.

* Email: mett@usal.es

Background

For a long time, it was considered that a healthy plant was a plant without microbes within its tissues. However, this view has started to change with new approaches to allow strains to grow for a longer time upon isolation as well as the use of NGS, which has permitted the identification of several strains present in the tissues of healthy plants, in particular several actinobacteria [1,2].

Endophytic microorganisms live inside plants for at least part of their life cycle. According to their life strategies, bacterial endophytes can be classified as "obligate" or "facultative". Obligate endophytes are strictly dependent on the host plant for their growth and survival while facultative endophytes have a stage in their life cycle during which they exist outside host plants [3]. These endophytes originate from soil, initially infecting the host plant by colonizing, for instance, the cracks formed at points of emergence of lateral roots from where they quickly spread to the intercellular spaces in the root [4]. Thus, a series of environmental and genetic factors is presumed to have a role in enabling a specific bacterium to become endophytic [5]. Conversely, Marchetti and

co-workers [6] recently showed how a pathogen can evolve in a few generations to become a symbiotic endophyte by losing specific transporters and regulators linked to pathogenesis.

Micromonospora is a genus of Gram-positive Actinobacteria that was first isolated from soil [7]. This bacterium has received a lot of attention during natural product screening programs, given its ability to produce a very rich array of secondary metabolites [8,9,10]. The distribution of members of *Micromonospora* is wide-ranging since these bacteria have been isolated from different geographical zones. In addition, its habitats are also diverse and include: soil, freshwater and marine sediments, mangrove soils, rocks, and nitrogen fixing nodules of both leguminous and actinorhizal plants [11,12,13]. The recent report [13] that *Micromonospora* inhabits nitrogen-fixing nodules in a systematic way, has opened up a question as to what is the potential ecological role of this bacterium in the plant and whether this bacterium is in a process of adaptation from a terrestrial to a facultative endophytic life style.

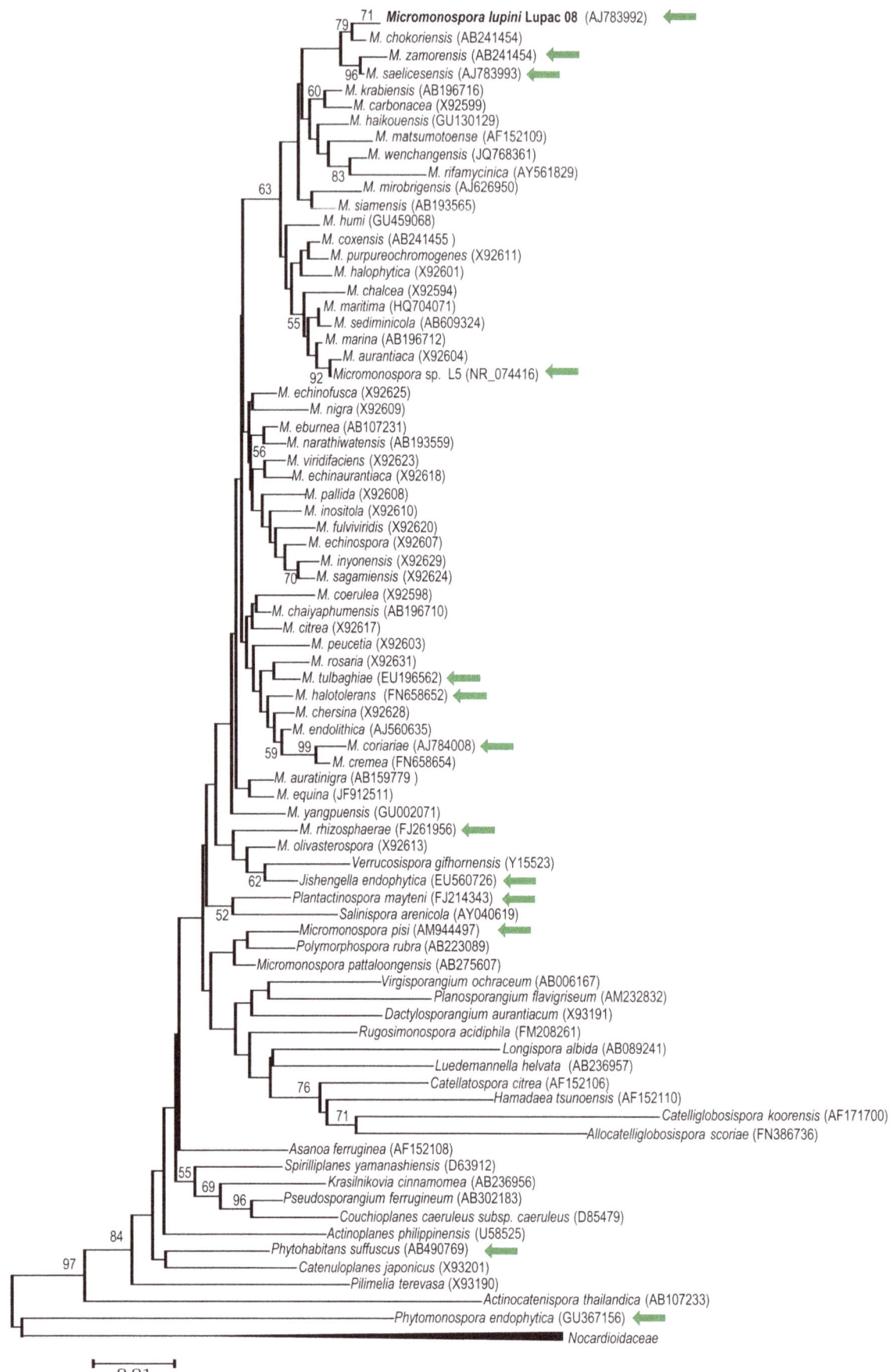

Figure 1. Neighbour-joining tree based on 16S rRNA gene sequences showing the relationship of *Micromonospora* **species and other members of the family** *Micromonosporaceae.* Strains isolated from plant related sources are indicated by a green arrow.

Taxonomically, *Micromonospora* belongs to the family *Micromonosporaceae* which currently contains 27 genera and includes aerobic, non-acid fast and mesophilic microorganisms. Many strains produce mycelial carotenoid pigments giving the colonies an orange to red appearance, but blue-green, brown or purple pigmented strains have also been isolated. The family *Micromonosporaceae* also harbors the genus *Salinispora*, which is widely distributed in tropical and sub-tropical marine sediments. This taxon was described as the first marine actinomycete given its inability to grow in low salinity medium. Indeed, genomic information obtained from the genomes of *Salinispora tropica* and *Salinispora arenicola* provide evidence of marine adaptation of *Salinispora* species [14]. Thus, it appears that *Salinispora* evolved from a terrestrial environment to a marine habitat. In the case of some *Micromonospora* lineages, the question is whether this bacterium has followed a comparable evolution process, changing from a terrestrial to an endophytic lifestyle.

Further examples of closely related actinobacteria with different lifestyles reflected in their genomes include, among others, the genera *Frankia*, *Mycobacterium* and *Streptomyces*. In the case of *Frankia*, comparative genomic analysis of three representative strains, differing by less than 2% in their 16S rRNA genes revealed significant differences in their genome sizes (5.4–9.0 Mb) suggesting that these differences (e.g. gene deletion, acquisition and duplication, etc.) reflect their rapid adaptation to contrasted host plants and to their environments [15]. Similarly, several mycobacterial genomes were analyzed both at the nucleotide and protein levels. One of the most striking features was lipid metabolism genes with marked expansions of the number of genes related to saturated fatty acid metabolism in the pathogenic mycobacteria compared to the soil-dwelling strains [16].

In an effort to identify the genomic traits which make possible adaptation from a soil dwelling way of life to an endophytic habitat, the aim of this work was to present the genome sequence analysis of a representative strain, *Micromonospora lupini* Lupac 08, isolated from a nitrogen fixing nodule of the legume *Lupinus angustifolius*. This strain is part of a collection of more than 2000 strains isolated from nitrogen fixing root nodules of diverse legume [17,18] and actinorhizal species [13]. Strain Lupac 08 was selected as it showed good plant growth promotion, was used previously for in situ localization studies *in planta* [11] and produced several new secondary metabolites [9,10]. The results presented here show that the genome of *M. lupini* Lupac 08 contains a diverse array of genes that may help its survival in soils or in plant tissues, while the high number of putative plant degrading enzyme genes identified in its genome is quite surprising since this bacterium is not considered a plant-pathogen and may instead reflect their ability to bind to plant structural compounds.

Results

Phylogenetic position of *M. lupini* Lupac 08 and general genome features

The phylogenetic position based on 16S rRNA gene sequence analysis of strain Lupac 08 with respect to currently described *Micromonospora* species and other members of the family *Micromonosporaceae* is presented in Figure 1. Those strains associated with plant/rhizosphere sources are highlighted. Strain Lupac 08 was clearly positioned within the genus *Micromonospora* and forms a subgroup together with the species *Micromonospora*

saelicesensis, *Micromonospora zamorensis* and *Micromonospora chokoriensis*. These strains were isolated from a nitrogen fixing nodule, the rhizosphere of a *Pisum sativum* plant and a sandy soil, respectively. Nevertheless, a clear picture based on the habitat cannot emerge from this analysis.

M. lupini Lupac 08 was shown to have a circular chromosome of 7,327,024 bp with a GC content of 71.96% and no plasmid. A total of 7158 genomic objects were identified: 7,054 protein-coding, 10 rRNAs, 77 tRNAs, and 12 miscRNAs genes. The average gene length was 964 bp with an average intergenic distance of 126 bp. After manual validation of the automatic annotation, 61.5% (4338 CDSs) of the genes were assigned a biological function while 38.5% were registered as open reading frames (ORFs) with an unknown function. Based on the G+C skew analysis and position of *dnaA*, the probable origin of replication (*oriC*), was mapped close to the ribosomal protein *rpmH*. A circular representation of the *M. lupini* chromosome is provided in Figure 2 indicating some of the features described above.

The genomic characteristics of strain Lupac 08 and three additional *Micromonospora* genomes deposited in the public databases including *Micromonospora* sp. strain L5 isolated from root nodules of *Casuarina equisetifolia* [19]; *M. aurantiaca* ATCC 27029[T] and *Micromonospora* sp. ATCC 39149 isolated from soil (Table 1) were compared. An important difference between the four strains was the number of tRNAs identified. *M. lupini* 08 contained by far the highest number with 77 tRNAs while the other strains had between 51 and 53. At present, *M. lupini* Lupac 08 contains one of the largest numbers of tRNAs reported among the actinobacteria sequenced. The number of rRNA and tRNA genes in a genome can be seen as an indication of positive selection. A high number of rRNA genes increases ribosome synthesis, which in turn increases the protein synthesis rate [20] and growth rate [21].

Comparative genome analysis

COG distribution. Nearly 70% of the CDS were classified into clusters of orthologous groups (COGs, Table S1). Thus, 4873 out of 7054 CDS were assigned to 24 different categories, including those for amino acid transport and metabolism (E, 12.7%), transcription (K, 10.8%), carbohydrate transport and metabolism (G, 9.7%), inorganic ion transport and metabolism (P, 8.7%), energy production and conversion (C, 5.5%), and signal transduction mechanisms (5.5%).

The COG distribution of *M. lupini* was similar to that observed in other bacteria in the family *Micromonosporaceae*, however various differences were detected such as the abundance of genes related to carbohydrate transport and metabolism. Among the *Micromonospora* genomes currently available, *M. lupini* Lupac 08 contained the highest percentage of genes (9.7%, 685) related to this category, followed by *Micromonospora* sp. L5 (8.9%, 598) and *M. aurantiaca* ATCC 27029 (8.5%, 576). The gene contents (in the same COG category) of other bacterial genomes classified in the family *Micromonosporaceae* were lower as in the case of *S. tropica* CNS-205 (7.4%, 391) and *S. arenicola* CNH-643 (6.4%, 374) two obligate marine actinomycetes. On the other hand, the overall COG profiles of *Verrucosispora maris* AB-18-032[T] (genome size 6.7 Mb) and *M. lupini* Lupac 08 were very similar and no clear differences were found. Although *V. maris* was isolated from a sea sediment, it does not require sea salts for growth and it is not considered an obligate marine microorganism

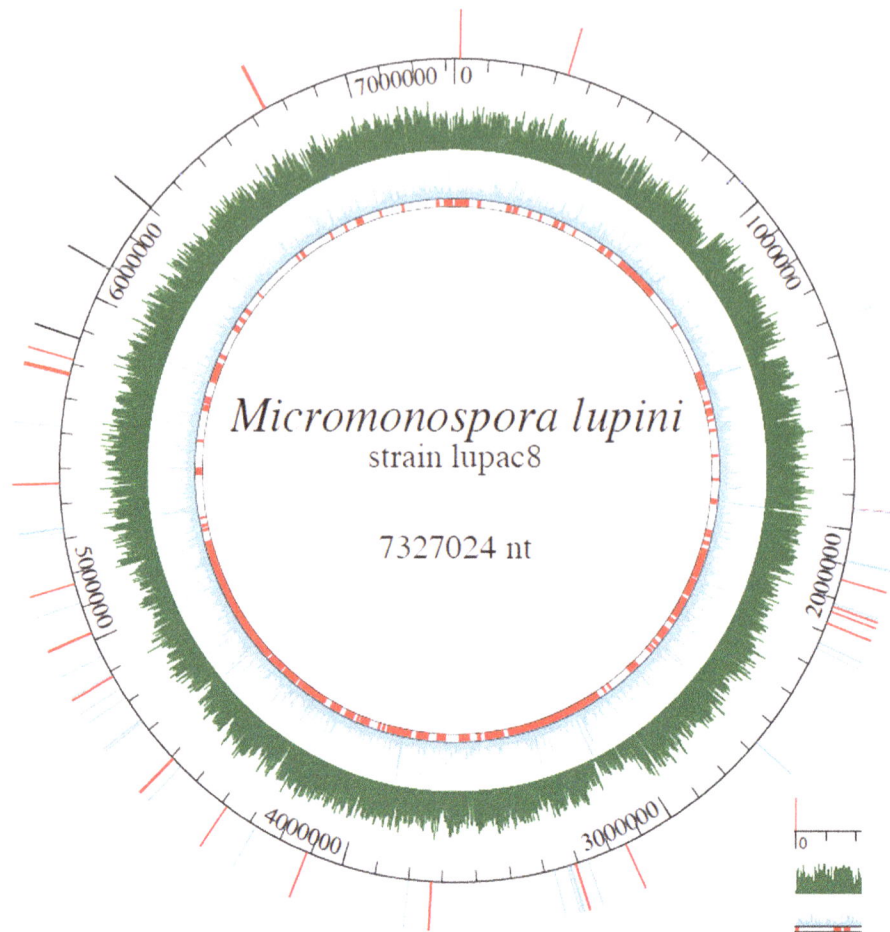

Figure 2. Circular representation of *Micromonospora lupini* **Lupac 08.** Circles displayed from the outside in: 1. Cellulose-binding genes in black, chitin-binding genes in red, lectin genes in lavender blue; 2. Genome coordinates; 3. MW; 4. GC% (linear range between 65 and 80%); 5. Regions of genome plasticity according to the RGP_Finder method (Mage platform) based on synteny breaks between the query genome (Lupac 08) and close genomes (*Micromonospora aurantiaca* ATCC 27029[T], *Micromonospora* sp. L5 and *Verrucosispora maris* AB-18-032[T]) correlated with HGT features (tRNA hotspot, DNA repeats, mobility genes), and compositional bias and GC deviation computation. C1 to C15 indicate the position of the 15 clusters of genes coding for secondary metabolites of Table 4.

unlike *S. tropica* and *S. arenicola*. Thus, its metabolism suggests a terrestrial life style. Micromonosporae are well known for their ability to degrade complex polysaccharides such as cellulose, chitin and lignin [22,23]. In particular, cellulose is frequently utilized as a carbon source [24,25]. Therefore the abundance of these genes in the genome of strain Lupac 08, at first glance may not seem surprising, however, the value of 9.7% is comparable to that of highly active cellulolytic microorganisms such as *Cellulomonas*

Table 1. Comparative genomic characteristics of *M. lupini* Lupac 08 and three *Micromonospora* genomes publicly available.

Feature	*M. lupini* **Lupac 08**	*M. aurantiaca* **ATCC 27029**[T]	*Micromonospora* **sp. L5**	*Micromonospora* **sp. ATCC 39149**
Size (Mb)	7.3	7.0	6.9	6.8
GC%	72	73	73	72
rRNA Operon	10	9	9	6
tRNA	77	52	53	51
CDS number	7054	6676	6617	5633
Average gene size (kb)	946	964	969	975
Protein-coding density (%)	90.1	90.4	90.4	89.9
Genes in COGs (%)	70.2%	68.3%	69%	nd

nd, not determined.

flavigena 134^T (9.5%) and *Thermobifida fusca* XY (7.9%), which are abundant in cellulose enriched environments such as soil, or plant tissues.

Synteny. The genome sequence of strain Lupac 08 was aligned with those of *Micromonospora* sp. L5, *M. aurantiaca* ATCC 27029^T and *Micromonospora* sp. ATCC 39149^T (Fig. 3). Although the four genomes share a significant amount of genetic characteristics, they have undergone various inversions and translocations and *M. lupini* Lupac 08 contains the highest number of non-conserved regions. In addition, this alignment shows a high homology between strains *Micromonospora* sp. L5 and *M. aurantiaca* ATCC 27029^T confirming their close phylogenetic relationship as suggested by 16S rRNA gene phylogeny (Fig. 1); nevertheless, strain L5 shows a large inversion event. Thus, although the four *Micromonospora* genomes share many common features, it is also evident that *M. lupini* contains unique genomic regions as compared to *M. aurantiaca* ATCC 27029^T or *Micromonospora* sp. L5.

Diversity of *Micromonosporae*: core vs. flexible gene pool

Using the *Micromonospora* genomes of strains M. *lupini* Lupac 08, *M. aurantiaca* ATCC 27029^T and that of *Micromonospora* sp. L5 available in the NCBI [19], the core genome was calculated using the SiLix software [26]. The core genome was composed of 2294 CDSs, which correspond to approximately 32% of the predicted proteome. In addition, *M. lupini* Lupac 08 contained the highest number of strain specific CDSs, 4702 (66.6%), which is a very high value when compared to *Micromonospora* sp. L5 and *M. aurantiaca* ATCC 27029^T (13–14%, Figure 4), which both share a high gene similarity (85–86%).

Horizontal gene transfer is universally recognized as an efficient mechanism for microorganisms to acquire functions that enable them to adapt to environments with different selective pressures. Therefore insertion elements, transposases, integrated phages, and plasmids can be related to the plasticity of a genome. Strain Lupac 08 contained 49 CDSs (0.7%, of total CDSs) related to gene exchange including eight integrases and eleven recombinases. Except for seven CDSs, most of these genes were grouped into 20 clusters. Interestingly, eight of these mobile element clusters were found near genes related to carbohydrate transport and metabolism.

Metabolic Features. A metabolic pathway reconstruction was performed between the genome of strain Lupac 08 and 20 additional strains among which plant pathogens, symbiotic and saprophytic bacteria were included. The distribution and grouping of the microorganisms analyzed using 798 metabolic routes are presented in Figure 5. A good correlation was obtained between the microorganisms, their life style and phylogeny. Two main groups were obtained, the proteobacteria and actinobacteria. Within the actinobacteria, three clusters were clearly identified: the first one contained strains that belonged to the family *Micromonosporaceae*, the second cluster corresponded to various streptomycetes and the third cluster included the three *Frankia* genomes. Surprisingly, *Micromonospora lupini* Lupac 08 showed a closer metabolic relationship with the three *Frankia* strains (ACN14a, CcI3 and EAN1pec) than with the other two *Micromonospora* genomes.

Plant/Soil-associated life style

Transport systems. Organisms living in endophytic associations need to share resources with their host. Membrane transport systems play essential roles in cellular metabolism and activities. Current data suggest a correlation of transporter profiles to both evolutionary history and the overall physiology and lifestyles of organisms [27].

A total of 631 CDSs were located in the genome of *M. lupini* coding for a large diversity of transporters, representing approximately 8.9% of the genome. The majority of CDSs were related

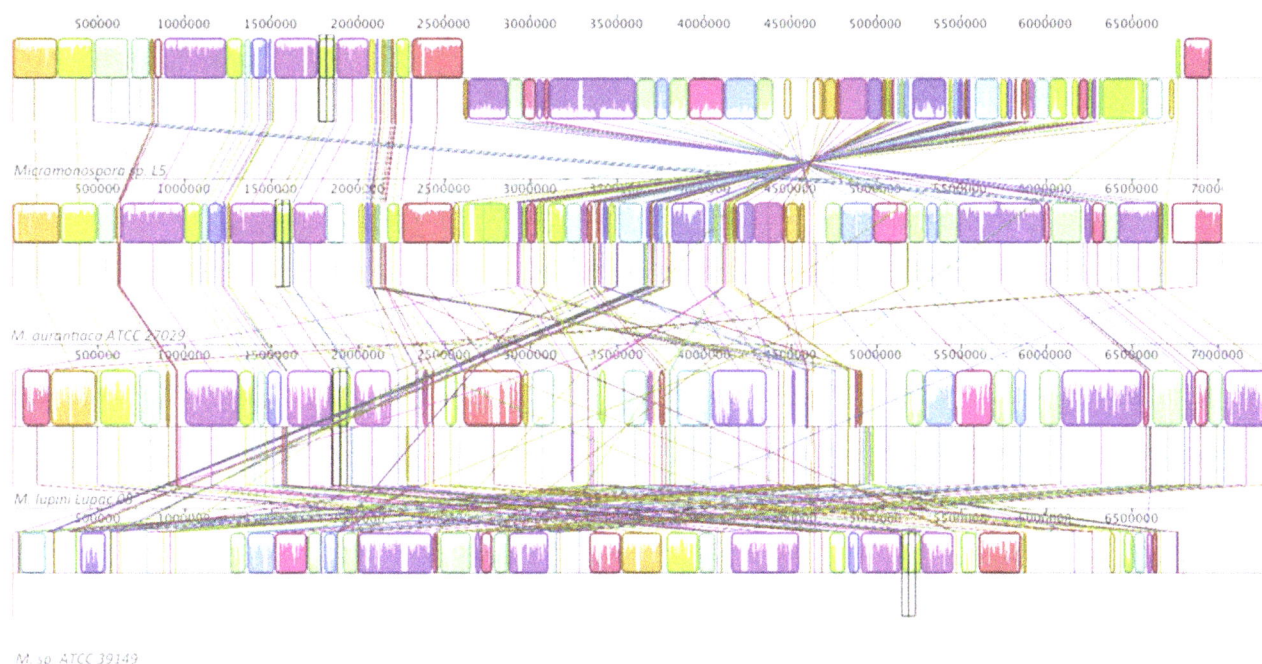

Figure 3. MAUVE alignment of the genome sequences of *Micromonospora lupini* **Lupac 08,** *Micromonospora* **sp. L5,** *Micromonospora aurantiaca* **ATCC 27029^T and** *Micromonospora* **sp. ATCC 39149.** When boxes have the same colour, this indicates syntenic regions. Boxes below the horizontal line indicate inverted regions. Rearrangements are shown by coloured lines. Scale is in nucleotides.

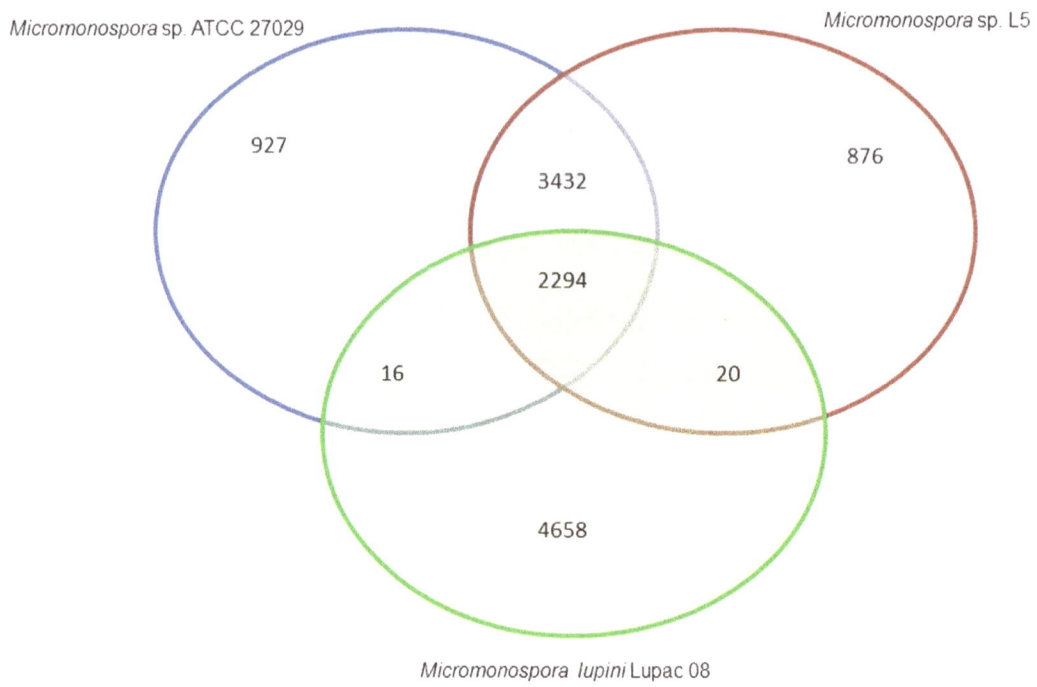

Figure 4. Venn diagram showing the number of clusters of orthologous genes, shared and unique, between *M. lupini* Lupac 08, *Micromonospora* sp. L5 and *M. aurantiaca* ATCC 27029[T].

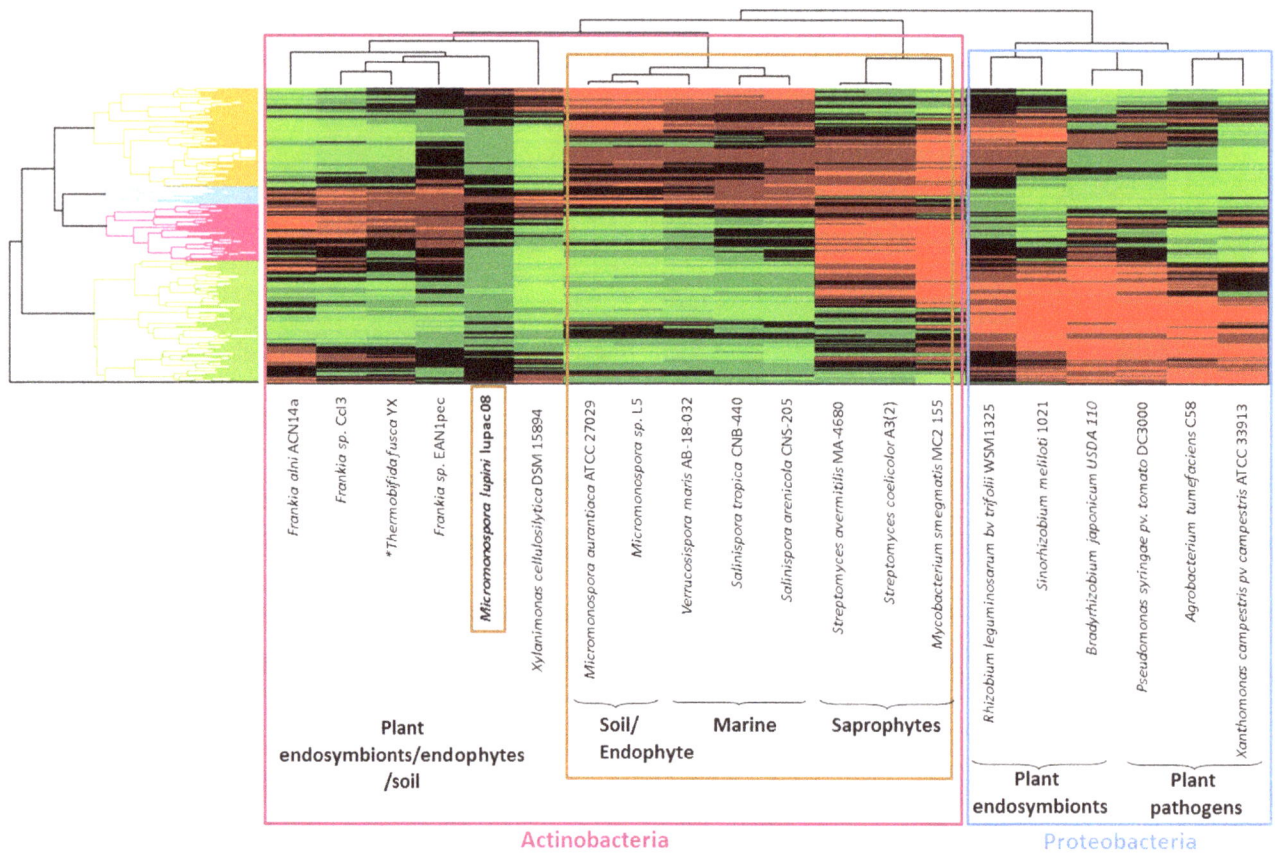

Figure 5. Bicluster plot of the metabolic profiles of *M lupini* Lupac 08 and 20 other bacterial genomes.

to ATP-binding (dependent) transporters of which 362 corresponded to ABC transporters; the next most abundant (215 CDSs) coded for secondary transporters, with 105 classified in the Major Facilitator Superfamily –MFS-); 17 transporters belonged to ion channels and 20 were unclassified. The number of transporters determined in *M. aurantiaca* ATCC 27029^T and *Micromonospora* sp. L5 were lower with 575 and 587, respectively (Table 2).

The number of transporters identified in the genome of *M. lupini* Lupac 08 is correlated with those in other bacteria with a plant/soil associated life style, which requires an efficient nutrient uptake system to obtain nutrients produced by the host plant, in addition to those found in the rhizosphere and the soil (Table 2) [27,28]. However, the number of transporters identified in strain Lupac 08 was lower than those present in other bacteria such as *Bradyrhizobium japonicum* USDA6^T (1138, 11.8%), *Mesorhizobium loti* MAF303099 (968, 14.2%), *Sinorhizobium meliloti* Sm1021 (1024, 16.4%) and *Rhizobium leguminosarium* bv. *trifolii* WSM 3125 (1087, 15.5%), which form a very close interaction with legumes. Nevertheless, the overall distribution (types) and the percentages of these values were similar. An additional difference was the absence of phosphotransferase system transporters (PTS) in *M. lupini* as compared to the strains mentioned above and other soil/plant bacteria included in Table 2. On the other hand, the overall profile of *M. lupini* Lupac 08 was very similar to those of *Frankia* sp. ACN14a and *Frankia* sp. CcI3 which also lack a PTS system.

Secretion systems. Secreted proteins play a number of essential roles in bacteria, including the colonization of niches and host–pathogen interactions. In Gram-positive bacteria, the majority of proteins are exported out of the cytosol by the conserved Sec translocase system or, alternately, by the twin-arginine translocation system. In addition, a unique protein export system, the type VII or ESX secretion system also exists in some Gram positive bacteria [29].

The genome of *M. lupini* Lupac 08 encodes for 537 (7.6%) secreted proteins including several protein secretion systems (Table 3). All genes related to the Sec-dependent pathway were located and included the SecY and SecE proteins which form the membrane channel and interact with the cytoplasmic membrane protein SecG; the auxiliary proteins SecD, YajC and the ATPase SecA. In addition, the heterodimer Ffh-FtsY (MiLup08_41486 and MiLup08_41460) was also present. As in other Gram-positive bacteria, *M. lupini* Lupac 08 lacks homologs of SecB, the chaperone that targets proteins to the Sec translocon for passage through the cytoplasmic membrane [30].

Genes related to the Sec-independent twin-arginine translocation pathway (TAT), which exports prefolded proteins across the cytoplasmic membrane using the transmembrane proton gradient as the main driving force for translocation were also located in strain Lupac 08 (Table 3). Homologs of TatA and TatC were identified, however no homolog for TatB was found. Similar to other actinobacteria (e.g. *Frankia* sp. ACN14a) the *tatA* gene was found next to *tatC*. Only an ORF encoding TatC was located in the genomes of *Micromonospora* sp. L5 and *M. aurantiaca* ATCC 27029^T while no copies of *tatA* or *tatB* were found.

A set of fifteen genes identified as part of the type VII secretion system were located in *M. lupini* Lupac 08 (Table 3). These are arranged in three different clusters and included the essential proteins for secretion EccC, EccD, EsxA and EsxB [31]. The first cluster contains eight genes: *eccC*, *esxA*, *esxB*, *eccD*, *eccB*, *eccE* and two copies of *mycP*, a subtilisin-like serine protease which also appears essential but the function of which is not yet known [32]. The second cluster includes a copy of *esxA* (MiLup08_40381), *esxB* (MiLup08_40380) and *mycP*, annotated as S8 S53 subtilin

kexin sedolisin (MiLup08_40382). Finally a third cluster contains the genes *eccC* (MiLup08_46744), *eccD* (MiLup08_46743) and *mycP* (MiLup08_46745).

Gram-negative bacteria use the type II secretion system to transport a large number of secreted proteins from the periplasmic space into the extracellular environment. Many of the secreted proteins are major virulence factors in plants and animals [33]. Type II secretion systems have been found in all completely sequenced plant pathogenic bacterial genomes, except in *Agrobacterium tumefaciens*. In addition, other bacteria have been shown to use secretion systems for the delivery of toxins, proteases, cellulases and lipases [34–37]. Genes coding for this system have also been reported for the three symbiotic strains *Frankia* [38].

Fifteen genes in *M. lupini* were annotated as components of the Type II secretion system, grouped into clusters of three to five genes (Table 3). Nine of these genes were annotated as Type II secretion system proteins including protein E and protein F; four were recorded as TadE family proteins and Milup08_40403 was annotated as an uncharacterized protein closest to one found in the *Frankia* symbiont of *Datisca glomerata*.

The secretion systems III and IV which are commonly related to plant-associated bacteria transport a wide variety of effector proteins into the extracellular medium or into the cytoplasm of eukaryotic host cells thus affecting the interaction [39]. In addition, a functional type IV system has been described in the plant symbiont *M. loti* strain R7A [40]. A gene annotated as *virB4* and related to secretion system IV was located in Lupac 08 (MiLup08_42651), this ORF is surrounded by proteins with unknown function related to those present in the genomes of *Micromonospora* sp. L5 and *M. aurantiaca* ATCC 27029^T.

Survival against plant defenses. Reactive oxygen species (ROS) play a major role in plant defense against pathogens. In response to attempted invasion, plants mount a broad range of defense responses, including the synthesis of ROS. *M. lupini* needs to survive under an oxidative environment in the rhizosphere before it can colonize plant roots and its genome revealed several genes encoding proteins to neutralize oxidative stress. The following genes were identified: three *sod* genes (MiLup08_45788, MiLup08_46012 and MiLup08_46604) that code for superoxide dismutases; a catalase HPII *katE* (MiLup08_44247); a catalase-peroxidase (*katG*, MiLup08_44435) and a catalase hydroperoxidase (*katA*, MiLup08_45857); four hydroperoxide reductases (MiLup08_40110, MiLup08_40293, MiLup08_41393, MiLup08_45407); a chloroperoxidase (MiLup08_44157) and a thiol peroxidase (MiLup08_43629).

In addition, a putative organic hydroperoxide resistance protein (Ohr, MiLup08_45098); a 4-hydroxyphenylpyruvate dioxygenase (Hpd, MiLup08_46664) and a homogentisate 1,2-dioxygenase (MiLup08_46677) were identified. Other enzymes include a glutathione peroxidase (MiLup08_45173); two glutathione transferases (MiLup08_46358 and MiLup08_41529) and four glutathione-S-transferases (*fdh*, MiLup08_42270, MiLup08_42834, MiLup08_44416 and 45648). Experimental data indicated that *M. lupini* indeed yields a catalase positive reaction [17] confirming the functionality of some of these genes. Therefore, to successfully reach the internal plant tissues, these genes may defend the bacterium against a ROS release by the plant.

Regulation as a means of adaptation

Lifestyle can be viewed as the set of biotopes an organism can thrive into and the relationships that it establishes with other species and its abiotic components. It is one of the driving forces that contribute to the overall characteristics of bacterial genomes [41].

Table 2. Transporters identified in the genome of *M. lupini* Lupac 08 and comparison with other bacteria with a plant/soil associated life styles.

	M. lupini Lupac 08	*M. aurantiaca* ATCC 27029T	*Micromonospora* sp. L5	*S. arenicola* CNS 205	*S. tropica* CNB 440	*F. alni* ACN14	*Frankia* sp. Ccl3	*F. symbiont Dastica glomerata*	*B. japonicum* USDA6T	*R. leguminosarum* Bv trifolii WSM 3125	*Enterobacter* sp. 368	*S. coelicolor* A32	*S. scabiei* 8722	*Pseudomonas syringae* pv *phaseolicola* 1448A
Genome size (Mb)	7.3	7.0	6.9	5.7	5.2	7.5	5.4	5.3	9.6	7.4	4.6	9.1	10	5.9
Total transport proteins	631	575	587	405	413	433	253	300	1138	1087	662	798	775	670
Transporters (%)	8.9	8.5	8.7	7.1	7.8	6.4	4.5	7.1	11.8	15.5	15.6	9.7	8.2	12.5
No. Transporters/Mb genome	0.08	0.08	0.08	0.07	0.08	0.06	0.05	0.06	0.12	0.15	0.14	0.09	0.08	0.11
ATP dependent (% of total)	379 (60.1%)	362 (63.0%)	366 (62.4%)	244 (60.2%)	247 (59.8%)	281 (64.9%)	146 (57.7%)	210 (70%)	684 (60.1%)	800 (73.6%)	317 (49.7%)	461 (57.8%)	480 (61.9%)	392 (58.5%)
ABC family*	362 (95.5%)	341 (94%)	342 (93.4%)	225 (92.2%)	228 (92.3%)	262 (93.2%)	127 (87%)	189 (90%)	645 (94.3%)	769 (96.1%)	287 (90.5%)	433 (93.9%)	455 (94.3%)	346 (88.3%)
Ion channels (% of total)	17 (2.7%)	14 (2.4%)	15 (2.6%)	9 (2.2%)	11 (2.7%)	12 (2.8%)	7 (2.8%)	7 (2.3%)	24 (2.1%)	26 (2.4%)	23 (3.5%)	19 (2.4%)	22 (2.8%)	32 (4.8%)
Phosphotransferase system (PTS)	–	–	–	4 (1%)	–	–	–	–	4 (0.4%)	6 (0.6%)	47 (7.1%)	10 (1.3%)	6 (0.8%)	5 (0.7%)
Secondary transporter	215 (34.1%)	185 (32.3%)	192 (32.7%)	139 (34.3%)	145 (35.1%)	130 (30%)	89 (35.2%)	75 (25%)	408 (35.9%)	234 (21.5%)	256 (38.7%)	286 (35.8%)	252 (32.5%)	226 (33.7%)
MFS family*	105 (48.8%)	68 (36.8%)	70 (36.5%)	65 (46.8%)	74 (51%)	64 (49.2%)	36 (40.4%)	33 (44%)	114 (27.9%)	69 (29.5%)	84 (32.8%)	120 (42%)	111 (44%)	72 (31.9%)
RND family	6 (3.2%)	6 (3.2%)	7 (3.6%)	7 (5%)	8 (5.5%)	9 (6.9%)	7 (7.9%)	7 (9.3%)	31 (7.6%)	13 (5.6%)	18 (7%)	15 (5.2%)	18 (7.1%)	16 (7.1%)
Unclassified	20 (3%)	13 (2.3%)	13 (2.2%)	8 (2%)	9 (2.2%)	10 (2.3%)	11 (4.3%)	8 (2.7%)	11 (1%)	16 (1.5%)	14 (2.1%)	21 (2.6%)	14 (1.8%)	11 (1.6%)

*Number and percentage in relation to the total number of ATP dependent and secondary transporters respectively.

Table 3. Secretion system genes present in the genome of *M. lupini* Lupac 08.

Secretion System	Gene (Milup08_X)	Product
Sec-dependent	*secY* (*prlA*)(46297)	Preprotein translocase, membrane component
	secE (46336)	Preprotein translocase subunit secE
	secG (44961)	Preprotein translocase SecG subunit
	secD (42464)	Protein-export membrane protein secD
	secF (42465)	Protein-export membrane protein secF
	yajC (42463)	Preprotein translocase, YajC subunit
	secA (41087)	Protein translocase subunit secA
	ffh (41468)	Signal recognition particle protein
	scRNA (misc_RNA-12)	SRP, Ribosome-nascent chain complex (RNC)
	yidC (30220)	Cytoplasmic insertase into membrane protein
	yidC-like (43138)	Membrane protein insertase, YidC/Oxa1 family
	yidC (45964)	Inner membrane protein translocase component YidC
	Milup_08_41485	Signal peptidase I
	Milup_08_41486	Signal peptidase I
	Milup_08_42560	Conserved protein of unknown fuction (probable signal peptidase I)
	lspA (45113)	Lipoprotein signal peptidase
	lgt (45071)	Prolipoprotein diacylglyceryltransferase
TAT-	*tatA* (43424)	Sec-independent protein translocase protein tatA/E homolog
	tatC (43425)	Sec-independent protein translocase protein tatC homolog
Type II- (T2SS)	Milup_08_40403	Similar to uncharacterized protein from *Frankia* symbiont of *Diastica glomerata*
	Milup_08_40405	Putative helicase/secretion neighbourhood TadE-like protein
	tadE (40223)	TadE Family protein
	tadE (40224)	TadE Family protein
	tadE (42690)	Similar to TadE family protein
	tadE (42691)	Similar to TadE family protein
	Milup_08_40226	Type II secretion system protein
	Milup_08_40227	Type II secretion system protein
	Milup_08_40228	Type II secretion system protein E
	Milup_08_40398	Type II secretion system protein E
	Milup_08_40399	Similar to Type II secretion system protein E
	Milup_08_40401	Similar to Type II secretion system protein
	Milup_08_42693	Type II secretion system protein F
	Milup_08_42694	Type II secretion system protein F
	Milup_08_42695	Type II secretion system protein
Type IV- (T4SS)	Milup_08_42651	VirB4 protein-like protein
Type VII/WXG100-	*eccB* (40554)	ESX-4 secretion system protein eccB4
	eccC (40438)	FtsK/SpoIIIE family protein
	eccC (40557)	ESX-4 secretion system protein/cell division protein ftsK/spoIIIE
	eccC (46744)	FtsK/SpoIIIE-like transmembrane protein
	eccD (40556)	ESX-4 secretion system protein eccD4/Putative secretion protein snm4
	eccD (46743)	FtsK/SpoIIIE family protein
	eccE (40555)	Putative uncharacterized protein
	esxA (40381)	Putative uncharacterized protein
	esxA (40559)	Putative uncharacterized protein
	esxB (40380)	Putative uncharacterized protein
	esxB (40558)	Putative uncharacterized protein
	mycP (40382)	Peptidase S8 and S53 subtilisin kexin sedolisin
	mycP (40560)	Peptidase S8 and S53 subtilisin kexin sedolisin
	mycP (40564)	Peptidase S8 and S53 subtilisin kexin sedolisin
	mycP (46745)	Peptidase S8 and S53 subtilisin kexin sedolisin

TAT, twin-arginine translocation; X, corresponds to the annotation gene numbers given in parenthesis.

The *M. lupini* genome shows a strong emphasis on regulation, with 643 proteins (~10%) predicted to have a regulatory function. This value is lower than that reported for the saprophytic strain *Streptomyces coelicolor* A3(2) with an exclusively terrestrial lifestyle (965 proteins; 12.3%) [42], but higher than the endosymbiotic strains *M. loti* MAFF303099 (542 proteins, 7.7%) [43]; *Frankia alni* ACN14a (515 proteins, 7.6%); *Frankia* sp. EAN1pec (555, 6.1%) and *Frankia* sp. CcI3 (244 proteins, 4.3%).

The genome codes for various regulator families such as TetR, AraC, LacI, ArsR, MerR, AsnC, MarR, DeoR, GntR and Crp. In addition, thirty-three ECF (extra-cytoplasmic function) sigma factors were located. Furthermore, 147 genes were related to two-component regulatory systems of which 34 were LuxR proteins. These two-component systems appear to play a crucial role in quorum sensing of Gram-positive bacteria and a positive correlation between plant-microbe interactions and the number of LuxR proteins has been suggested [44,45].

Many regulatory genes (~18%) were located near polysaccharide related loci including those involved in plant cell wall degradation. Specifically, 63% of cellulose degradation or cellulose binding genes had a nearby regulator (proximity ranged from 2–4 genes up or downstream). In the case of xylan metabolism, regulators were identified for 50% of the genes, while 43% of pectin metabolism genes also had a regulator nearby. An extended overview of the regulators and their associated carbohydrate genes is presented in Table S2.

An endophytic bacterium highly equipped with an array of plant cell wall degrading enzymes

The ability of *M. lupini* lupac 08 to assimilate a wide range of sugars was previously reported [19] and this is clearly reflected in its genome. The range of simple and complex saccharides assimilated by this strain include cellobiose, cellulose, glucose, mannitol, starch, sucrose, trehalose, xylan and xylose among others. Genomic analyses confirmed the presence of a large number of genes devoted to the metabolism of carbohydrates, including many compounds of plant origin. Plant-polymer degrading enzymes such as cellulases, xylanases and pectinases have been suspected to play a role in internal plant colonization [46]. In the case of plant pathogenic bacteria and fungi, these gain access by actively degrading plant cell wall compounds using glycoside hydrolases including cellulases and endoglucanases. However, genomic analyses show that non-pathogenic endophytic microorganisms such as *Enterobacter* sp. 638 [47], *Azoarcus* BH72 [39] or the symbiotic actinobacterium *Frankia sp.* [48] have only a reduced set of cell-wall degrading enzymes.

The genome of *M. lupini* Lupac 08 revealed a significant number of genes encoding enzymes potentially involved in plant-polymer degradation but also an important number of cellulose-binding related genes. Overall, about 10% of the genome coded for genes related to carbohydrate metabolism of which 192 had a hydrolytic function. At least 79 genes putatively involved in interactions with plants and with the potential to hydrolyze plant polymers were identified (Table S1). These genes were placed into the glycosyl hydrolase families GH5, GH6, GH9, GH10, GH11, GH18, GH20, GH43, GH44 and GH62, or into the carbohydrate binding modules CBM2, CBM13, CBM33, CBM3, CBM46, CBM42, CBM5, CBM4, CBM6 and CBM32. The CBM2 family was the most abundant appearing in 46 of the 79 genes identified.

Fourteen genes were further identified as lectins or proteins with lectin binding domains, which presumably bind to and interact with carbohydrates. Some of these loci (e.g. Milup_42969, Milup_42975, Milup_44484, and Milup_44962) appear to be related to cellulases and xylanases, respectively. These proteins are

important as they serve as a means of attachment between a bacterium and its host (animal or plant) and are produced by either of the two interacting organisms [48].

Compared to the 45 enzymes predicted to act on oligo- and/or polysaccharides reported for *T. fusca* XY [49], the number of these enzymes present in the genome of *M. lupini* is significantly higher.

Cellulose metabolism. Aerobic cellulolytic actinobacteria have been shown to use a system for cellulose degradation consisting of sets of soluble cellulases and hemicellulases. Most of these independent cellulolytic enzymes contain one or more carbohydrate binding domains [50].

A total of 46 genes were found to present a hydrolytic or binding fuction towards cellulose (Table S1). Several endoglucanases were detected in strain Lupac 08 (e.g. C1, C2 C10 and C14), these enzymes hydrolyze internal bonds at random positions of amorphous regions of cellulose and generate chain ends for the processive action of cellobiohydrolases (exoglucanases). A copy of the exoglucanase gene *cbhA* (C16) was also located in the genome. Exoglucanases act on the ends of cellulose polysaccharide chains, liberating cellobiose as the major product. β-D-glucosidases such as M108 and M109 which would further hydrolyze cellobiose were also identified. In addition, several extracellular cellulase coding genes were identified including *celA* (C3 and C6), *celB* (C5) and *celD* (C13). These results strongly suggest that strain Lupac 08 is potentially capable of completely degrading cellulose.

Strain Lupac 08 was tested for *in vitro* production of cellulases. Very high cellulase activity was detected in minimal agar supplemented with carboxymethylcellulose (CMC, 0.5%) (Fig. 6A). When the culture medium was supplemented with glucose (1%) similar results were obtained indicating that this sugar did not repress nor derepress the expression of the genes responsible for the production of cellulases.

Hemicellulosic substrates. Genome analysis also revealed the ability of *M. lupini* to convert various hemicellulosic substrates to sugars. Twelve putative genes related to the metabolism of xylan included several copies of extracellular xylanases (X1, X3, X4, X5, X6, X7, X9, X10 and X12; see Table S1); an extracellular bifunctional xylanase/deacetylase (X8); and an arabinofuranosidase (X2) which work synergistically with xylanases to degrade xylan to its component sugars. Genes for several α-arabinofuranosidases were also identified (C17, M33, and M39); these are exo-acting enzymes which hydrolyze nonreducing arabinofuranose residues from arabinoxylan, pectins, and shorter oligosaccharides.

In vitro xylanase activity was detected in strain Lupac 08 when tested in a minimal medium supplemented with xylan (1%). Production of xylanases was detected after incubation for 4 days increasing significantly after 14 days (Fig. 6D). The substrate was assayed with and without glucose with similar results.

Starch degradation. Starch is a ubiquitous and easily accessible source of energy. In plant cells it is usually deposited as large granules in the cytoplasm. Several genes coding for amylo-α-1,6-glucosidases (e.g. M26, M32, M44, M63, M111 and M121; Table S1) were located in addition to two *amyE* homologs that code for an extracellular α-amylase. Furthermore, strain Lupac 08 was able to degrade this polymer under laboratory conditions (Fig. 6C) and it was previously shown that Lupac 08 can utilize this substrate as a carbon source [17].

Pectin degradation. Pectinolytic enzymes can degrade pectic substances either through hydrolysis (hydrolases) or trans-elimination (lyases) [51] and are important virulence mechanisms in many soft-rotting and macerating pathogens [52]. Six pectate lyases (P1, P3, P4, P5 and P6; Table S1) were located in the

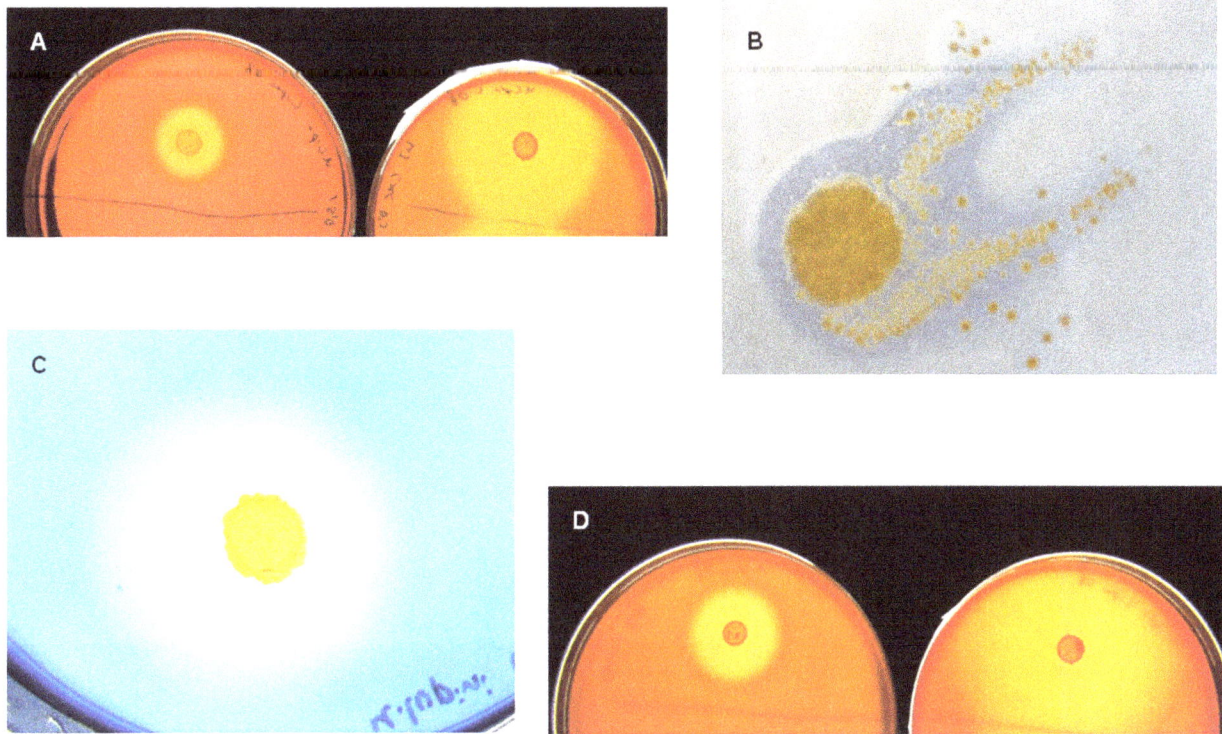

Figure 6. Expression of cellulose, starch, xylan and chitin degrading genes in *Micromonospora lupini* **Lupac 08.** (A) carboxymetheylcellulose hydrolysis at 4 (left) and 14 (right) days after inoculation. (B), starch hydrolysis at 4 days after inoculation. (C), chitin degradation at 7 days after inoculation. (D), xylan degradation at 4 (left) and 14 (right) days after inoculation.

genome of *M. lupini*, two of which were annotated as virulence factors (P5 and P6). In addition, an extracellular pectin methylesterase gene, *pmeA*, and a gene coding for a pectate lyase involved in D-galacturonic acid hydrolysis (P7) were also identified. Interestingly, *T. fusca* XY contains two pectin lyase homologs but does not appear to possess a pectin methylesterase or a pectin acetylesterase gene. Pectinase-encoding genes are reported to be absent in other endophytic microorganisms such as *Azoarcus* sp. BH72 or *Enterobacter* sp. 638 [39,47]. Production of pectinases was observed under laboratory conditions and activity was visualized after 8 days of incubation (Fig. 7D).

Expansin-like proteins. Expansins are proteins that were first described from plants [53]. These molecules function as cell wall loosening proteins by disrupting the noncovalent binding of matrix polysaccharides to cellulose [54], resulting in physical effects, such as polymer creep and stress relaxation of extended cell walls [55,56]. Many plant-associated microorganisms including several pathogenic actinobacterial species have been shown to contain proteins with expansin-like domains [57].

Two genes (MiLup_41274 and MiLup_45306) were identified in the genome of strain Lupac 08 that encode for a secreted protein showing 42% and 48% sequence similarity to the corresponding *celA* genes of *Clavibacter michiganensis* subsp. *michiganensis* and *Clavibacter michiganensis* subsp. *sepedonicus*, respectively. This gene corresponds to a secreted β-1,4-endoglucanase (CelA) that is required for virulence and contains a C-terminal α-expansin like domain [58,59]. In the case of *C. michiganesis* subsp. *michiganensis* CelA, this expansin-like domain is essential for development of wilting symptoms [58]. It is suggested that microbial expansins function to promote microbe-plant interactions, both harmful and beneficial ones [60].

Plant growth promotion traits of *Micromonospora* lupini Lupac 08

Our current knowledge of plant-microbe interactions indicates that populations inhabiting a host plant are not restricted to a single microbial species but comprise several genera and species. Few reports are available regarding the presence of other microorganisms (associated or endophytic) in nitrogen fixing nodules, in spite of the fact that nodules are much richer in nutrients as compared to roots [61]. The recent reports on the isolation of large *Micromonospora* populations from nitrogen fixing nodules clearly suggest that this bacterium plays an important role which has yet to be defined.

Effect of *M. lupini* **Lupac 08 on** *Trifolium***.** *Micromonospora lupini* Lupac 08 clearly produced a plant growth enhancing effect when it was co-inoculated with *Rhizobium* sp. E11 under laboratory conditions on clover plantlets. In general, the number of nodules was higher in those plants co-inoculated (18–24 nodules) with both bacteria as compared to the plants inoculated only with *Rhizobium* sp. E11 (11–15 nodules). Overall, the co-inoculated plants showed better growth and were larger in size as compared to the other two treatments (Fig. 7C). Similar results were previously observed when strain Lupac 08 was inoculated in its original host, *Lupinus* [62].

Nitrogen fixing capacity. Indirect evidence of nitrogen fixing genes was obtained by partial amplification of *nifH*-like gene fragments in strains *Micromonospora* sp. L5 [12] and *M. lupini* Lupac 08 [11]. In the present work the genomes of *Micromonospora lupini* Lupac 08 and *Micromonospora sp.* L5 were screened for the presence of nitrogen fixing genes to confirm this earlier finding. After thorough analysis of the complete genome, no sequences related to this biological process were detected, supporting the results reported for strain

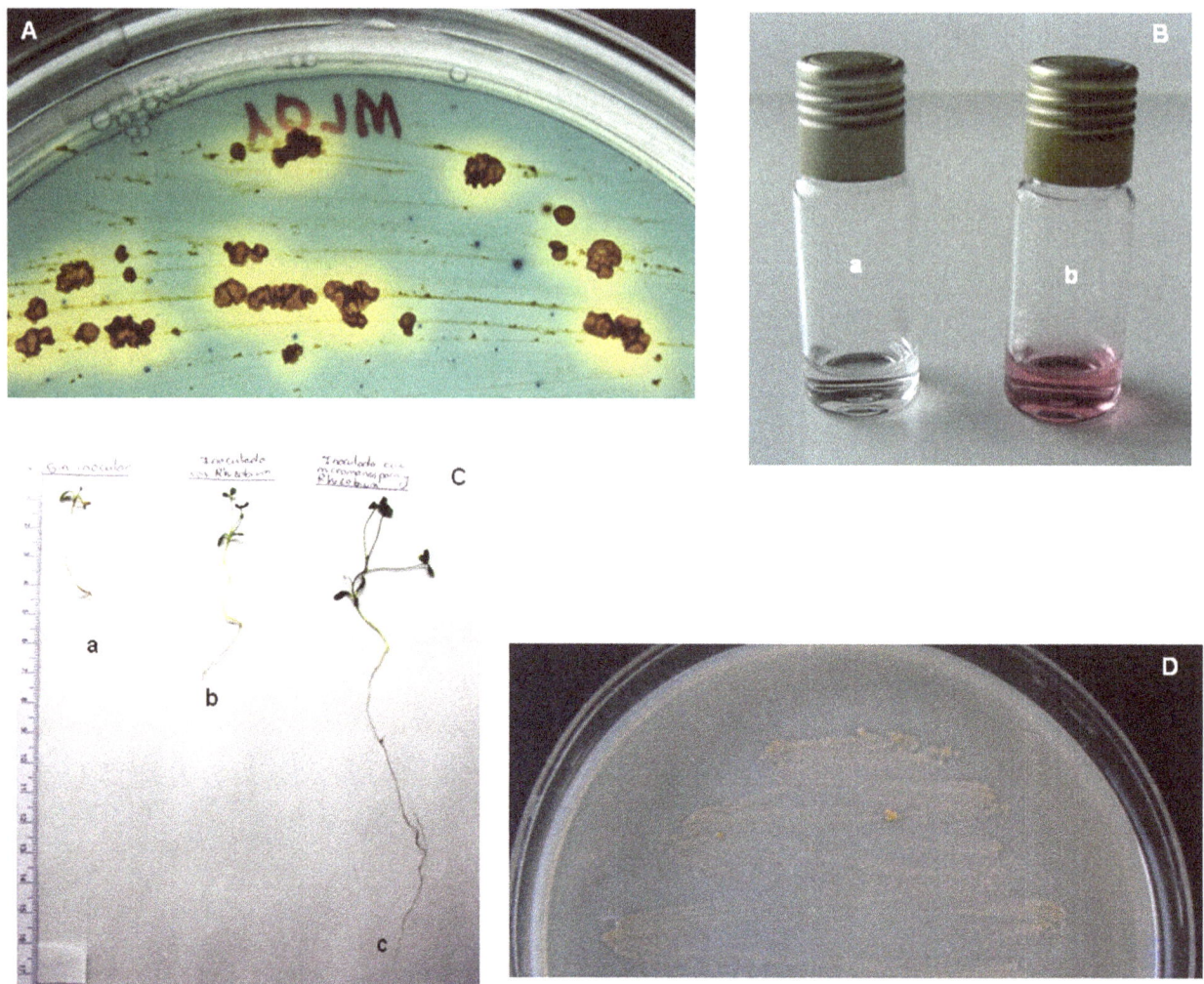

Figure 7. Plant growth promotion and biological control features of *M. lupini* **Lupac 08.** (A) Siderophore, (B) indole-3-acetic acid [a, negative control *E. coli* DH5α; b, Lupac 08] and pectinase production (D) by *M. lupini* strain Lupac 08;. (C) Plant growth promoting effect of *M. lupini* Lupac 08 on clover plantlets. a) control; b) inoculated with *Rhizobium* sp. E11; c) co-inoculated with *Rhizobium* sp. E11 and *M. lupini* Lupac 08.

Micromonospora sp. L5 [19]. Nitrogenase activity detection by acetylene reduction assays carried out with strain Lupac 08 over a period of two weeks were negative. A positive result was reported for strain L5 [12.].

Trehalose and its role in nodulation and bacteroid survival. Trehalose is a common reserve disaccharide in the root nodules of legumes, present at high concentrations in bacteroids at the onset of nitrogen fixation [63]. It has been reported that in the interaction between *Phaseolus vulgaris* and *Rhizobium,* enhanced germination, quality and grain yield have been correlated with trehalose content, and a higher tolerance to abiotic stress [64,65]. On the other hand, the trehalose content appears to be regulated by trehalase, a nodule stimulated plant enzyme [66,67]. Although trehalose metabolism in leguminous plants is still poorly understood, it has been shown that in senescent nodules, trehalose becomes the most abundant non-structural carbohydrate [68] and it is proposed that trehalose, a stress protectant accumulated in bacteria, could offset membrane injuries and/or serve as an intermediate energy reserve. Indeed, Müller *et al.* [68] showed that during terminal senescence of nodules an appreciable part of the bacteria maintained their trehalose pools and survived.

Eight genes related to the metabolism of trehalose were detected in the genome of Lupac 08; seven genes were related with trehalose synthesis (Mlup08_40949, Mlup08_43225, Mlup08_43226, Mlup_45189, Mlup_45758, Mlup08_45759 and Mlup08_45961) and one (*treA*, Mlup08_45961) with the enzyme trehalase. Barraza *et al.* [67] proposed that modification of the trehalose content in the nodules could trigger physiological alterations that would enhace carbon and nitrogen metabolism, as well as bacteroid fitness (greater survival) and nitrogen fixation, which in turn would positively impact on symbiotic interactions. *Micromonospora* may contribute to the survival of rhizobia by helping to maintain high levels of trehalose.

Chitin degradation and protection against pathogens. Plant β-1,3-glucanases are directly involved in defense by hydrolyzing the cell walls of fungal pathogens most commonly in combination with chitinases. Nine chitin-related ORFs were identified in *M. lupini*. Specifically, six code for a chitooligosaccharide deacetylase, several extracellular endo- and exo-chitinases and a β-N-acetyl-hexosaminidase (MiLup08_41789, MiLup08_41912, MiLup08_43481, MiLup08_44343, MiLup08_45172, MiLup08_45568), while three CDS code for putative chitin-binding domain proteins (MiLup08_41110, MiLup08_41724, MiLup08_41729).

Chitinases often work synergistically with chitin-binding proteins (CBPs). The biological roles of bacterial chitinases and carbon binding proteins are easily understood in an environmental context, especially in soil (that harbour fungi and insects) and marine (shellfish) habitats and their impact on chitin cycling. However, there is an increasing amount of direct or indirect evidence suggesting that some chitinases and CBPs additionally serve as virulence factors for bacterial pathogens during infection of non-chitinous substrates [69].

Experimental data confirmed *in vitro* chitin degradation of strain Lupac 08 (Figure 6B). As with other endophytic bacteria *Micromonospora* may produce chitinases to inhibit fungal pathogens, or may produce these molecules to elicit the plant defense mechanism. Either way, it seems that *Micromonospora* would provide a benefit to its host.

Siderophores (Iron-transport) and other secondary metabolites

Iron is an element essential for every living organism, as a cofactor of numerous proteins. Siderophores produced by plant growth promoting bacteria may reduce the growth of phytophathogens by depriving them of iron. Thus, an efficient iron uptake system can contribute to protect the host plant against pathogens. Interestingly, siderophores can also act as important virulence determinants for both plant and animal pathogens [70].

The genome of strain Lupac 08 revealed several siderophore related genes including specific iron uptake transporters, secretion of different siderophores and synthesis of siderophore receptors. Namely, a zinc/iron permease (MiLup_40258), a ferrous iron permease FTR1 (efeU, MiLup_41076) and eight iron ABC transporters (MiLup_42281-MiLup_42285). The number of the latter transporters is similar to the number of those found in the genome of the endophytic bacterium *Enterobacter* sp. 638 [47] while the plant pathogen *Erwinia amylovora* CFBP 1430 presents only three such transporters [71].

Several gene clusters related to the biosynthesis and transport of the siderophores enterobactin (MiLup_44069-MiUp_44071), aerobactin (iucA/iucC family protein, MiLup_44063 and MiLup_44064; MiLup_40326) and alcaligin (MiLup_44065) were also located. In the case of aerobactin, the gene *iucA* is highly correlated with virulence in avian pathogenic *E. coli* strains [72].

Two siderophore-interacting proteins (MiLup_40648 and MiLup_45559) were also found. One of these genes (MiLup_40648) was located next to a siderophore transporter of the RhtX/FptX family; RhtX from *S. meliloti* 2011 and FptX from *Pseudomonas aeruginosa* appear to be single polypeptide transporters from the major facilitator family for import of siderophores as a means to import iron [73]. In addition, a thiazolinyl imide reductase involved in siderophore biosynthesis was also identified (MiLup_43551).

The genome of Lupac 08 also contained several regulators including an iron-dependent repressor (IdeR, MiLup_41668), two ferric uptake regulation proteins (MiLup_40794 and MiLup_44436) and a putative iron-regulated membrane protein which suggests that these systems are highly regulated. Production of siderophores was detected experimentally (Fig. 7A).

Actinobacteria are well known to be capable of producing a vast diversity of natural secondary metabolite compounds with applications in medicine, agriculture, and other biotechnological areas [74]. Endophytic bacteria are currently of significant interest as an untapped resource of novel bioactive small molecules because their metabolites are speculated to affect the physiological conditions of host plants including growth and disease resistance. *Micromonospora* strains are well known for their capacity to produce many secondary metabolites and *M. lupini* Lupac 08 was previously screened for the production of novel compounds with antitumoral activity and the results obtained confirmed the production of a new family of molecules named Lupinacidins A, B and C [9,10].

Fifteen clusters involved in the biosynthesis of secondary metabolites were identified in the genome of *M. lupini* Lupac 08. These included siderophores (see above), terpenes, butyrolactones, polyketides (PKS), nonribosomal peptides (NRPS), chalcone synthases and bacteriocins (Table 4). A DNA stretch of 544 kb was estimated to code predominantly for secondary metabolites, accounting for about 7.4% of the genome. This percentage is lower than that reported for the marine actinobacterium *S. tropica* (9.9%, [75]) but it is within the range of other actinobacteria e.g. *S. coelicolor* (8.2%, [43]). Interestingly, *Frankia* strains ACN14a and EAN1pec dedicate about 5% of their genomes to natural product assembly while the potential of CcI3, which has the smallest genome of the three *Frankia* strains, has a much reduced host range and is absent from most soils is significantly smaller (~3%) [76].

Several clusters identified in the genome of *M. lupini* were also located in other genomes of phylogenetically related bacteria, especially in *S. tropica* CNB-440, *S. arenicola* CNS-205 and *V. maris* AB-18-032. Nevertheless other clusters were unique to *M lupini* (Table 4). Eight of the 15 clusters identified were located in the region between coordinates 4,000 kb and 5,000 kb of the genome, close to the terminus of replication. This area of the genome also contains a high density of genes coding for the biosynthesis of various plant cell wall degrading enzymes and several transposases.

Terpene related enzymes present in the genome of *M. lupini* are involved in the synthesis of carotenoids, sugar-binding lipids and the production of pentalenolactone type antibiotics. Similar molecules have also been predicted from the genomes of the three sequenced *Frankia* strains [77]. Various polyketide biosynthetic and non-ribosomal peptide synthase pathways were also identified specifically as PKI, PKII (2 clusters), PKIII types, NRPS (2 clusters) and hybrid PKS/NRPS clusters (2 clusters). The presence of these gene clusters suggests that *M. lupini* is capable of producing a vast diversity of secondary metabolites such as the antitumor anthraquinone derivative lupinacidins reported earlier [9,10]. Some of these metabolites may perform specialized functions in ecological niches and recent studies have reported on the importance of PKS and NRPS molecules and their potential role in communication during root colonization [78,79]. In addition, cluster 10 contains genes that putatively code for the production of granaticin. Granaticins are antibiotics of the benzoisochromanequinone class of aromatic polyketides, the best known member of which is actinorhodin produced by *S. coelicolor* A3(2). Production of granaticins has mainly been reported from *Streptomyces* strains [80]. NRPS cluster 11 (see Table 4), appeared to be unique to strain Lupac 08 as this group of genes was not detected in any of the other genomes compared except in *S. tropica* CNB-440 where it seems to be only partially conserved.

The PKS type III cluster corresponds to several genes that code for the production of naringenin, a central precursor of many flavonoids. It has recently been proposed that flavonoids play an important role in the establishment of plant root endosymbioses. In the case of legume-*Rhizobium* interactions, flavonoids released by plant roots induce genes involved in nodulation [81]. In a similar way it has also been suggested that these molecules play an important role during the early stages of the symbiotic association between *Frankia* and actinorhizal plants [82]. *Micromonospora* flavonoids may contribute to support communication between the nitrogen fixing bacteria and their host plants.

Table 4. Comparison of secondary metabolite clusters found in the genome of *M. lupini* Lupac 08 and other related microorganisms.

Cluster	Type	*M. lupini* **Lupac 08** (Milup08_X)	*Micromonospora* sp. **L5**	*M. aurantiaca* **ATCC 27029**[T]	*Verrucosispora maris* **AB-18-032**	*Salinispora tropica* **CNB-440**	*Salinispora arenicola* **CNS-205**	*Streptomyces coelicolor* **A3(2)**
1	Terpene	40204–40210	Conserved	Conserved	Conserved	Conserved	Conserved	Conserved
2	Terpene	40306–40320	Conserved	Conserved	Conserved	Absent	Absent	Present
3	Butyrolactone	40602–40668	Absent	Absent	Absent	Conserved	Conserved	Conserved
4	Type I PKS	41995–42009	Conserved	Conserved	Conserved	Conserved	Conserved	Absent
5	Terpene	43134–43144	Conserved	Conserved	Conserved	Conserved	Conserved	Conserved
6	NRPS+PKS	43546–43581	Absent	Absent	Conserved	Conserved	Absent	Partially conserved
7	Type II PKS	43804–43844	Conserved	Conserved	Conserved	Conserved	Conserved	Partially conserved
8	Siderophore	44063–44071	Conserved	Conserved	Conserved	Conserved	Conserved	Conserved
9	NRPS-Type I PKS	44386–44405	Conserved	NRPS Absent	NRPS Absent	NRPS Absent	NRPS Absent	NRPS Absent
10	Type II PKS	44613–44624	Partially conserved	Partially conserved	Partially conserved	Partially conserved	Partially conserved	Partially conserved
11	NRPS	44684–44691	Absent	Absent	Absent	Partially conserved	Absent	Absent
12	Bacteriocin	44929–44933	Conserved	Conserved	Conserved	Conserved	Conserved	Partially conserved
13	Terpene	45087–45093	Conserved	Conserved	Conserved	Absent	Absent	Conserved
14	NRPS	45439–45446	Conserved	Conserved	Conserved	Conserved	Conserved	Conserved
15	Type III PKS	46684–46700	Conserved	Conserved	Conserved	Conserved	Conserved	Absent

PKS, polyketide synthases; NRPS, non-ribosomal peptide synthases.

Genomic information also revealed that *M. lupini* has the potential to produce bacteriocins (cluster 12) as suggested by the presence of a putative short-chain dehydrogenase/reductase.

Phytohormones

Indole-3 acetic acid. Diverse bacterial species have the ability to produce auxinic phytohormones such as indole-3-acetic acid (IAA) and a few can also produce phenyl-acetate (PAA) such as *Frankia alni* [83,84]. Different biosynthesis pathways have been identified and redundancy for IAA biosynthesis is widespread among plant-associated bacteria [83]. Interactions between IAA-producing bacteria and plants may lead to several outcomes, from pathogenesis to phytostimulation [85]. The genome of *M. lupini* Lupac 08 contains a gene (Milup_45687) potentially involved in the biosynthesis of IAA via the indole-3-acetonitrile pathway. This gene corresponds to the conversion of indole-3-acetonitrile to indole-3-acetic acid. Nitrilases with specificity for indole-3-acetonitrile have been reported in *Alcaligenes faecalis* [86]. In *A. tumefaciens* and *Rhizobium* spp. nitrile hydratase and amidase activity could be identified, indicating the conversion of indole-3-acetonitrile to indole-3-acetic acid via indole-3-acetamide [87]. Analysis of IAA production by strain Lupac 08 was carried out, yielding a positive result (Fig. 7B).

Acetoin and 2,3-butanediol. The volatile compounds acetoin and 2,3-butanediol produced by bacteria such as *Bacillus subtilis* GB03 and *Bacillus amyloliquefaciens* IN937a have been reported as plant growth promoting hormones [88]. Several genes were located in the genome of strain Lupac 08 which could be involved in the production of these compounds. Two copies of the gene *pdhB* (MiLup_40114 and MiLup_43782) that encode the enzyme pyruvate dehydrogenase were identified. This enzyme transforms pyruvate to acetaldehyde and in this process a small fraction of pyruvate is converted to acetoin as a by-product. In addition, three acetolactate synthases involved in the synthesis of acetolactate from pyruvate are present (MiLup_41336, MiLup_41383 and MiLup_41384). Under aerobic conditions, acetolactate is converted to acetoin by the enzyme acetoin dehydrogenase (MiLup_41670).

Discussion

The most extensively studied bacteria interacting with plants are Gram-negative proteobacteria because they are readily isolated from plant tissues and genetically handled for interaction studies. However, the impact of Gram-positive bacteria on plants should not be underestimated as has been done for many years mainly due to their slow growth. *M. lupini* Lupac 08, a Gram-positive actinobacterium was isolated from the internal root nodule tissues of *Lupinus angustifolius* but it is only a representative of a large collection of more than 2000 *Micromonospora* strains isolated from diverse legumes and actinorhizals from different geographical locations. So far, several genomes of root symbionts and soil saprophytes have been studied; therefore we decided to focus on an intermediary category, that of facultative endophytes.

Lupac 08 was isolated from lupine nodules, and shown to produce the anticancer agents Lupinadicin A, B and C. The genome of strain Lupac 08 was sequenced to obtain information about the potential ecological role of *Micromonospora* in interaction with legumes and actinorhizal plants. Genomic analysis revealed several strategies which are probably necessary to lead a successful lifestyle as a saprophyte in the rhizosphere, a competitive and harsh environment, and as an endophyte capable of colonizing the internal plant tissues. *Micromonospora* species have less than 3% distance in their 16S rRNA genes, which can be

roughly translated to a time of 150MY according to the equivalence proposed by Ochman and Wilson [89]. In the current phylogeny (Fig. 1), *M. lupini* has as closest neighbours *M chokoriensis M. saelicesensis* and *M. zamorensis*, isolated from sandy soil, root nodules of *L. angustifolius* and the rhizosphere of *P. sativum* respectively. *Micromonospora* sp. L5 and *M. aurantiaca* are located further away, with a distance of 1.2% that would translate to 60 MY for the emergence of a group of species that interact with plants, a date that would be close to the postulated time of emergence of *Fabaceae* and that of many actinorhizal plant families [90]. The separation from the *Salinispora* and *Verrucosispora* lineages would constitute two independent events that would have occurred slightly earlier at 160MY and 170MY, while the emergence of the *Actinoplanes* and that of the *Dactylosporangium* would have occurred 250MY ago, a time when dicotyledons had not yet appeared but when continents and thus soils had appeared that did permit the growth of primitive plants such as the gymnosperms.

The size of the *Micromonospora* genomes analyzed is quite uniform, with that of Lupac 08 slightly larger. The chromosome size of *M. lupini* Lupac 08 appears to reflect a wealth of genes allowing for adaptation to a complex saprophytic/endophytic lifestyle, which means adapting to a wider range of environmental conditions with the ability to metabolize a large variety of nutrient sources. Considering that *Micromonospora* sp. L5 and *M. lupini* Lupac 08 were both isolated from nitrogen fixing nodules (actinorhizal and legume plants, respectively), it would be expected that the genomes of these strains be more similar to each other than to *M. aurantiaca* ATCC 27029[T] which was originally isolated from soil. Surprisingly this was not the case as confirmed by the high number of strain specific genes identified in the genome of Lupac 08, suggesting a high capacity of adaptation to a fluctuating environment by this microorganism. On the other hand, *Micromonospora* sp. L5 and *M. aurantiaca* ATCC 27029[T] share a high number of orthologous genes (86%) suggesting that the niche of origin is not crucial.

An interesting result was the distribution of the metabolic profiles of 20 bacteria representing different living environments (Fig. 5). There was a clear proximity between *M. lupini* Lupac 08 and the three *Frankia* genomes. This result suggests that strain Lupac 08 contains metabolic functions similar to those found in *Frankia* strains that are probably useful for its interaction with plants. This metabolic versatility combined with a diverse transport system make Lupac 08 an organism fit to adapt to a soil/plant environment.

The emergence and evolution of nitrogen fixation ability among the domains *Bacteria* and *Archaea* is complex and has not yet been fully elucidated. Although it was previously reported that *Micromonospora* strains isolated from legume and actinorhizal root nodules contained *nifH*-like gene fragments [11,12], we could not confirm these results. In a similar approach based on PCR-amplification, other authors reported the presence of *nif*-H like sequences for bacterial isolates obtained from legumes collected in arid zones including *Microbacterium*, *Agromyces*, *Mycobacterium* and *Ornithinicoccus* [91]. One recurrent problem with the use of a PCR-based approach is that it is limited to a single gene amplified billions of times, which may provide false-positive results [92] and for this reason must always be confirmed by an independent approach.

Plant-polymer degrading enzymes such as cellulases and pectinases have been suspected to play a role in internal colonization. Most plant pathogens secrete cellulases, pectinases, xylanases, or other enzymes to hydrolyze plant cell wall polymers, while a lack of secreted hydrolases has been proposed to be favourable for microorganisms that form beneficial association with plants. Examples of endophytic plant growth promoting bacteria that lack large amounts of cell wall degrading enzymes include *Frankia* [38], *Enterobacter* sp. 638 [47], *Azoarcus* sp. BH72 [60] and *Herbaspirillum seropedicae* [93]. An *Azospirillum* sp. that does not colonize root tissues proper, but only the rhizosphere, has a genome containing a large number of putative cellulases similar to soil cellulolytic bacteria with 26–34 glycosyl hydrolases [94], as compared to the 37 present in *T. fusca*, a highly cellulolytic actinobacterium isolated from soil.

The genome of *M. lupini* revealed a high number of putative genes that encode for hydrolytic enzymes and specifically cellulolytic, xylanolytic, chitinolytic and pectinolytic activities were confirmed in the laboratory, indicating the capacity of *Micromonospora* to degrade plant polymers in a way similar to that of plant pathogen microorganisms. In the case of *Micromonospora*, there seems to be a paradox since strain *M. lupini* Lupac 08 shows a very high *in vitro* activity for cellulases and xylanases, however, preliminary inoculation experiments in our laboratory indicate that the microorganism does not behave as a pathogen, on the contrary, *Micromonospora* appears to interact in a tripartite relationship stimulating nodulation and plant growth (Fig. 7c). Therefore the question arises as to what is the function of these enzymes when *Micromonospora* interacts with the host plant. Alternatively some of the genes present, especially those related to the metabolism of cellulose may not necessarily imply that the bacterium is involved in plant cell wall degradation but have a different role, yet to be defined [95]. In addition many of the cellulose-related genes contain binding-domains suggesting that these may be related to the adhesion of the bacterium to the plant. These genes could also help *Micromonospora* digest plant cell walls upon senescence of the nodules.

M. lupini Lupac 08 contains several secondary metabolite gene clusters, many of which appear to be involved in the synthesis of siderophores and also of antibiotics. These would also in all likelihood be involved in the synthesis of the antitumor anthraquinone molecules described previously [9,10]. *Micromonospora*, like many other endophytic bacteria is a facultative plant colonizer that must compete with other microorganisms in the rhizosphere before entering the plant. In this sense, the NRPS and PKS gene clusters identified in the genome of *M. lupini* Lupac 08 may be involved in defense as well as in interaction and communication with its host plant. Thus, it will be necessary to identify these compounds and their functional attributes to further expand our knowledge of this plant-microbe interaction.

Conclusions

We have provided experimental data which supports the hypothesis that *M. lupini* Lupac 08 is a plant growth promoting bacterium. *Micromonospora lupini* Lupac 08 clearly produces a plant growth enhancing effect as observed in laboratory experiments. The localization of several genes involved in plant growth promotion traits such as the production of siderophores, phytohormones, the degradation of chitin (biocontrol) and the biosynthesis of trehalose may all contribute to the welfare of the host plant. *Micromonospora* appears as a new candidate in plant-microbe interactions with an important potential in agriculture and other biotechnological applications. The current data is promising but it is still too early to determine which specific roles are played by this microorganism in interactions with nitrogen fixing plants.

Methods

Genome sequencing, annotation and analysis

The genome sequence of *M. lupini* Lupac 08 was determined using the 454 FLX system and Titanium platform (454 Life Sciences) as previously reported [96]. Sequences were assembled into 50 contigs and four scaffolds ranging from 583 to 7,083,659 nucelotides using the MaGe (Magnifying Genomes) interface [97]. This Whole Genome Shotgun project has been deposited at European Nucleotide Archive under accession number NZ_CAIE00000000.01.

16S rRNA gene phylogeny

Sequences obtained from public databases (Genbank/EMBL) were manually aligned using clustal X software [98]. Phylogenetic distances were calculated with the Kimura 2-parameter model [99] and the tree topologies were inferred using the maximum-likelihood method [100]. All analyses were carried out using the MEGA5 program [101].

Comparative genome analysis

Genome rearrangement of the *Micromonospora* strains *M. lupini* Lupac 08, *Micromonospora* sp. L5, *M. aurantiaca* ATCC 27029[T] and *Micromonospora* sp. ATCC 39149 were carried out using MAUVE software [102]. The number of shared and unique genes present in the respective genomes were calculated and represented by a Venn-diagram using the EDGAR software [103]. Potential horizontally transferred genes were predicted using the "Region of Genomic Plasticity Finder" method implemented on the MicroScope platform. First we selected genomes included in the PkGDB and NCBI RefSeq databases that presented high synteny percentages with the Lupac 08 strain. Automatic results were manually curated according to several features such as base composition, DNA repetitions, presence of near mobile elements and information provided by SIGI and IVON programs [104].

Comparative analysis of metabolic profiles

A bicluster plot of the metabolic profiles for *M lupini* Lupac 08 and 20 other bacterial genomes was performed with Multibiplot [105]. A comparison of 798 MicroCyc metabolic pathways was made using MaGe. This comparison is based on the calculation of 'pathway completion' values, scaled from 0 to 1, where 0 means that a particular organism does not contain any enzyme for a given pathway and 1 that it contains all the reactions of the pathway. These values were transformed applying row standardization and a JK-Biplot was constructed after performing a PCA (Singular Value Decomposition estimation method). The heatmap was then obtained with the expected values computing the Euclidian distance and average linkage for rows and columns.

Transport proteins identification and classification

Information about transport proteins of genomes was obtained from the TransportDB relational database when available (http://www.membranetransport.org/). The identification and classification of the transporters of strain Lupac 08 was performed using the TransAAP tool based on TransportDB [106] followed by manual validation.

Cellulose, starch and xylan degradation

Strain Lupac 08 was cultivated on yeast-malt agar for 5 days and subsequently transferred to M3 agar [107] with and without glucose and supplemented with one of the substrates in the following way: carboxymethylcellulose (CMC, 0.5%), starch (1%)

and xylan (1%). A bacterial suspension of 10^6 per ml was prepared in saline solution (0.85%) and 200 µl were inoculated on the different plates which were then incubated at 28°C and results were recorded at 4, 7 and 14 days after inoculation. Xylan and CMC plates were stained with Congo red [108] while starch plates were flooded with iodine solution [109].

Pectin degradation

Pectinolytic acitivity was determined as described in Williams et al. [109]. Agar plates supplemented with pectin (0.5%) were streaked with strain Lupac 08 and incubated at 28°C. Hydrolysis zones were detected after 14 days incubation by flooding plates with an aqueous solution of cetyltrimethyl ammonium bromide (CTAB, 1%) and examining them after 30 min.

Chitin degradation

Chitinolytic acitivity was determined as described in Murthy et al. [110]. Agar plates supplemented with colloidal chitin at 0.5% (Gift of France-Chitine, Orange, http://france-chitine.com/), partly hydrolysed by stirring in 0.5 M HCl for 2h, were inoculated with strain Lupac 08 and incubated at 28°C. Hydrolysis zones were detected as cleared zones after 14 days incubation.

Siderophore production

Siderophore production was assessed using a modified chrome azurol S (CAS) assay [111]. Strain Lupac 08 was cultured on yeast-malt agar and incubated for 7 days, subsequently it was streaked onto CAS agar plates and incubated at 28°C for 7–10 days. A positive result was indicated by an orange halo around bacterial colonies.

Indole-3-acetic acid production

Production of indole acetic acid was assayed following the method of Glickmann and Dessaux [112]. Strain Lupac 08 was inoculated in 5 ml of yeast-malt medium supplemented with L-tryptophan (0.2%) and incubated at 28°C at 150 rpm during 7 days. The culture was then centrifuged at 12,000 x g for 10 min and 1 ml of the supernatant was mixed with 2 ml of Salkowski's reagent [113] and incubated at room temperature for 30 min. IAA

production was measured spectrophotometrically at 530 nm to assess the development of a pink colour.

Plant growth

Surface-sterilized seeds of *Trifolium* sp. were germinated axenically in Petri dishes on 1.4% w/v agar. Seedlings were transferred to sterile square plastic plates that contained a nitrogen-free nutrient solution [114]. Fifteen plantlets were inoculated in the following manner (5 per treatment): 500 µl (each) of bacterial suspensions (10^6 cells per ml) of *M. lupini* Lupac 08 and *Rhizobium* sp. E11 for coinoculation treatment; inoculation with *Rhizobium* sp. E11; and uninoculated plants as negative controls.

Acetylene reduction activity

Nitrogenase activity was measured using acetylene reduction [115] in sterile 150 ml plasma flasks with a rubber stopper. Cells of Lupac 08 were cultured in liquid minimal glucose medium without nitrogen at 28°C with shaking. The air in flasks was replaced with mixture of air and acetylene (ration 90:10 v/v). One mililiter of mixture was sampled for each measure using gas chromatography with a flame ionization detector (Girdel 30, France). *Mesorhizobium melitolti* Sm1021 was used as positive control.

Author Contributions

Conceived and designed the experiments: MET PN. Performed the experiments: MET PP PB RR. Analyzed the data: MET RB PP PN CM. Contributed reagents/materials/analysis tools: MET PN PP YI CM. Contributed to the writing of the manuscript: MET PN RB PP.

References

1. Conn VM, Franco CM (2004) Analysis of the endophytic actinobacterial population in the roots of wheat (*Triticum aestivum* L.) by terminal restriction fragment legth polymorphism and sequencing of 16S rRNA clones. Appl Environ Microbiol 70: 1787–94.

2. Kaewkla O, Franco CM (2013) Rational approaches to improving the isolation of endophytic actinobacteria from Australian native trees. Microb Ecol 65: 384–393.

3. Hardoim PR, van Overbeek LS, Elsas DJ (2008) Properties of bacterial endophytes and their proposed role in plant growth. Trends Microbiol 16: 463–471.

4. Chi F, Shen SH, Cheng HP, Jing YX, Yanni YG, et al. (2005) Ascending migration of endophytic rhizobia, from roots to leaves, inside rice plants and assessment of benefits to rice growth physiology. Appl Environ Microbiol 71: 7271–7278.

5. Reinhold-Hurek B, Hurek T (1998) Life in grasses: diazotrophic endophytes. Trends Microbiol 6: 139–144.

6. Marchetti M, Capela D, Glew M, Cruvellier S, Chane-Woon-Ming B, et al. (2010) Experimental evolution of a plant pathogen into a legume symbiont Plos Biol 8: e1000280. doi: 10.1371/journal.pbio.1000280.

7. Orskov J (1923) Investigations into the morphology of the ray fungi. Copenhagen.

8. Igarashi Y, Ogura H, Furihata K, Oku N, Indananda C, et al. (2011a) Maklamicin, an antibacterial polyketide from an endophytic *Micromonospora* sp. J Nat Prod 74: 670–674.

9. Igarashi Y, Trujillo ME, Martinez-Molina E, Yanase S, Miyanaga S, et al. (2007) Antitumor anthraquinones from an endophytic actinomycete *Micromonospora lupini* sp. nov. Bioorg Med Chem Lett 17: 3702–3705.

10. Igarashi Y, Yanase S, Sugimoto K, Enomoto M, Miyanaga S, et al. (2011b) Lupinacidin C, an inhibitor of tumor cell invasion from *Micromonospora lupini*. J Nat Prod 74: 862–865.

11. Trujillo ME, Alonso-Vega P, Rodriguez R, Carro L, Cerda E, et al. (2010) The genus *Micromonospora* is widespread in legume root nodules: the example of *Lupinus angustifolius*. ISME J 4: 1265–1281.

12. Valdés M, Perez NO, Estrada-de Los Santos P, Caballero-Mellado J, Pena-Cabriales JJ, et al. (2005) Non-*Frankia* actinomycetes isolated from surface-sterilized roots of *Casuarina equisetifolia* fix nitrogen. Appl Environ Microbiol 71: 460–466.

13. Carro L, Pujic P, Trujillo ME, Normand P (2013) *Micromonospora* is a normal inhabitant of actinorhizal nodules. J Biosci 38: 685–693.

14. Penn K, Jensen PR (2012) Comparative genomics reveals evidence of marine adaptation in *Salinispora* species. BMC Genomics 13: 86.

15. Normand P, Lapierre P, Tisa LS, Gogarten JP, Alloisio N, et al. (2007) Genome characteristics of facultatively symbiotic *Frankia* sp. strains reflect host range and host plant biogeography. Genome Res 17: 7–15.

16. Smith SE, Corneli-Showers P, Dardenne CN, Harpending HH, Martin DP, et al. (2012) Comparative genomic and phylogenetic approaches to characterize the role of genetic recombination in mycobacterial evolution. PLoS ONE 7: e50070.

17. Trujillo ME, Kroppenstedt RM, Fernandez-Molinero C, Schumann P, Martinez-Molina E (2007) *Micromonospora lupini* sp. nov. and *Micromonospora saelicesensis* sp. nov., isolated from root nodules of *Lupinus angustifolius*. Int J Syst Evol Microbiol 57: 2799–2804.

18. Carro L, Spröer C, Alonso P, Trujillo ME (2012) Diversity of *Micromonospora* strains isolated from nitrogen fixing nodules and rhizosphere of *Pisum sativum* analyzed by multilocus sequence analysis. Syst Appl Microbiol 35: 73–80.

19. Hirsch AM, Alvarado J, Bruce D, Chertkov O, De Hoff PL, et al. (2013) Complete genome sequence of *Micromonospora* strain L5, a potential plant-growth regulating actinomycete, originally isolated from *Casuarina equisetifolia* root nodules. Genome Announc 1: 2–00759–13.

20. Lethlefsen L, Schmidt TM (2007) Performance of the translational apparatus varies with the ecological strategies of bacteria J Bacteriol 189: 3237–3245.

21. Yano K, Wada T, Suzuki S, Tagami K, Matsumoto T, et al. (2013) Multiple rRNA operons are essential for efficient cell growth and sporulation as well as outgrowth in *Bacillus subtilis*. Microbiol 159: 2225–2236.

22. McCarthy AJ, Broda P (1984) Screening for lignin degrading actinomycetes and characterization of their activity against ^{14}C-lignin labelled wheat lignocellulose. J Gen Microbiol 130: 905–2913.

23. Jendrossek D, Tomasi G, Kroppenstedt R (1997) Bacterial degradation of natural rubber: a privilege of actinomycetes? FEMS Microbiol Lett 150: 179–188.

24. Sandrak NA (1977) Degradation of cellulose by micromonospores. Mikrobiologiia 46 478–481.

25. de Menezes AB, Lockhart RJ, Cox MJ, Allison HE, McCarthy AJ (2008) Cellulose degradation by micromonosporas recovered from freshwater lakes and classification of these actinomycetes by DNA gyrase B gene sequencing. Appl Environ Microbiol 74: 7080–7084.

26. Miele V, Penel S, Duret L (2011) Ultra-fast sequence clustering from similarity networks with SiLiX. BMC Bioinformatics 12: 116 doi: 10.1186/1471-2105-12-116.

27. Ren Q, Paulsen IT (2007) Large-scale comparative genomic analyses of cytoplasmic membrane transport systems in prokaryotes. J Mol Microbiol Biotechnol 12: 165–179.

28. Ren Q, Paulsen IT (2005) Comparative analyses of fundamental differences in membrane transport capabilities in prokaryotes and eukaryotes. PLoS Comp Biol 1: e27.

29. Sutcliffe I (2011) New insights into the distribution of WXG100 protein secretion systems. Antonie van Leeuwenhoek 99: 127–131.

30. Scott JR, Barnett TC (2006) Surface proteins of Gram-positive bacteria and how they get there. Annu Rev Microbiol 60: 397–423.

31. Fyans JK, Bignell D, Loria R, Toth IT, Palmer T (2013) The ESX7type VII secretion system modulates development, but not virulence, of the plant pathogen *Streptomyces scabies* Mol Plant Pathol 14: 119–130.

32. Abdallah AM, Gey van Pittius NC, DiGiuseppe Champion PA, Cox J, Luirink J, et al. (2007) Type VII secretion – mycobacteria show the way. Nat Rev 5: 883–891.

33. Johnson TL, Abendroth J, Hol WGJ, Sandkvist M (2006) Type II secretion: from structure to function. FEMS Microbiol Lett 255: 175–186.

34. Dow JM, Daniels MJ, Dums F, Turner PC, Gough C (1989) Genetic and biochemical analysis of protein export from *Xanthomonas campestris*. J Cell Sci Suppl 11: 59–72.

35. Filloux A, Bally M, Ball G, Akrim M, Tommassen J, et al. (1990) Protein secretion in gram-negative bacteria: transport across the outer membrane involves common mechanisms in different bacteria. EMBO J 9: 4323–4329.

36. Reeves PJ, Whitcombe D, Wharam S, Gibson M, Allison G, et al. (1993) Molecular cloning and characterization of 13 out genes from *Erwinia carotovora* subspecies *carotovora*: genes encoding members of a general secretion pathway (GSP) widespread in Gram-negative bacteria. Mol Microbiol 8: 443–456.

37. DeShazer D, Brett PJ, Burtnick MN, Woods DE (1999) Molecular characterization of genetic loci required for secretion of exoproducts in *Burkholderia pseudomallei*. J Bacteriol 181: 4661–4664.

38. Mastronunzio JE, Tisa LS, Normand P, Benson DR (2008) Comparative secretome analysis suggests low plant cell wall degrading capacity in *Frankia* symbionts. BMC Genomics 9: 47 doi: 10.1186/1471-2164-9-47.

39. Krause A, Ramakumar A, Bartels D, Battistoni F, Bekel T, et al. (2007) Complete genome of the mutualistic, N$_2$-fixing grass endophyte *Azoarcus* sp. strain BH72. Nat Biotechnol 24: 1385–1391.

40. Hubber A, Vergunst AC, Sullivan JT, Hooykaas PJJ, Ronson CW (2004) Symbiotic phenotypes and translocated effector proteins of the *Mesorhizobium loti* strain R7A VirB/D4 type IV secretion system. Mol Microbiol 25: 561–574.

41. Cases I, de Lorenzo V, Ouzounis CA (2003) Transcription regulation and environmental adaptation in bacteria. Trends Microbiol 11: 248–253.

42. Bentley SD, Chater KF, Cerdeño-Tárraga AM, Challis GL, Thomson NR, et al. (2002) Complete genome sequence of the model actinomycete *Streptomyces coelicolor* A3(2). Nature 417: 141–147.

43. Kaneko T, Nakamura Y, Sato S, Asamizu E, Kato T et al. (2000) Complete genome structure of the nitrogen-fixing symbiotic bacterium *Mesorhizobium loti*. DNA Res 7: 331–338.

44. Lopes-Santos C, Correia-Neves M, Moradas-Ferreira P, Vaz-Mendes M (2009) A walk into the LuxR regulators of actinobacteria: phylogenomic distribution and functional diversity. Plos One 7: e46758.

45. Patankar AV, Gonzalez JE (2009) Orphan LuxR regulators of quorum sensing. FEMS Microbiol Rev 33: 739–756.

46. Compant S, Duffy B, Nowak J, Clément C, Barka EA (2005) Use of plant-growth-promoting bacteria for biocontrol of plant diseases: principles, mechanisms of action, and future prospects. Appl Environ Microbiol 71: 4951–4959.

47. Taghavi S, van der Lelie D, Hoffman A, Zhang YB, Walla MD, et al. (2010) Genome sequence of the plant growth promoting endophytic bacterium

48. *Enterobacter* sp. 638. Plos Genet 6 e1000943. doi:10.1371/jounal.pgen.1000943.

48. Pujic P, Fournier P, Alloisio N, Hay AE, Maréchal J, et al. (2012) Lectin genes in the *Frankia alni* genome. Arch Microbiol 194: 47–56.

49. Lykidis A, Mavromatis K, Ivanova N, Anderson I, Land M, et al. (2007) Genome sequence and analysis of the soil cellulolytic actinomycete *Thermobifida fusca* YX. J Bacteriol 189: 2477–2486.

50. Anderson I, Abt B, Lykidis A, Klenk HP, Kyrpides N, et al. (2012) Genomics of aerobic cellulose utilization sytems in actinobacteria. PlosOne 7 e39331.

51. Jayani RS, Saxena S, Gupta R (2005) Microbial pectinolytic enzymes: a review. Process Biochem 40: 2931–2944.

52. Jakob K, Kniskern J, Bergelson MJ (2007) The role of pectate lyase and the jasmonic acid defense response in *Pseudomonas viridiflava* virulence. Mol Plant Microb Interact 20: 146–158.

53. McQueen-Mason S, Durachko DM, Cosgrove DJ (1992) Two endogenous proteins that induce cell wall extension in plants. Plant Cell 4: 1425–1433.

54. Georgelis N, Tabuchi A, Nikolaidis N, Cosgrove DJ (2011) Structure-function analysis of the bacterial expansin EXLX1. J Biol Chem 286: 16814–16823.

55. McQueen-Mason SJ, Cosgrove DJ (1995) Expansin Mode of Action on Cell Walls - Analysis of wall hydrolysis, stress relaxation, and binding. Plant Physiol107: 87–100.

56. Cosgrove DJ (2000) Loosening of plant cell walls by expansins. Nature 407: 321–326.

57. Bignell DR, Huguet-Tapia JC, Joshi MV, Pettis GS, Loria R (2010) What does it take to be a pathogen: genomic insights from *Streptomyces* species. Antonie van Leeuwenhoek 98: 179–194.

58. Jahr H, Dreider J, Meletzus D, Bahro R, Eichenbalub R (2000) The endo-beta-1,4-glucanase CelA of *Clavibacter michiganensis* subsp. *michiganensis* is a pathogenicity determinant required for induction of bacterial wilt of tomato. Mol Plant Microb Interact 13: 703–14.

59. Laine MJ, Haapalainen M, Wahlroos T, Kankare K, Nissinen R, et al. (2000) The cellulase encoded by the native plasmid of *Clavibacter michiganensis* ssp. *sepedonicus* plays a role in virulence and contains an expansin-like domain. Physiol Mol Plant Pathol 57: 221–233.

60. Kerff F, Amoroso A, Herman R, Sauvage E, Petrella S, et al. (2008) Crystal structure and activity of *Bacillus subtilis* YoaJ (EXLX1), a bacterial expansin that promotes root colonization. Proc natl Acad Sci USA 105: 16876–16881.

61. Dudeja SS, Giri R, Saini R, Suneja-Madan P, Kothe E (2012) Interaction of endophytic microbes with legumes. J Basic Microbiol 52: 248–60.

62. Cerda ME (2008) Aislamiento de *Micromonospora* de nódulos de leguminosas tropicales y análisis de su interés como promotor del crecimiento vegetal. Ph.D. Thesis. Universidad de Salamanca, Spain.

63. Streeter JG (1985) Accumulation of alpha, alpha-trehalose by *Rhizobium* bacteria and bacteroids. J Bacteriol 164: 78–84.

64. Farías-Rodriguez R, Mellor RB, Arias C, Peña-Cabriales JJ (1998) The accumulation of trehalose in nodules of several cultivars of common bean (*Phaseolus vulgaris*) and its correlation with resistance to drought stress. Physiol Plant 102 353–359.

65. Altamirano-Hernández J, López MG, Acosta-Gallegos JA, Farías-Rodríguez R, Peña-Cabriales JJ (2007) Influence of soluble sugars on seed quality in nodulated common bean (*Phaseolus vulgaris* L.): the case of trehalose. Crop Sci 47: 1193–1205.

66. Aeschbacher RA, Müller J, Boller T, Wiemken A (1999) Purification of the trehalase GMTRE1 from soybean nodules and cloning of its cDNA. GMTRE1 is expressed at a low level in multiple tissues. Plant Physiol 119: 489–495.

67. Barraza AG, Estrada-Navarrete G, Rodriguez-Alegria ME, Lopez-Munguia A, Merino E, et al. (2013) Down-regulation of PvTRE1 enhances nodule biomass and bacteroid number in the common bean. New Phytol 197: 194–206.

68. Müller J, Boller T, Wiemken A (2001) Trehalose becomes the most abundant non-structural carbohydrate during senescence of soybean nodules. J Experimental Bot 52 943–947.

69. Frederiksen RF, Paspaliari DK, Larsen T, Storgaard BG, Larsen MH, et al. (2013) Bacterial chitinases and chitin-binding proteins as virulence factors. Microbiology 159: 833–847.

70. Taguchi F, Suzuki T, Inagaki Y, Toyoda K, Shiraishi T, et al. (2010) The siderophore pyoverdine of *Pseudomonas syringae* pv. tabaci 6605 is an intrinsic virulence factor in host tobacco infection. J Bacteriol 192: 117–126.

71. Smits THM, Jaenicke S, Rezzonico F, Kamber T, Blom J, et al. (2010) Complete genome sequence of the fire blight pathogen *Erwinia amylovora* CFBP 1430 and comparsion to other *Erwinia* spp. Mol Plant Microbe Interact 23: 384–393.

72. Tivendale KA, Allen JL, Ginns CA, Crabb BS, Browning GF (2004) Association of *iss* and *iucA*, but not *tsh*, with plasmid-mediated virulence of avian pathogenic *Escherichia coli*. Infect Immun 72: 6554–6560.

73. Cuiv PO, Clarke P, Lynch D, O'Connell M (2004) Identification of *rhtX* and *fptX*, novel genes encoding proteins that show homology and function in the utilization of the siderophores rhizobactin 1021 by *Sinorhizobium meliloti* and pyochelin by *Pseudomonas aeruginosa*, respectively. J Bacteriol 186: 2996–3005.

74. Genilloud O (2014) The re-emerging role of microbial natural products in antibiotic discovery. Antonie van Leeuwenhoek 106: 173–178.

75. Udwary DW, Zeigler L, Asolkar RN, Vasanth S, Alla L, et al. (2007) Genome sequencing reveals complex secondary metabolome in the marine actinomycete *Salinispora tropica*. Pro Natl Acad Sci 104: 10376–10381.

76. Nett M, Ikeda H, Moore BS (2009) Genomic basis for natural product biosynthetic diversity in the actinomycetes. Nat Prod Rep 26: 1362–1384.

77. Udwary DW, Gontang EA, Jones AC, Jones CS, Schultz AW, et al. (2011) Significant natural product biosynthetic potential of actinorhizal symbionts of the genus *Frankia*, as revealed by comparative genomic and proteomic analyses. Appl Environ Microbiol 77: 3617–3625.

78. Velázquez-Robeldo R, Contreras-Cornejo H, Macías-Rodriguez LI, Hernández-Morales A, Aguirre J, et al. (2011) Role of the 4-phosphopantetheinyl transferase of *Trichoderma virens* in secondary metabolism, and induction of plant defense responses. Mol Plant Microb Interact 24: 1459–1471.

79. Mukherjee PK, Buensanteai N, Moran-Diez ME, Druzhinina IS, Kenerley CM (2012) Functional analysis of non-ribosomal peptide synthetases (NRPs) in *Trichoderma virens* reveals a polyketide synthase (PKS)/NRPS hybrid enzyme involved in the induced systemic resistance response in maize. Microbiol 158: 155–165.

80. Tornus D, Floss HG (2001) Identification of four genes from the granaticin biosynthetic gene cluster of Stretpomyces violaceoruber Tü22 involved in the biosynthesis of L-rhodisone. J Antibiot 54: 91–101.

81. Tadra-Sfeir MZ, Souza EM, Faoro H, Müller-Santos M, Baura VA, et al. (2011) Naringenin regulates expression of genes involved in cell wall synthesis in *Herbaspirillum seropedicae* Appl Environ Microbiol 77: 2180–2183.

82. Abdel-Lateif K, Boguz D, Hocher V (2012) The role of flavonoids in the establishment of plant roots endosymbioses with arbuscular mycorrhiza fungi, rhizobia and *Frankia* bacteria. Plant Signal Behav 7: 636–641.

83. Hammad Y, Nalin R, Marechal J, Fiasson K, Pepin R, et al. (2003) A possible role for phenyl acetic acid (PAA) on *Alnus glutinosa* nodulation by *Frankia* bacteria. Plant and Soil 254: 193–206.

84. Duca D, Lorv J, Patten CL, Rose D, Glick BR (2014) Indole-3-acetic acid in plant-microbe interactions. Antonie van Leeuwenhoek 106: 85–125.

85. Spaepen S, Vanderleyden J, Remans R (2007) Indole-3-acetic acid in microbial and microorganism-plant signaling. FEMS Microbiol Rev 31: 425–448.

86. Kobayashi M, Izui H, Nagasawa T, Yamada H (1993) Nitrilase in biosynthesis of the plant hormone indole-3-acetic acid from indole-3-acetonitrile: cloning of the *Alcaligenes* gene and site-directed mutagenesis of cysteine residues. Proc Natl Acad Sci 90: 247–251.

87. Kobayashi M, Suzuki T, Fujita T, Masuda M, Shimizu S (1995) Occurrence of enzymes involved in biosynthesis of indole-3-acetic acid from indole-3-acetonitrile in plant-associated bacteria, *Agrobacterium* and *Rhizobium*. Proc Natl Acad Sci 92: 714–718.

88. Ryu CM, Farag MA, Hu CH, Reddy MS, Wei HX, et al. (2003) Bacterial volatiles promote growth in Arabidopsis. Proc Natl Acad Sci 100: 4927–4933.

89. Ochman H, Wilson AC (1987) Evolution in bacteria: evidence for a universal substitution rate in cellular genomes. J Mol Evol 26: 74–86.

90. Bell CD, Soltis DE, Soltis PS (2010) The age and diversification of the angiosperms re-revisited. Amer J Bot 97: 1–8.

91. Zakhia F, Jeder H, Willems A, Gillis M, Dreyfus B, et al. (2006) Diverse bacteria associated with root nodules of spontaneous legumes in Tunisia and first report for nifH-like gene within the genera Microbacterium and Starkeya. Microb Ecol 51: 375–393.

92. Gtari M, Ghodhbane-Gtari F, Nouioui I, Beauchemin N, Tisa LS (2012) Phylogenetic perspectives of nitrogen-fixing actinobacteria. Arch Microbiol 194: 3–11.

93. Pedrosa FO, Monteiro RA, Wassem R, Cruz LM, Ayub RA, et al. (2011) Genome of *Herbaspirillum seropedicae* strain SmR1, a specialized diazotrophic endophyte of tropical grasses. PLoS Genet 7: e1002064. doi:10.1371/journal.pgen.1002064.

94. Wisniewski-Dyé F, Borziak K, Khalsa-Moyers G, Alexandre G, Sukharnikov LO, et al. (2011) *Azospirillum* genomes reveal transition of bacteria from aquatic to terrestrial environments. Plos Genetics 7: e1002430. doi: 10.1371/journal.pgen.1002430.

95. Medie FM, Davies GJ, Drancourt M, Henrissat B (2012) Genome analyses highlight the different biological roles of cellulases. Nat Rev Microbiol 10: 227–234.

96. Alonso-Vega P, Normand P, Bacigalupe R, Pujic P, Lajus A, et al. (2012) Genome sequence of *Micromonospora lupini* Lupac 08, isolated from root nodules of *Lupinus angustifolius*. J Bacteriol 194: 4135.

97. Vallenet D, Labarre L, Rouy Z, Barbe C, Bocs S, et al. (2006) MaGe–a microbial genome annotation system supported by synteny results. Nucleic Acids Res 34: 53–65.

98. Thompson JD, Gibson TJ, Plewniak F, Jeanmougin F, Higgins DG (1997) The CLUSTAL_X windows interface: flexible strategies for multiple sequence alignment aided by quality analysis tools. Nucleic Acids Res 25: 4876–4882.

99. Kimura M (1980) A simple method for estimating evolutionary rates of base substitutions through comparative studies of nucleotide sequences. J Mol Evol 16: 111–120.

100. Felsenstein J (1981) Evolutionary trees from DNA sequences: a maximum likelihood approach. J Mol Evol 17: 368–376.

101. Tamura K, Peterson D, Peterson N, Stecher G, Nei M, et al. (2011) MEGA5: molecular evolutionary genetics analysis using maximum likelihood, evolutionary distance, and maximum parsimony methods. Mol Biol Evol 10: 2731–2739.

102. Darling AE, Mau B, Perna NT (2010) progressiveMauve: multiple genome alignment with gene gain, loss and rearrangement. PLoS ONE 5: e11147. doi: 10.1371/journal.pone.0011147.

103. Blom J, Albaum S, Doppmeier D, Pühler A, Vorhölter FJ, et al. (2009) EDGAR: a software framework for the comparative analysis of prokaryotic genomes. BMC Bioinformatics 10: 154. doi: 10.1186/1471-2105-10-154.

104. Vernikos GS, Parkhill J (2006) Interpolated variable order motifs for identification of horizontally acquired DNA: revisiting the *Salmonella* pathogenicity islands. Bioinformatics 22: 2196–2203.

105. Vicente-Villardón JL (2010) MULTBIPLOT: A package for Multivariate Analysis using Biplots. Departamento de Estadística. Universidad de Salamanca. Available: http://biplot.usal.es/ClassicalBiplot/index.html.

106. Ren Q, Chen K, Paulsen IT (2007) TransportDB: a comprehensive database resource for cytoplasmic membrane transport systems and outer membrane channels. Nucleic Acids Res 35: 274–279.

107. Rowbotham TJ, Cross T (1977) Ecology of *Rhodococcus coprophilus* and associated actnomycetes in fresh water and agricultural habitats. J Gen Microbiol 100: 231–240.

108. Mateos PF, Jimenez-Zurdo JI, Chen J, Squartini AS, Haack SK, et al. (1992) Cell-associated pectinolytic and cellulolytic enzymes in *Rhizobium leguminosarum* biovar trifolii. Appl Environ Microbiol 58: 1816–1822.

109. Williams ST, Goodfellow M, Alderson G, Wellington EMH, Sneath PHA, et al. (1983) Numerical classification of *Streptomyces* and related genera J Gen Microbiol 129: 1743–1813.

110. Murthy N, Bleakley B (2012) *Simplified method of preparing colloidal chitin used for screening of chitinase-producing microorganisms*. Internet J Microbiol 10: DOI:10.5580/2bc3.

111. Milagres AMF, Machuca A, Napoleão D (1999) Detection of siderophore production from several fungi and bacteria by a modification of chrome azurol S (CAS) agar plate assay. J Microbiol Meth 37: 1–6.

112. Glickmann E, Dessaux Y (1995) A critical examination of the specificity of the salkowski reagent for indolic compounds produced by phytopathogenic bacteria. Appl Environ Microbiol 61: 793–796.

113. Gordon SA, Weber RP (1951) Colorimetric estimation of indole acetic acid. Plant Physiol 26: 192–195.

114. Rigaud J, Puppo A (1975) Indole-3-acetic acid catabolism by soybean bacteroids J Gen Microbiol 88: 223–228.

115. Hardy RWF, Hoilsten RD, Jackson EK, Burns RC (1968) The acetylene-ethylene assay for N_2 fixation: laboratory and field evaluation. Plant Physiol 43: 1185–1207.

Biological Instability in a Chlorinated Drinking Water Distribution Network

Alina Nescerecka[1,2], Janis Rubulis[1], Marius Vital[2], Talis Juhna[1], Frederik Hammes[2]*

1 Department of Water Engineering and Technology, Riga Technical University, Riga, Latvia, **2** Department of Environmental Microbiology, Eawag, Swiss Federal Institute for Aquatic Science and Technology, Dübendorf, Switzerland

Abstract

The purpose of a drinking water distribution system is to deliver drinking water to the consumer, preferably with the same quality as when it left the treatment plant. In this context, the maintenance of good microbiological quality is often referred to as biological stability, and the addition of sufficient chlorine residuals is regarded as one way to achieve this. The full-scale drinking water distribution system of Riga (Latvia) was investigated with respect to biological stability in chlorinated drinking water. Flow cytometric (FCM) intact cell concentrations, intracellular adenosine tri-phosphate (ATP), heterotrophic plate counts and residual chlorine measurements were performed to evaluate the drinking water quality and stability at 49 sampling points throughout the distribution network. Cell viability methods were compared and the importance of extracellular ATP measurements was examined as well. FCM intact cell concentrations varied from 5×10^3 cells mL^{-1} to 4.66×10^5 cells mL^{-1} in the network. While this parameter did not exceed 2.1×10^4 cells mL^{-1} in the effluent from any water treatment plant, 50% of all the network samples contained more than 1.06×10^5 cells mL^{-1}. This indisputably demonstrates biological instability in this particular drinking water distribution system, which was ascribed to a loss of disinfectant residuals and concomitant bacterial growth. The study highlights the potential of using cultivation-independent methods for the assessment of chlorinated water samples. In addition, it underlines the complexity of full-scale drinking water distribution systems, and the resulting challenges to establish the causes of biological instability.

Editor: Jose Luis Balcazar, Catalan Institute for Water Research (ICRA), Spain

Funding: The authors acknowledge the financial support of the EU project "TECHNEAU" (Nr. 018320) and the NMS-CH project "BioWater: Assessment of biological stability in drinking water distribution networks with chlorine residuals" (Sciex-N-7 12.265). The funders had no role in study design, data collection and analysis, decision to publish, or preparation of the manuscript.

Competing Interests: The authors have declared that no competing interests exist.

* E-mail: frederik.hammes@eawag.ch

Introduction

The goal of public drinking water supply systems is to produce water of acceptable aesthetic and hygienic quality and to maintain that quality throughout distribution until the point of consumption. From a microbiological perspective, the quality of treated water can deteriorate as a result of excessive bacterial growth, which can lead to problems such as a sensory deterioration of water quality (e.g. taste, odor, turbidity, discoloration) as well as pathogen proliferation [1–10]. To avoid this, biological stability during distribution can be achieved by maintaining sufficient residual disinfectants in the water, and/or through nutrient limitations [3,7,11,12]. However, drinking water systems should not be viewed as sterile; complex indigenous bacterial communities have been shown to inhabit both chlorinated and non-chlorinated drinking water distribution systems [5,13–17].

The concept of biological stability and its impact on a system's microbiology has been discussed extensively in the framework of non-chlorinated drinking water distribution systems [3,7,17–20]. However, many treatment plants worldwide employ a final disinfection step to ensure that no viable bacteria enter the distribution system. The latter is often achieved by oxidative disinfection, usually by chlorination [21]. Disinfection has a number of implications for a biological system. During chlorination, one can expect that a considerable fraction of bacteria in the

water are killed or damaged, while some residual chlorine may remain in the water (Figure 1). This could be visible through numerous microbial monitoring methods. For example, the number of cultivable bacteria, measured with heterotrophic plate counts, would reduce dramatically [22,23]. Secondly, bacteria cells are likely to display measurable membrane damage irrespective of their cultivability [24], though the rate and extent of damage may differ between different communities. This would be detectable with several staining techniques coupled with epifluorescence microscopy or flow cytometry (FCM). Also, adenosine tri-phosphate (ATP), often used as a cultivation-independent viability method [19,22,25] will be severely affected. Based on data from Hammes and co-workers [4] one may reasonably expect increased levels of extracellular ATP (so-called free ATP) and decreased concentrations of intracellular ATP (bacterial ATP) following oxidative disinfection. Irrespective of the detection method, the overall consequence of disinfection is a considerable decrease in the viable biomass, potentially opening a niche for microorganisms to occupy downstream of the treatment process. Following initial disinfection, residual chlorine might provoke undesirable changes during drinking water distribution. Disinfectants target not only bacteria, but it also react with natural organic matter, pipe surfaces and particles in the network, thus potentially forming/releasing assimilable organic carbon (AOC) [26–30]. AOC can easily be consumed by bacteria, and is therefore seen as a main contributor to biological instability. Moreover, chlorine decay within the network negatively affects its ability to inhibit microbial growth at the far ends of the network [12]. If all factors were considered, the presence of nutrients, a reduction in the number of competing bacteria, and the lack of residual disinfectant would potentially lead to biological instability in the distribution network, manifesting in a subsequent bacterial growth (Figure 1). Besides the importance of nutrients, the extent of bacterial growth will be influenced by a number of factors. For example, increased water temperature can accelerate chlorine decay and favor bacteria growth [19,31], while changes in hydraulic conditions can alter nutrient supply for microorganisms in biofilms and/or bacteria detachment from the pipe surfaces [32,33]. Finally, the quality of materials in contact with drinking water, as well as the presence of sediments and loose deposits, can both affect the general microbial quality of the water [6,34,35].

In the present study we examined some of the above-discussed concepts in a full-scale, chlorinated distribution system in the city of Riga (Latvia) with a number of microbiological methods. The purpose was a detailed investigation of the entire city's distribution network, asking the basic question whether evidence of spatial and/or temporal biological instability exists, and if so, to which degree. Additional goals were to evaluate the use of fluorescent staining coupled with FCM, as well as ATP analysis, for the assessment of chlorinated drinking water in a distribution network with disinfectant residuals.

Materials and Methods

Ethics statement

Permission for sampling at all locations in the present study was obtained from the local water utility (Rīgas Ūdens).

Description of study site

Sampling was performed in the full-scale distribution network of Riga (Latvia) with a total length of about 1400 km. The city is supplied with drinking water from six water treatment plants (WTP) produced from both surface and groundwater (150 000 m^3 d^{-1}). Only the three major WTP, which are continuously operated, were included in the sampling campaign. Average WTP effluent water quality parameters for each treatment plant are shown in Table 1. The distribution network mainly consists of cast iron (80%) and unlined iron (15%) pipes as old as 50 years.

The diameters of pipes ranged from 100 to 1200 mm. Three reservoirs are operated in the network to compensate for fluctuations in the daily water demand, while four high-pressure zones are maintained in some distal areas of the network. The high-pressure zones were excluded from the present study. A total of 49 sampling sites were selected across the city to cover the network broadly and to include both proximal and distal zones relative to the treatment plants. The sampling sites were selected according to the approximate water retention times obtained from a validated hydraulic model made in EPANET 2.0 [36,37] based on a total length of 538 km (39% of the total length of the network). Apart from the effluents of the three treatment plants, the sampling sites were in all cases fire hydrants in order to attain some degree of reproducibility between sampling and to avoid localized effects (e.g. household growth). The exact locations of sampled fire hydrants can be obtained from the authors after agreement from the local water utility.

Sampling protocol

A specific sampling protocol was designed and followed in order to avoid artifacts due to water stagnation in unused fire hydrants. Each hydrant was pre-flushed at a high velocity (never exceeding 1.6 m s^{-1}) for no more than 60 s, then immediately adjusted to a low velocity of 0.015–0.25 m s^{-1} and connected to an online system for monitoring pH, temperature, redox potential, electro-conductivity and turbidity. The low sampling velocity was specifically used to ensure a minimal possible impact of cell wall erosion and detachment from biofilms on the samples and measurements. Readings of all parameters were taken at 5–10 minute intervals, and water was only sampled for microbiological analysis once all of the parameters stabilized. The impact of this hydrant flushing is demonstrated in an example in Figure 2 and discussed in detail in the results section. Samples were kept in cold storage ($\approx 5°C$) and analyzed within four hours of sampling.

Chemical analysis

Determination of free chlorine was performed according to standard method EN ISO 7393-1, based on the direct reaction with N,N-diethyl-1,4-phenylenediamine (DPD) and subsequent formation of a red compound at pH 6.2–6.5. Afterwards titration by means of a standard solution of ammonium iron (III) sulfate until disappearance of the red color was performed. Determination of total chlorine was performed according to EN ISO 7393-1,

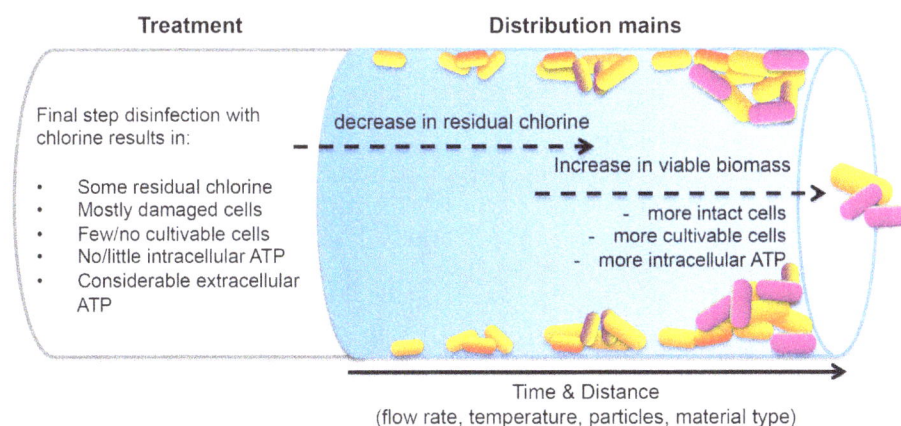

Treatment   **Distribution mains**

Final step disinfection with chlorine results in:

- Some residual chlorine
- Mostly damaged cells
- Few/no cultivable cells
- No/little intracellular ATP
- Considerable extracellular ATP

decrease in residual chlorine

Increase in viable biomass
- more intact cells
- more cultivable cells
- more intracellular ATP

Time & Distance
(flow rate, temperature, particles, material type)

Figure 1. A worst-case-scenario in an unstable, chlorinated distribution network. Prediction of changes in the microbiological state of the water due to the depletion of residual chlorine and the concomitant growth of bacteria, potentially resulting in hygienic and sensory deterioration of the water quality.

Table 1. Average water quality parameters for the final effluents of the the three main treatment plants of Riga (Latvia).

	WTP 1	WTP 2	WTP 3
Source water	surface water	artificially recharged groundwater	groundwater
Final treatment step[a]	Cl_2 (0.5–3 mg L^{-1})	Cl_2 (ca. 1.5 mg L^{-1})	N.A.
Residual chlorine (mg L^{-1})[b]	0.44±0.11	0.51±0.01	0.42±0.26
Total organic carbon (TOC) (mg L^{-1})[a]	6±1	9±3	3
Assimilable organic carbon (AOC) (μg L^{-1})[a]	213±37	209±59	N.A.
Total cell concentration (cells mL^{-1})[b]	$5.31\pm0.97\times10^5$	$5.45\pm0.47\times10^5$	$1.69\pm0.18\times10^5$
Intact cell concentration (cells mL^{-1})[b]	$1.83\pm1.18\times10^4$	$1.4\pm0.86\times10^4$	$1.03\pm0.68\times10^4$
Total ATP (nM)[b]	0.015±0.005	0.029±0.004	0.011±0.002
Intracellular ATP (nM)[b]	0.007±0.003	0.000±0.004	0.001±0.002
HPC 22°C (CFU mL^{-1})[b]	23±24	4±2	4
HPC 36°C (CFU mL^{-1})[b]	16±16	4±2	1
Conductivity (μS cm^{-1}), 25°C[a]	468±101	625±4	272±25
pH[a]	6.63±0.18	7.41±0.04	7.5±0.05

[a]Data supplied by the water utility or measured in previous sampling campaigns.
[b]Data from present study.

based on the reaction with DPD in the presence of an excess of potassium iodide, and then titration as described above.

Fluorescent staining and flow cytometry (FCM) of water samples

Staining and FCM analysis was done as described previously [4,38]. In short, for a working solution, SYBR® Green I (SG) stock (Invitrogen AG, Basel, Switzerland) was diluted 100x in anhydrous dimethylsulfoxide (DMSO) and propidium iodide (PI; 30 mM) was mixed with the SYBR® Green I working solution to a final PI concentration of 0.6 mM. This working solution was stored at $-20°C$ until use. From every water sample, 1 mL was stained with SGPI at 10 μL mL^{-1}. Before analysis, samples were incubated in the dark for 15 minutes. Prior to FCM analysis, the water samples were diluted with 0.22 μm filtered bottled water (Evian, France) to

10% v/v of the initial concentration. FCM was performed using a Partec CyFlow SL instrument (Partec GmbH, Münster, Germany), equipped with a blue 25 mW solid state laser emitting light at a fixed wavelength of 488 nm. Green fluorescence was collected at 520±10 nm, red fluorescence above 630 nm, and high angle sideward scatter (SSC) at 488 nm. The trigger was set on the green fluorescence channel and data were acquired on two-parameter density plots while no compensation was used for any of the measurements. The CyFlow SL instrument is equipped with volumetric counting hardware and has an experimentally determined quantification limit of 1000 cells mL^{-1} [4].

Adenosine tri-phosphate (ATP) analysis

Total ATP was determined using the BacTiter-Glo reagent (Promega Corporation, Madison, WI, USA) and a luminometer

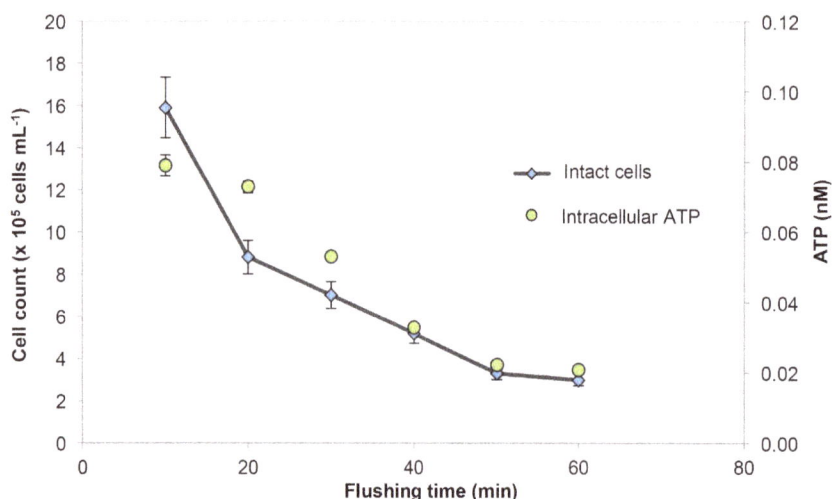

Figure 2. The impact of low velocity flushing on the water quality in a newly-opened fire hydrant. Intracellular adenosine tri-phosphate (ATP) data points were derived from duplicate measurements of extracellular and total ATP concentrations. FCM intact cells (after SYBR Green I and propidium iodide staining) were single measurements, with a relative standard deviation of 9% calculated from all data in the present study.

(Glomax, Turner Biosystems, Sunnyvale, CA, USA) as described elsewhere [25]. A water sample (500 µl) and the ATP reagent (50 µl) were warmed to 38°C simultaneously in separate sterile Eppendorf tubes. The sample and the reagent were then combined and then the luminescence was measured after 20 s reaction time at 38°C. The data were collected as relative light units (RLU) and converted to ATP (nM) by means of a calibration curve made with a known ATP standard (Promega). For extracellular ATP analysis, each sample was filtered through a 0.1 µm sterile syringe filter (Millex-GP, Millipore, Billerica, MA, USA), followed by analysis as described above. The intracellular ATP was calculated by subtracting the extracellular ATP from the total ATP for each individual sample. ATP was measured in duplicate for all samples.

Heterotrophic plate counts

To obtain heterotrophic plate counts (HPC), samples were serially diluted in sterile distilled water and then inoculated onto nutrient yeast agar plates using the spread plate technique. All plates were incubated in dark at 22°C or 36°C for 3 and 7 days, respectively. Results were expressed as colony forming units (CFU) per ml of water sample.

Statistical analysis

Statistical data evaluation was performed with the MS Excel Data Analysis tool (Descriptive statistics, Regression). The reproducibility for indirect/calculated data (e.g., intracellular ATP) was calculated by a propagation-of-uncertainty method. FCM data was not always measured in duplicate, due to practical constraints. In these cases, a 9% error (average coefficient of variation (CV) (n = 39)) was applied for representing FCM data. The residual chlorine concentration distribution box plot was created using on-line calculator on http://www.physics.csbsju.edu/stats/.

Results and Discussion

The importance of correct sampling

Sample collection during this study elucidated some of the problems specific for this network and highlighted the broader importance of correct sampling procedures. Fire hydrants were selected as sampling points to enable direct access to the distribution network and avoid potential household effects [39]. We opted for a low velocity water flow in combination with online monitoring to achieve comparable samples. In some cases, the water initially emerging from the fire hydrants were visibly turbid and/or discolored (data not shown). Turbid water is clearly unwanted and serves as a first visual confirmation of some form of system failure. In this regard, a recent study in the Netherlands has established an important link between suspended solids and microbial growth and biological instability [6]. Hence in some instances continuous low velocity flushing of up to 60 minutes was required before stable values for chemical and physical parameters as well as microbiological parameters were obtained (Figure 2; Table S1). The data in Figure 2 demonstrate clearly the need for a carefully planned sampling protocol when assessing full-scale systems. It should be noted that Figure 2 represents an example of some of the worst sampling points in the system. Data from other hydrants often showed less fluctuation during flushing (Figure S1). One potential problem during the sampling procedure is the re-suspension of sediments/particles and sloughing of biofilms from the pipes, causing artifacts in the measurements. In this respect, we specifically employed a low velocity (0.015–0.25 m s^{-1}) pre-sampling flushing procedure. The latter differs from extreme flushing applied for network cleaning, which is operated with high velocities of 1.5–1.8 m s^{-1} [40,41]. According to Antoun and co-workers [40] low-velocity flushing (below 0.3 m s^{-1}) does not cause any scouring actions. However, it should be considered that part of the samples, especially during the first minutes of the flushing, can cointain biofilm bacteria detatched in a result of pre-flushing [35].

The concept of detecting instability: a single point in the distribution network

In the introduction we proposed the straightforward hypothesis that biological parameters would show an increase between the point of treatment and a point during distribution in case of biological instability (Figure 1). Before the relation between different parameters and the impact on the entire network are discussed in detail below, a single sampling point is compared to its source water as an example to illustrate the concept (Figure 3A). The point was selected on the basis of (1) hydraulic data linking it with a specific WTP, (2) its medial distance from WTP (neither too close and nor too far from the WTP) and, (3) the fact that all microbiological parameters (FCM, ATP and HPC) as well as residual chlorine measurements were performed on this sample. For the purpose of clarity, the data was normalized to the values of the treated water and expressed as the relative change (the raw data and standard deviations for the data in Figure 3A are shown in Figure S2). Evidently the data from Figure 3A supports the basic hypothesis. The microbial parameters such as intact cell concentration, ATP and colony forming units all show a considerable increase in their values. Simultaneously, only 12% (0.06 mg L^{-1}) of the initial residual chlorine concentration (0.5 mg L^{-1}) was left in the water sample. The data suggests that the residual chlorine in the network was not sufficient to inhibit microbial growth, concurring with earlier report from Prévost and colleagues [42] showing increased HPC, total direct and direct viable bacteria counts in a distribution network coinciding with chlorine depletion. Other studies also showed the presence of viable bacteria in water with chlorine concentration lower than 0.1 mg L^{-1} [23] and that residual chlorine levels below 0.07 mg L^{-1} allows bacterial growth [12]. Data of residual chlorine concentrations in the drinking water network is summarized in Figure S3. Evidently a considerable fraction of samples (18%) had residual chlorine concentrations below 0.1 mg L^{-1}.

Staining of bacteria with fluorescent dyes was previously suggested as a way to distinguish between viable and damaged bacteria in real water samples [43,44], and the application of this approach has been successfully demonstrated in laboratory scale chlorination studies [24,45]. One focus point of the present study was to determine whether FCM combined with viability staining can be used for a fast and meaningful assessment of viable bacteria in chlorinated drinking water systems. The same samples from Figure 3A, stained with SYBR Green and propidium iodide (SGPI), are shown as density plots obtained with FCM (Figure 3B). The theory behind the staining method and the interpretation of such data are discussed in detail elsewhere [20,24,38,43,46]. In the treatment plant sample, where the water was recently exposed to chlorine, 98% of all cells were measured as membrane compromised, seen by absence of events inside the gated area of the plot (Figure 3B). In the distribution network (DN) sample, a high concentration of intact cells appeared (Figure 3B). Since these intact cells were clearly not present in the influent, the plausible conclusion is that the bacterial growth occurred during distribution.

A

B

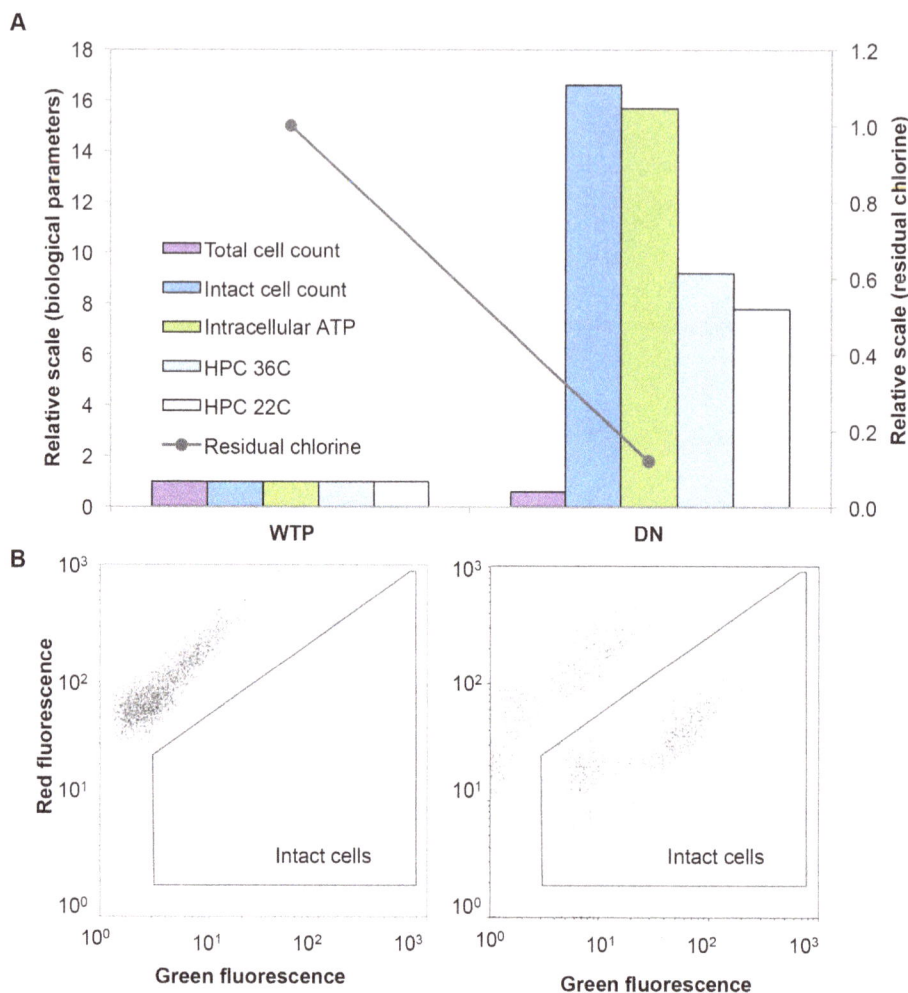

Figure 3. Changes in bacterial parameters between water treatment plant (WTP) and distribution network (DN) sampling points. (A) For comparison, all values at the WTP were set to 1, and values in the DN were expressed relative to their values at the WTP. The original raw data for these samples are shown in Figure S2. Data points are average values for duplicate FCM and ATP measurements and triplicate HPC measurements; (B) Flow cytometric density plots of samples stained with SYBR Green I and propidium iodide, showing the intact cell concentration at the plant and in the specific network point.

Detailed assessment of dynamic changes in a single point

High frequency monitoring of a single sampling point revealed temporal instability in the distribution network. We monitored the effluent of one treatment plant and one point in the network with 1-hour intervals during a day (ca. 21 h). The sampling was arranged in such a way that the network sampling started 15 hours after the treatment plant sampling, which corresponded with the estimated water residence time (WRT) for this location. Figure 4 displays the changes of intracellular ATP and intact cell concentrations in the network and the water treatment plant. Values for both parameters were low in the water samples from the treatment plant (n = 19): intracellular ATP varied from 0.0025 nM to 0.0096 nM (mean = 0.0061±0.002 nM) and the intact cell concentration amongst 19 samples varied from 7.5×10^3 to 6.3×10^4 cells mL^{-1} (mean = $1.6 \times 10^4 \pm 1.2 \times 10^4$ cells mL^{-1} in average). In turn, the values from the distribution network point (n = 23) were significantly higher: intact cell concentrations ranged from 1.37×10^5 to 4.66×10^5 cells mL^{-1} (mean = $2.5 \times 10^5 \pm 9.9 \times 10^4$ cells mL^{-1}), and the ATP concentrations from 0.021 to 0.063 nM (mean = 0.038±0.012 nM). More-

over, a distinct pattern was apparent in the distribution network data, with values peaking at about 05:00–07:00 and again at 12:00–13:00. During both these events, the intracellular ATP data followed a similar pattern as the intact cell concentration data, with a good overall correlation ($R^2 = 0.81$; p<0.005). Although it is not evident exactly why the bacterial concentrations peaked at these specific time periods, a plausible explanation is a change in the flow velocity due to diurnal changes in water consumption by both industrial and domestic consumers. It was previously shown in laboratory scale experiments that increased flow velocity could lead to increased bacterial detachment from biofilms and a re-suspension of loose deposits, thus leading to an increase in suspended cell concentrations [32,33,47]. In addition, it is possible that lower water consumption overnight resulted in considerably reduced flow rates, and consequently a faster decay of chlorine and increased bacterial growth [42,48].

Detailed data sets of diurnal changes in the microbial quality of water mains, such as Figure 4, are particularly scarce in literature. Importantly, this clearly demonstrated temporal instability in the network for which the exact cause remains uncertain. Moreover, it

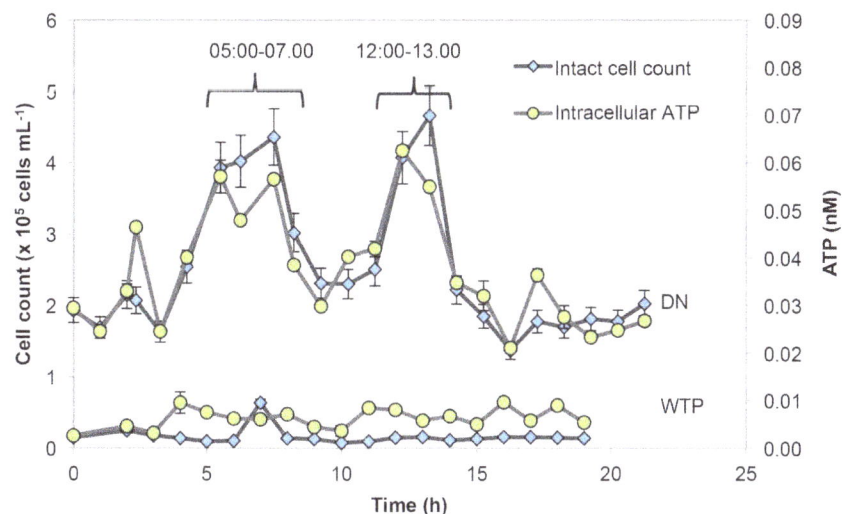

Figure 4. Diurnal changes in bacterial parameters of WTP and DN points. Intensive sampling of one WTP (n = 19) and one point in the DN (n = 23) during 21 hours reveals steady cell concentrations at the treatment plant but clear variations in the distribution network. Intracellular adenosine tri-phosphate (ATP) data points were derived from duplicate measurements of extracellular and total ATP concentrations. FCM intact cells (after SYBR Green I and propidium iodide staining) were single measurements, with a relative standard deviation of 9% calculated from all data in the present study.

shows that the absolute cell concentrations at any sampling point may be influenced by the time of sampling.

Instability data for the entire network

Full-scale distribution networks are complicated systems, not restricted to a single source or a straight distribution line [17]. The Riga distribution network is supplied with drinking water from several separate treatment plants (Table 1). One plant treats surface water from the Daugava River (WTP 1) and the others supply natural groundwater (WTP 3) and artificially recharged groundwater (WTP 2). Chlorination is applied as the final disinfection step at all plants, resulting in low concentrations of intact cells, intracellular ATP and cultivable bacteria in the effluents (Table 1). A large fraction of the active chorine is rapidly consumed due to relatively high levels of organic matter. Despite the fact that the purpose of chlorination and residual chlorine is to limit microbial growth during distribution, a considerable increase in the concentration of intact cells was detected throughout the distribution network. Figure 5A shows the range of intact cell concentrations arranged in ascending order. Treated water contained between 1.84×10^5–5.63×10^5 total cells mL^{-1} and between 9.7×10^3–2.13×10^4 intact cells mL^{-1} (hence 2–5% intact cells) depending on WTP. The data confirms effective final disinfection in all treatment plants. The total cell concentration values of the drinking water samples from the distribution network (n = 49) varied from 1.62×10^5 cells mL^{-1} to 1.07×10^6 cells mL^{-1} and the range of the intact cell concentration was from 5.28×10^3 cells mL^{-1} to 4.66×10^5 cells mL^{-1} (3–59% intact cells). Notably, 50% of all samples contained more than 1.06×10^5 intact cells mL^{-1} corresponding to an increase of at least one order of magnitude in those samples compared to effluent water, which clearly shows that bacterial growth in the distribution network was not an isolated occurrence. The observed increase in intact cell concentration is likely related to the presence of assimilable organic carbon (AOC) in the distributed water. While AOC was not measured in the present study, previous data for two of the treatment plants were high (in the range of 200 µg L^{-1}; Table 1), and nutrient availability in the water is generally regarded as a key

factor that promotes microbial growth [29,49]. It cannot be excluded that some variability in the data resulted from bacteria detached from biofilms or re-suspended from sediments during the fire hydrant sampling procedure. However, the potential adverse impact of this was minimized by the low velocity sampling protocol (see above), while the systematic increase in cell concentrations in the network clearly suggests the occurrence of biological instability rather than sampling artifacts. In contrast to these findings, several studies analyzing drinking water distribution systems without any additional residual disinfectants showed no (or only minute) changes in bacterial parameters during distribution [17,19,20]. These distributions systems rely on nutrient limitation to achieve biological stability, and while intact cell concentrations are often relatively high (ca. 1×10^5 cells mL^{-1}) [17,20], changes during distribution tend to be negligible.

To examine the spatial distribution of the growth/instability in the network, the data was divided into four broad categories based on the extent of growth (Figure 5A). These were visualized on the sampling map (Figure 5B). The sampling points with the lowest intact cell concentration (less than 5×10^4 cells mL^{-1}) are marked with green bullets. Yellow and orange colored bullets indicate higher concentrations, while the points with the highest values (over 2×10^5 cells mL^{-1}) are shown as red bullets. As could be expected, the map shows that the points with the lowest cell concentrations are mostly concentrated in areas close to the water treatment plants. Low intact cell concentrations in those areas could be ascribed to (1) disinfection during treatment and (2) growth inhibition from sufficient residual chlorine. Also the flow rate in the outgoing pipes closest to the treatment plants is high, which likely prevents water stagnation, sedimentation and cell adhesion on the pipe surface, and, consequently, biofilm formation and further bacterial growth. A different situation is observed in the distant areas from the water treatment plants and particularly in the so-called mixing zones, where the water from three different water treatment plants potentially mix. The map displays different color points spread in these zones without any visible order. The prevalence of the samples with higher cell concentrations there compared to the areas close to WTPs also corroborates the

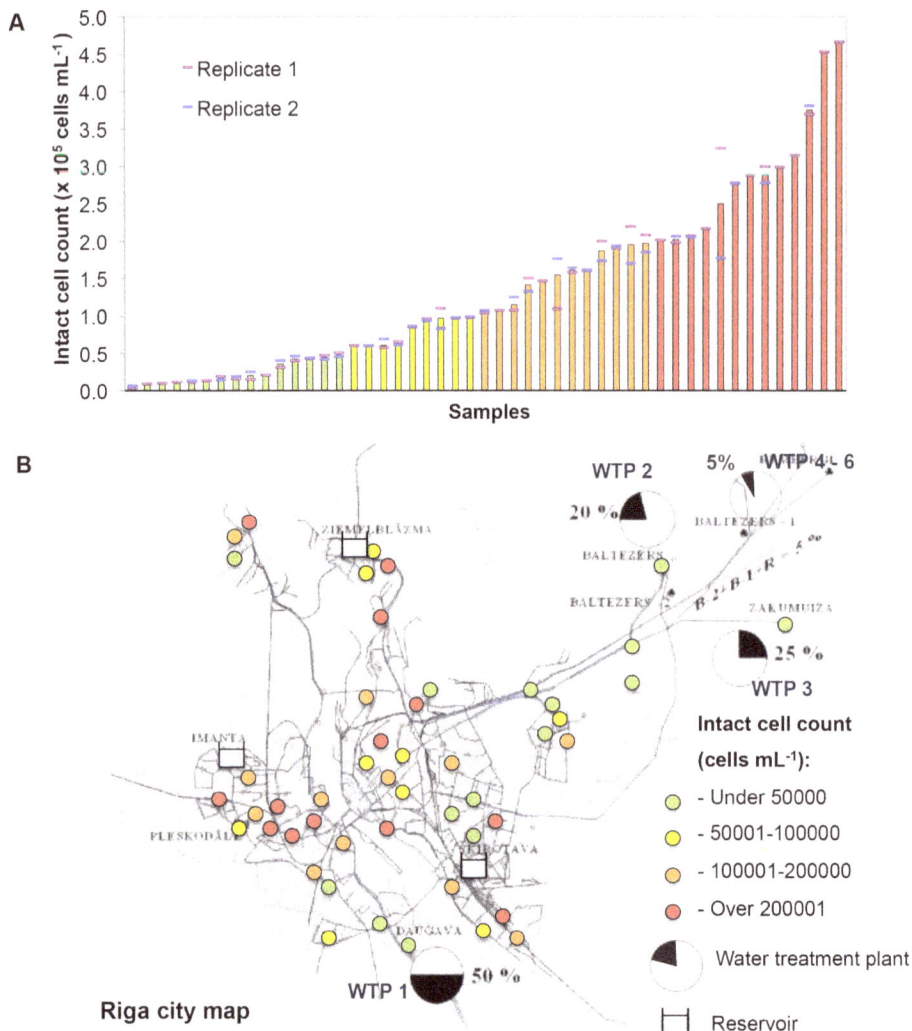

Figure 5. Intact cell concentrations of all samples measured from the distribution network (n = 49). (A) Intact cell concentrations arranged in ascending order and categorized into four main classes (colored bars) according to increasing concentrations. Data points are average values of duplicate measurements. Blue and purple stripes above and below data bars show the measured values. (B) Actual distribution of the classes of intact cells (colored circles) throughout the drinking water distribution network. WTP 1, WTP 2, WTP 3 represent location and productivity of the main water treatment plans supplying the city: WTP 1 operates using surface water, WTP 2 – artificially recharged ground water, WTP 3 – natural groundwater. WTP 4 – 6 indicates on other three pump stations with less significance for the city water supply.

argument that increasing distance and water residence time could lead to chlorine decay with concomitant oxidation of dissolved organic matter; both these events would favor bacterial growth. Moreover, mixing zones are potential hot-spots for bacterial growth, as one water might well contain the nutrients that are growth limiting in the other.

The uneven spatial distribution of the samples with different intact cell concentrations is noteworthy, highlighted for example by the three points in upper-left corner of the map. Based on the long distance from the WTPs, high intact cell concentrations were expected, but the samples taken from the hydrants located in this small area rather show variability (respectively 1.82×10^4, 1.87×10^5, 2.51×10^5 intact cells mL^{-1}). Such different intact cell concentrations could be due to several reasons: the time the samples were taken, which is linked to water consumption and the potential impact of which is shown in Figure 4, the condition of the pipes in this specific area (unknown), the way water flows from the treatment plant, and/or the relative proximity of these sample

points to one of the reservoirs (Figure 5B), etc. Other authors showed a decrease in AOC [13] and ATP [19] in the some distal points of the distribution networks. Decrease of AOC concentration was explained by its consumption by bacteria within the network. These authors argued that an insufficient amount of nutrients led to starvation and a decrease in bacterial parameters at the end of the pipelines. However, it is an unlikely reason in the present study, because this phenomenon seems more occasional than systematic.

The combined data demonstrates clearly biological instability throughout the distribution network. However, despite the relative simplicity of the concept (Figure 1; Figure 3A), a complex interplay of chemical, physical and biological parameters and hydraulic conditions should be taken into account for characterization of each particular case of instability.

113

Comparison of different microbiological parameters

FCM and ATP data showed clear correlations, but these data did not correlate well with conventional HPC data. A total amount of 49 different samples was measured in duplicate with ATP (total and extracellular) and FCM (total and intact cell concentration) analyses, while 38 of those samples were further analyzed with HPC. The significant linear correlation ($R^2 = 0.77$; n = 49) between intracellular ATP and FCM intact cell concentration is shown in Figure 6A. This corroborates previous studies that showed good results comparing total ATP with total cell concentration [19,50] and intracellular ATP with intact cell count as well [20,25]. The strong correlation is encouraging, since FCM and ATP analysis are independent viability parameters – integrity of the cell membrane (FCM) and cellular energy (ATP). A correlation between these parameters during disinfection is not necessarily a given fact. The membrane integrity based PI staining method implies that PI positive cells are damaged and thus considered as inactive, yet extreme examples where living cells became permeable for propidium iodide have been described [46]. In turn, Nocker and co-workers [51] showed that after UV-C exposure cells became inactivated, while their membranes remained essentially intact. Discrepancies between intracellular ATP and intact cell concentration can also result from cell morphology, bacterial species and physiological state, that was discussed in detail previously [25]. The results provided by FCM provide information on single cell level, whereas during ATP analyses the values are evaluated per volume. Hence, intracellular ATP-per-cell was calculated for characterization of biomass activity. In the present study intracellular ATP-per-cell ranges from zero (no cell-bound ATP observed) to 5.92×10^{-10} nM cell^{-1} ($= 3 \times 10^{-17}$ g cell^{-1}) with the average value of 1.68×10^{-10} nM cell^{-1} ($= 8.52 \times 10^{-18}$ g cell^{-1}) (stdev $= 9.58 \times 10^{-11}$ nM cell^{-1}, n = 49). The result is in the same range as ATP-per-cell values obtained from various water sources, which were analyzed with the same methods [20,25]. This suggests that bacterial activity (ATP values) in the intact cells was not affected by any remaining chlorine residuals, and that membrane damage (SGPI values) was in this case reflective of viability in the sample. The good correlation between these two independent parameters is an optimistic prospect for applying these methods for chlorinated water analyses in future studies.

The conventional HPC results were compared with the FCM intact cell concentration values. A weak correlation ($R^2 = 0.18$, n = 38) was observed between HPC (at 22°C) and intact cell concentrations (Figure 6B), similar to reports in previous studies [38,50]. It could be explained by the often described phenomenon, that less than 1% of drinking water bacteria are cultivable on conventional agar plates [25,50,52]. In addition, Mezule and co-workers [53] demonstrated evidence of the presence of so called viable-but-not-cultivable (VNBC) bacterial state, in both drinking water and biofilms for the network investigated here, thus indicating further limitations in the HPC method. Since intracellular ATP showed a good correlation with intact cell concentration, but intact cell count correlated weakly with HPC, it was expected that intracellular ATP and HPC would not correlate well (Figure 6C; $R^2 = 0.11$, n = 38). Various studies were performed to compare ATP and HPC parameters from water samples, but good correlations were never observed e.g., $R^2 = 0.20$ [19], $R^2 = 0.36$ [22] and $R^2 = 0.31$ [50]. Our results combined with those from previous studies cast further doubts on the value of using the HPC method for general microbiological drinking water quality control. In our opinion, the clear correlation between two methodologically independent viability parameters (intracellular ATP and FCM intact cell counts), and the absence of any correlations with

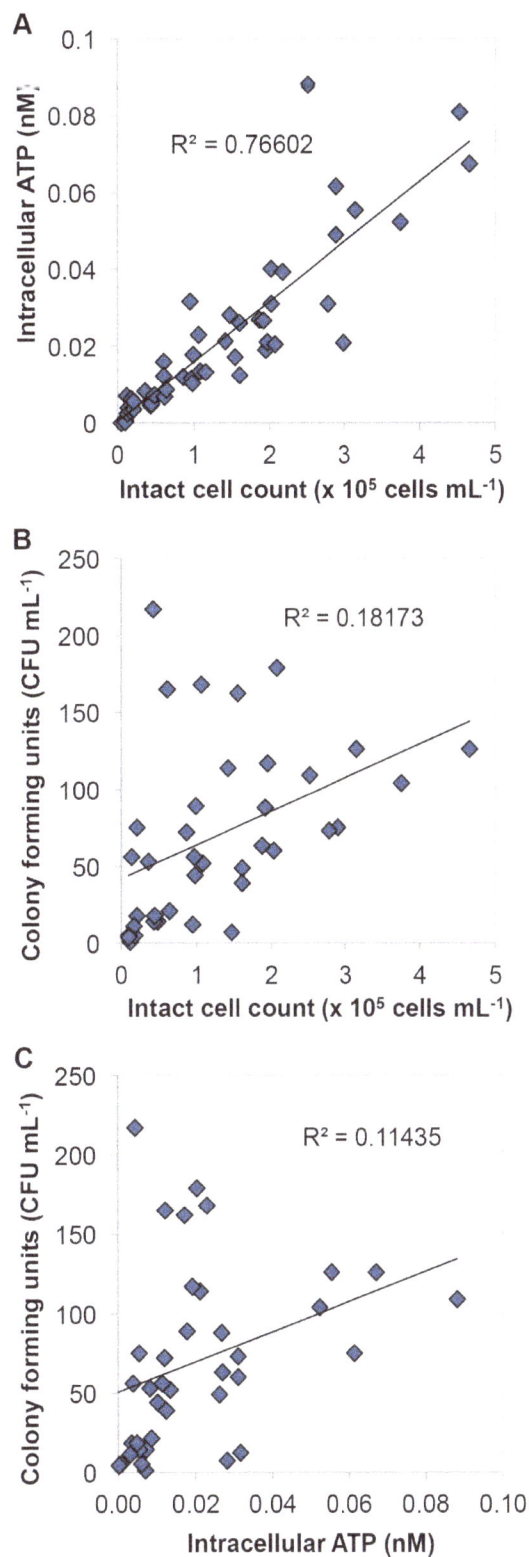

Figure 6. Comparison of the various microbiological parameters. Clear correlations were observed between intact cells and intracellular ATP (n = 49) (A), but no obvious correlations between these two parameters and heterotrophic plate counts at 22°C (n = 38) (B) (C).

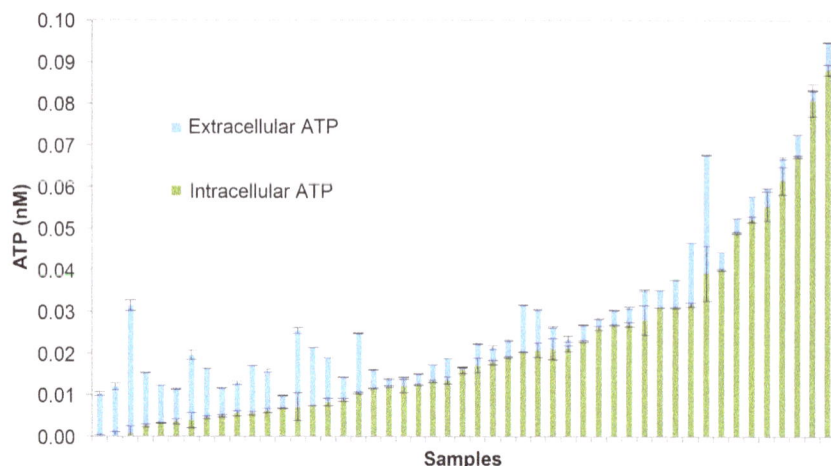

Figure 7. Distribution of intracellular and extracellular ATP in the water samples. In general, higher concentrations and relative percentages of extracellular ATP were measured in samples that exhibited lower intracellular ATP concentrations (n = 49).

two different HPC methods, renders the former methods more meaningful for assessing and understanding biological instability, particularly in chlorinated environments.

Importance of measuring extracellular ATP

Arguments for and against the concept and importance of measuring extracellular ATP have been made [4,19,20,25,54,55]. To understand this better, we arranged our data according to increasing intracellular ATP concentrations, after which the measured extracellular ATP values were added to each corresponding sample (Figure 7). It is evident that extracellular ATP constitutes a considerable fraction of the total ATP amount in some samples – varying from 3% up to 100% – with an average contribution of 36% (n = 49). Moreover, 33% of the samples contain more that 50% of extracellular ATP. This data supports other studies, where analyses showed high extracellular ATP in drinking water samples from the distribution networks [20,25]. Interestingly, the highest extracellular ATP ratio is mostly observed in the samples with relatively low intracellular ATP, in this case samples with close proximity to the treatment plant. In the case of chlorinated water, this could potentially be explained by the oxidative effect of chlorine on bacterial cells. Previous studies have shown extensive damage to bacterial membranes during chlorination [24,45], after which a release of extracellular ATP from the damaged bacteria can occur. This membrane damage was also clearly detected in the present study (e.g., Figure 3B). Although, there is lack of detailed data considering the release of extracellular ATP in water samples affected by chlorination, strong evidence of ATP release during oxidation was presented in previous studies [4,20]. Both these works showed a significant decrease in cell concentrations and intracellular ATP after ozonation, whereas extracellular ATP comprised 83–100% of the total ATP. Moreover, Figure 7 shows that samples with increased intracellular ATP concentrations, which we linked to bacterial growth during distribution, often had considerably less extracellular ATP in relation to total ATP. This could be due to the fact that extracellular ATP can be biodegraded by bacteria or extracellular enzymes in the network [54,56–58]. However, it cannot be excluded that a decrease in extracellular ATP during distribution occurs due to oxidation by residual chlorine present in the network.

Conclusions

- An investigation of a full-scale chlorinated drinking water distribution network with various microbiological methods clearly demonstrated both spatial and temporal biological instability in the network.

- Fluorescent staining with SGPI in combination with ATP measurements provided reliable and descriptive information about bacterial density and viability in chlorinated drinking water samples.

- A good correlation was observed between intracellular ATP and intact cell counts ($R^2 = 0.77$), whereas HPC showed poor correlations with both parameters ($R^2 = 0.18$ with intact cell concentration and $R^2 = 0.11$ with intracellular ATP).

- Extracellular ATP constituted on average 36% of total ATP in the present study, which confirms the necessity of extracellular ATP subtraction from total ATP measurements during chlorinated drinking water analyses.

- Overall the results raise questions with respect to the offset between increased biological safety gained from disinfection opposed to increased risk from instability (uncontrolled bacterial growth). While an improvement of the chlorination procedure could be a solution, the data suggests looking beyond only disinfection for achieving biological stability of drinking water.

Supporting Information

Figure S1 Additional examples of hydrant flushing. Changes in intact cell concentration and intracellular ATP during flushing in 6 newly-opened fire hydrants. Intact cell concentration values are shown as solid lines with blue markers, whereas intracellular ATP results displayed as single green bullets.

Figure S2 Actual data for Figure 3A. Changes in various bacterial parameters between one water treatment plant and a randomly selected point in the distribution network (actual values for Figure 3A).

Figure S3 Residual chlorine concentration in the distri-

bution network. 50% of residual chlorine concentration in the network was between 0.12 (first quartile) and 0.23 (third quartile) mg mL^{-1}, with a mean value of 0.17 mg mL^{-1} (n = 27). The whiskers indicate on minimum and maximum values, whereas bullets show outliers of the population.

Table S1   Physical and chemical parameters of water measured on-line during low velocity flushing of newly-opened fire hydrant. Some measurements were omitted during the first 20 minutes of flushing due to the high fluctuation in measuring tools readings.

Acknowledgments

The authors thank Stefan Kötzsch for critical input, Arturs Briedis, Edgars Grundbergs and Kaspars Neilands for assistance in the sampling campaigns, and sampling/information support from Rigas Udens Ltd.

Author Contributions

Conceived and designed the experiments: FH MV JR TJ. Performed the experiments: FH MV JR. Analyzed the data: FH MV JR TJ AN. Contributed reagents/materials/analysis tools: FH TJ. Wrote the paper: FH MV JR TJ AN.

References

1. Bartram J, Cotruvo J, Exner M, Fricker C, Glasmacher A (2003) Heterotrophic plate count and drinking-water safety: The significance of HPCs for water quality and human health. World Health Organization. 244 p.

2. Boe-Hansen R, Albrechtsen H-J, Arvin E, Jørgensen C (2002) Bulk water phase and biofilm growth in drinking water at low nutrient conditions. Water Res 36: 4477–4486. doi:10.1016/S0043-1354(02)00191-4.

3. Hammes F, Berger C, Köster O, Egli T (2010) Assessing biological stability of drinking water without disinfectant residuals in a full-scale water supply system. J Water Supply Res Technol 59: 31. doi:10.2166/aqua.2010.052.

4. Hammes F, Berney M, Wang Y, Vital M, Köster O, et al. (2008) Flow-cytometric total bacterial cell counts as a descriptive microbiological parameter for drinking water treatment processes. Water Res 42: 269–277. doi:10.1016/j.watres.2007.07.009.

5. Juhna T, Birzniece D, Larsson S, Zulenkovs D, Sharipo A, et al. (2007) Detection of Escherichia coli in biofilms from pipe samples and coupons in drinking water distribution networks. Appl Environ Microbiol 73: 7456–7464. doi:10.1128/AEM.00845-07.

6. Liu G, Lut MC, Verberk JQJC, Van Dijk JC (2013) A comparison of additional treatment processes to limit particle accumulation and microbial growth during drinking water distribution. Water Res 47: 2719–2728. doi:10.1016/j.watres.2013.02.035.

7. Van der Kooij D (2000) Biological stability: A multidimensional quality aspect of treated water. In: Belkin S, editor. Environmental Challenges. Springer Netherlands. pp. 25–34.

8. Vital M, Stucki D, Egli T, Hammes F (2010) Evaluating the growth potential of pathogenic bacteria in water. Appl Environ Microbiol 76: 6477–6484. doi:10.1128/AEM.00794-10.

9. Vital M, Füchslin HP, Hammes F, Egli T (2007) Growth of Vibrio cholerae O1 Ogawa Eltor in freshwater. Microbiology 153: 1993–2001. doi:10.1099/mic.0.2006/005173-0.

10. Vital M, Hammes F, Egli T (2008) Escherichia coli O157 can grow in natural freshwater at low carbon concentrations. Environ Microbiol 10: 2387–2396. doi:10.1111/j.1462-2920.2008.01664.x.

11. LeChevallier MW, Schulz W, Lee RG (1991) Bacterial nutrients in drinking water. Appl Environ Microbiol 57: 857–862.

12. Niquette P, Servais P, Savoir R (2001) Bacterial dynamics in the drinking water distribution system of Brussels. Water Res 35: 675–682.

13. Liu W, Wu H, Wang Z, Ong SL, Hu JY, et al. (2002) Investigation of assimilable organic carbon (AOC) and bacterial regrowth in drinking water distribution system. Water Res 36: 891–898.

14. Eichler S, Christen R, Höltje C, Westphal P, Bötel J, et al. (2006) Composition and dynamics of bacterial communities of a drinking water supply system as assessed by RNA- and DNA-based 16S rRNA gene fingerprinting. Appl Environ Microbiol 72: 1858–1872. doi:10.1128/AEM.72.3.1858-1872.2006.

15. Hong P-Y, Hwang C, Ling F, Andersen GL, LeChevallier MW, et al. (2010) Pyrosequencing analysis of bacterial biofilm communities in water meters of a drinking water distribution system. Appl Environ Microbiol 76: 5631–5635. doi:10.1128/AEM.00281-10.

16. Pinto AJ, Xi C, Raskin L (2012) Bacterial community structure in the drinking water microbiome Is governed by filtration processes. Environ Sci Technol 46: 8851–8859. doi:10.1021/es302042t.

17. Lautenschlager K, Hwang C, Liu W-T, Boon N, Köster O, et al. (2013) A microbiology-based multi-parametric approach towards assessing biological stability in drinking water distribution networks. Water Res 47: 3015–3025. doi:10.1016/j.watres.2013.03.002.

18. Rittmann BE, Snoeyink VL (1984) Achieving biologically stable drinking water. J - Am Water Works Assoc 76: 106–114.

19. Van der Wielen PWJJ, van der Kooij D (2010) Effect of water composition, distance and season on the adenosine triphosphate concentration in unchlorinated drinking water in the Netherlands. Water Res 44: 4860–4867. doi:10.1016/j.watres.2010.07.016.

20. Vital M, Dignum M, Magic-Knezev A, Ross P, Rietveld L, et al. (2012) Flow cytometry and adenosine tri-phosphate analysis: Alternative possibilities to evaluate major bacteriological changes in drinking water treatment and distribution systems. Water Res 46: 4665–4676. doi:10.1016/j.watres.2012.06.010.

21. LeChevallier MW, Au K-K (2004) Water treatment and pathogen control: Process efficiency in achieving safe drinking-water. IWA Publishing. 136 p.

22. Delahaye E, Welté B, Levi Y, Leblon G, Montiel A (2003) An ATP-based method for monitoring the microbiological drinking water quality in a distribution network. Water Res 37: 3689–3696. doi:10.1016/S0043-1354(03)00288-4.

23. Francisque A, Rodriguez MJ, Miranda-Moreno LF, Sadiq R, Proulx F (2009) Modeling of heterotrophic bacteria counts in a water distribution system. Water Res 43: 1075–1087. doi:10.1016/j.watres.2008.11.030.

24. Ramseier MK, von Gunten U, Freihofer P, Hammes F (2011) Kinetics of membrane damage to high (HNA) and low (LNA) nucleic acid bacterial clusters in drinking water by ozone, chlorine, chlorine dioxide, monochloramine, ferrate(VI), and permanganate. Water Res 45: 1490–1500. doi:10.1016/j.watres.2010.11.016.

25. Hammes F, Goldschmidt F, Vital M, Wang Y, Egli T (2010) Measurement and interpretation of microbial adenosine tri-phosphate (ATP) in aquatic environments. Water Res 44: 3915–3923. doi:10.1016/j.watres.2010.04.015.

26. LeChevallier MW, Welch NJ, Smith DB (1996) Full-scale studies of factors related to coliform regrowth in drinking water. Appl Environ Microbiol 62: 2201–2211.

27. Polanska M, Huysman K, van Keer C (2005) Investigation of assimilable organic carbon (AOC) in flemish drinking water. Water Res 39: 2259–2266. doi:10.1016/j.watres.2005.04.015.

28. Ramseier MK, Peter A, Traber J, von Gunten U (2011) Formation of assimilable organic carbon during oxidation of natural waters with ozone, chlorine dioxide, chlorine, permanganate, and ferrate. Water Res 45: 2002–2010. doi:10.1016/j.watres.2010.12.002.

29. Van der Kooij D (1990) Assimilable organic carbon (AOC) in drinking water. In: McFeters GA, editor. Drinking Water Microbiology. New York, NY: Springer New York. pp. 57–87.

30. Weinrich LA, Jjemba PK, Giraldo E, LeChevallier MW (2010) Implications of organic carbon in the deterioration of water quality in reclaimed water distribution systems. Water Res 44: 5367–5375. doi:10.1016/j.watres.2010.06.035.

31. Jjemba P (2010) Guidance document on the microbiological quality and Biostability of reclaimed water following storage and distribution. WateReuse Research Foundation.

32. Lehtola MJ, Laxander M, Miettinen IT, Hirvonen A, Vartiainen T, et al. (2006) The effects of changing water flow velocity on the formation of biofilms and water quality in pilot distribution system consisting of copper or polyethylene pipes. Water Res 40: 2151–2160. doi:10.1016/j.watres.2006.04.010.

33. Manuel CM, Nunes OC, Melo LF (2007) Dynamics of drinking water biofilm in flow/non-flow conditions. Water Res 41: 551–562. doi:10.1016/j.watres.2006.11.007.

34. Bucheli-Witschel M, Kötzsch S, Darr S, Widler R, Egli T (2012) A new method to assess the influence of migration from polymeric materials on the biostability of drinking water. Water Res 46: 4246–4260. doi:10.1016/j.watres.2012.05.008.

35. Douterelo I, Husband S, Boxall JB (2014) The bacteriological composition of biomass recovered by flushing an operational drinking water distribution system. Water Res 54: 100–114. doi:10.1016/j.watres.2014.01.049.

36. Rossman LA (2000) EPANET 2 Users manual. National Risk Management Research Laboratory. U.S. Environmental Protection Agency, Cincinatti, Ohio.

37. Rubulis J, Dejus S, Meksa R (2011) Online measurement usage for predicting water age from tracer tests to validate a hydraulic model American Society of Civil Engineers. pp. 1488–1497. doi:10.1061/41203(425)133.

38. Berney M, Vital M, Hülshoff I, Weilenmann H-U, Egli T, et al. (2008) Rapid, cultivation-independent assessment of microbial viability in drinking water. Water Res 42: 4010–4018. doi:10.1016/j.watres.2008.07.017.

39. Lautenschlager K, Boon N, Wang Y, Egli T, Hammes F (2010) Overnight stagnation of drinking water in household taps induces microbial growth and changes in community composition. Water Res 44: 4868–4877. doi:10.1016/j.watres.2010.07.032.

40. Antoun EN, Dyksen JE, Hiltebrand DJ (1999) Unidirectional flushing: A powerful tool. J - Am Water Works Assoc 91: 62–71.

41. Friedman M, Kirmeyer GJ, Antoun E (2002) Developing and implementing a distribution system flushing program. J - Am Water Works Assoc 94: 48–56.

42. Prévost M, Rompré A, Coallier J, Servais P, Laurent P, et al. (1998) Suspended bacterial biomass and activity in full-scale drinking water distribution systems: Impact of water treatment. Water Res 32: 1393–1406. doi:10.1016/S0043-1354(97)00388-6.

43. Berney M, Hammes F, Bosshard F, Weilenmann H-U, Egli T (2007) Assessment and interpretation of bacterial viability by using the LIVE/DEAD BacLight Kit in combination with flow cytometry. Appl Environ Microbiol 73: 3283–3290. doi:10.1128/AEM.02750-06.

44. Grégori G, Citterio S, Ghiani A, Labra M, Sgorbati S, et al. (2001) Resolution of viable and membrane-compromised bacteria in freshwater and marine waters based on analytical flow cytometry and nucleic acid double staining. Appl Environ Microbiol 67: 4662–4670. doi:10.1128/AEM.67.10.4662-4670.2001.

45. Lisle JT, Pyle BH, McFeters GA (1999) The use of multiple indices of physiological activity to access viability in chlorine disinfected Escherichia coli O157:H7. Lett Appl Microbiol 29: 42–47. doi:10.1046/j.1365-2672.1999.00572.x.

46. Shi L, Günther S, Hübschmann T, Wick LY, Harms H, et al. (2007) Limits of propidium iodide as a cell viability indicator for environmental bacteria. Cytometry A 71A: 592–598. doi:10.1002/cyto.a.20402.

47. Tsai Y-P (2005) Impact of flow velocity on the dynamic behaviour of biofilm bacteria. Biofouling 21: 267–277. doi:10.1080/08927010500398633.

48. Srinivasan S, Harrington GW, Xagoraraki I, Goel R (2008) Factors affecting bulk to total bacteria ratio in drinking water distribution systems. Water Res 42: 3393–3404. doi:10.1016/j.watres.2008.04.025.

49. Van der Kooij D (1992) Assimilable organic carbon as an indicator of bacterial regrowth. J - Am Water Works Assoc 84: 57–65.

50. Siebel E, Wang Y, Egli T, Hammes F (2008) Correlations between total cell concentration, total adenosine tri-phosphate concentration and heterotrophic plate counts during microbial monitoring of drinking water. Drink Water Eng Sci Discuss 1: 71–86. doi:10.5194/dwesd-1-71-2008.

51. Nocker A, Sossa KE, Camper AK (2007) Molecular monitoring of disinfection efficacy using propidium monoazide in combination with quantitative PCR. J Microbiol Methods 70: 252–260. doi:10.1016/j.mimet.2007.04.014.

52. Van der Kooij D, Vrouwenvelder JS, Veenendaal HR (2003) Elucidation and control of biofilm formation processes in water treatment and distribution using the Unified Biofilm Approach. Water Sci Technol J Int Assoc Water Pollut Res 47: 83–90.

53. Mezule L, Larsson S, Juhna T (2013) Application of DVC-FISH method in tracking Escherichia coli in drinking water distribution networks. Drink Water Eng Sci 6: 25–31. doi:10.5194/dwes-6-25-2013.

54. Cowan DA, Casanueva A (2007) Stability of ATP in Antarctic mineral soils. Polar Biol 30: 1599–1603. doi:10.1007/s00300-007-0324-9.

55. Venkateswaran K, Hattori N, La Duc MT, Kern R (2003) ATP as a biomarker of viable microorganisms in clean-room facilities. J Microbiol Methods 52: 367–377. doi:10.1016/S0167-7012(02)00192-6.

56. Azam F, Hodson RE (1977) Dissolved ATP in the sea and its utilisation by marine bacteria. Nature 267: 696–698. doi:10.1038/267696a0.

57. Mempin R, Tran H, Chen C, Gong H, Ho KK, et al. (2013) Release of extracellular ATP by bacteria during growth. BMC Microbiol 13: 301. doi:10.1186/1471-2180-13-301.

58. Riemann B (1979) The occurrence and ecological importance of dissolved ATP in fresh water. Freshw Biol 9: 481–490. doi:10.1111/j.1365-2427.1979.tb01532.x.

Colonic Immune Suppression, Barrier Dysfunction, and Dysbiosis by Gastrointestinal *Bacillus anthracis* Infection

Yaíma L. Lightfoot[1,2,9], Tao Yang[1,2,9], Bikash Sahay[1,2], Mojgan Zadeh[1,2], Sam X. Cheng[3], Gary P. Wang[4], Jennifer L. Owen[5], Mansour Mohamadzadeh[1,2]*

1 Department of Infectious Diseases and Pathology, University of Florida, Gainesville, Florida, United States of America, 2 Division of Gastroenterology, Hepatology and Nutrition, Department of Medicine, University of Florida, Gainesville, Florida, United States of America, 3 Division of Gastroenterology, Department of Pediatrics, University of Florida, Gainesville, Florida, United States of America, 4 Division of Infectious Diseases and Global Medicine, Department of Medicine, University of Florida, Gainesville, Florida, United States of America, 5 Department of Physiological Sciences, College of Veterinary Medicine, University of Florida, Gainesville, Florida, United States of America

Abstract

Gastrointestinal (GI) anthrax results from the ingestion of *Bacillus anthracis*. Herein, we investigated the pathogenesis of GI anthrax in animals orally infected with toxigenic non-encapsulated *B. anthracis* Sterne strain (pXO1$^+$ pXO2$^-$) spores that resulted in rapid animal death. *B. anthracis* Sterne induced significant breakdown of intestinal barrier function and led to gut dysbiosis, resulting in systemic dissemination of not only *B. anthracis*, but also of commensals. Disease progression significantly correlated with the deterioration of innate and T cell functions. Our studies provide critical immunologic and physiologic insights into the pathogenesis of GI anthrax infection, whereupon cleavage of mitogen-activated protein kinases (MAPKs) in immune cells may play a central role in promoting dysfunctional immune responses against this deadly pathogen.

Editor: Nupur Gangopadhyay, University of Pittsburgh, United States of America

Funding: This work was supported by the National Institute of Allergy and Infectious Diseases RO1 AI093370 to MM. The funders had no role in study design, data collection and analysis, decision to publish, or preparation of the manuscript.

Competing Interests: The authors have declared that no competing interests exist.

* Email: m.zadeh@ufl.edu

9 These authors contributed equally to this work.

Introduction

Gastrointestinal (GI) anthrax, named for its primary route of infection, is an acute infectious disease resulting from the ingestion of the spore-forming, Gram-positive bacterium, *Bacillus anthracis* [1]. Anthrax can also be contracted via inhalation or cutaneous exposure, with inhalation anthrax having the highest mortality rate of the three clinical subtypes [2]. Disease-causing *B. anthracis* spores primarily infect grazing animals, but humans may be exposed to anthrax through the handling of infected animals and animal products, the consumption of tainted meat, or through intentional exposure [1]. Independent of the route of entry, unchecked infection rapidly becomes systemic and death occurs due to septicemia and/or toxemia [3].

Within fully virulent *B. anthracis* strains, two large plasmids, pXO1 and pXO2, are composed of the genes needed for toxin production and capsule formation, respectively, and both plasmids are necessary for complete virulence [4,5]. The pXO1 encodes protective antigen (PA), lethal factor (LF), and edema factor (EF); lethal toxin (LT) comprises PA+LF, while edema toxin (ET) comprises PA+EF. Via these two toxins, *B. anthracis* evades and inhibits critical signals of the innate and adaptive immune systems [6]. The poorly immunogenic anthrax capsule is encoded on pXO2 and consists of poly-γ-D-glutamic acid, which protects *B. anthracis* from phagocytosis and complement binding [7,8]. Several therapeutic strategies have targeted specific *B. anthracis* virulence factors [9,10]; however, development of next generation vaccines and therapeutics against *B. anthracis* requires a better understanding of disease pathogenesis in humans. In particular, insufficient data exist regarding the pathogenesis of GI anthrax [11–13]. GI *B. anthracis* infection is not only a persistent and major problem in developing countries, but also poses a threat in biological warfare, whereby intentional contamination of food sources may occur [1].

Here, we report that GI *B. anthracis* spore infection results in swift morbidity and mortality and is associated with pathogen dissemination throughout visceral organs by induction of leakage in the intestinal barrier and significant changes in the gut's microbial composition, all of which may orchestrate dysfunctional immune responses. A greater understanding of the pathogenesis of GI anthrax and molecular studies of the "microorganism-mammalian immune defense interface" [14] is imperative and may result in improvement of a protective vaccine in man.

Materials and Methods

Mice and Ethics Statement

A/J mice were purchased from the Jackson Laboratory and bred in-house in the animal facility at the College of Veterinary Medicine, University of Florida. For microbiota composition experiments, mice were tested after a minimum of two generations of in-house breeding. Mice were used at 6–8 weeks of age in accordance with the Animal Welfare Act and the Public Health

Policy on Humane Care. All procedures were approved by the Institutional Animal Case and Use Committee (IACUC) at the University of Florida under protocol number 201107129, and all efforts were made to minimize animal suffering. Infected mice were monitored every 24 hours and were humanely euthanized when signs of advanced infection (e.g., difficulty breathing) were noted; in some cases, mice died as a direct result of the infection before euthanasia could take place. Euthanasia was performed by prolonged inhalation of isoflurane and confirmed by cervical dislocation.

B. anthracis Spore Preparation and Mouse Infections

Spores were prepared with a toxigenic non-encapsulated strain of *B. anthracis* (Sterne), as described previously [15] with the approval of the Institutional Biosafety Committee (IBC) at the University of Florida. To calculate final concentrations, serial dilutions were grown in triplicate on lysogeny broth agar plates and colonies counted. For survival studies, mice were orally infected with 10^5 spores (n = 10), 10^7 spores (n = 10), or 10^9 spores (n = 20) in a final volume of 100 μL with a reusable, 30 mm, 20 gauge, barrel-tipped feeding needle after fasting for 4 hours; infected mice were monitored, and deaths recorded. For immunologic and microbiota composition studies, A/J mice (n = 10/group) were orally infected with Sterne spores (10^9 spores/100 μL PBS/mouse) for the specified time points. Groups of A/J mice (n = 10/group) were also either orally gavaged or injected intraperitoneally (i.p.) with 125 μg LT (PA+LF) and monitored for morbidity and death.

Histopathology

Sterne-infected A/J mice were sacrificed at various days post-infection and the colon, spleen, liver, kidney, and lungs surgically excised for analyses. Tissues were fixed, sectioned, and stained with hematoxylin and eosin (H&E) by Histology Tech Services (Gainesville, FL). Histopathology and bacterial dissemination in infected mice were analyzed by a boarded veterinary pathologist (JLO). In some cases, cytocentrifugation of single cell suspensions from spleen, mesenteric lymph nodes (MLNs), and bronchoalveolar lavage (BAL) fluid were evaluated. Neither vegetative bacilli nor spores could be detected in the lungs prior to three days post-infection, confirming that spores were not accidently introduced into the respiratory tract during oral gavage.

Ex vivo Trans Epithelial Electrical Resistance (TEER), Short-circuit Current (I_{SC}) and Trans-epithelial Conductance (G_T) Measurements of Intestinal Tissues

Differences in TEER and electrogenic ion transport in the colons of Sterne-infected versus uninfected mice were quantified by measuring the short circuit current responses of isolated colonic tissues mounted in modified Ussing chambers, as previously described [16]. Briefly, Sterne-infected (day 3) and uninfected A/J mice were euthanized and colons quickly isolated. Segments were cut open along the mesenteric border into a flat sheet and flushed with ice-cold HEPES-Ringer solution; intact, full-thickness segments containing all of the layers of colons were used. The intestinal sheets were mounted between two halves of a modified Ussing chamber (Physiologic Instruments, San Diego, CA) and short-circuited by a voltage clamp (VCC MC6, Physiologic Instruments) with correction for solution resistance. The exposure area was 0.3 cm². The mucosal and serosal surfaces of the tissues were bathed in reservoirs with 3 mL HEPES-Ringer solution containing 100 mM NaCl, 25 mM HCO_3, 1.5 mM $CaCl_2$, 1.5 mM $MgCl_2$, 5 mM KCl, 10 mM glucose, and 22 mM

HEPES, pH 7.4, maintained at 37°C and continuously bubbled with 95% O_2 and 5% CO_2. Tissues and solutions were maintained at 37°C by surrounding water jackets. Tissues were allowed a 30 min stabilization period before measurements were recorded at basal and challenged conditions. The delta current (ΔI_{SC}) before and after challenge was used to estimate electrogenic current stimulation by external stimuli; the post-challenge secretory current was measured 10 minutes after challenge as basolateral bumetanide-sensitive I_{SC}. Throughout the experiments, tissues were constantly short-circuited by clamping the transepithelial potential at 0 mV, except for 1 second intervals every 20 seconds when tissues were clamped at ±3 mV and G_T determined. Data were acquired via DATAQ instruments (Akron, OH) and processed using the Acquire & Analyze software (Warner Instruments, Hamden, CT).

Real-time PCR

RNA was isolated from distal colons of mice with Aurum Total RNA Kit (Bio-Rad). iScript Select cDNA Synthesis Kit (Bio-Rad) was used for reverse transcription and cDNA used for quantitative PCR by SYBR Green Dye gene expression assay on a Bio-Rad CFX96 Real time system. mRNA levels are shown as fold-increase over uninfected mice; n = 10/group. For fecal bacteria detection, PowerSoil DNA Isolation Kit (MO BIO Laboratories, Carlsbad, CA) was used to extract total DNA according to the manufacturer's instructions. Real-time PCR analyses were performed on 2 ng of total DNA template (SsoAdvanced SYBR Green Supermix, Bio-Rad) to target bacteria group-specific 16S rDNA sequences, as previously described [17]. Groups tested include Enterobacteriaceae family (Proteobacteria phyla) and *Bifidobacterium* genus (Actinobacteria phyla). Specific groups were normalized to the housekeeper Eubacteria group; n = 10/group. For bacterial dissemination, total DNA from murine organs was extracted by using gDNA MiniPrep kit (Zymoresearch, Irvine, CA). Ten ng of total DNA was used to assess the dissemination of *Bifidobacterium* and Enterobacteriaceae; n = 6/group. A list of primers used and their sequences can be found in Table S1. For statistical analyses of gene array data, unpaired *t*-tests were performed using the relative expression levels of each gene at the specified time point compared to the relative expression level of the same gene in uninfected mice.

16S Ribosomal DNA Sequencing

For microbiome analyses, fecal DNA samples were amplified by Illumina Miseq compatible primers, targeting the 16s rDNA V4–V5 region. Amplicons were purified by QIAquick Gel extraction kit (Qiagene, Madison, WI) and quantified by Qubit 2.0 Fluorometer (Invitrogen, Grand Island, NY) and Kapa SYBR fast qPCR kit (Kapa Biosystems, Inc., Woburn, MA). Equal amounts of amplicons were pooled with 10% of Phix control. Miseq v2 reagent kit (Illumina, Inc., San Diego, CA) was used to run the pooled samples on the Illumina Miseq machine. The Q score of this run was 86.59% and cluster density was 975±61. Data were analyzed, as previously described [18]. Primers used are found in Table S2.

Colonoscopy of Uninfected and Sterne-infected A/J Mice

Colons of uninfected or day 3 Sterne-infected A/J mice were imaged with a Multi-Purpose Rigid Telescope attached to a TELE PACK X (Karl Storz–Endoscope, Germany). A/J mice were fasted for 6h prior to visualization of the colons of live animals under appropriate anesthetic conditions.

Propria (LP) Leukocyte Preparation

Colons surgically excised from Sterne-infected and uninfected control A/J mice were flushed with ice-cold PBS to remove fecal contents, cut open longitudinally, washed, and cut into 1 cm pieces. Tissue pieces were incubated with agitation at 37°C for 25 min in 30 mL of RPMI 1640 Glutamine (GIBCO, Life Technologies, Grand Island, New York) containing 5 mM EDTA (Ambion, Life Technologies), 1 mM DTT (Sigma-Aldrich, St. Louis, MO), 10 mM HEPES (GIBCO, Life Technologies), and supplemented with 5% heat inactivated FBS (HyClone, Thermo-Fisher Scientific, Waltham, MA) to remove intraepithelial lymphocytes and epithelial cells. After incubation, colonic tissues were washed with ice-cold PBS, minced, and incubated at 37°C with pre-warmed DMEM (GIBCO, Life Technologies) containing 0.25 mg/mL collagenase type VII (Sigma-Aldrich), 0.125 U/mL Liberase TM Research Grade (Roche Applied Science, Indianapolis, IN), 10 mM HEPES, 0.1 M $CaCl_2$ (Sigma-Aldrich), and 5% FBS (3×10 min digestions). After each digestion, cell suspensions were passed through a strainer, spun down, and resuspended in DMEM supplemented with 5% FBS. Cells obtained from the three digestions were combined and immediately counted for staining and flow cytometry-based analyses.

Flow Cytometry and Antibodies

To exclude dead cells, single cell suspensions obtained from processed spleens and LPLs were stained with LIVE/DEAD Aqua Dead Cell Stain Kit (Molecular Probes, Life Technologies). Cells were subsequently washed and incubated with Mouse Fc Blocking Reagent (Miltenyi Biotec, Auburn, CA) prior to staining with combinations of the following antibodies or their corresponding isotype controls purchased from eBioscience (San Diego, CA), Biolegend (San Diego, CA), BD Pharmingen, R&D Systems (Minneapolis, MN), or Cell Signaling Technology, Inc. (Danvers, MA): CD45 (30-F11), CD11c (N418), CD11b (M1/70), CD11b (M1/70), F4/80 (BM8), GR1 (RB6-8C5), I-A/I-E MHCII (2G9), CD3 (145-2C11), CD4 (RM4–5), CD8 (53–607), PD-1/Rat IgG2a, κ, Pro-IL-1β (NJTEN3)/Rat IgG1, κ, TNFα (MP6-XT22)/Rat IgG1, κ, IL-6 (MP5-20F3)/Rat IgG1, κ, IFNγ (XMG1.2)/Rat IgG1, κ, FoxP3 (FJK-16A)/Rat IgG2a, κ, phospho-p38 MAPK (28B10)/Mouse IgG1, phospho-p44/42 MAPK (Erk1/2) (D13.14.4E)/Rabbit. Prior to intracellular staining, cells were fixed and permeabilized with BD Cytofix/Cytoperm (BD Biosciences). Colonic T cells were stimulated with phorbol 12-myristate 13-acetate [(PMA) 50 ng/mL] and ionomycin (1 μg/mL) for 2.5 hrs for the detection of intracellular cytokines. After staining, a BD LSRFortessa (BD Biosciences) cell analyzer was used to acquire fixed cells. Data were analyzed with FlowJo software (Tree Star, Ashland, OR).

Ex vivo Evaluation of MAPKs

Colonic cells were isolated as described above. Equilibrated LP cells were incubated with 1 multiplicity of infection (MOI) of spores or left untreated for 1, 3 and 6 h. Cells were stained and analyzed, as described above.

Sera Analyses

Cytokines in the sera of Sterne-infected and uninfected A/J mice were measured using the Bio-Plex Pro Mouse Cytokine 23-plex immunoassay kit (Bio-Rad, Hercules, CA). Magnetic beads were acquired with a MAGPIX system (Luminex, Austin, TX) and data analyzed with Bio-Plex Data Pro Software (Bio-Rad).

Statistical Analyses

Unless stated otherwise, representative data indicate mean ± SEM. Significance was determined by two-tailed unpaired t-tests for two group comparisons (GraphPad Prism 5 for Mac OS X, La Jolla, CA).

Results

Lethality and Systemic Dissemination of GI *B. anthracis* Infection

To explore GI anthrax pathogenesis while circumventing the difficulties of working with biosafety level 3 (BSL3) *B. anthracis* strains, we employed an extensively used mouse model, A/J mice, which is highly susceptible to *B. anthracis* Sterne [19]. Given the route of infection utilized in our studies, and the potential excretion of gavaged Sterne spores, doses that have been previously shown to be lethal in experimental models of inhalational anthrax [20] resulted in little to no death of A/J mice when administered orally. The lethal dose 50 (LD_{50}) for this bacterium via the respiratory route of infection has been reported to be as low as 1×10^3 CFU of spores/mouse [21], and the LD_{50} of the vegetative bacteria was previously found to be 2.3×10^7 CFU [13]. However, to better mimic the likely form of *B. anthracis* when ingested, we fed A/J mice varying doses of Sterne spores. We found that 10^5 spores given orally did not kill any mice and inoculation with 10^7 spores resulted in 80% survival, likely because of the myriad of enzymes and the extreme pH levels of the gut milieu. The oral LD_{50} was approximately 5×10^8 spores/mouse (Fig. 1A), and 80% of infected mice died when orally gavaged with 10^9 Sterne spores (Fig. 1A). These mice were lethargic and showed signs of dyspnea as early as 2 days post-infection. After 3 days of infection, long chains of vegetative bacilli with the characteristic "bamboo-like appearance" were observed (marked with asterisk for ease of visualization) not only in the colon (Fig. 1B), but also systemically (Fig. 1C–I). Vegetative bacilli were seen in numerous organs, including the mesenteric lymph nodes (MLNs) (Fig. 1C), spleen (Fig. 1D, E), liver (Fig. 1F), kidneys (Fig. 1G), lungs (Fig. 1H), and in BAL fluid collected from the lungs (Fig. 1I). In some mice, vegetative bacilli were found in the liver after only 1 day of infection (Fig. S1A), suggesting a hematogenous component in the dissemination process, considering the portal vein carries blood from the GI tract to the liver for nutrition and detoxification. Anthrax-induced death in A/J mice is thought to be due to bacteremia and toxemia resulting from the high number of *B. anthracis* microbes in the periphery [19]. Indeed, we noted focal and diffuse areas of lympholysis in the spleens of infected mice (Fig. 1D, E), consistent with the known necrotizing effects of the toxins. Establishment of active infection is required for morbidity and mortality in our model, as A/J mice that were given LT (125 μg) orally did not show any signs of illness, likely due to proteolytic enzymes within the GI tract (Fig. S1B). In contrast, the same LT dose, when given i.p., resulted in the death of over 50% of the mice (Fig. S1B). Nonetheless, both routes of injection resulted in decreased expression of interleukin (IL)-1β in the colonic tissues (Fig. S1C).

GI Epithelial Barrier Dysfunction and Dysbiosis in Sterne-infected A/J Mice

Observing the dissemination of *B. anthracis* Sterne in various mouse organs led us to investigate the intestinal barrier function of infected mice. We tested the epithelial barrier integrity of colons isolated from mice infected for 3 days. Infected colons showed a significant breakdown of intestinal barrier integrity, as evidenced by a significantly lower transepithelial electrical resistance (TEER)

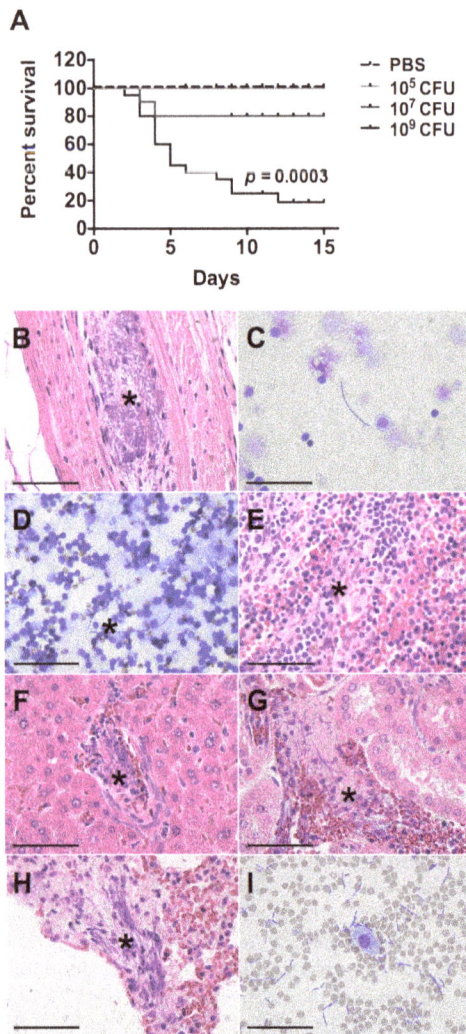

Figure 1. Lethality and Systemic *B. anthracis* Sterne Dissemination in A/J Mice. A. A/J mice were orally gavaged with 10^5, 10^7, or 10^9 spores of the Sterne strain of *B. anthracis*. Lethal infection was established within 3 days in A/J mice receiving 10^9 spores; $n = 10$ mice/group (10^5 and 10^7), $n = 20$ mice/group (10^9). Experiments were performed a minimum of three times. Statistical significance was calculated using the log-rank test. After 3 days of infection, A/J mice were sacrificed; both spores and vegetative bacilli (marked with *) were observed in the colon (**B**), MLNs (**C**), spleen (**D, E**), liver (**F**), kidneys (**G**), lungs (**H**), and in bronchoalveolar lavage (BAL) fluid (**I**). Bar = .

A mutualistic relationship exists between the intestinal microbiota and the epithelial cells that comprise the single cell barrier between the host and the intestinal lumen, and both populations can modulate the other [22]. Thus, we examined the composition of the gut microbiota of infected mice. In fact, significant changes in the composition of the microbiota were noted in the Sterne-infected mice (Fig. 3). Global changes in microbial community distribution were analyzed by the UniFrac method [23]. In principle coordinate analyses (PCoA), the gut microbes before and 3 days post-infection clustered separately (Fig. 3A), indicating that GI *B. anthracis* Sterne significantly promoted microbial dysbiosis, in which a significant reduction in species richness and changes in the dominant phyla were observed (Fig. 3B, C). Several algorithms were used to determine the species richness and diversity within our samples. For instance, Chao Richness represents an estimate of the total number of species in the samples analyzed as it considers the sequences obtained a subsample of the entire community. Pielou evenness describes how close in numbers species in an environment are; a value nearing 1 indicates an even distribution within the community. Finally, Shannon Diversity takes into account both abundance and evenness within the community.

Subsequently, the algorithm linear discriminant analysis (LDA) effect size (LEfSe) [24] was used to determine which bacteria taxa were differentially represented with infection. Differentially depleted and enhanced genera were mostly composed of unculturable bacteria (Fig. 3D); however, infected mice were enriched for the novel genus *Anaerotruncus*, of which *A. colihominis* is the only described species and has been previously shown to cause bacteremia [25]. Additionally, to overcome PCR primer bias leading to underrepresentation of *Actinobacteria*, in particular bifidobacteria [26], real-time PCR was used to quantify *Bifidobacterium* in the colons. We noted a significant decrease in the relative abundance of *Bifidobacterium* with infection (Fig. 3E); the relative abundance of Enterobacteriaceae was also decreased (Fig. 3E). Because of the breach in intestinal epithelial integrity with infection, we then determined the presence of Enterobacteriaceae systemically. Indeed, members of this bacterial family were detected in the spleen, liver, and MLNs of infected mice (Fig. 3F). However, *Bifidobacterium* was not detected in these tissue samples (data not shown), indicating that its reduction in the colon may represent an actual depletion subsequent to infection. Not surprisingly, fecal shedding of Sterne was observed in all infected mice after one day of infection (Fig. 3G).

Innate Immune Responses in *B. anthracis* Sterne-infected Mice

It was previously demonstrated that innate cells, including dendritic cells (DCs) are the first targets for *B. anthracis* toxins such as LT [27,28]. The internalized toxin complexes in the cytosol function as Zn^{+2}-dependent metalloproteases that cleave members of the mitogen-activated protein kinase kinase (MAPKK) family, consequently blocking critical signals for cell activation [29–33], all of which could contribute to a failure in microbial clearance [28,34,35]. In line with these findings, it was recently observed that both toxins, LT and ET, significantly impair protective innate immune responses [28,34–36].

To demonstrate whether GI anthrax infection induces the same effects in DCs derived from orally infected mice, colonic DCs were isolated and studied. Indeed, the activation of colonic $CD45^+MHCII^{hi}CD11c^+F4/80^-CD11b^+$ DCs (Fig. 4A), co-stimulatory molecules, CD80/CD86, and co-inhibitory, B7-H1, were unchanged or significantly depressed during infection (Fig. 4B), suggesting inhibition of immune functions. It is worth noting that

(Fig. 2A). Consistent with barrier disruption, infected mice also exhibited higher transepithelial conductance (G_T), and short-circuit current (I_{SC}) (Fig. 2B, C). Additionally, infection altered transcellular ion transport, which is manifested as abnormal electrogenic ion transport responses to cholinergic/muscarinic and histamine challenges (Fig. 2D, E, respectively) and abnormally high post-challenge secretory currents (Fig. 2F). Despite these signs of colonic mucosal damage, colonoscopies performed 3 days post-infection indicated no signs of hemorrhagic lesions within the colons of the majority of the mice studied (Fig. 2G). However, a very small number of mice (<10%) did show gross intestinal hemorrhage (Fig. 2H), vascular congestion, and trace microscopic evidence of hemorrhage in the colons (Fig. 2I) and in the small intestines of infected mice (Fig. 2J).

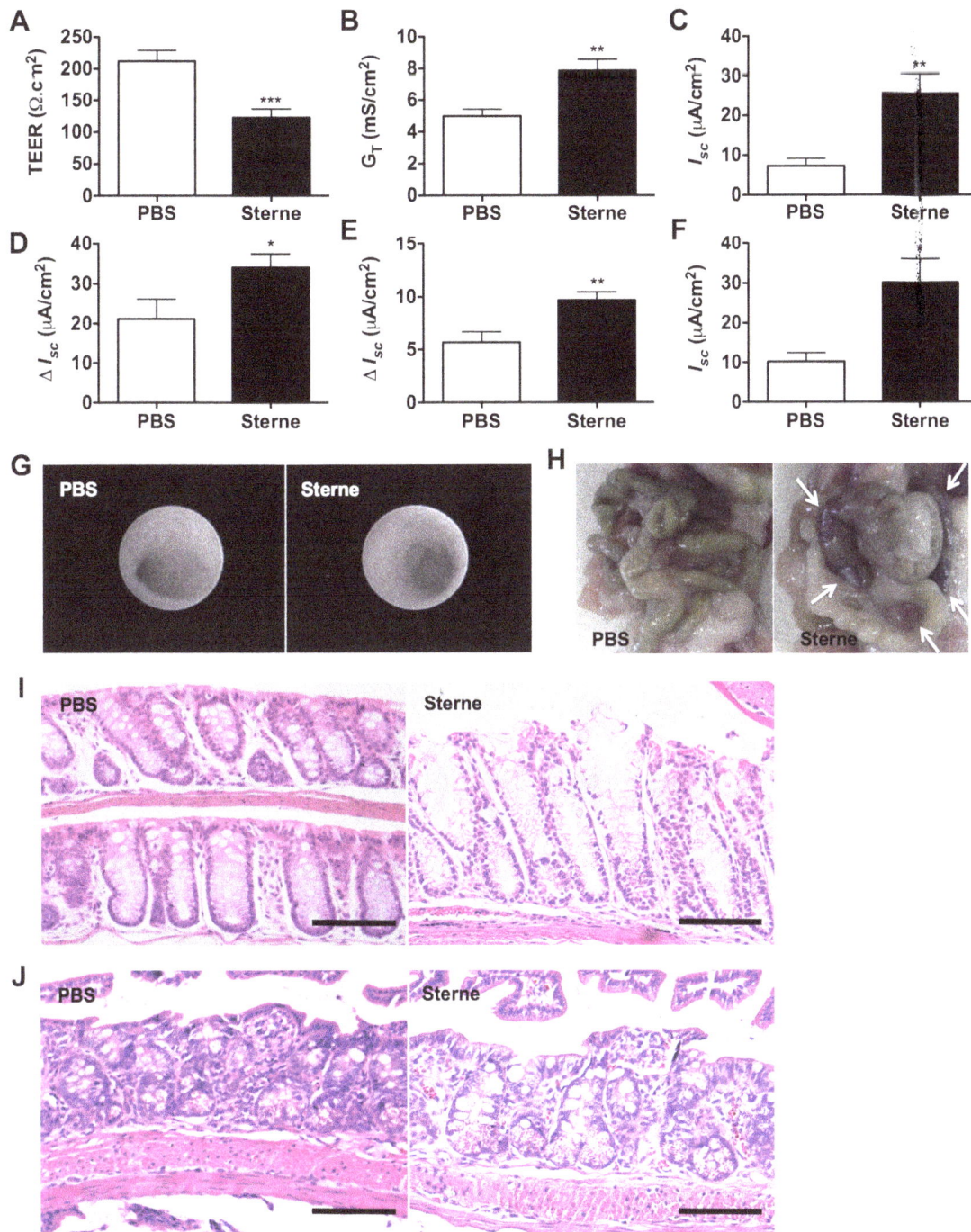

Figure 2. GI Epithelial Barrier Dysfunction Induced by Sterne Infection. A/J mice were orally gavaged with 10^9 spores of the Sterne strain of *B. anthracis* and intestinal barrier integrity was analyzed *ex vivo* three days post-infection; n = 5 mice/group. TEER (**A**), trans-epithelial conductance (**B**), and short-circuit current (**C**) of Sterne-infected versus uninfected A/J mice. **D and E.** Delta current (ΔI_{SC}) before and after cholinergic (**D**) and histamine (**E**) challenges. **F.** Post-challenge secretory current of Sterne-infected versus uninfected A/J mice. Data are shown as mean +/− SEM. *P< 0.05, **P<0.01, ***P<0.001 compared with PBS. **G.** Colonoscopies were performed in groups of uninfected and 10^9 Sterne spores-infected A/J mice three days post infection with a Multi-Purpose Rigid Telescope attached to a TELE PACK X. **H.** Gross hemorrhage in the small intestines. **I and J.** Hemorrhagic lesions in the colon (**I**) and small intestine (**J**) of Sterne-infected A/J mice. Bar = 200 µm.

B7-H1 is tightly regulated by ERK/p38 MAPK signaling [37], which is known to be disrupted by LF [38]. Accordingly, we addressed the question whether GI Sterne infection induces the cleavage of MAPKs in colonic DCs by evaluating the activation of downstream kinases, p38 and Erk1/2, *in vivo* and *ex vivo*. Data clearly show that infection with Sterne resulted in the impairment of p38

and Erk1/2 phosphorylation in colonic DCs of mice that were orally gavaged with 1×10^9 spores (Fig. 4C). To strengthen this observation, we then isolated and infected colonic cells with 1 MOI of *B. anthracis* Sterne spores for 1, 3, or 6 hours. A transient increase in p38 and Erk1/2 phosphorylation in total colonic cells (Fig. S2) and more specifically phospho-p38 in DCs was observed at earlier time points;

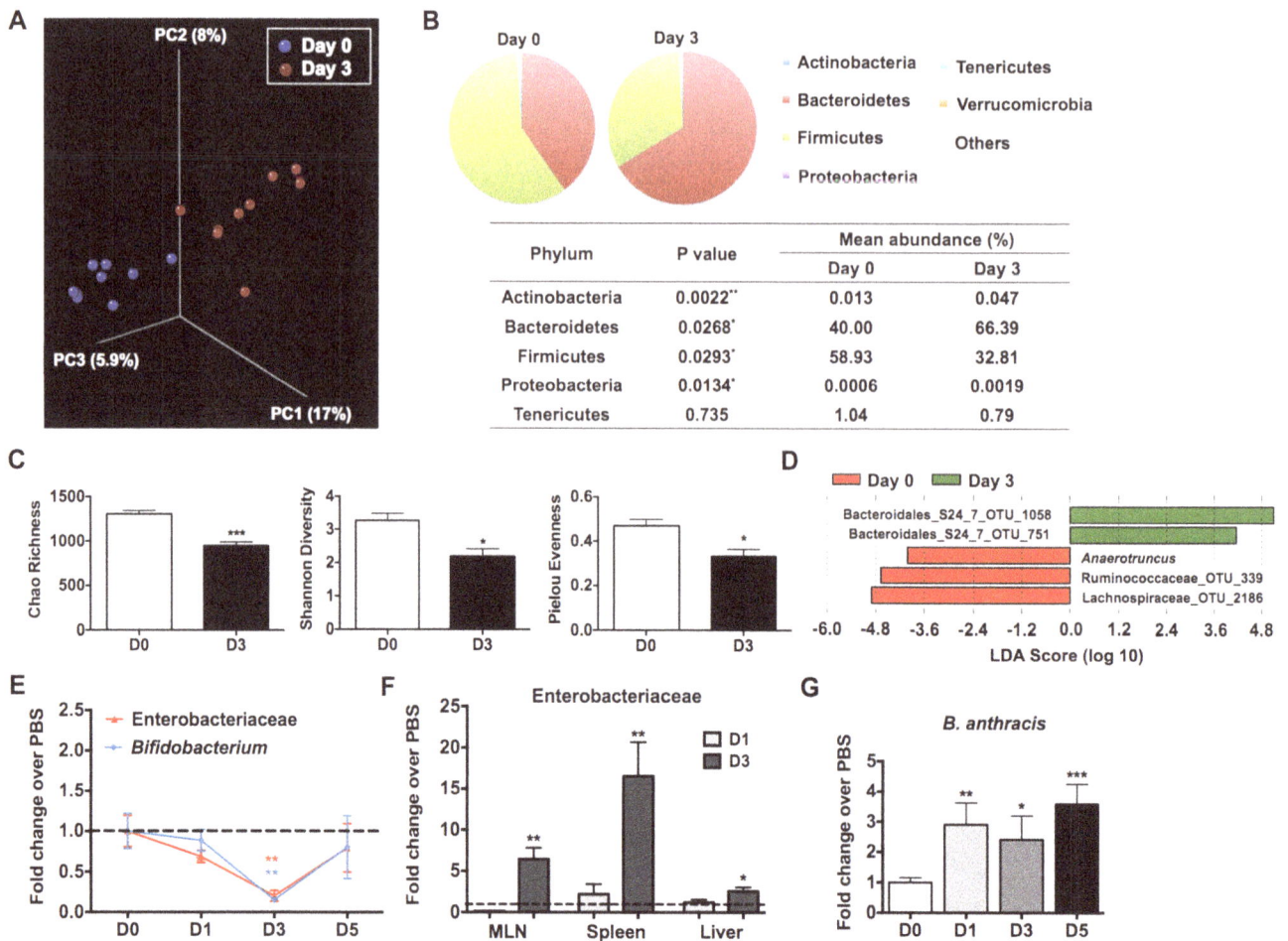

Figure 3. GI Dysbiosis Subsequent to Sterne Infection. A/J mice were orally gavaged with 10^9 spores of the Sterne strain of *B. anthracis* and changes in microbiota composition during the course of Sterne infection were monitored. **A**. Unweighted UniFrac analyses were used to calculate distances between samples obtained from Sterne-infected A/J mice before infection and three days post-infection and three dimensional scatterplots were generated by using principal coordinate analysis (PCoA); n = 9 mice/group. **B**. Average abundance values of indicated phyla. **C**. Decreased microbial diversity, evenness, and species richness. Left: The Chao richness index was used as a measure of species richness. Middle: The Shannon diversity index was used to estimate microbial diversity for each group. Right: The species evenness index was calculated using the formula J' = H'/ H'_{max}, where H' is the Shannon diversity index and H'_{max} is the maximal value of H'. Data are shown as mean +/− SEM. *P<0.05, ***P<0.001 compared with PBS-treated or day 0 mice. **D**. Bacteria genera most enriched or depleted in Sterne-infected mice at day 3 versus day 0, as measured by linear discriminant analysis (LDA). **E**. Reduced relative abundance of Enterobacteriaceae and *Bifidobacterium* in Sterne-infected A/J mice. **F**. Presence of Enterobacteriaceae in MLNs, spleens, and livers of Sterne-infected mice. **G**. Persistence of *B. anthracis* spores in the feces of Sterne-infected mice.

however, after 6 hours, cleavage of p38 and Erk1/2 phosphorylation was measured (Fig. 4D, and Fig. S2). Furthermore, *B. anthracis* Sterne also downregulated IL-1β, tumor necrosis factor (TNF)-α, and IL-6 by DCs (Fig. 5A), a consequence that is attributed to cleavage of MAPKs [28,34,35].

Having noted inhibition of colonic DC function, transcriptional changes in genes encoding pattern recognition receptors and inflammatory mediators were analyzed in the colonic tissues of infected mice as a measure of local induced inflammation and immune activation by *B. anthracis* Sterne. Inflammation-associated genes were found to be transiently upregulated early during Sterne infection (day 1, Fig. 5B); however, this gene upregulation was rapidly downregulated on days 3 and 5 post-infection, possibly due to the release of bacterial toxins (Fig. 5B). Therefore, the GI milieu mirrored the phenomenon of immune dysfunction observed in colonic DCs. Similar to the lack of colonic innate immune activation observed, systemic innate responses were also suppressed in infected mice, as both splenic DCs and macrophages

produced less pro-inflammatory cytokines than those of uninfected mice (Fig. S3A, B). Not unexpectedly, given the aforementioned immune defects, with the exception of IL-1β at day 1 post-infection, no major increases in pro-inflammatory cytokine levels were detected in the sera of infected mice compared to control animals (Fig. S4); instead, we found that circulating levels of interferon (IFN)γ were reduced after 3 days of infection (Fig. S4).

Local and Systemic T Cell Immune Responses in Sterne-infected A/J Mice

To better understand the impact of deteriorated innate cell function resulting from GI anthrax on local and systemic T cell responses, we evaluated Th1, Th17, and regulatory (Treg) T cell activation in the colonic LP (Fig. 6A) and in the spleens of infected mice. Colonic and splenic T cell responses in infected mice were mostly limited to the CD4+ T cell subset. No significant changes were observed in day 1 and day 3-infected mice (Data Not

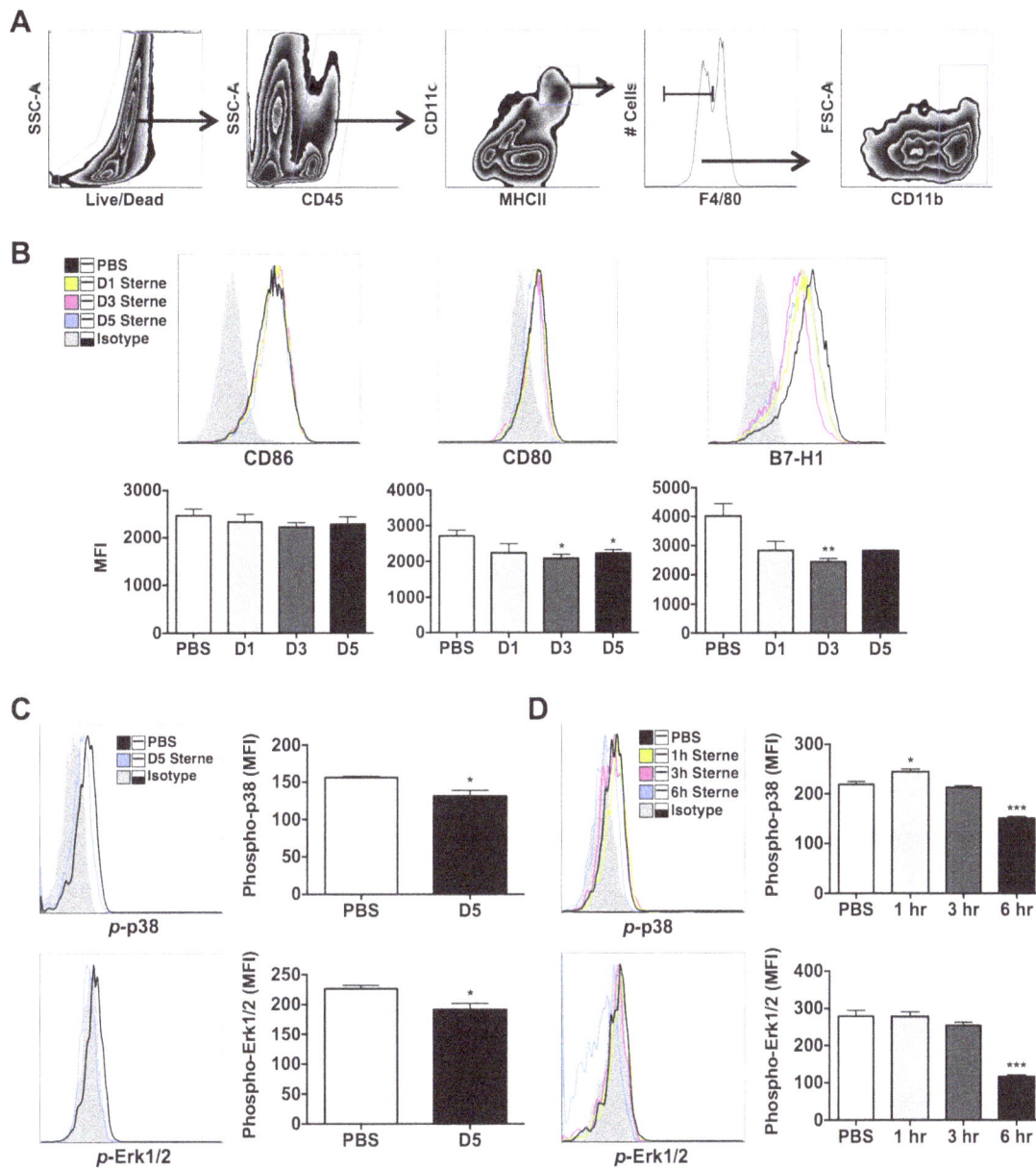

Figure 4. Cleavage of MAPKs in DCs of Sterne-Infected A/J Mice. A/J mice were orally gavaged with 10^9 spores of the Sterne strain of *B. anthracis* and DCs analyzed at various time points. **A**. Gating strategy for the analysis of CD45⁺MHCII^hiCD11c⁺F4/80⁻CD11b⁺ colonic DCs. **B**. Cell surface expression of CD86, CD80, and B7-H1 by colonic DCs was analyzed by flow cytometry. Gray tinted line = isotype control; black line = PBS group; green line = day 1 Sterne-infected A/J mice; magenta line = day 3 Sterne-infected A/J mice; blue line = day 5 Sterne-infected mice. **C**. Activity of MAPKs in colonic DCs of infected versus uninfected mice was analyzed by flow cytometry. Gray tinted line = isotype control; black line = PBS group; blue line = day 5 Sterne-infected mice. **D**. Colonic LP cells were isolated from uninfected A/J mice and incubated with 1 MOI of *B. anthracis* spores for 1, 3, or 6 hours. Activity of p38 and Erk1/2 was subsequently analyzed in colonic DCs. Gray tinted line = isotype control; black line = PBS group; green line = 1 hour treatment; magenta line = 3 hour treatment; blue line = 6 hour treatment. Data represent observations from three independent experiments and are shown as mean +/− SEM. *P<0.05, **P<0.01, ***P<0.001 compared with PBS.

Shown). However, a significant increase in the frequency of IL-17A⁺CD4⁺ T cells (day 5) was noted in the colons of Sterne infected mice (Fig. 6B, top). Systemically, splenic CD4⁺ T cells from infected mice showed increased intracellular IFNγ and a transient increase in IL-17A⁺CD4⁺ T cells (Fig. S5A, B). Recently, it has been shown that depending on the levels of ET, *B. anthracis* can suppress T cell proliferation, skew CD4⁺ T cells toward Th17 differentiation, and/or potentiate Th2 polarization [39]. Moreover, concurrent with increased colonic Th17 responses, Tregs were also enhanced in day 5-infected mice (Fig. 6B, bottom),

suggesting that these induced levels of Tregs might potentially regulate protective Th17 immunity, which has been demonstrated to be critical for protection in inhalational anthrax infection [40].

Increased levels of Programmed Death-1 (PD-1) on T cells has recently been shown to be a critical "molecular signature" of T cell exhaustion in several models of chronic viral infection; this "molecular signature" also includes increased mRNA for cell-surface receptors suspected or known to have inhibitory function, such as 2B4, Ly49 family members, and GP49B [41]. Consistent with induction of immune dysfunction by GI anthrax infection,

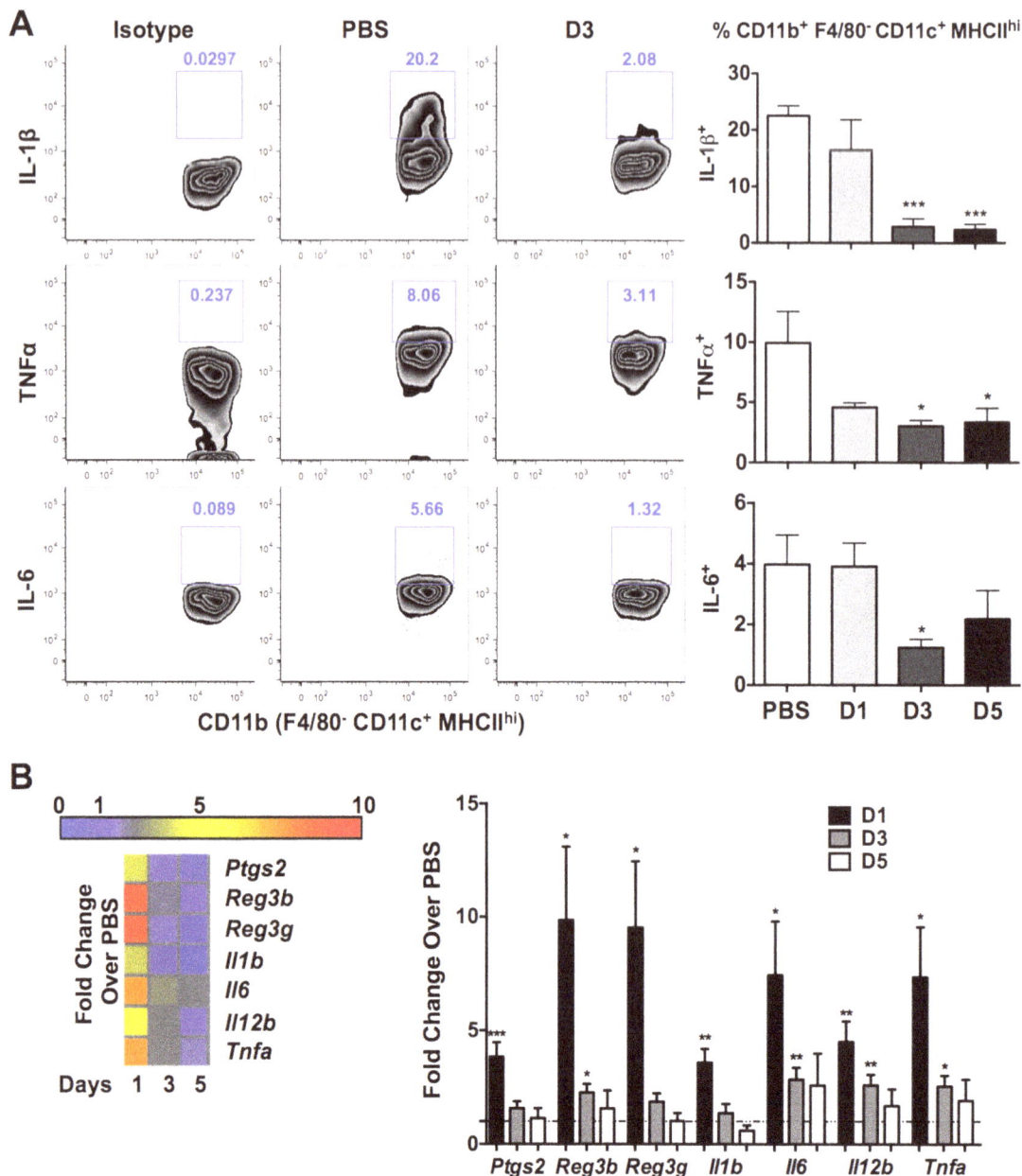

Figure 5. Suppression of Innate Immune Responses in Sterne-Infected A/J Mice. A/J mice were orally gavaged with 10^9 spores of the Sterne strain of *B. anthracis* and innate immune responses analyzed at various time points. **A**. DCs isolated from the colons of Sterne-infected A/J mice were analyzed by flow cytometry for the production of the pro-inflammatory cytokines IL-1β, TNF-α, and IL-6. Representative plots indicate cytokine production of uninfected and day 3 infected mice. Corresponding isotype controls were utilized for gating of intracellular cytokines. **B**. Gene expression profile of the distal colon of Sterne-infected A/J mice. Data represent observations from four independent experiments and are shown as mean +/− SEM. *P<0.05, **P<0.01, ***P<0.001 compared with PBS.

PD-1 receptor expression was significantly augmented on colonic CD4+ and CD8+ T cells of infected mice (Figure 6C), the transcription of which was also confirmed by gene expression analyses (Fig. 6D). We also observed increased gene expression of *Gp49b*, *Ly49*, and the *2B4* short isoform (Fig. 6D) in colonic tissues derived from *B. anthracis* infected mice. Moreover, colonic expression of *Cxcr3*, a chemokine receptor that is preferentially expressed on Th1 cells and promotes their recruitment to sites of inflammation [42,43], was significantly increased (~3 fold) in day 3-infected mice (Fig. 6D); however, expression of this critical

chemokine receptor was significantly down-modulated by day 5 of infection, highlighting the change in the inflammatory status of the gut upon GI anthrax progression.

Discussion

Contact with *B. anthracis*-infected animals or consumption of tainted, undercooked meat may result in GI anthrax in humans [13,44]. In recent years, advances have been made toward the development of experimental models to study the pathogenesis of

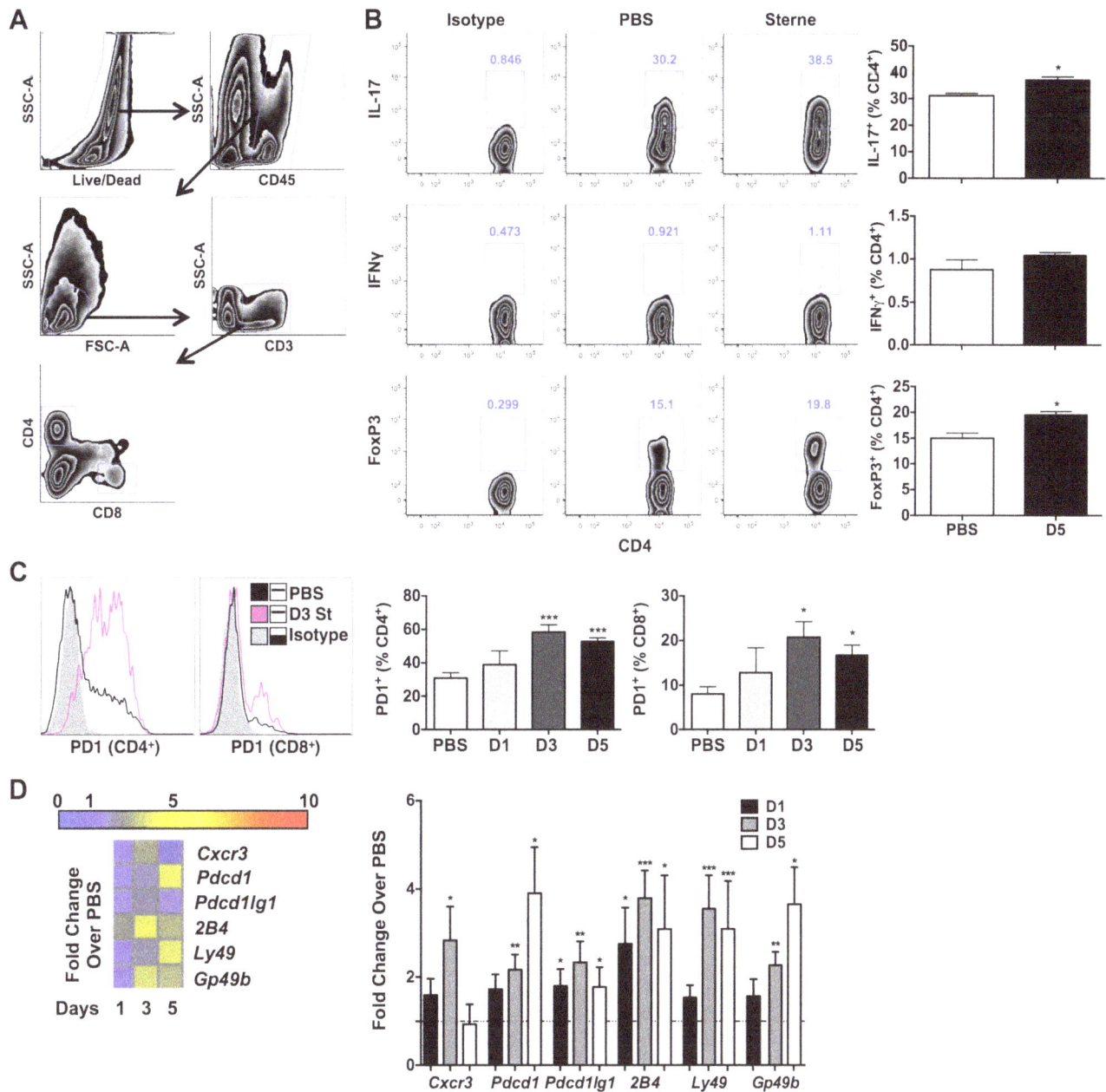

Figure 6. T Cell Responses in Sterne-Infected A/J Mice. A/J mice were orally gavaged with 10^9 spores of the Sterne strain of *B. anthracis* and adaptive immune responses in the colon analyzed at various time points by flow cytometry. **A**. Gating strategy for the analysis of colonic T cells. **B**. Th1, Th17, and regulatory T cells responses were tested by flow cytometry. Representative plots indicate cytokine production of uninfected and day 5-infected mice. **C**. Surface expression of PD1 in colonic T cells. **D**. Gene expression profile of the distal colon of Sterne-infected A/J mice. Data represent observations from four independent experiments and are shown as mean +/− SEM. *P<0.05, **P<0.01, ***P<0.001 compared with PBS.

GI *B. anthracis* infection and the effects of its toxins on GI health [11–13,45]. Unlike direct intragastric infection with *B. anthracis* spores [11,12], oral gavage with Sterne spores did not induce severe pathology in the small intestine, and the Peyer's Patch did not appear to be the primary site of *B. anthracis* growth. Moreover, intragastric infection with vegetative bacilli resulted in rapid morbidity and mortality of the mice [13]. Such morbidity leads to significant nutritional deficiencies in the animals, which have known negative immunologic consequences [46]. Therefore, our model of GI anthrax provides a unique window of moderate

disease, whereby the immunologic status of the animals better indicates responses to the pathogen itself. To gain a better understanding of GI anthrax pathogenesis, A/J mice were orally infected with 10^9 *B. anthracis* Sterne spores. Bacterial dissemination was observed in various visceral organs as early as one day post-infection (liver), which potentially contributed to further systemic bacterial pathogenesis resulting in impaired immunity, whereupon mice succumbed to infection as early as day 2 post-infection. Consequently, our findings provide in-depth immunologic studies as well as detailed investigations of the intestinal health of infected mice as it relates to its barrier function and microbial composition.

Emerging data suggest that *B. anthracis* toxins suppress host immunity, allowing the pathogen to significantly replicate, resulting in septic shock [47]. We posit that, as is the case with many other microbial pathogens, the first event is the ability of *B. anthracis* to disrupt epithelial barrier function via synergy of LT with *B. anthracis* S-layer protein A (BslA) [32], as this protein has been identified as an important surface adhesin of this deadly pathogen [48]. Disruption of intestinal barrier function undoubtedly contributes to bacterial dissemination locally and systemically, a mechanism that may change the composition of the gut microbiota, as significant decreases in the relative abundance of *Enterobacteriaceae* and *Bifidobacterium* in the feces of Sterne infected mice were observed. We also detected *Enterobacteriaceae* members in two important peripheral immune organs, the MLNs and the spleen. The observed bacterial translocation and dissemination could be seen as a potential cause for the lower numbers in the feces of infected mice; however, further studies are warranted to ascertain this possibility. On the other hand, *Bifidobacterium*, typically recognized as a beneficial bacterial group in the host gut, was not detected in any other organ, suggesting that this group was potentially out-competed for survival in the altered intestinal microenvironment upon *B. anthracis* Sterne infection. Indeed, this has been shown to be the case with many patients with colorectal inflammatory diseases, since *Bifidobacteria* protect the mucosae from damaging inflammation [49] and enhance barrier function [50]. When considering the concomitant intestinal barrier damage caused by *B. anthracis* infection, the dissemination of *Enterobacteriaceae* may have contributed to pathogenic inflammation, while elimination of the beneficial bacterium, *Bifidobacterium*, may have left the host with less regulatory mucosal responses involved in healthy gut homeostasis.

PCoA demonstrated a significantly altered gut microbial composition in mice after 3 days of infection, indicating a global change in the gut flora at the phylum level. Uncultured *operational taxonomic units* (OTU) 751 and OTU 1058 were shown to be dramatically increased, along with *Anaetruncus*, while uncultured OTU 339 and OTU 2186 were decreased with infection. Both unculturable OTU 751 and OTU 1058 are classified as part of the Bacteroidales S24-7 family, which has been found to be a dominant uncultured family in the gut whose function is likely the predominant contributor to the Bacteroidetes phylum expansion. Unculturable OTU 339 and OTU 2186 can be classified into the Ruminococcaceae family and Lachnospiraceae family, respectively. Depletion of both of these two families has consistently been associated with colonic inflammation [51]. Most of these two families' members belong to *Clostridium* cluster IV and XIVa, which have the potential to induce butyrate production [52], Tregs [53], maintenance of barrier function [52], and competition with other pathogens in the colonization of the gut. Therefore, the collective depletion of these commensal bacteria in the gut may have negatively affected the local immunity by promoting pathogenic inflammation and disrupting mucosal homeostasis.

Controlled inflammatory responses are necessary to promote protective immunity and overcome pathogen challenge after infection. However, certain pathogens have evolved to evade immune recognition and clearance through NF-κB and MAPK signaling inhibition [54]. Accordingly, one day post-infection we found a robust increase in the transcription of genes involved in innate immune recognition (Data Not Shown); however, expression of these genes rapidly decreased to basal levels, and in certain cases, was even lower than those of the controls, suggesting suppression of critical sensing molecules to mobilize protective

immunity against pathogens. Most of the genes whose expression were upregulated are MAPK-dependent, indicating that within one day endospores germinate into the vegetative form to secrete LT, which likely causes immune suppression. Indeed, infected mice exhibited global immune suppression, as infection did not elicit protective immune activation in either colonic or splenic DCs. This is consistent with previous reports showing that LT suppresses IL-6 and TNF-α in DCs by disrupting MAPK signaling [28]. Suppression is thought to occur early in infection, while in later stages, the toxin exerts inflammatory effects contributing to toxic shock and bacteremia [28]. Our studies clearly demonstrate the inhibition of MAPK signaling in intestinal immune cells (e.g., DCs) upon GI anthrax infection.

The gut microenvironment is tightly controlled by innate cells that potentially impact T cell function by regulating PD-1/B7-H1 interactions, which may result in T cell "exhaustion" as a consequence of infection [55]. Importantly, it has been shown that the MAPK-dependent transcription factor, T-bet [56], is a negative regulator of PD-1 [57]. Thus, inhibition of MAPK-activity by LT may have downregulated T-bet activation, resulting in the increased expression of PD-1 on T cells. While the expression of B7-H1 was reduced on DCs [58], B7-H1 transcription was globally upregulated in colonic cells other than DCs. Furthermore, MAPK-dependent cytokines (e.g., IL-1β, TNF-α, and IL-6) were downregulated in DCs and macrophages as a result of Sterne infection. In the colon, expression of *Cxcr3*, which is a signature chemokine receptor for IFNγ+ Th1 cell recruitment, was transiently increased with a significant reduction by day 5, indicating a potential lack of supporting immune mechanisms to induce microbial protective immunity in the GI tract. Conversely, in the spleen, the immune environment was shifted toward a mixed Th1/Th17 response, with increased IFNγ and IL-17A-producing CD4+ T cells, a hallmark of bacterial infection [59].

In summary, GI infection with *B. anthracis* is quite different and more complex than infection via the respiratory route. Respiratory surfaces maintain an anti-inflammatory environment [60], as does the GI tract [61]; however, the microbial load of the GI tract is considerably larger and more complex. Respiratory infection with *B. anthracis* has a higher fatality rate and requires smaller inoculums than GI infection, perhaps due to the presence of mitigating factors in the latter, including proteolytic enzyme activity, the extensive gut microbiota, and peristaltic expulsion of spores from the GI tract. Additionally, the breach in gut barrier function caused by *B. anthracis* and its gene products leads to translocation of *B. anthracis* and other gut microbes, which may engage complex signaling events, locally and systemically. Translocation of the gut microbiota could be beneficial to the host by mounting a Th1/Th17 response to clear the bacteria; conversely, it may cause bacteremia and septic shock. The precise contribution of this shift in the composition of the gut microbiome to the pathobiology of GI anthrax has yet to be examined. An early distinction between mice that survive and those that succumb to infection and a better understanding of early innate immune cell activation that bypasses suppressive attempts by *B. anthracis* will be critical in the prevention of fatal systemic disease in animals and in humans.

Supporting Information

Figure S1 Morbidity and Mortality in GI Anthrax is Dependent on Active Infection and Involves Hematogenous Spread of Infection. A. A/J mice were orally gavaged with 10^9 spores of the Sterne strain of *B. anthracis*. One day post-infection, *B. anthracis* Sterne bacilli could be found within the liver of some mice Bar = 50 μm. **B**. A/J mice were orally gavaged with 125 μg LT (PA+LF) or injected intraperitoneally (i.p.) and monitored for morbidity and death. **C**. IL-1β expression profile of the distal colons of A/J mice that were given LT (125 μg) by the oral route versus i.p. injection.

Figure S2 *Ex vivo* **Inhibition of MAPKs in Immune Cells by Sterne.** Colonic LP cells were isolated from uninfected A/J mice and incubated with 1 MOI of *B. anthracis* spores for 1, 3, or 6 hours. Activity of p38 and Erk1/2 was subsequently analyzed with phosphorylation-specific antibodies by Western blot.

Figure S3 **Splenic Innate Immune Responses in Sterne-infected A/J Mice.** A/J mice were orally gavaged with 10^9 spores of the Sterne strain of *B. anthracis* and innate immune responses analyzed at various time points by flow cytometry. Splenic DC (**A**) and macrophage (**B**) functions were inhibited after infection as measured by IL-1β and TNF-α production. n = 10 mice/group. Data represent observations from four independent experiments and are shown as mean +/− SEM. **P<0.01, ***P< 0.001 compared with PBS.

Figure S4 **Sera Cytokine Levels of Sterne-infected A/J Mice.** A/J mice were orally gavaged with 10^9 spores of the Sterne strain of *B. anthracis* and sera collected. Cytokines in the sera of Sterne-infected and uninfected A/J mice were measured using the Bio-Plex Pro Mouse Cytokine 23-plex immunoassay kit. Data are

shown as mean +/− SEM; each symbol represents one mouse. *P<0.05, ***P<0.001 compared with PBS.

Figure S5 **Splenic T Cell Responses in Sterne-infected A/J Mice.** A/J mice were orally gavaged with 10^9 spores of the Sterne strain of *B. anthracis* and Th1 (**A**) and Th17 (**B**) responses analyzed at various time points by flow cytometry. n = 10 mice/group. Data represent observations from four independent experiments and are shown as mean +/− SEM. *P<0.05 compared with PBS.

Table S1 List of primer sequences for Real-Time PCR analyses.

Table S2 List of primer sequences for 16S rDNA analyses.

Acknowledgments

We would like to extend our gratitude to Dr. Arthur Friedlander and Dr. Timothy Hoover for fruitful discussions, as well as to Dr. Lieqi Tang and Eric Li for technical support.

Author Contributions

Conceived and designed the experiments: YLL TY SXC GPW JLO MM. Performed the experiments: YLL TY BS MZ JLO. Analyzed the data: YLL TY BS MZ SXC GPW JLO MM. Wrote the paper: YLL TY SXC GPW JLO MM.

References

1. Beatty ME, Ashford DA, Griffin PM, Tauxe RV, Sobel J (2003) Gastrointestinal anthrax: review of the literature. Arch Intern Med 163: 2527–2531.
2. Mock M, Fouet A (2001) Anthrax. Annu Rev Microbiol 55: 647–671.
3. Turnbull PC (1991) Anthrax vaccines: past, present and future. Vaccine 9: 533–539.
4. Mikesell P, Ivins BE, Ristroph JD, Dreier TM (1983) Evidence for plasmid-mediated toxin production in Bacillus anthracis. Infect Immun 39: 371–376.
5. Green BD, Battisti L, Koehler TM, Thorne CB, Ivins BE (1985) Demonstration of a capsule plasmid in Bacillus anthracis. Infect Immun 49: 291–297.
6. Tonello F, Zornetta I (2012) Bacillus anthracis factors for phagosomal escape. Toxins (Basel) 4: 536–553.
7. Makino S, Watarai M, Cheun HI, Shirahata T, Uchida I (2002) Effect of the lower molecular capsule released from the cell surface of Bacillus anthracis on the pathogenesis of anthrax. J Infect Dis 186: 227–233.
8. Scorpio A, Chabot DJ, Day WA, O'Brien D K, Vietri NJ, et al. (2007) Poly-gamma-glutamate capsule-degrading enzyme treatment enhances phagocytosis and killing of encapsulated Bacillus anthracis. Antimicrobial agents and chemotherapy 51: 215–222.
9. Mohamadzadeh M, Duong T, Sandwick SJ, Hoover T, Klaenhammer TR (2009) Dendritic cell targeting of Bacillus anthracis protective antigen expressed by Lactobacillus acidophilus protects mice from lethal challenge. Proc Natl Acad Sci U S A 106: 4331–4336.
10. Tournier JN, Ulrich RG, Quesnel-Hellmann A, Mohamadzadeh M, Stiles BG (2009) Anthrax, toxins and vaccines: a 125-year journey targeting Bacillus anthracis. Expert Rev Anti Infect Ther 7: 219–236.
11. Glomski IJ, Piris-Gimenez A, Huerre M, Mock M, Goossens PL (2007) Primary involvement of pharynx and peyer's patch in inhalational and intestinal anthrax. PLoS Pathog 3: e76.
12. Tonry JH, Popov SG, Narayanan A, Kashanchi F, Hakami RM, et al. (2013) In vivo murine and in vitro M-like cell models of gastrointestinal anthrax. Microbes Infect 15: 37–44.
13. Xie T, Sun C, Uslu K, Auth RD, Fang H, et al. (2013) A New Murine Model for Gastrointestinal Anthrax Infection. PLoS One 8: e66943.
14. Baldari CT, Tonello F, Paccani SR, Montecucco C (2006) Anthrax toxins: A paradigm of bacterial immune suppression. Trends in immunology 27: 434–440.
15. Welkos SL, Keener TJ, Gibbs PH (1986) Differences in susceptibility of inbred mice to Bacillus anthracis. Infect Immun 51: 795–800.
16. Cheng SX (2012) Calcium-sensing receptor inhibits secretagogue-induced electrolyte secretion by intestine via the enteric nervous system. Am J Physiol Gastrointest Liver Physiol 303: G60–70.
17. Barman M, Unold D, Shifley K, Amir E, Hung K, et al. (2008) Enteric salmonellosis disrupts the microbial ecology of the murine gastrointestinal tract. Infect Immun 76: 907–915.
18. Antharam VC, Li EC, Ishmael A, Sharma A, Mai V, et al. (2013) Intestinal dysbiosis and depletion of butyrogenic bacteria in Clostridium difficile infection and nosocomial diarrhea. J Clin Microbiol 51: 2884–2892.
19. Goossens PL (2009) Animal models of human anthrax: the Quest for the Holy Grail. Mol Aspects Med 30: 467–480.
20. Loving CL, Kennett M, Lee GM, Grippe VK, Merkel TJ (2007) Murine aerosol challenge model of anthrax. Infect Immun 75: 2689–2698.
21. Welkos SL, Keener TJ, Gibbs PH (1986) Differences in susceptibility of inbred mice to Bacillus anthracis. Infection and immunity 51: 795–800.
22. Salzman NH, Hung K, Haribhai D, Chu H, Karlsson-Sjoberg J, et al. (2010) Enteric defensins are essential regulators of intestinal microbial ecology. Nature immunology 11: 76–83.
23. Lozupone C, Lladser ME, Knights D, Stombaugh J, Knight R (2011) UniFrac: an effective distance metric for microbial community comparison. ISME J 5: 169–172.
24. Segata N, Izard J, Waldron L, Gevers D, Miropolsky L, et al. (2011) Metagenomic biomarker discovery and explanation. Genome Biol 12: R60.
25. Lau SK, Woo PC, Woo GK, Fung AM, Ngan AH, et al. (2006) Bacteraemia caused by Anaerotruncus colihominis and emended description of the species. J Clin Pathol 59: 748–752.
26. Hill JE, Fernando WM, Zello GA, Tyler RT, Dahl WJ, et al. (2010) Improvement of the representation of bifidobacteria in fecal microbiota metagenomic libraries by application of the cpn60 universal primer cocktail. Appl Environ Microbiol 76: 4550–4552.
27. Hu H, Leppla SH (2009) Anthrax toxin uptake by primary immune cells as determined with a lethal factor-beta-lactamase fusion protein. PLoS one 4: e7946.
28. Agrawal A, Lingappa J, Leppla SH, Agrawal S, Jabbar A, et al. (2003) Impairment of dendritic cells and adaptive immunity by anthrax lethal toxin. Nature 424: 329–334.
29. Duesbery NS, Webb CP, Leppla SH, Gordon VM, Klimpel KR, et al. (1998) Proteolytic inactivation of MAP-kinase-kinase by anthrax lethal factor. Science 280: 734–737.
30. Vitale G, Bernardi L, Napolitani G, Mock M, Montecucco C (2000) Susceptibility of mitogen-activated protein kinase kinase family members to proteolysis by anthrax lethal factor. The Biochemical journal 352 Pt 3: 739–745.
31. Ebrahimi CM, Sheen TR, Renken CW, Gottlieb RA, Doran KS (2011) Contribution of lethal toxin and edema toxin to the pathogenesis of anthrax meningitis. Infection and immunity 79: 2510–2518.
32. Xie T, Auth RD, Frucht DM (2011) The effects of anthrax lethal toxin on host barrier function. Toxins 3: 591–607.
33. Xu L, Frucht DM (2007) Bacillus anthracis: a multi-faceted role for anthrax lethal toxin in thwarting host immune defenses. The international journal of biochemistry & cell biology 39: 20–24.

34. Park JM, Greten FR, Li ZW, Karin M (2002) Macrophage apoptosis by anthrax lethal factor through p38 MAP kinase inhibition. Science 297: 2048–2051.

35. During RL, Li W, Hao B, Koenig JM, Stephens DS, et al. (2005) Anthrax lethal toxin paralyzes neutrophil actin-based motility. The Journal of infectious diseases 192: 837–845.

36. Chou PJ, Newton CA, Perkins I, Friedman H, Klein TW (2008) Suppression of dendritic cell activation by anthrax lethal toxin and edema toxin depends on multiple factors including cell source, stimulus used, and function tested. DNA Cell Biol 27: 637–648.

37. Karakhanova S, Meisel S, Ring S, Mahnke K, Enk AH (2010) ERK/p38 MAP-kinases and PI3K are involved in the differential regulation of B7-H1 expression in DC subsets. Eur J Immunol 40: 254–266.

38. Chopra AP, Boone SA, Liang X, Duesbery NS (2003) Anthrax lethal factor proteolysis and inactivation of MAPK kinase. J Biol Chem 278: 9402–9406.

39. Paccani SR, Benagiano M, Savino MT, Finetti F, Tonello F, et al. (2011) The adenylate cyclase toxin of Bacillus anthracis is a potent promoter of T(H)17 cell development. The Journal of allergy and clinical immunology 127: 1635–1637.

40. Datta SK, Sabet M, Nguyen KP, Valdez PA, Gonzalez-Navajas JM, et al. (2010) Mucosal adjuvant activity of cholera toxin requires Th17 cells and protects against inhalation anthrax. Proc Natl Acad Sci U S A 107: 10638–10643.

41. Wherry EJ, Ha SJ, Kaech SM, Haining WN, Sarkar S, et al. (2007) Molecular signature of CD8+ T cell exhaustion during chronic viral infection. Immunity 27: 670–684.

42. Bonecchi R, Bianchi G, Bordignon PP, D'Ambrosio D, Lang R, et al. (1998) Differential expression of chemokine receptors and chemotactic responsiveness of type 1 T helper cells (Th1s) and Th2s. J Exp Med 187: 129–134.

43. Lacotte S, Brun S, Muller S, Dumortier H (2009) CXCR3, inflammation, and autoimmune diseases. Ann N Y Acad Sci 1173: 310–317.

44. Swartz MN (2001) Recognition and management of anthrax–an update. N Engl J Med 345: 1621–1626.

45. Sun C, Fang H, Xie T, Auth RD, Patel N, et al. (2012) Anthrax lethal toxin disrupts intestinal barrier function and causes systemic infections with enteric bacteria. PLoS One 7: e33583.

46. Chandra RK (1996) Nutrition, immunity and infection: from basic knowledge of dietary manipulation of immune responses to practical application of ameliorating suffering and improving survival. Proc Natl Acad Sci U S A 93: 14304–14307.

47. Coggeshall KM, Lupu F, Ballard J, Metcalf JP, James JA, et al. (2013) The sepsis model: an emerging hypothesis for the lethality of inhalation anthrax. Journal of cellular and molecular medicine 17: 914–920.

48. Kern J, Schneewind O (2010) BslA, the S-layer adhesin of B. anthracis, is a virulence factor for anthrax pathogenesis. Mol Microbiol 75: 324–332.

49. Manichanh C, Borruel N, Casellas F, Guarner F (2012) The gut microbiota in IBD. Nature reviews Gastroenterology & hepatology 9: 599–608.

50. Ewaschuk JB, Diaz H, Meddings L, Diederichs B, Dmytrash A, et al. (2008) Secreted bioactive factors from Bifidobacterium infantis enhance epithelial cell barrier function. American journal of physiology Gastrointestinal and liver physiology 295: G1025 1034.

51. Biddle A SL, Blanchard J, Leschine S (2013) Untangling the Genetic Basis of Fibrolytic Specialization by Lachnospiraceae and Ruminococcaceae in Diverse Gut Communities. Diversity 5: 627 640.

52. Thibault R, Blachier F, Darcy-Vrillon B, de Coppet P, Bourreille A, et al. (2010) Butyrate utilization by the colonic mucosa in inflammatory bowel diseases: a transport deficiency. Inflammatory bowel diseases 16: 684–695.

53. Atarashi K, Tanoue T, Shima T, Imaoka A, Kuwahara T, et al. (2011) Induction of colonic regulatory T cells by indigenous Clostridium species. Science 331: 337–341.

54. Mohamadzadeh M, Chen L, Schmaljohn AL (2007) How Ebola and Marburg viruses battle the immune system. Nat Rev Immunol 7: 556–567.

55. Gianchecchi E, Delfino DV, Fierabracci A (2013) Recent insights into the role of the PD-1/PD-L1 pathway in immunological tolerance and autoimmunity. Autoimmunity reviews.

56. Bachmann M, Dragoi C, Poleganov MA, Pfeilschifter J,M (2007) Interleukin-18 directly activates T-bet expression and function via p38 mitogen-activated protein kinase and nuclear factor-Œ/B in acute myeloid leukemia, Aiderived predendritic KG-1 cells. Molecular Cancer Therapeutics 6: 723–731.

57. Kao C, Oestreich KJ, Paley MA, Crawford A, Angelosanto JM, et al. (2011) Transcription factor T-bet represses expression of the inhibitory receptor PD-1 and sustains virus-specific CD8+ T cell responses during chronic infection. Nature immunology 12: 663–671.

58. Qian Y, Deng J, Geng L, Xie H, Jiang G, et al. (2008) TLR4 signaling induces B7-H1 expression through MAPK pathways in bladder cancer cells. Cancer investigation 26: 816–821.

59. Pepper M, Linehan JL, Pagan AJ, Zell T, Dileepan T, et al. (2010) Different routes of bacterial infection induce long-lived TH1 memory cells and short-lived TH17 cells. Nature immunology 11: 83–89.

60. Balhara J, Gounni AS (2012) The alveolar macrophages in asthma: a double-edged sword. Mucosal immunology 5: 605–609.

61. Tsuji NM, Kosaka A (2008) Oral tolerance: intestinal homeostasis and antigen-specific regulatory T cells. Trends in immunology 29: 532–540.

Erythrocytic Mobilization Enhanced by the Granulocyte Colony-Stimulating Factor Is Associated with Reduced Anthrax-Lethal-Toxin-Induced Mortality in Mice

Hsin-Hou Chang[1,2], Ya-Wen Chiang[1], Ting-Kai Lin[1], Guan-Ling Lin[2], You-Yen Lin[2], Jyh-Hwa Kau[3], Hsin-Hsien Huang[4], Hui-Ling Hsu[4], Jen-Hung Wang[5], Der-Shan Sun[1,2]*

1 Department of Molecular Biology and Human Genetics, Tzu-Chi University, Hualien, Taiwan, 2 Institute of Medical Sciences, Tzu-Chi University, Hualien, Taiwan, 3 Department of Microbiology and Immunology, National Defense Medical Center, Taipei, Taiwan, 4 Institute of Preventive Medicine, National Defense Medical Center, Taipei, Taiwan, 5 Department of Medical Research, Tzu Chi General Hospital, Hualien, Taiwan

Abstract

Anthrax lethal toxin (LT), one of the primary virulence factors of *Bacillus anthracis*, causes anthrax-like symptoms and death in animals. Experiments have indicated that levels of erythrocytopenia and hypoxic stress are associated with disease severity after administering LT. In this study, the granulocyte colony-stimulating factor (G-CSF) was used as a therapeutic agent to ameliorate anthrax-LT- and spore-induced mortality in C57BL/6J mice. We demonstrated that G-CSF promoted the mobilization of mature erythrocytes to peripheral blood, resulting in a significantly faster recovery from erythrocytopenia. In addition, combined treatment using G-CSF and erythropoietin tended to ameliorate *B. anthracis*-spore-elicited mortality in mice. Although specific treatments against LT-mediated pathogenesis remain elusive, these results may be useful in developing feasible strategies to treat anthrax.

Editor: Nupur Gangopadhyay, University of Pittsburgh, United States of America

Funding: This work was supported by grants of National Science Council http://web1.nsc.gov.tw/mp.aspx (NSC 96-2311-B-320-005-MY3 and NSC 99-2311-B-320-003-MY3) and Tzu-Chi University http://www.tcu.edu.tw/ (610400130). The funders had no role in study design, data collection and analysis, decision to publish, or preparation of the manuscript.

Competing Interests: The authors have declared that no competing interests exist.

* Email: dssun@mail.tcu.edu.tw

Introduction

Infection with *Bacillus anthracis*, a gram-positive spore-forming bacterium, can lead to life-threatening anthrax [1]. Anthrax lethal toxin (LT) is comprised of a protective antigen (PA, 83 kDa) and lethal factor (LF, 90 kDa) [2–4], and is one of the primary virulence factors of *B. anthracis*. LF is a specific metalloprotease for mitogen-activated protein kinase (MAPK) kinases (MKKs/MEKs) [5], and can thus disrupt MAPK signaling cascades including p38 MAPK, p42/44 extracellular signal-regulated kinase (ERK), and c-Jun N-terminal kinase (JNK) [6,7]. All of these 3 MAPK pathways are critical in maintaining fundamental cellular homeostasis, including cell proliferation, differentiation, and apoptosis [8]. LF can be delivered into cells by PA, a cell-receptor binding component [3,9]. Although experimental LT treatments may not reproduce the full pathogenesis of anthrax, LT studies in cell or animal models have revealed certain pathogenic progressions. Various cell types, which include macrophages [10,11], dendritic cells [12], lymphocytes [13,14], erythrocytes [15], and megakaryocytes [16], cardiomyocytes [17], and smooth muscle [17], are sensitive to LT treatment. LT has been shown to suppress the differentiation and maturation of the progenitors of macrophages, megakaryocytes, and erythrocytes [15,16,18]. In addition, blood cell count analyses have indicated that LT treatment significantly reduced levels of circulating red blood cells (RBCs) and platelets in mice [15,16], suggesting multiple targets of

LT in hematopoietic lineage cells. Deficiencies of platelets and RBCs may lead to hemorrhage and lethal hypoxic damage [19,20]. Because high levels of LT accumulate in the body when anthrax enters the bacteremia stage, death is typically inevitable even after aggressive antibiotic treatments. This suggests that a specific treatment to overcome the toxic effect is crucial in controlling the disease [21]. Unfortunately, an effective therapeutic approach against LT remains elusive.

Cytokine treatments, particularly hematopoietic growth factors, have been used in various clinical settings to rescue pathological defects [22]. Our previous demonstration was the first to indicate that thrombopoietin (TPO), a megakaryopoiesis-enhancing cytokine [23], can ameliorate LT-induced thrombopoiesis suppression, thrombocytopenia, and likely reduce the mortality in mice [16]. Our data also revealed that erythropoietin (EPO), a potent erythropoiesis-stimulating cytokine [24], ameliorated LT-induced erythropoiesis suppression (particularly those precursors in early erythropoiesis stages), erythrocytopenia, and reduced mortality rates from 100% to 50% after lethal-dose LT challenges in experimental mice [15]. Bone marrow is the primary stem niche supporting erythropoiesis, displaying technically-divided 4-differentiation stages of erythroblasts based on the expression levels of surface markers CD71 and TER-119 [25]. The transferrin receptor (CD71) is first expressed on early erythroblasts, such as erythroid burst-forming units (BFU-Es) and erythroid colony-forming units (CFU-Es) cells. Erythrocytic CD71 is downregulated by more mature erythroblasts [26]. By contrast, TER-119 is primarily expressed on relatively mature erythroblasts, reticulocytes, and mature erythrocytes [27]. Accordingly, 4 cell populations (CD71^{high}TER-119^{med}, CD71^{high}TER-119^{high}, CD71^{med}-TER-119^{high}, and CD71^{low}TER-119^{high}) can be defined, which are morphologically equivalent to proerythroblasts (flow cytometry-gated region 1; R1), basophilic erythroblasts (R2), late basophilic and polychromatophilic erythroblasts (R3), and orthochromatophilic erythroblasts (R4), from the early to late stages of erythroid differentiation, respectively [25]. Following these approaches, we are thus able to characterize the mechanism to use hematopoietic cytokines/growth factors as ameliorative agents to rescue anthrax LT-induced mortality [15,16].

The granulocyte colony-stimulating factor (G-CSF) has been found to regulate granulopoiesis [28], and is a multifunctional cytokine. For example, it has been found to stimulate cell proliferation, differentiation, enhance hematopoiesis, mobilize hematopoietic stem cells, and induce anti-apoptotic and anti-inflammatory effects [29–31]. Both G-CSF and EPO are U.S. FDA-approved drugs. Although the mechanism remains unclear, combined treatments using G-CSF and EPO were shown to ameliorate aplastic anemia in patients with myelodysplastic syndrome [32–36]. Given that EPO treatments are beneficial for LT-challenged mice [15], we hypothesized that combining G-CSF and EPO may be useful in treating anthrax. Consequently, we used mouse models to discuss the combined treatments of G-CSF and EPO on reduced anthrax LT and spore-induced mortality. In addition, we also discussed the differential erythropoietic regulation in response to G-CSF and EPO treatments.

Materials and Methods

Ethics Statement

Our research approaches involving experimental mice were approved by the Institutional Animal Care and Use Committee of Tzu Chi University (Approval ID: 98104) and the National Defense Medical Center (Approval ID: AN-100-04).

Toxins and spores

B. anthracis-derived LT was purified according to previously described procedures [49]. LT was delivered in a 1:5 ratio of LF and PA [16]. Spores derived from the *B. anthracis*-nonencapsulated mutant strain (pXO1$^+$, pXO2$^-$) were purchased from the American Type Culture Collection (Manassas, VA, USA) (ATCC 14186).

Erythroid colony-forming cell assay

The erythroid colony-forming cell assay was conducted according to the manufacturer's instructions (MethoCult M3334, StemCell Technologies). For the *in vitro* erythroid colony-forming cell assay, bone marrow cells were collected from the femurs and tibiae of C57BL/6J mice. C57BL/6J mice (males, 8–10 wk of age) were obtained from the National Laboratory Animal Center (Taipei, Taiwan) and kept in a specific pathogen-free (SPF) environment in the experimental animal center of Tzu Chi University. For the *ex vivo* erythroid colony-forming cell assay, C57BL/6J mice were retro-orbitally injected with 55 μg/kg/d of recombinant human G-CSF (Filgrastim, Kirin, Tokyo, Japan) in 250 μl saline, once daily for 5 d, initiated 5 d before the challenges of a lethal dose of LT (1.5 mg/kg in 250 μl saline, retro-orbitally injected). Treatments using saline, G-CSF, and LT alone were used as comparison controls. Bone marrow cells were collected at 69 h after LT treatment and flushed with Roswell Park Memorial Institute medium (RPMI)-1640 containing 20% anticoagulant acid

citrate dextrose formula A (ACD-A: 38 mM citric acid, 75 mM trisodium citrate, 139 mM D-glucose, 12.5 mM EDTA [15]). After depleting RBCs by adding a hypotonic buffer (153 mM NH$_4$Cl and 17 mM Tris-HCl) at room temperature for 10 min, 100 μl of remaining cells (9×10^5/ml) were resuspended in Iscove's Modified Dulbecco's Medium (IMDM) (StemCell Technologies) and mixed with 1 ml of semisolid methylcellulose-based medium containing 3 units of EPO. Finally, each 1.1 ml of methylcellulose-cell suspension was mixed with or without a dose of G-CSF (20 ng/ml or 764 ng/ml) and duplicate seeded in 35-mm dishes. A G-CSF dose of 20 ng/ml was used in the colony-forming cell assay for hematopoietic cells [50]. Because the volume of mice blood is 70–80 ml/kg [51], a dose of 764 ng/ml approximated the dose used to ameliorate LT-induced mortality in the experiments. Two doses of G-CSF (20 ng/ml and 764 ng/ml) were added to the medium supplement of the erythroid colony-forming cell assay. The cultures were incubated at 37°C for 14 d. Dynamic changes in colony number were measured on Days 3, 7, and 14 after initiating the colony assay. The erythroid colonies were separated into 3 groups by size: small (8–50 cells), medium (more than 50, but less than 200 cells), and large (more than 200 cells).

Analysis of erythropoiesis in bone marrow

Bone marrow cells were purified and blocked with 5% bovine serum albumin in RPMI medium at 37°C for 1 h and incubated in 500 μl RPMI-1640 medium with 1 μl of fluorescein (FITC)-conjugated rat anti-mouse CD71 antibody (BioLegend) and 3 μl of R-Phycoerythrin (R-PE)-conjugated rat anti-mouse TER-119 antibody (BD Immunocytometry System) at 37°C for 1 h. After washing with phosphate-buffered saline (PBS), the cells were measured and analyzed using a FACSCalibur flow cytometer and the CellQuest^{TM} Pro program (Becton-Dickinson).

Flow cytometry analysis of peripheral blood cells of G-CSF-treated EGFP mice

EGFP mice [C57BL/6J-Tg (Pgk1-EGFP) 03Narl, males, 10–12 wk of age] were obtained from the National Laboratory Animal Center (Taipei, Taiwan) and maintained in the aforementioned SPF environments. EGFP mice were retro-orbitally injected with G-CSF (55 μg/kg/d in 250 μl saline) once daily for 4 d. To detect the erythrocytes' specific surface markers, 50 μl of retro-orbital blood samples were obtained 22, 44, 66, and 94 h after the initial G-CSF injection, and subsequently mixed with 450 μl of anticoagulant ACD-A (1:9). Cells were incubated in 300 μl of RPMI-1640 medium with 3 μl of the R-Phycoerythrin (R-PE)-conjugated rat anti-mouse TER-119 antibody (BD Immunocytometry System) at 37°C for 1 h. After washing with PBS, the cells were analyzed using a FACSCalibur flow cytometer and the CellQuest^{TM} Pro program.

G-CSF treatment to reduce LT-induced mortality

C57BL/6J mice (males, 8–10 wk of age) were retro-orbitally injected with 55 μg/kg/d of recombinant human G-CSF in 250 μl saline, once daily for 5 d, initiated 5 d before or 1 d after the challenges of a lethal dose of LT (1.5 mg/kg in 250 μl saline, retro-orbitally injected). Treatments using saline, G-CSF, and LT alone were used as comparison controls. Because no suitable potential predictor of death/survival exists for LT-challenged mice, we used death as an endpoint for the survival experiment. The survival time and mortality of mice were recorded after the LT challenges. LT treatment in mice did not induce obvious discomfort and body weight loss, except for reducing activities. The experimental mice were continually monitored up to 250 h

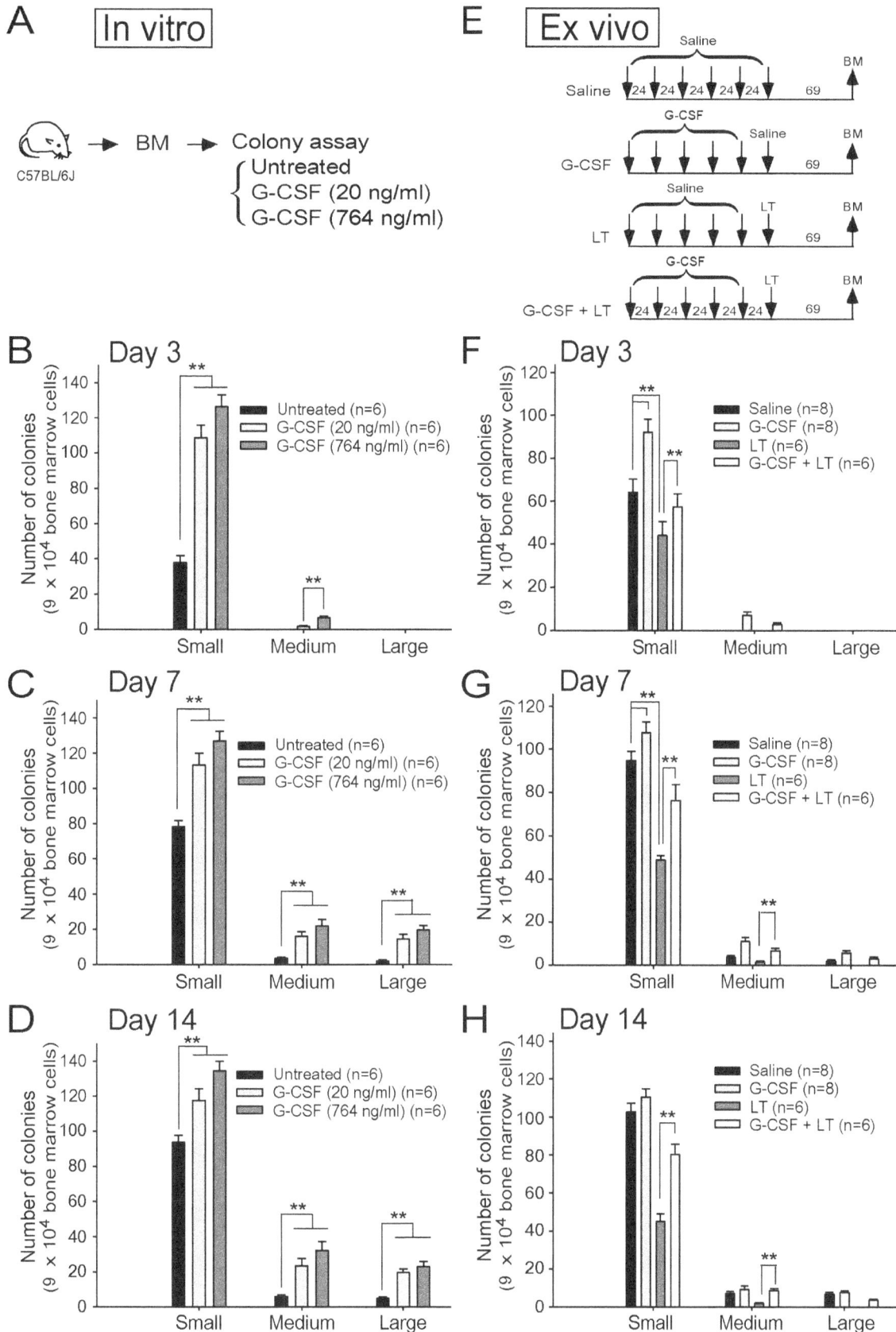

Figure 1. Erythroid colony-forming cell assays to measure the effect of G-CSF on erythropoiesis. The experimental outlines of *in vitro* (A) and *ex vivo* (E) analyses are shown. An *in vitro* assay was performed using murine bone marrow (BM) cells that were incubated with [20 ng/ml (n = 6) or 764 ng/ml (n = 6)], or without G-CSF (n = 6). The colonies were quantified on Days 3 (B), 7 (C), and 14 (D). Untreated bone marrow cells were used

as a control. Colony numbers of bone marrow cells from mice, which were treated with G-CSF (n = 8), LT (n = 6), or G-CSF and LT (n = 6), were measured on Days 3 (F), 7 (G), and 14 (H) following the *ex vivo* colony assay. Bone marrow cells from saline treated mice (n = 8) served as controls. **P<0.01 was compared between the indicated groups. Data are shown as mean ± standard deviation (SD) and represent results from 2 independent experiments. The mouse drawing used in this and all following figures was originally published in *Blood*. Huang, H. S., Sun, D. S., Lien, T. S. and Chang, H. H. Dendritic cells modulate platelet activity in IVIg-mediated amelioration of ITP in mice. *Blood*. 2010; 116: 5002–5009. © the American Society of Hematology.

for every 4–6 h. All surviving mice were monitored each day for 2 subsequent mo. For hematopoietic parameters, 50 μl of retro-orbital blood samples were collected at 22, 44, and 66 h after LT challenges and analyzed by an automated hematology analyzer (KX-21, Sysmex Corporation).

Combined treatments with G-CSF and EPO to reduce anthrax-spore-induced mortality

C57BL/6J mice were retro-orbitally injected with G-CSF (55 μg/kg/d in 250 μl saline) daily for 5 consecutive d or injected

with a combination of recombinant human EPO (rhEPO, Neorecormon, Roche, Mannheim, Germany) (2 IU/g, in 250 μl saline) twice at 24 and 48 h after injecting spores (1×10^7 in 1 ml saline, intraperitoneal injection). The survival times and mortality of mice were recorded. Because no suitable potential predictor of death/survival exists for spore-challenged mice, we used death as an endpoint for the survival experiment. Spore treatment in mice did not induce obvious discomfort and body weight loss, except for reducing activities. The experimental mice were continually

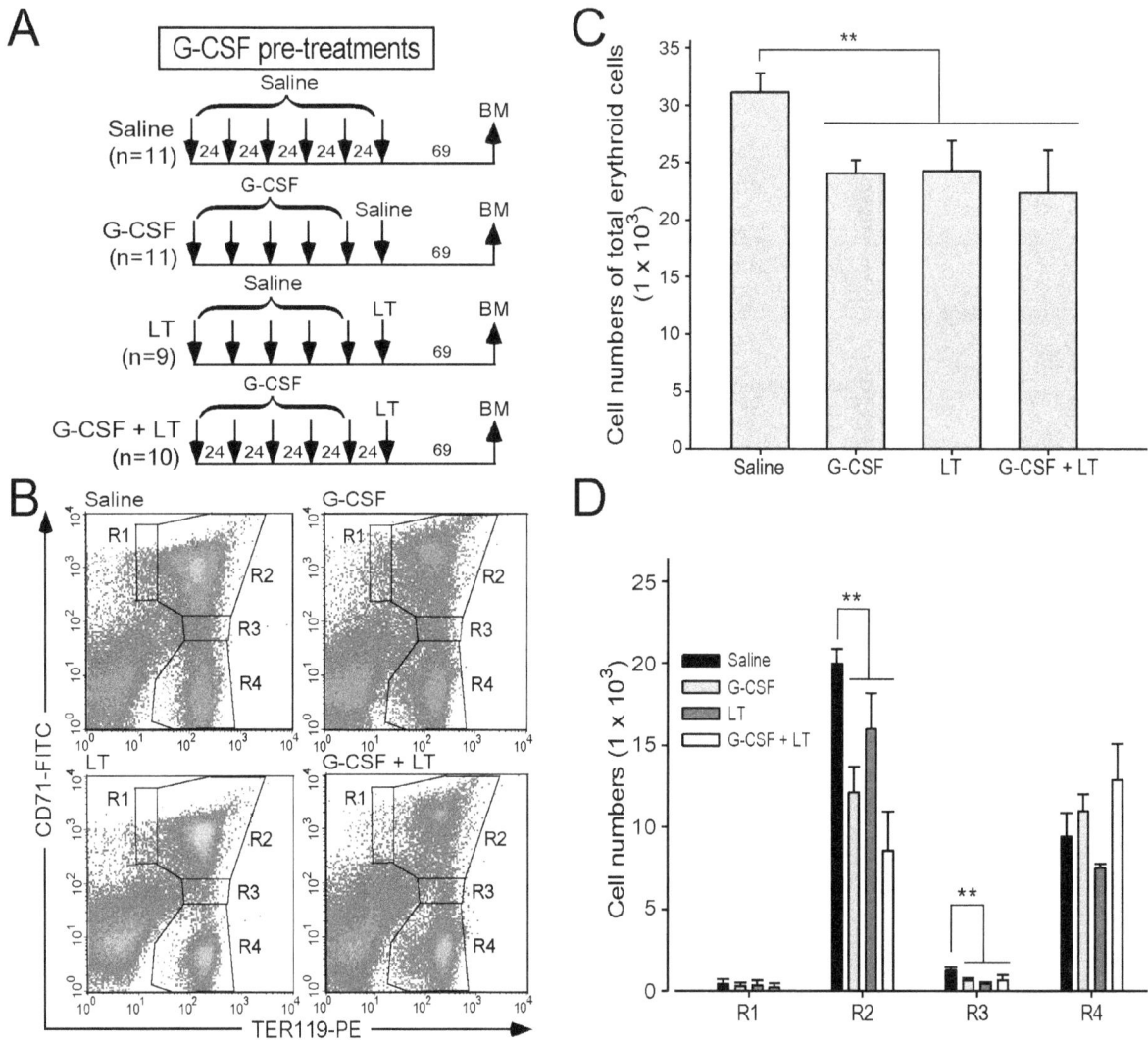

Figure 2. Regulation of G-CSF on bone marrow erythroblast populations. The experimental outlines are illustrated (A). Mice were treated with saline (n = 11), G-CSF (n = 11), LT (n = 9), or G-CSF plus LT (n = 10). Flow cytometry analysis analyzed erythroblast populations of BM cells at 69 h after LT challenges. The erythroblast cells were gated as R1 (CD71^{high}, TER-119^{med}), R2 (CD71^{high}, TER-119^{high}), R3 (CD71^{med}, TER-119^{high}), and R4 (CD71^{low}, TER-119^{high}) in all groups (B) as described [25]. The cell numbers of all erythroblast cells (sum of R1 to R4) (C) and individual erythroblast (R1, R2, R3, and R4) (D) in each group were quantified. **P<0.01 was compared between indicated groups. Data are showed as mean ± SD and represent the results from 2 independent experiments.

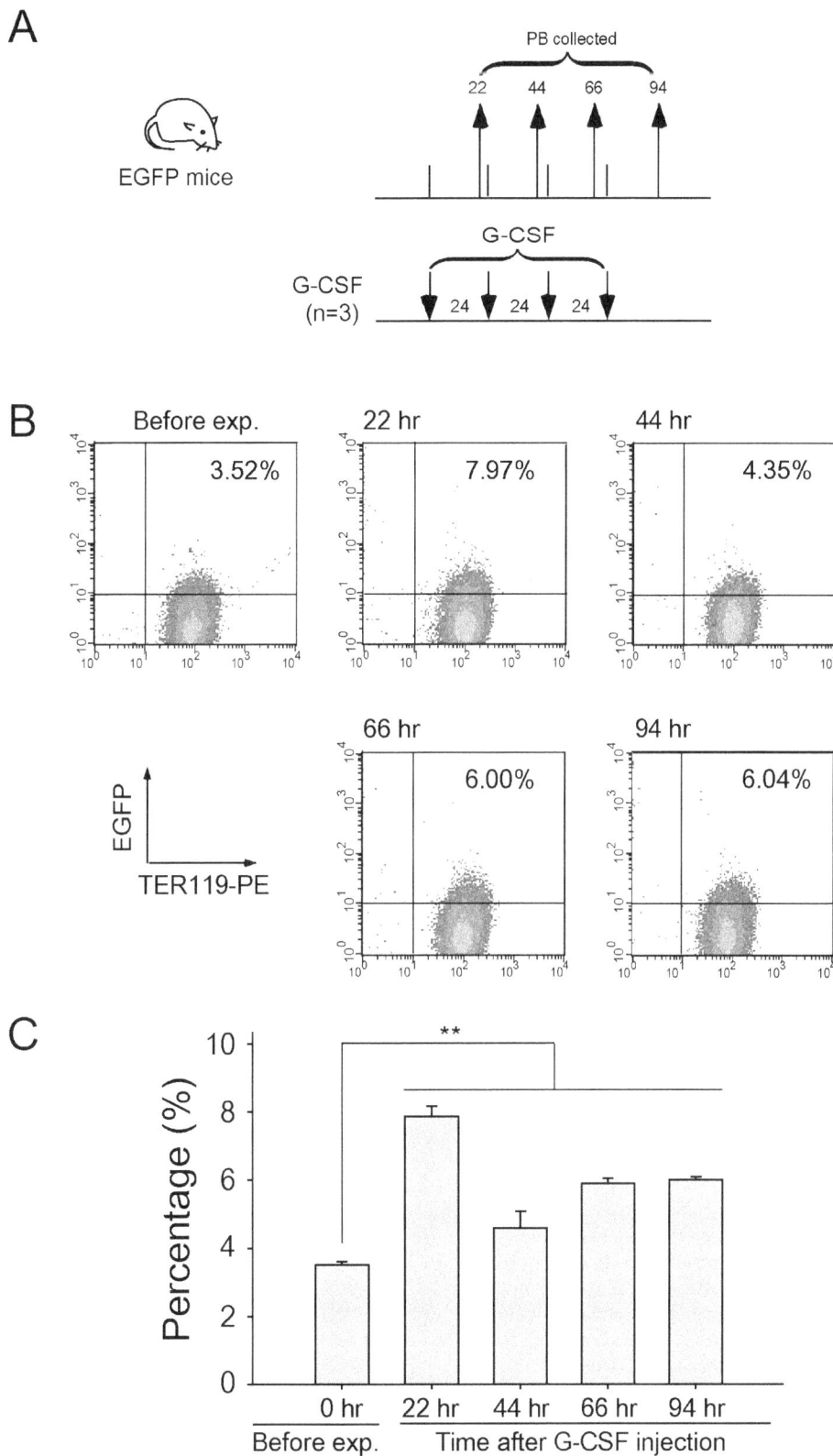

Figure 3. Mobilization of newly synthesized erythrocytes into peripheral blood by G-CSF. The experimental outline is illustrated (A). The EGFP mice were injected with G-CSF (n = 3). The percentage of EGFP⁺/TER119⁺ cells in peripheral blood (PB) was analyzed by flow cytometry (B) and quantified (C) at 22, 44, 66, and 94 h after G-CSF injection. PB collected from mice before G-CSF injection served as the negative control. **$P < 0.01$ was compared to the negative control. Data are shown as mean ± SD.

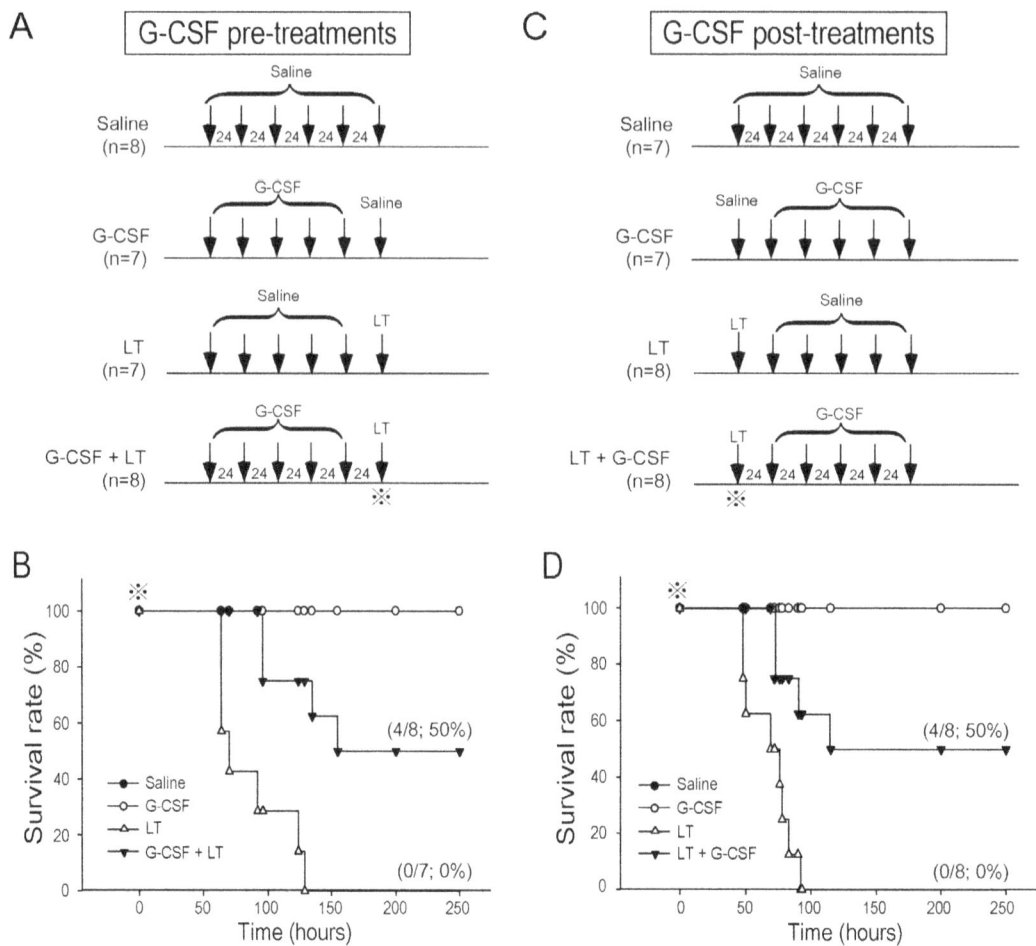

Figure 4. G-CSF treatments ameliorated LT-elicited mortality in mice. The experimental timetable (A), (C), and the survival rates of mice pre-treated (B) and post-treated (D) with G-CSF, LT, or G-CSF and LT are indicated. Saline treated mice served as negative controls. The symbol (※) in (A) to (D) indicates the onset time point for recording survival rates.

monitored up to 15 d for every 4–6 h. All surviving mice were monitored each day for 2 subsequent mo.

Statistics

All results are presented as the mean ± SD (standard deviation) for each group. Data significance was examined by one-way ANOVA followed by the post-hoc Bonferroni-corrected t-test. Univariate Kaplan-Meier analysis was used to compare the difference in survival rate between groups with various treatments. P-values were calculated and log-rank tests were performed to determine statistical significance. A probability of type 1 error $\alpha = 0.05$ was recognized as the threshold of statistical significance. Statistical analysis was conducted using the statistical software SPSS, version 17.0 (SPSS Inc., Chicago, IL, USA).

Results

G-CSF treatment promoted erythrocytic differentiation and proliferation *in vitro* and *ex vivo*

To elucidate the role of G-CSF on erythrocytic differentiation and proliferation, an *in vitro* erythroid colony-forming cell assay was performed to quantify BFU-Es and CFU-Es. Control groups without using filgrastim G-CSF supplements formed only medium-sized colonies by Day 7 (Figure 1C). By contrast, G-CSF

treatments accelerated the formation of medium-sized colonies, which appeared earlier by Day 3 (Figure 1B). Following G-CSF treatment, the numbers of colonies in all colony sizes were greater than those in the untreated control groups (Figure 1B–1D). Based on the traditional concept that G-CSF primarily regulates granulopoiesis [28], those colonies to be affected by G-CSF treatment may not be exclusively erythroid-origin cells. Consequently, 3, 3′-diaminobenzidine tetrahydrochloride (DAB) [37] was used to identify erythroid colonies, and the pseudoperoxidase activity of erythroid cells was stained on Day 14. Compared with the untreated groups, the number of DAB⁺ colonies was greater in G-CSF-treated groups (Figure S1). This data indicated that G-CSF treatment enhances the proliferation and differentiation of erythrocytes *in vitro*. Further experiments were performed to investigate the effect of G-CSF and LT treatments *ex vivo* (Figure 1E). The number of erythroid colonies sharply decreased with LT treatment on Days 3, 7 and 14 (Figure 1F–1H). Medium-sized colonies first appeared on Day 3 in G-CSF treated groups (Figure 1F), compared with saline and LT-treated groups, whereas medium-sized colonies only appeared on Day 7 (Figure 1G). In addition, G-CSF treatments ameliorated LT-induced suppression on erythropoiesis in the *ex vivo* erythroid colony-forming cell assay (Figure 1F–1H). These results indicated that G-CSF promoted erythrocytic proliferation and differentiation *in vitro* and *ex vivo*

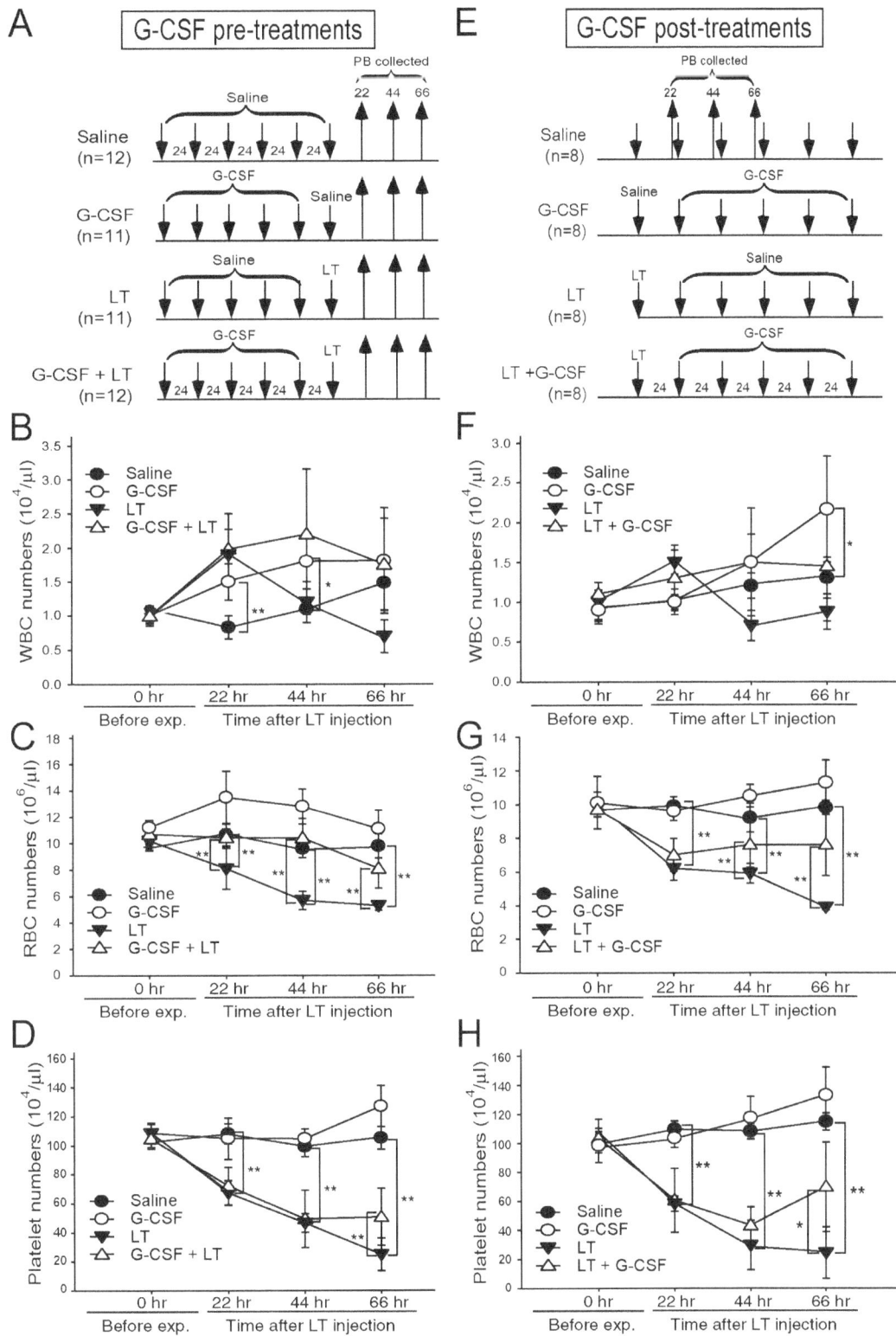

Figure 5. Amelioration of LT-induced erythrocytopenic response by G-CSF. The experimental outlines are indicated (A), (E). Mice were treated with saline (n = 12), G-CSF (n = 11), LT (n = 11), and G-CSF plus LT (n = 12) before (A) or saline (n = 8), G-CSF (n = 8), LT (n = 8), and LT plus G-CSF (n = 8) after (E) the LT challenges; their WBC, RBC, and platelet counts were subsequently analyzed at 22, 44, and 66 h after the LT challenges. Saline-treated mice were used as negative controls. *$P < 0.05$, **$P < 0.01$ were compared between the indicated groups. Data are shown as mean $\pm$ SD and represent the results from 2 independent experiments.

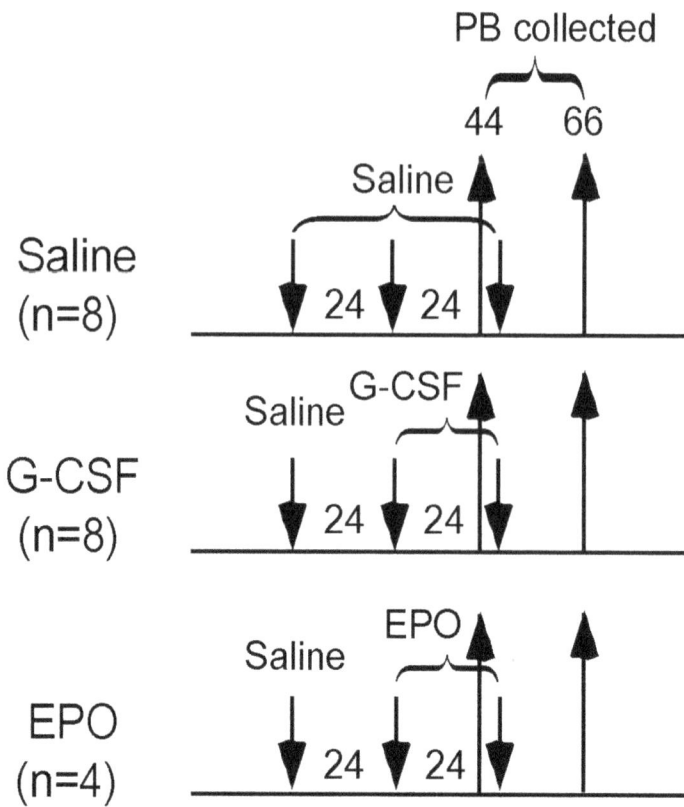

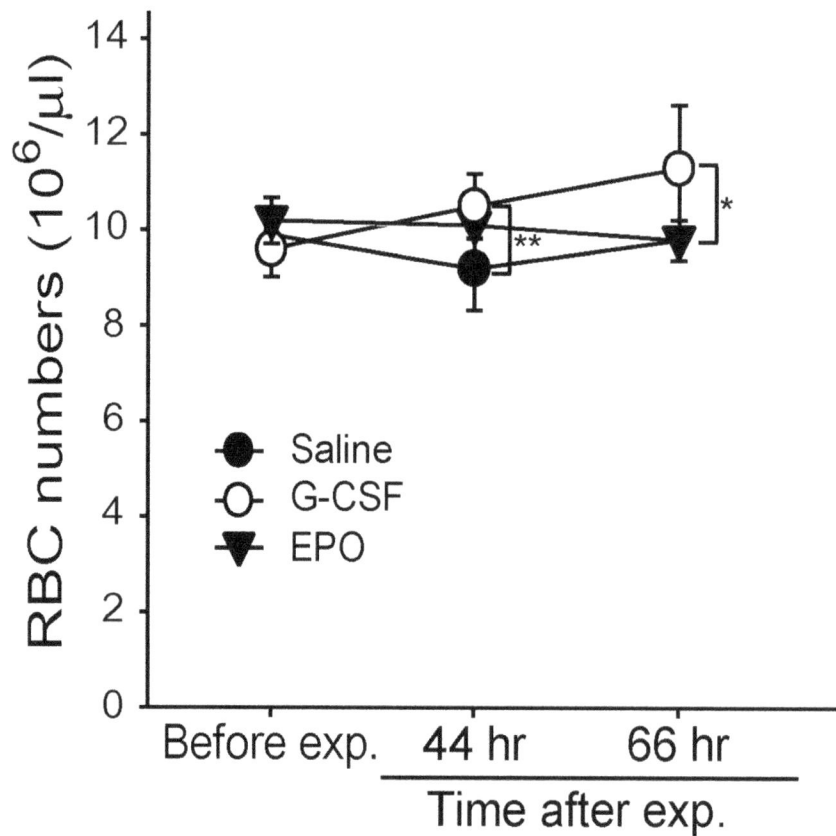

Figure 6. Fast mobilization of erythrocytes into peripheral blood by G-CSF versus EPO treatments. Experimental outline for measuring the PB-RBC counts of mice treated with G-CSF (n = 8) or EPO (n = 4) for 2 consecutive d (A). The PB RBC counts were measured before the experiments and at 44 and 66 h after the first saline injection (B). Data are shown as mean ± SD and represent the results from 2 independent experiments. Saline treated groups (n = 8) were used as the negative control. *P<0.05, **P<0.01 were compared to the negative control.

and ameliorated LT-induced erythropoiesis suppression in the *ex vivo* erythroid colony-forming cell assay.

G-CSF treatment promoted mobilization of newly synthesized RBC to peripheral blood

After the promising analyses *in vitro* and *ex vivo*, this study investigated the erythropoietic progression *in vivo*. Our previous report revealed that LT suppressed erythropoiesis in bone marrow [15]. Following similar approaches [15,25], we used surface expression of CD71 and TER-119 to verify the maturation status of various bone marrow erythroblasts under the G-CSF treatments with or without anthrax LT challenges. Although we found that G-CSF treatment rescued LT-induced erythrocytopenia (please see the following section), G-CSF pre-treatments could not overcome LT-mediated suppression on the cell numbers of both total erythroblast and individual subpopulations of erythroblast (R1 to R4 populations) (Figure 2C and 2D). This prompted us to verify whether G-CSF could mobilize mature erythrocytes into peripheral blood; we employed C57BL/6J mice with the whole-body-expressing enhanced-green-fluorescence-protein (EGFP) transgene. Prior to G-CSF analyses, we found that only a small fraction of EGFP$^+$ RBCs was detectable in the peripheral blood of normal control groups (Figure 3, 3.5% cells, before exp. groups). This is likely because mature erythrocytes do not have a nucleus, and that newly differentiated RBCs, rather than aged RBCs, express detectable EGFP. We employed an acute hemorrhage model, in which 35% of total blood was removed, to provoke the natural induction of erythropoiesis to investigate whether newly synthesized erythrocytes contain additional fluorescence. Our data revealed that EGFP$^+$/TER-119$^+$ erythrocytes increased consistently by Days 2, 4, and 6 after acute anemia (Figure S2). This suggested that EGFP$^+$/TER-119$^+$ cells are newly synthesized erythrocytes. Using the same strategy, analyses revealed that G-CSF treatment can mobilize newly synthesized erythrocytes to peripheral blood in mice (Figure 3, beginning at 22 h after G-CSF injections).

G-CSF treatment reduced LT-mediated mortality, erythrocytopenia, and thrombocytopenia

To investigate the ameliorative effect of G-CSF on LT, C57BL/6J mice were treated with G-CSF according to the manufacturer's instructions (once daily for 5 d). Treatments of G-CSF were initiated 5 d before (Figure 4A) or 1 d after the challenges of a lethal dose of LT (Figure 4C). LT initiated mortality within 48 to 129 h (Figure 4B and 4D, LT groups). Administration of 5 doses of G-CSF before (Figure 4A) and after (Figure 4C) the LT challenges significantly improved survival rates (Figure 4B and 4D) (P<0.01). Treatments using saline, G-CSF, and LT alone served as the controls (Figure 4B and 4D). The peripheral white blood cell (WBC) counts of the G-CSF-treated groups increased approximately 2-fold at 22 and 44 h (Figure 5A and 5B), as well as at 66 h (Figure 5E and 5F) following the implementation of differing G-CSF regimens; this is in consistent with a previous G-CSF report [38]. Notably, both G-CSF treatments significantly ameliorated LT-induced erythrocytopenia (Figure 5C, G-CSF + LT vs. LT; Figure 5G, LT + G-CSF vs. LT). Compared with RBC counts, the ameliorative effect of G-CSF on LT-induced thrombocytopenia was somewhat later and was observed at

approximately 66 h after LT treatment (Figure 5D and 5H). These results indicated that G-CSF positively regulated both RBC and platelet counts.

G-CSF treatment induced erythrocytes to mobilize into peripheral blood faster than EPO

Our previous study showed that EPO up-regulated RBC counts in peripheral blood [15]. To compare the EPO and G-CSF treatments in their efficiency at increasing RBC counts, mice were injected with 2 doses of either G-CSF or EPO. The circulating RBC counts were measured (Figure 6). Our data revealed that G-CSF induced a faster increase of RBC counts, within 20 h of the first G-CSF administration, than the EPO treatment, in which no increased RBC counts were observed (Figure 6B, G-CSF vs. EPO). These results suggested that G-CSF induced a faster mobilization of erythrocytes into peripheral blood than that of EPO.

Combined G-CSF and EPO had an ameliorative effect on anthrax-spore-induced mortality in C57BL/6J mice

Our previous report suggested that EPO ameliorates LT-mediated erythrocytopenia by enhancing erythropoiesis [15]. Because G-CSF increases the erythrocyte supply through a diverse mechanism by enhancing the mobilization of erythrocytes into peripheral blood, these results prompted us to investigate whether combined treatments using G-CSF and EPO may be more effective than respective single treatments alone. The analysis indicated that EPO treatment did not exert a protective effect on anthrax-spore-challenged mice (Figure 7B, Spore + EPO vs. Spore only). This is consistent with another line of evidence; the survival rates of anthrax LT-challenged mice increased only 25% following EPO post-treatment (Figure S3, LT + EPO vs. LT, P = 0.101). By contrast, G-CSF treatments with or without EPO effectively increased the survival rate of anthrax spore-challenged mice from 18.75% to 37.5% (Figure 7, Spore + G-CSF vs. Spore only). Post-treatments combining G-CSF and EPO prolonged the survival period of anthrax-spore-challenged mice (Figure 7, Spore + G-CSF + EPO vs. Spore + G-CSF). Statistical analysis revealed that the P value is marginally significant (Figure 7, P = 0.094, Spore + G-CSF + EPO vs. Spore; P = 0.088, Spore + G-CSF + EPO vs. Spore + EPO). These results suggested that combined treatment using G-CSF and EPO tended to ameliorate anthrax-spore-induced mortality in mice.

Discussion

This study demonstrated that G-CSF, a stimulating factor for granulopoiesis, enhanced erythrocytic mobilization, by which it enhanced RBC counts in peripheral blood. This likely therefore rescued anthrax LT-induced anemia and mortality in mice.

Our *in vitro* (Figure 1A–1D) and *ex vivo* (Figure 1E–1H) evidence suggested that G-CSF may promote erythropoiesis. The *in vivo* analyses revealed that G-CSF ameliorated LT-induced erythrocytopenia in peripheral blood (Figure 5C and 5G), but did not increase erythroblast cell numbers in bone marrow (Figure 2). Therefore, the effects of G-CSF on erythropoiesis *in vivo* require clarification. One study demonstrated that G-CSF had a negative effect on bone marrow erythropoiesis in mice [39].

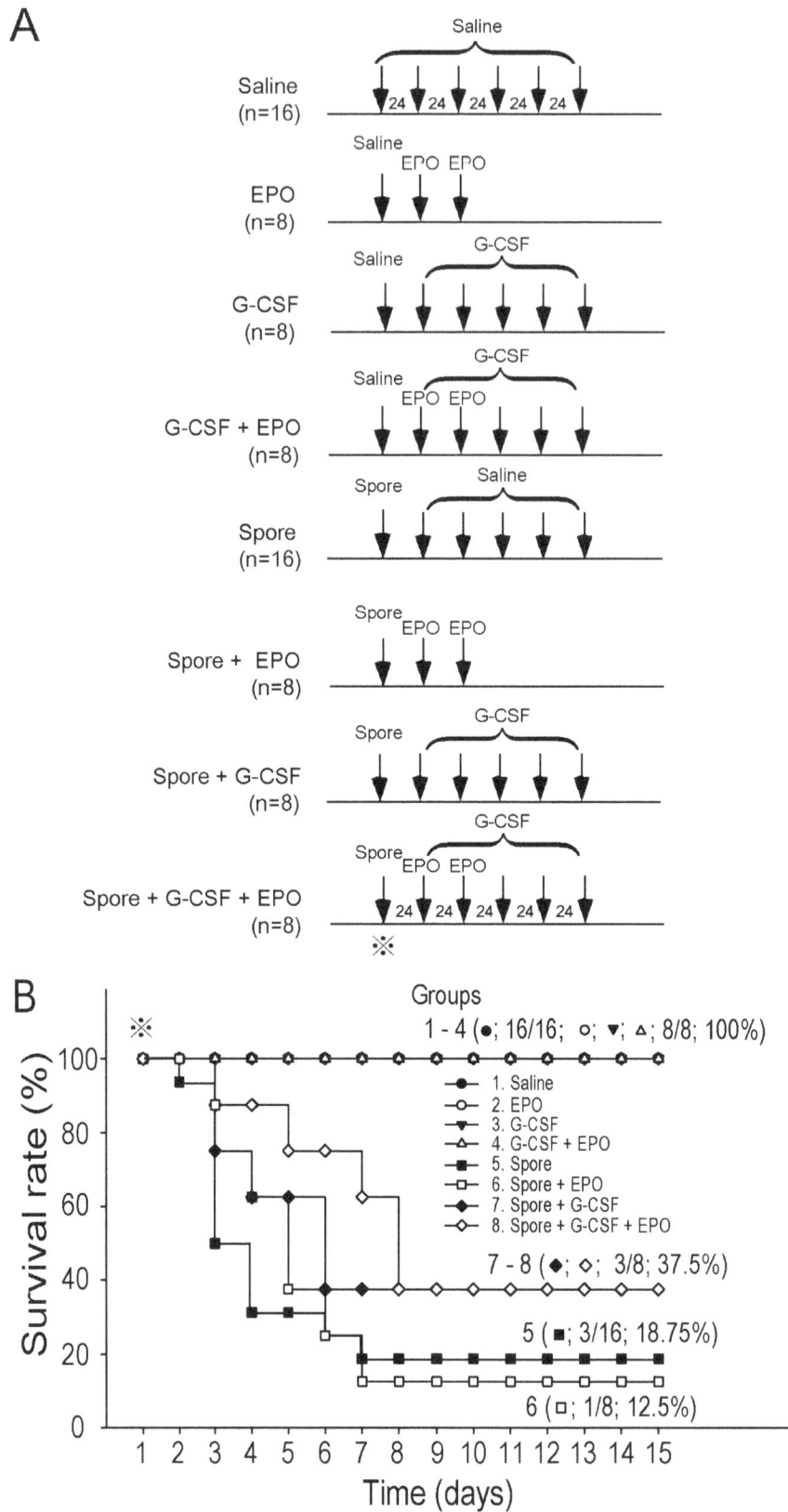

Figure 7. Post-treatments of G-CSF and EPO tended to ameliorate anthrax spore elicited-lethality in mice. Experimental outlines (A) and the survival rates of mice treated with G-CSF (n = 8), EPO (n = 8), and G-CSF combined with EPO (n = 8) after anthrax spore injection (B). Saline (n = 16), EPO (n = 8), G-CSF (n = 8), G-CSF and EPO (n = 8), and anthrax spore (n = 16) injected groups were used for comparisons. The symbol (※) in (A) and (B) indicates the onset time point for recording survival rates. Data represent the results from 2 independent experiments.

However, another study demonstrated that G-CSF treatments in humans increased immature reticulocytes in peripheral blood [40]. Clinical observations also found that the reticulocyte fraction, an assessment of immature erythroid cells in peripheral blood, was an early surrogate marker for the rise of CD34$^+$ hematopoietic stem cells during G-CSF mobilization [41]. This evidence suggests that G-CSF is involved in regulating erythroid precursor cells in humans.

G-CSF induced fast mobilization of RBCs to peripheral blood within 20 h of the first G-CSF administration, compared with the EPO treatment (Figure 6). This data is consistent with the EGFP mice experiment, regarding the time in which G-CSF treatment increased the percentage of newly synthesized erythrocytes in peripheral blood (Figure 3). Thus, a single post-treatment dose of G-CSF was sufficient to save approximately 40% of mice within 24 h (Figure 4D, 100% and 60% survival rates in LT + G-CSF and LT, 48 h groups, respectively). The advantage of fast mobilization of RBCs of G-CSF can also be confirmed by the higher survival rate of G-CSF compared with EPO in LT-induced (Figure 4D vs. Figure S3B) and LT-spore-induced (Figure 7B) mortality in mice.

G-CSF is a multi-function cytokine that stimulates the proliferation and differentiation of myeloid precursors and modulates mature cells [29]. G-CSF is also used to prevent or shorten neutropenia in chemotherapy-induced or primary congenital neutropenia [42], and mobilize hematopoietic stem cells to peripheral blood for transplantation [43]. Probably because of the anti-apoptotic and anti-inflammatory effects, G-CSF has also been used to treat nonhematopoietic targets including cerebral ischemia [44], spinal cord ischemia [45], infarct heart [46], and end stage liver disease [47]. Consequently, the effects of G-CSF on other cell types (e.g., macrophage, lymphocyte, endothelial, dendritic cells, cardiomyocytes, and smooth muscle) may not be completely excluded from the rescue mechanism of G-CSF. Although the mechanism is unclear, combination therapy using G-CSF and EPO has been used to treat myelodysplastic anemia in clinical settings [32–36,48]. This suggests that the medical community has empirically recognized the RBC-enhancing effect of the combination treatment using G-CSF and EPO. One critical aspect of G-CSF is the rapid induction of peripheral RBCs, a property superior to EPO, which may be applied to anthrax or other diseases with urgent RBC and oxygen demands. Further detailed and well-designed clinical studies are required to explore the therapeutic potential of combination treatment using G-CSF and EPO.

In this study, we demonstrated that G-CSF mobilized erythrocytes into peripheral blood. In addition, combined treatments of G-CSF and EPO tended to ameliorate anthrax spore-induced mortality. An optimized rescue protocol may provide new perspectives and assist the development of a feasible therapeutic strategy against anthrax.

References

1. Mock M, Fouet A (2001) Anthrax. Annu Rev Microbiol 55: 647–671.
2. Brossier F, Mock M (2001) Toxins of Bacillus anthracis. Toxicon 39: 1747–1755.
3. Collier RJ, Young JA (2003) Anthrax toxin. Annu Rev Cell Dev Biol 19: 45–70.
4. Mourez M (2004) Anthrax toxins. Rev Physiol Biochem Pharmacol 152: 135–164.
5. Bardwell AJ, Abdollahi M, Bardwell L (2004) Anthrax lethal factor-cleavage products of MAPK (mitogen-activated protein kinase) kinases exhibit reduced binding to their cognate MAPKs. Biochem J 378: 569–577.
6. Hagemann C, Blank JL (2001) The ups and downs of MEK kinase interactions. Cell Signal 13: 863–875.
7. Wada T, Penninger JM (2004) Mitogen-activated protein kinases in apoptosis regulation. Oncogene 23: 2838–2849.

Supporting Information

Figure S1 G-CSF treatments enhanced bone marrow erythroid colony numbers. An *in vitro* erythroid colony-forming cell assay was performed using murine bone marrow (BM) cells incubated with [20 ng/ml (n = 6) or 764 ng/ml (n = 6)] or without G-CSF (n = 6). The erythroid colonies were confirmed by 3, 3′-diaminobenzidine tetrahydrochloride (DAB) staining (A) and quantified on Day 14 (B). Untreated BM cells were used as the control. **$P<0.01$ was compared to the untreated groups. Scale bar: 500 μm. Data are shown as mean ± SD.

Figure S2 Mobilization of newly synthesized erythrocytes (EGFP$^+$/TER-119$^+$) into peripheral blood by acute anemia in EGFP transgenic mice. Experimental timetable used in acute anima assay (A). During acute anima (after aspirating 35% of total blood), the population of EGFP$^+$/TER-119$^+$ cells in peripheral blood (PB) of EGFP mice was gated as R1 and quantified. The total cell number was defined as 100%. The percentage of R1 was analyzed by flow cytometry (B) on Day 2, 4, and 6 after the removal of 35% of total blood. PB samples from mice before acute anemia were used as negative controls. Data were collected from 2 representative EGFP mice.

Figure S3 Post-treatments of EPO increased the survival rates of LT-challenged mice. Experimental timetable (A). The survival rates of mice challenged with EPO (n = 4), LT (n = 4), or LT plus EPO (n = 4) are shown (B). Saline-treated mice were used as controls (n = 4). C57BL/6J mice were retro-orbitally injected with recombinant human EPO (rhEPO, Neorecormon, Roche, Mannheim, Germany) (2 IU/g, in 250 μl saline) twice every 24 h after the challenges of a lethal dose of LT (1.5 mg/kg in 250 μl saline, retro-orbitally injected). The symbol (※) in (A) and (B) indicates the onset time point for recording the survival rates of mice.

Acknowledgments

We wish to thank Professor Yu MS for providing 3, 3′-diaminobenzidine tetrahydrochloride (DAB) tablets. We also wish to thank Professor Wang MH and his team, the Experimental Animal Center of Tzu-Chi University for animal care.

Author Contributions

Conceived and designed the experiments: HHC DSS. Performed the experiments: YWC TKL GLL YYL JHK HHH HLH. Analyzed the data: HHC JHW DSS. Contributed reagents/materials/analysis tools: JHK HHH HLH JHW. Wrote the paper: HHC DSS.

8. Raman M, Chen W, Cobb MH (2007) Differential regulation and properties of MAPKs. Oncogene 26: 3100–3112.
9. Moayeri M, Leppla SH (2004) The roles of anthrax toxin in pathogenesis. Curr Opin Microbiol 7: 19–24.
10. Kau JH, Sun DS, Huang HS, Lien TS, Huang HH, et al. (2010) Sublethal doses of anthrax lethal toxin on the suppression of macrophage phagocytosis. PLoS One 5: e14289.
11. Muehlbauer SM, Evering TH, Bonuccelli G, Squires RC, Ashton AW, et al. (2007) Anthrax lethal toxin kills macrophages in a strain-specific manner by apoptosis or caspase-1-mediated necrosis. Cell Cycle 6: 758–766.
12. Alileche A, Serfass ER, Muehlbauer SM, Porcelli SA, Brojatsch J (2005) Anthrax lethal toxin-mediated killing of human and murine dendritic cells impairs the adaptive immune response. PLoS Pathog 1: e19.

13. Comer JE, Chopra AK, Peterson JW, Konig R (2005) Direct inhibition of T-lymphocyte activation by anthrax toxins in vivo. Infect Immun 73: 8275–8281.

14. Fang H, Xu L, Chen TY, Cyr JM, Frucht DM (2006) Anthrax lethal toxin has direct and potent inhibitory effects on B cell proliferation and immunoglobulin production. J Immunol 176: 6155–6161.

15. Chang HH, Wang TP, Chen PK, Lin YY, Liao CH, et al. (2013) Erythropoiesis suppression is associated with anthrax lethal toxin-mediated pathogenic progression. PLoS One 8: e71718.

16. Chen PK, Chang HH, Lin GL, Wang TP, Lai YL, et al. (2013) Suppressive effects of anthrax lethal toxin on megakaryopoiesis. PLoS One 8: e59512.

17. Liu S, Zhang Y, Moayeri M, Liu J, Crown D, et al. (2013) Key tissue targets responsible for anthrax-toxin-induced lethality. Nature 501: 63–68.

18. Kassam A, Der SD, Mogridge J (2005) Differentiation of human monocytic cell lines confers susceptibility to Bacillus anthracis lethal toxin. Cell Microbiol 7: 281–292.

19. Kau JH, Sun DS, Tsai WJ, Shyu HF, Huang HH, et al. (2005) Antiplatelet activities of anthrax lethal toxin are associated with suppressed p42/44 and p38 mitogen-activated protein kinase pathways in the platelets. J Infect Dis 192: 1465–1474.

20. Moayeri M, Haines D, Young HA, Leppla SH (2003) Bacillus anthracis lethal toxin induces TNF-alpha-independent hypoxia-mediated toxicity in mice. J Clin Invest 112: 670–682.

21. Rainey GJ, Young JA (2004) Antitoxins: novel strategies to target agents of bioterrorism. Nat Rev Microbiol 2: 721–726.

22. Wadhwa M, Thorpe R (2008) Haematopoietic growth factors and their therapeutic use. Thromb Haemost 99: 863–873.

23. Debili N, Wendling F, Katz A, Guichard J, Breton-Gorius J, et al. (1995) The Mpl-ligand or thrombopoietin or megakaryocyte growth and differentiative factor has both direct proliferative and differentiative activities on human megakaryocyte progenitors. Blood 86: 2516–2525.

24. Fisher JW (2003) Erythropoietin: physiology and pharmacology update. Exp Biol Med (Maywood) 228: 1–14.

25. Socolovsky M, Nam H, Fleming MD, Haase VH, Brugnara C, et al. (2001) Ineffective erythropoiesis in Stat5a(-/-)5b(-/-) mice due to decreased survival of early erythroblasts. Blood 98: 3261–3273.

26. Trowbridge IS, Lesley J, Schulte R (1982) Murine cell surface transferrin receptor: studies with an anti-receptor monoclonal antibody. J Cell Physiol 112: 403–410.

27. Kina T, Ikuta K, Takayama E, Wada K, Majumdar AS, et al. (2000) The monoclonal antibody TER-119 recognizes a molecule associated with glycophorin A and specifically marks the late stages of murine erythroid lineage. Br J Haematol 109: 280–287.

28. Hubel K, Dale DC, Liles WC (2002) Therapeutic use of cytokines to modulate phagocyte function for the treatment of infectious diseases: current status of granulocyte colony-stimulating factor, granulocyte-macrophage colony-stimulating factor, macrophage colony-stimulating factor, and interferon-gamma. J Infect Dis 185: 1490–1501.

29. Morstyn G, Burgess AW (1988) Hemopoietic growth factors: a review. Cancer Res 48: 5624–5637.

30. Metcalf D (2008) Hematopoietic cytokines. Blood 111: 485–491.

31. Xiao BG, Lu CZ, Link H (2007) Cell biology and clinical promise of G-CSF: immunomodulation and neuroprotection. J Cell Mol Med 11: 1272–1290.

32. Bessho M, Hirashima K, Asano S, Ikeda Y, Ogawa N, et al. (1997) Treatment of the anemia of aplastic anemia patients with recombinant human erythropoietin in combination with granulocyte colony-stimulating factor: a multicenter randomized controlled study. Multicenter Study Group. Eur J Haematol 58: 265–272.

33. Greenberg PL, Sun Z, Miller KB, Bennett JM, Tallman MS, et al. (2009) Treatment of myelodysplastic syndrome patients with erythropoietin with or without granulocyte colony-stimulating factor: results of a prospective randomized phase 3 trial by the Eastern Cooperative Oncology Group (E1996). Blood 114: 2393–2400.

34. Hellstrom-Lindberg E, Gulbrandsen N, Lindberg G, Ahlgren T, Dahl IM, et al. (2003) A validated decision model for treating the anaemia of myelodysplastic syndromes with erythropoietin + granulocyte colony-stimulating factor: significant effects on quality of life. Br J Haematol 120: 1037–1046.

35. Hellstrom-Lindberg E, Negrin R, Stein R, Krantz S, Lindberg G, et al. (1997) Erythroid response to treatment with G-CSF plus erythropoietin for the anaemia of patients with myelodysplastic syndromes: proposal for a predictive model. Br J Haematol 99: 344–351.

36. Negrin RS, Stein R, Doherty K, Cornwell J, Vardiman J, et al. (1996) Maintenance treatment of the anemia of myelodysplastic syndromes with recombinant human granulocyte colony-stimulating factor and erythropoietin: evidence for in vivo synergy. Blood 87: 4076–4081.

37. Ogawa M, Parmley RT, Bank HL, Spicer SS (1976) Human marrow erythropoiesis in culture. I. Characterization of methylcellulose colony assay. Blood 48: 407–417.

38. Kikuta T, Shimazaki C, Ashihara E, Sudo Y, Hirai H, et al. (2000) Mobilization of hematopoietic primitive and committed progenitor cells into blood in mice by anti-vascular adhesion molecule-1 antibody alone or in combination with granulocyte colony-stimulating factor. Exp Hematol 28: 311–317.

39. Nijhof W, De Haan G, Dontje B, Loeffler M (1994) Effects of G-CSF on erythropoiesis. Ann N Y Acad Sci 718: 312–324; discussion 324–315.

40. Park K, Im T, Sasaki A, Yamane T, Nakao Y, et al. (1991) Positive effect of granulocyte-colony stimulating factor on erythropoiesis in humans. Osaka City Med J 37: 123–132.

41. Remacha AF, Martino R, Sureda A, Sarda MP, Sola C, et al. (1996) Changes in reticulocyte fractions during peripheral stem cell harvesting: role in monitoring stem cell collection. Bone Marrow Transplant 17: 163–168.

42. Page AV, Liles WC (2011) Colony-stimulating factors in the prevention and management of infectious diseases. Infect Dis Clin North Am 25: 803–817.

43. Rankin SM (2012) Chemokines and adult bone marrow stem cells. Immunol Lett 145: 47–54.

44. Abe K, Yamashita T, Takizawa S, Kuroda S, Kinouchi H, et al. (2012) Stem cell therapy for cerebral ischemia: from basic science to clinical applications. J Cereb Blood Flow Metab 32: 1317–1331.

45. Chen WF, Jean YH, Sung CS, Wu GJ, Huang SY, et al. (2008) Intrathecally injected granulocyte colony-stimulating factor produced neuroprotective effects in spinal cord ischemia via the mitogen-activated protein kinase and Akt pathways. Neuroscience 153: 31–43.

46. Baldo MP, Davel AP, Damas-Souza DM, Nicoletti-Carvalho JE, Bordin S, et al. (2011) The antiapoptotic effect of granulocyte colony-stimulating factor reduces infarct size and prevents heart failure development in rats. Cell Physiol Biochem 28: 33–40.

47. Gaia S, Smedile A, Omede P, Olivero A, Sanavio F, et al. (2006) Feasibility and safety of G-CSF administration to induce bone marrow-derived cells mobilization in patients with end stage liver disease. J Hepatol 45: 13–19.

48. Jadersten M, Montgomery SM, Dybedal I, Porwit-MacDonald A, Hellstrom-Lindberg E (2005) Long-term outcome of treatment of anemia in MDS with erythropoietin and G-CSF. Blood 106: 803–811.

49. Kau JH, Lin CG, Huang HH, Hsu HL, Chen KC, et al. (2002) Calyculin A sensitive protein phosphatase is required for Bacillus anthracis lethal toxin induced cytotoxicity. Curr Microbiol 44: 106–111.

50. Sarma NJ, Takeda A, Yaseen NR (2010) Colony forming cell (CFC) assay for human hematopoietic cells. J Vis Exp.

51. Harkness JE, Wagner JE (1995) Clinical procedures: Biology and Medicine of Rabbits and Rodents 93.

Draft Genome Sequence Analysis of a *Pseudomonas putida* W15Oct28 Strain with Antagonistic Activity to Gram-Positive and *Pseudomonas* sp. Pathogens

Lumeng Ye[1], Falk Hildebrand[1], Jozef Dingemans[1], Steven Ballet[2], George Laus[2], Sandra Matthijs[3], Roeland Berendsen[4], Pierre Cornelis[1]*

[1] Department of Bioengineering Sciences, Research group Microbiology, Vrije Universiteit Brussel and VIB Structural Biology Brussels, Brussels, Belgium, [2] Chemistry Department, Vrije Universiteit Brussel, Pleinlaan 2, 1050 Brussels, Belgium, [3] Institut de Recherches Microbiologiques - Wiame, Campus du CERIA, Brussels, Belgium, [4] Plant-Microbe Interactions, Utrecht University, Utrecht, The Netherlands

Abstract

Pseudomonas putida is a member of the fluorescent pseudomonads known to produce the yellow-green fluorescent pyoverdine siderophore. *P. putida* W15Oct28, isolated from a stream in Brussels, was found to produce compound(s) with antimicrobial activity against the opportunistic pathogens *Staphylococcus aureus*, *Pseudomonas aeruginosa*, and the plant pathogen *Pseudomonas syringae*, an unusual characteristic for *P. putida*. The active compound production only occurred in media with low iron content and without organic nitrogen sources. Transposon mutants which lost their antimicrobial activity had the majority of insertions in genes involved in the biosynthesis of pyoverdine, although purified pyoverdine was not responsible for the antagonism. Separation of compounds present in culture supernatants revealed the presence of two fractions containing highly hydrophobic molecules active against *P. aeruginosa*. Analysis of the draft genome confirmed the presence of putisolvin biosynthesis genes and the corresponding lipopeptides were found to contribute to the antimicrobial activity. One cluster of ten genes was detected, comprising a NAD-dependent epimerase, an acetylornithine aminotransferase, an acyl CoA dehydrogenase, a short chain dehydrogenase, a fatty acid desaturase and three genes for a RND efflux pump. *P. putida* W15Oct28 genome also contains 56 genes encoding TonB-dependent receptors, conferring a high capacity to utilize pyoverdines from other pseudomonads. One unique feature of W15Oct28 is also the presence of different secretion systems including a full set of genes for type IV secretion, and several genes for type VI secretion and their VgrG effectors.

Editor: Stefan Bereswill, Charité-University Medicine Berlin, Germany

Funding: This work was supported by the European Research Council (ERC) [ERC Advanced Grant agreement number 269072 – PLANTIMMUSYS]. LY received a CSC-VUB fellowship for her PhD. The funders had no role in study design, data collection and analysis, decision to publish, or preparation of the manuscript.

Competing Interests: The authors have declared that no competing interests exist.

* Email: pcornel@vub.ac.be

Introduction

Pseudomonas putida is a gram-negative rod-shaped γ-proteobacterium found throughout various environments. *P. putida* strains show a diverse spectrum of metabolic activities, including their ability to tolerate heavy metals and to degrade organic solvent, which enables them to survive in highly polluted environments. *P. putida* strains are also known to interact with the rhizosphere and for their plant-growth promoting activities [1–7]. Bacteria belonging to the *Pseudomonas* genus produce different bioactive secondary metabolites, but their exploitation is not as developed compared to the situation in Gram-positive bacteria such as *Bacillus* sp. and *Streptomyces* sp. strains [8]. Nowadays, with the advent of the next generation sequencing methods, together with the increased accuracy of gene annotations, new avenues are open for the discovery of secondary metabolite genes clusters in order to gather more information about the different molecules produced and their biological activity. A recent example

is the identification, next to the already described antimicrobial compounds pyrolnitrin, pyoluteorin and phloroglucinol, of rhizoxin analogs and orfamides from the well-studied plant-promoting rhizobacterium *P. protegens* Pf-5 (previously named *P. fluorescens* Pf5) through genome-mining [9]. The majority of secondary metabolites with a biological acitivity have so far been described in different strains belonging to the *P. fluorescens* group (although few of these have been renamed, such as *P. protegens* and *P. brassicacearum*) [10], but not much is known about the production of bioactive compounds by *P. putida*. *Pseudomonas putida* W15Oct28 was isolated from the Woluwe River, Belgium [11]. This strain showed some unique phenotypic characters such as the production of a new pyoverdine siderophore with a large peptide chain [12] and a capacity to produce compound(s) with anti-microbial activity against *P. aeruginosa* and *Staphylococcus aureus*. Consequently, its potential to produce new secondary metabolites drew our attention, which motivated us to acquire its whole genome sequence by Illumina Miseq technology. In this study, we focus on the secondary metabolite biosynthesis pathways of *P. putida* W15Oct28, which includes pyoverdine, putisolvins biosurfactants, and a novel antimicrobial molecule with broad spectrum inhibitory activity (against *P. aeruginosa*, *P. syringae*, *P. entomophila*, and *S. aureus*, including MRSA), and a partial safracin biosynthetic gene cluster. In addition, we found that this strain presents an extensive repertoire of iron uptake systems with 56 TonB-dependent receptors. W15Oct28 is also unique among pseudomonads because it has a full set of genes for a type IVb (Dot/Icm) secretion system, and several loci for type VI secretion together with five VgrG effector proteins.

Experimental Procedures

Bacterial strains and cultivation

P. putida W15Oct28 was isolated and purified from samples of the Woluwe river surface water in the frame of a project funded by the Belgian Federal Government who granted us the right to isolate bacteria from the Woluwe and the Senne rivers [11]. With the exception of *P. aeruginosa*, which were cultivated at 37°C, all the other *Pseudomonas* sp. strains were grown at 28°C. *Staphylococcus aureus* strains were cultivated at 37°C. All bacteria were grown in solid or liquid LB medium, except for the production of secondary metabolites (mentioned below). The list of strains and plasmids used in this study, as well as the primers list is presented in **Table S1 in file S1**.

Secondary metabolites production and purification

For pyoverdine production, *P. putida* W15Oct28 was grown at 28°C in 1 l of iron-poor CAA medium (Bacto Casamino Acid, BD, 5g l^{-1}; K_2HPO_4 1.18 g l^{-1}; $MgSO_4 \cdot 7H_2O$ 0.25 g l^{-1}) in 2 l Erlenmeyer flasks, at a shaking speed of 160 rpm for 48 hours. Bacterial cells were removed by centrifugation at 7,000g during 15 min. After filtration the supernatant was passed on a C-18 column that was activated with methanol and washed with distilled water. Elution was done with acetonitrile/H_2O (70/30%). Samples were lyophilized after most of the acetonitrile was evaporated [13]. The pyoverdines from different *Pseudomonas* sp. strains used for growth stimulation tests were purified by the same protocol, but using smaller scale cultures (20 ml).

For putisolvins and antimicrobial molecules, *P. putida* W15Oct28 was grown at 28°C in 1 l of iron poor M9 minimal medium (12.8 g of Na_2HPO_4, 3.0 g of KH_2PO_4, 0.5 g of NaCl, 1.0 g of NH_4NO_3, 100 μl of 1M $CaCl_2$, 2 ml of 1M $MgSO_4$, 10 ml of 20% W/V glucose, for 1 liter, pH 7.0) in 2 l Erlenmeyer flasks, at a shaking speed of 160 rpm for 48 hours. After 48 hours of culture, another 10 ml of 20% W/V glucose was supplemented to the culture again which was left at 4–8°C for 5 more days. Bacterial cells were separated by centrifugation at 7,000 g during 15 min. The supernatant was extracted by 40% volume of ethyl acetate for the extraction of the antimicrobial molecule. Cells were mixed with 30 ml of ethyl acetate and sonicated for 5 minutes. This extraction contained most of the putisolvins and a partial fraction of the antimicrobial molecule(s).

DNA extraction and whole genome sequencing

Genomic DNA of W15Oct28 was extracted by Puregen Yeast/ Bact Kit B (Qiagen, Cat. No. 158567). Four genomic DNA extraction samples were combined, further purified and concentrated by the DNA Clean & Concentrator Kit (ZYMO research, Cat. No. D4003S). The genome of *P. putida* W15Oct28 was sequenced at the VIB nucleomics core using the Illumina Miseq system. The library was constructed by the Nextera kit, yielding reads lengths of 150 bp paired end.

Genome assembly, annotation and analysis

The final genome coverage was about 62 times and the quality filtered sequences from the MiSeq run were *de novo* assembled using Velvet version 1.2.08 [14]. 138 contigs were further combined in scaffolds using SSPACE basic version 2.0, with 99 contigs representing the draft genome. The draft genome was uploaded and annotated by using the RAST website [15]. Each contig was uploaded to antiSMASH to detect the pyoverdine and other secondary metabolites biosynthetic genes clusters [16,17]. To predict the substrate of the different adenylation domains, antiSMASH gives 4 predictions based on the combination of NRPSPredictor2 SVM, Stachelhaus code, Minowa, and consensus. When the 4 predictions gave identical results, then the prediction was accepted and considered as valid. When the 4 predictions were different, the amino acids sequences of the adenylation domain were further analyzed by PKS-NRPS to compare with the non-ribosomal Stachelhaus code [18]. The circulated draft genome figure and whole genome comparison with *P. putida* GB-1, BIRD-1, and NBRC 14164 were done by CGView [19]. Genomic islands were searched for by submitting the draft genome to Islandviewer [20], and conserved IS elements were identified by IS Finder database (https://www-is.biotoul.fr/). The draft genome of *P. putida* W15Oct28 was examined by CRISPRFinder to detect the presence of CRISPR elements, which confer immunity against incoming DNA, including bacteriophages [21].

Random transposon mutagenesis and in-frame gene knock-out deletion

Random mutagenesis library was constructed by conjugating the *E. coli* strain containing the plasposon pTn*Mod*OTc [22] into wild type strain W15Oct28 via biparental mating. Mutants were harvested after antibiotic selection and were selected by functional screening. Information of the insertion flanking regions was acquired by sequencing the plasmid which was rescued from mutants genomes [22] with PITC-F/R primers. In frame genes deletions were facilitated by a yeast recombination method as described by Shanks *et al.* [23]. In summary, PCR primers to amplify upstream and downstream DNA fragments were designed around the region to delete. The two fragments were separately amplified, and mixed with double-digested (*Eco*RI and *Bam*HI) shuttle vector pMQ30, single-stranded carrier DNA (deoxyribonucleic acid sodium salt type III from salmon testes, Sigma, D1626), and engineered yeast strain *S. cerevisiae* InvSc1 in Lazy Bone solution (40% PEG 4000, 0.1 M LiAc, 10 mM Tris-HCl, pH 7.5, 1mM EDTA). After incubation at room temperature overnight, the solution was subjected to 42°C, 12 min for heat shock. In the end this transformation product was washed with TE buffer to remove the polyethylene glycol (PEG 4,000), resuspended in 600 μl of TE buffer before plating on SD-Ura medium (MP biomedicals, 4813065) for selection. The correctly constructed deletion vector was transferred to the recipient strain via conjugation using a donor *E. coli* strain. The merodiploids were selected by antibiotic resistance (Gm) and confirmed by colony PCR. The shuttle vector was removed by culturing the merodiploids in LB+10% sucrose, and positive mutants were selected by PCR screening.

High-performance liquid chromatography (HPCL) and mass spectrometry (MS)

HPCL analysis was carried out with an Agilent 1100 Series HPCL System with auto-sampler, degaser UV detector and a thermostated column compartment with an operating temperature range from ambient to 105°C. Preparative-scale purification was done using a Gilson 712 semi-preparative HPCL system with a 322 pump, an UV-VIS 156 detector, a manual injector and 206 Fraction Collector. ESI-LC/MS was done by means of a Waters 600E HPCL Pump, a Waters 2487 Dual Absorbance Wavelength Detector and a Fisons VG II Quattro Mass Spectrometer (ESP ionisation). Operating temperature is from ambient to 80°C. The mobile phase consists of a water/acetonitrile/TFA mixture with a gradient going from a mixture water/AcN (97:3) containing 0.1% TFA to a mixture water/AcN (0:100) containing 0.1% TFA in 30 min followed by 10 min isocratic run at these conditions, and with a flow rate of 20 ml min^{-1}. High resolution mass spectrometry and collision-induced dissociation tandem fragmentation MS were done on a QTof Micro mass spectrometry (positive ion mode). Under standard measurement conditions the sample was dissolved in CH_3CN/H_2O (1:1) containing 0.1% TFA. GC/MS spectra were recorded on a Trace MS Plus (Thermo). Separation was done on a J&W Scientific DBxXLB (30 m, 0.25 mm ID, 0.25 μm film thickness).

Antagonism tests

Wild-type and plasposon mutants W15Oct28 strains were pre-cultured in 3 ml of LB till and OD of 0.8 was reached. Then 10 μl of each culture was spotted on M9 minimal medium plates. After overnight cultivation, the producing strain was killed by exposure to UV for 2 minutes. Then soft agar containing the different indicator strains (10^5–10^6 CFU) was overlaid on the plates. Results were observed after 18 hours of cultivation at the appropriate temperature. To detect the antimicrobial activity of pre-HPCL isolated fractions, each fraction was dissolved in 50% methanol and 10 μl spotted on the plate with indicator strain overlay.

Pyoverdine utilization test

Pyoverdine cross feeding test was done on a W15Oct28 pyoverdine null mutant (7G11, transposon inserted in NRPS) on CAA +600μM 2, 2'- dipyridyl. Each type of pyoverdine was added as 10 μl of 8 mM pyoverdine on a paper filter. After incubation at 28°C for 48 hours, the presence of a growth stimulation zone surrounding the filter was considered as positive. Wild type strain and 7G11 were grow on the same medium without any pyoverdine supplement as positive and negative control, respectively.

Phylogenic analysis of *Pseudomonas* sp. strain and clustering of TonB-dependent receptors

Sequence alignments and trees were generated using CLC Main Workbench 6.7.2 (CLC bio, Aarhus, Denmark). Phylogenic analysis of W15Oct28 was done according to the multilocus sequence analysis (MLSA) of four concatenated housekeeping genes method [10]. Average Nucleotide Identity based on BLAST (ANIb) values was calculated using Jspecies [24]. Clustering of TonB-dependent receptors was done by comparing the amino acid sequences of W15Oct28 to the reference genes described by Hartney *et al.* [25]. Sequences were aligned with Gap open cost 15 and Gap extension cost 0.3. Neighbor joining trees were created by performing bootstrap analysis with 1000 replicates.

Results

P. putida W15Oct28 whole genome sequence analysis and comparison with other sequenced *P. putida* strains

The genome assembly resulted in 138 contigs assembled in 99 scaffolds representing the draft genome (longest scaffold size:

283,647 bp, 22 scaffolds longer that 100 kb, mean scaffold size 4,5726 bp, N50 scaffold size: 105,817 bp); the genome is 6,331,075 bps in length with average GC content of 62,8%. This Whole Genome Shotgun project has been deposited at DDBJ/EMBL/GenBank under the accession JENB00000000. The version described in this paper is version JENB01000000, biosample SAMN02644482. The genome contains 5,540 predicted coding sequences (CDS), a total of 116 RNA genes, including 6 rRNA operons (8 copies of 5S rRNAs, 6 copies of 16S rRNAs, and 6 copies of 23S rRNAs). In addition, 71 tRNA genes were identified. The W15Oct28 strain was clearly confirmed to be *Pseudomonas putida* by analysis of different housekeeping genes sequences (16s rDNA, *gyrB*, *rpoB*, *rpoD*) according to the phylogeny method described by Mulet *et al.* [10,26] with 99%, 97%, 98%, and 99% DNA sequence identity, respectively with the type strain of *P. putida* NBRC14164^T (**Figure 1**). It is interesting to notice that the closest relatives of W15Oct28 are the *P. putida* type strain NBRC 14164^T, and the recently sequenced strain H8234, a clinical isolate [27], confirming that our strain probably belongs to the species *P. putida* [28]. It appears also from the phylogenetic tree of Figure 1 that the other strains designated as *P. putida* probably represent other species. As a confirmation of this preliminary taxonomical assignment, **Table 1** shows the Average Nucleotide Identity based on BLAST (ANIb) between the genomes of all the *P. putida* group strains represented in Figure 1. The table confirms that *P. putida* W15Oct28 shares about 94% of its genome with the recently sequenced NBRC14164^T strain and 92% with H8234, totally in line with the previous phylogenetic assignment presented in Figure 1. The percentage of identity with other representatives of the *P. putida* group is lower, again confirming the previous multi-locus analysis-based taxonomic assignment. *P. putida* KT2440, BIRD 1, and DOT T1E form another cluster of related strains (**Table 1**). The circular map of the *P. putida* W15Oct28 genome is presented in **Figure S1**.

Genes corresponding to secondary metabolism pathways

A. Biosynthetic pathway of a new type pyoverdine and the presence of 56 TonB-dependent receptors. In our previous work, we reported that the only form of pyoverdine produced by *P. putida* W15Oct28 contains α-keto-glutaric acid as acyl side chain, a dihydropyoverdine chromophore and a 12 amino acid peptide chain [12]. One unique aspect of this pyoverdine is the incorporation of L-homoserine in peptide chain which was only previously reported in the pyoverdine produced by *Azotobacter vinelandii* DJ [12]. The pyoverdine biosynthesis gene clusters are localized in three regions in the genome. The first pyoverdine genomic region shares high identity (99%) with other *P. putida* type strains and contains the genes which catalyze the formation of the chromophore precursor (PvdL/G/Y/H) while the last part of the cluster contains genes involved in the transport of ferripyoverdine (*fpvC*, *fpvD*, *fpvE*, *fpvF*) [29] and in pyoverdine modification (*pvdA*, *pvdQ*) [30,31]. The second cluster of genes shares about 75% identity with other whole genome sequenced *P. putida* strains and contains the genes *pvdM*, *pvdN*, *pvdO*, *pvdP* encoding periplasmic enzymes which are involved in the final maturation of the chromophore in the periplasm [32]. The third cluster contains one thioesterase gene, the genes for four non-ribosomal peptide synthetases involved in the peptide chain synthesis (*pvdD*, *pvdI*, *pvdJ*, *pvdK*) and the *fpvA* gene corresponding to the TonB-dependent ferri-pyoverdine receptor. These last genes share less than 50% identity with the corresponding genes from other *P. putida* strains [12]. This third pyoverdine cluster was considered by IslandViewer to correspond to a

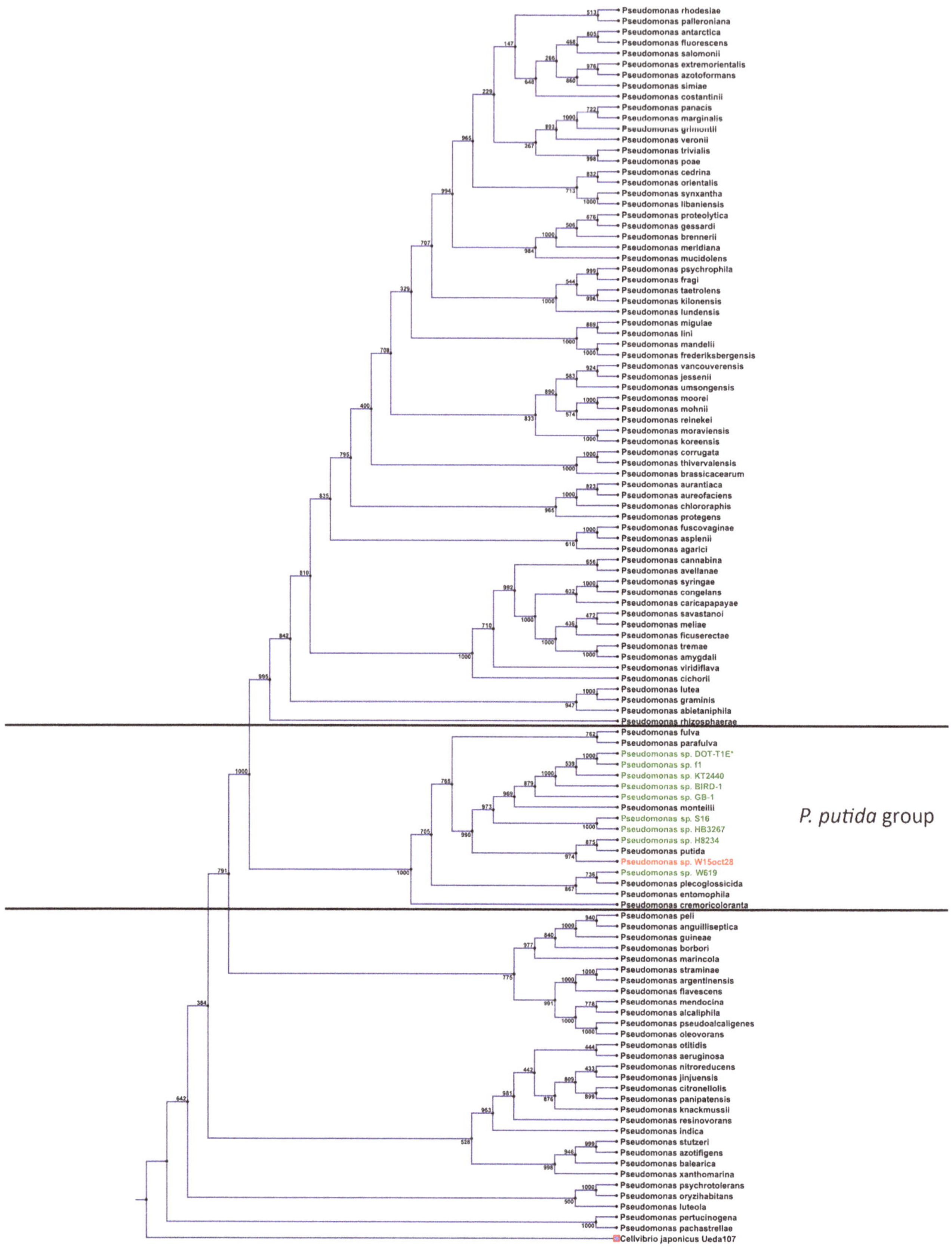

Figure 1. Pseudomonas phylogenetic tree. A phylogenetic tree of different Pseudomonas species based on the comparison of four different housekeeping genes sequences (16s rDNA, *gyrB*, *rpoB*, *rpoD*). The *P. putida* cluster is highlighted and the strain W15Oct28 is indicated in red. W15Oct28 closest relative is the *P. putida* type strain NBRC14164[T].

Table 1. Whole genome comparison (Average Nucleotide Identity based on BLAST [ANIb]) between different *P. putida* group strains, including W15Oct28.

ANIb	*Pseudomonas putida* NBRC 14164	*Pseudomonas* sp. W15oct28	*Pseudomonas* sp. H8234	*Pseudomonas* sp. GB 1	*Pseudomonas* sp. KT2440	*Pseudomonas* sp. BIRD 1	*Pseudomonas* sp. F1	*Pseudomonas* sp. DOT T1E	*Pseudomonas* sp. S16	*Pseudomonas* sp. HB3267	*Pseudomonas* sp. W619	*Pseudomonas entomophila* L48	*Pseudomonas* sp. UW4	*Pseudomonas fulva* 12 X
P. putida NBRC 14164[T]	–	94.2	93.3	90.4	89.7	89.6	89.5	89.5	89.4	89.5	85.2	84.2	77.3	76.2
Pseudomonas sp. **W15oct28**	94.2	–	92.5	90.4	89.5	89.6	89.4	89.4	89.5	89.4	85.0	84.1	77.2	76.0
Pseudomonas sp. H8234	93.1	92.4	–	89.7	89.2	88.9	88.8	88.9	88.4	88.6	84.8	83.9	77.1	75.7
Pseudomonas sp. GB 1	90.4	90.6	89.9	–	90.4	90.4	90.5	90.4	89.4	89.7	85.1	84.2	77.2	76.0
Pseudomonas sp. KT2440	89.7	89.7	89.2	90.4	–	97.2	96.7	96.9	88.9	89.2	84.9	84.0	77.2	76.0
Pseudomonas sp. BIRD 1	89.7	89.8	89.2	90.6	97.3	–	96.7	96.8	89.1	89.2	84.6	84.1	77.3	76.1
Pseudomonas sp.F1	89.6	89.5	89.1	90.5	96.7	96.6	–	98.3	88.9	89.2	84.8	84.0	77.1	76.0
Pseudomonas sp. DOT T1E	89.4	89.4	89.1	90.2	96.7	96.5	98.1	–	88.8	88.8	84.5	84.0	76.9	75.7
Pseudomonas sp. S16	89.5	89.6	89.0	89.5	88.9	89.1	89.0	89.0	–	96.9	85.4	84.9	77.7	76.7
Pseudomonas sp. HB3267	89.6	89.6	89.0	89.8	89.2	89.1	89.2	89.0	96.9	–	85.8	84.9	77.7	76.7
Pseudomonas sp. W619	85.3	85.3	85.1	85.2	84.9	84.6	84.8	84.7	85.4	85.7	–	84.0	76.8	75.8
P. entomophila L48	84.2	84.2	84.2	84.3	84.1	84.0	84.1	84.2	84.9	85.0	83.9	–	77.8	77.1
Pseudomonas sp. UW4	77.2	77.2	77.1	77.2	77.2	77.1	77.1	77.0	77.5	77.6	76.8	77.5	–	75.7
P. fulva 12 X	76.31	76.45	76.07	76.26	76.17	76.18	76.19	76.13	76.73	76.72	75.83	77	75.94	–

genomic island (GIs), because of its lower GC content compared to the average. Details about the genes involved in W15Oct28 pyoverdine biosynthesis and uptake have been described in our previous work [12]. *Pseudomonas putida* W15Oct28 genome contains 56 genes encoding TonB-dependent outer-membrane proteins receptors, 20 of them being probably involved in ferri-pyoverdine uptake, a situation similar to what we observed in *P. fluorescens* ATCC17400 which has 55 TonB-dependent receptors, including 17 that are involved in the uptake of different ferric pyoverdines [33]. Analysis of these receptors compared to other characterized *Pseudomonas* TonB-dependent receptors, revealed the presence of homologs of FpvAI, FpvAII, FpvAIII, and FpvU/V/W/X/Y/Z already described in *P. protegens* Pf-5 [25] (**Figure 2**). Pyoverdine feeding experiment confirmed the utilization of *P. aeruginosa* type I, II and III pyoverdines and the pyoverdines produced by *P. protegens*, and *P. fluorescens* species (**Table 2**). Other receptors predicted to recognize siderophores produced by other bacteria were found as well (enterobactin, achromobactin). The W15Oct28 FpvA receptor for its own pyoverdine utilization showed the best blast hit with the FpvA of *P. syringae* (42% identity), and it clusters with the *P. syringae* FpvA receptors in the tree presented in Figure 2, confirming the unique structure of the pyoverdine we recently disclosed [12]. The list of TonB-dependent receptors is presented in **Table S2 in file S1**.

B. Gene clusters for the biosynthesis of putisolvins.

The culture supernatant of *P. putida* W15Oct28 has a very low surface tension, which is likely to be the result of biosurfactant production, such as putisolvins. Putisolvins are lipopeptides biosurfactants produced by *P. putida* strains, such as *P. putida* PCL1445 and 267 [34–39]. Putisolvins were reported to increase motility and prevent biofilm formation of the producing strain, and can cause existing biofilms disruption. The structure of putisolvin I is C6-Leu-Glu-Leu-Ile-Gln-Ser-Val-Ile-c(Ser-Leu-Val-Ser), and its molecular weight is 1,379 Da. Putisolvin II has 14 Da more in mass due to the replacement of a valine to leucine or iso-leucine on the 11th amino acid position. **Figure 3a** is showing the organization of the putisolvin genes cluster of W15 Oct28 and *P. putida* PCL1445. Both genes clusters are highly similar and antiSMASH predicted the same amino acids to be activated by the different adenylation (A) domains. Although the three NRPS genes *psoABC* are in one cluster in *P. putida* PCL1145, the antiSMASH analysis found two different clusters for *P. putida* W15Oct28, one containing three NRPS genes while the remaining NRPS gene is found in another cluster. In both instances, the NRPS genes are flanked on one side by a LuxR regulator gene and by *macA* and *macB* genes encoding a transporter. On the other side of the PCL1145 cluster and in the second cluster of *P. putida* W15 Oct28, there is another *luxR* gene (*psoR*) and a gene coding for an outer membrane efflux protein (OprM). Although the amino acid sequences of the three non-ribosomal peptide synthetases (PsoA/B/C) only share an average of 80% identity between W15Oct28 and the sequences reported in PCL1445 (access no. DQ151887.2), only few differences in the signature residues of 3 adenylation domains for substrate recognition and activation in PsoB were found: the second A domain for leucine activation (W15Oct28: DAWSLGNV; PCL1445: DAWFLGNV), the fifth A domain for valine activation (W15Oct28: DALWMGGT; PCL1445: DALWIGGT), and the seventh A domain for serine activation (W15Oct28: DVWHXXXX; PCL1445: DVWHLSLV). However, the antiSMASH software analysis predicted the same amino acids to be activated by the different putisolvin NRPS domains in PCL1145 and W15Oct28 (**Figure 3a**). The biosurfactants produced by *P. putida* W15Oct28 were also characterized as

putisolvins by high-resolution and CID mass spectrometry (both molecular weight and amino acid sequence are identical to the reported data). In the LC/MS results presented in **Figure S2a and b** for putisolvins I and II, respectively, reveal one product with a mass of 1,380 (putisolvin I) and one with a mass of 1,394 (1,380 plus 14 Da for putisolvin II). These values are very close to those found for the two *P. putida* PCL1445 putisolvins I and II (1,379 and 1,393, respectively).

The *ppuI-rsaL-ppuR* quorum sensing system which regulates putisolvins production in PCL1445 [36] is apparently missing in the genome of W15Oct28.

C. Production of a broad-spectrum antimicrobial molecule.

One of the most interesting features of *P. putida* W15Oct28 is its strong antagonistic activity against several bacterial human pathogens (*P. aeruginosa*, *S. aureus*, *S. epidermidis*), the entomopathogenic *P. entomophila*, several plant pathogens (different pathovars of *P. syringae*, *Xanthomonas translucens* pv. *cerealis* LMG679, *Curtobacterium flaccumfaciens* [Gram-positive], and the yeast *Saccharomyces cerevisiae*) (**Table 3 and Figure 4a**). The production of the antimicrobial molecules could be detected after growing the strain W15Oct28 on M9 minimal medium supplemented with different carbon sources (glucose, citric acid or glycerol). Interestingly, the production of the antimicrobial molecule(s) was only detected when the cells were grown in media low in iron and organic nitrogen sources. The production of the antimicrobial compound(s) was detected both on solid and liquid aerated media (160 rpm, at 26–28°C). When screening a transposon mutant library for antagonism-negative mutants of W15Oct28, the majority of them turned out to be pyoverdine null mutants with transposon insertions in different pyoverdine biosynthesis genes (**Figure 4b**), which is in agreement with the fact that the antagonistic activity was lost in medium supplemented with iron, due to the suppression of pyoverdine production. The production of the antimicrobial molecule(s) was not observed when the cells were grown the in iron-poor and organic nitrogen sources-rich casamino acid medium, which allows high levels of pyoverdine production, suggesting that the antimicrobial compound is not pyoverdine. We also constructed an in-frame deletion mutant in the *pvdL* gene encoding an NRPS gene for the pyoverdine chromophore precursor synthesis [40]. As expected, the *pvdL* mutant lost its antagonism, confirming the link between pyoverdine production and the antagonistic activity. We also observed that when the CbrA/B two component regulation system [41] was deactivated, pyoverdine production was unchanged but no antimicrobial activity could be detected. These results suggest that the production of the antibiotic compound(s) by *P. putida* W15Oct28 is somehow linked to pyoverdine production, but is not due to pyoverdine itself. Indeed, purified pyoverdine only showed some limited growth inhibitory activity against gram-positive bacteria, probably as a result of iron deprivation (results not shown). To rule out the possibility that the antagonism could be due, totally or partially, to the production of putisolvins, we constructed a mutant with an in-frame partial deletion in the *psoB* homologue of W15Oct28 and tested it for its antagonistic activity. This mutant still maintained an inhibitory activity, albeit slightly reduced, excluding the involvement of the sole putisolvins in the antimicrobial activity of W15Oct28.

Two antimicrobial molecules fractions, termed fraction 1 and 4, were obtained from 18 L of culture in M9 minimal medium with yields of 9 mg and 6 mg, respectively (**Figure 5**). Antimicrobial molecule fraction 1 is a colorless oily like compound while the second antimicrobial molecule present in fraction 4 is a white powder. These two molecules are both highly hydrophobic and eluted from the HPCL column only at 100% acetonitrile.

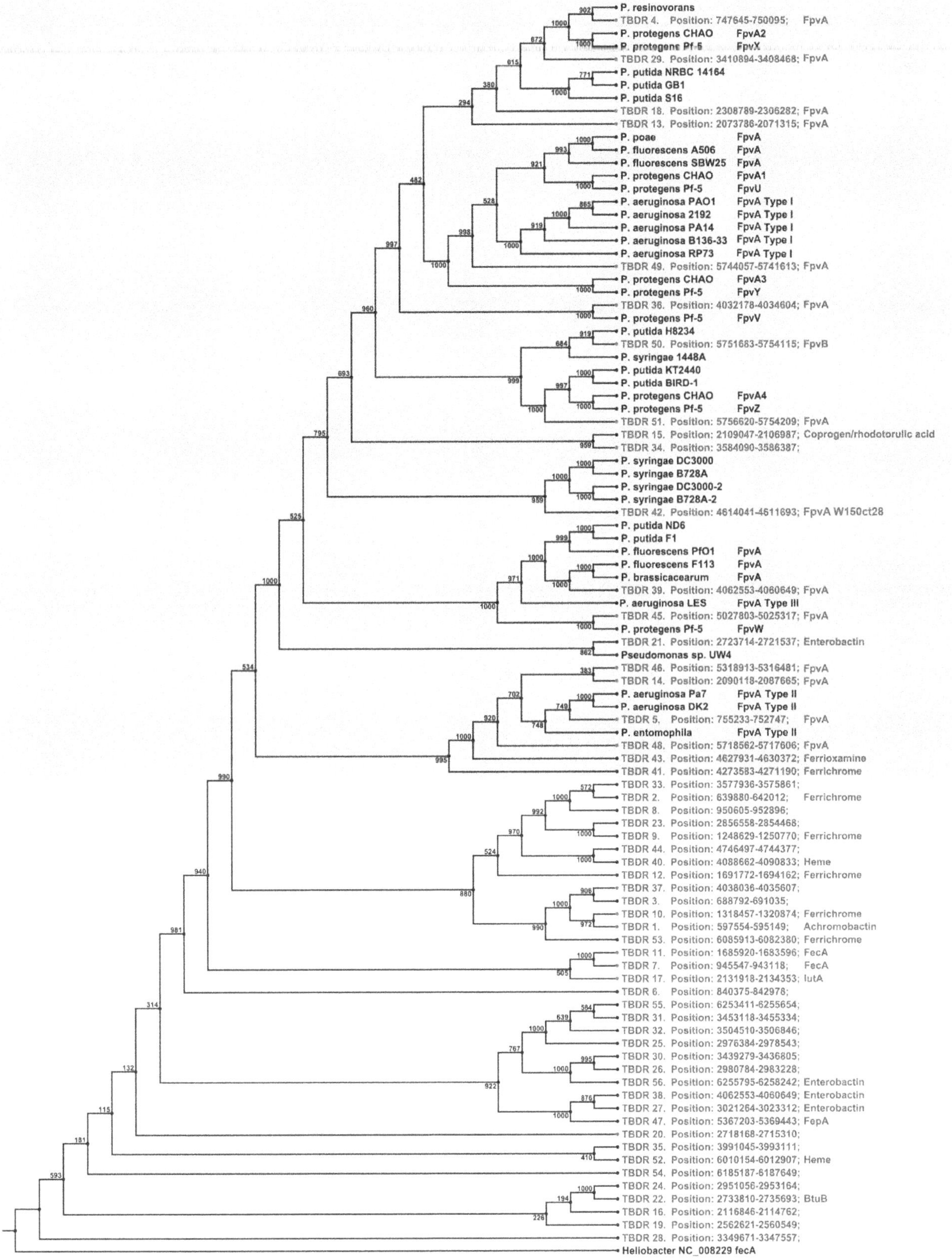

Figure 2. Phylogeny of TonB-dependent receptors. Neighbor joining tree based on the alignment of amino acid sequences of the 56 TonB dependent receptors (TBDR) detected in the genome of W15oct28 (green font) and a selection of known ferric-pyoverdine receptors from different sequenced Pseudomonas genomes (black font). W15oct28 TBDRs that are regulated through sigma-anti-sigma factors are indicated with a red node. Grey surface indicates part of the tree that contains all ferric-pyoverdine receptors that form a separate cluster.

Table 2. Pattern of pyoverdines utilization by the *pvdL*-pyoverdine-negative mutant.

Producing strain	MW/Da	Peptide chain	Utilization	Other producing strain
Pseudomonas sp. W2Aug36	989	εlys-OHAsp-Ala-aThr-Ala-cOHOrn	yes	*Pseudomonas* sp. B10
P. umsongensis LMG 21317T	1046	Ala-Lys-Thr-Ser-AcOHOrn-cOHOrn	yes	*P. fluorescens* Ps4a
P. aeruginosa PAO1 (type I)	1333	Ser-Arg-Ser-FOHOrn-(Lys-FOHOrn-Thr-Thr)	yes	
P. aeruginosa W15Oct32 (type II)	1091	Ser-FOHOrn-Orn-Gly-aThr-Ser-cOHOrn	yes	*P. aeruginosa* Pa 27853
P. aeruginosa 59.20 (type III)	1173	(Ser-Dab)-FOHOrn-Gln-Gln-FOHOrn-Gly	yes	*P. aeruginosa* Pa6/(R)
P. rhodesiae Lille 25	1421	Ser-Lys-FOHOrn-Ser-Ser-Gly-(FOHOrn-Ser-Ser)	yes	
P. protegens Pf-5	1287	Asp-FOHOrn-Lys-(Thr-Ala-Ala-FOHOrn-Lys)	yes	*P. protegens* CHA0
P. lurida LMG 21995T	1364	Ser-Ser-FOHOrn-Ser-Ser-(Lys-FOHOrn-Lys-Ser)	yes	*P.fluorescens* 95–275
P. salomonii LMG 22120T	1263	Ser-Orn-FOHOrn-Ser-Ser-(Lys-FOHOrn-Ser)	yes	*Pseudomonas* sp. 96–318
P. brenneri LMG 23068T	1187	Ser-Dab-Gly-Ser-OHAsp-Ala-Gly-Ala-Gly-cOHOrn	yes	*P.fluorescens* PflW
P. aureofciens	1277	Ser-AOHOrn-Gly-aThr-Thr-Gln-Gly-Ser-cOHOrn	yes	*P.aureofaciens* P. au
P. citronellolis LMG 18378T		structure unknown	yes	
Pseudomonas sp. W15Feb38	1046	Ser-AOHOrn-Ala-Gly-aThr-Ala-cOHOrn	yes	*P. fluorescens* PL7
Pseudomonas sp. W2Feb31B	1246	structure unknown	yes	
Pseudomonas sp. W2Jun14	1159	structure unknown	yes	
P. chlororaphis W2Apr9		structure unknown	yes	
P. putida F1	1370	Asp-OHbutOHOrn-Dab-Thr-Gly-Ser-Ser-OHAsp-Thr	yes (slightly)	*P. putida* PutC
P. fluorescens ATCC 17400	1299	Ala-Lys-Gly-Gly-OHAsp-(Gln-Dab)-Ser-Ala-cOHOrn	yes (slightly)	
P. fluorescens ATCC 17926	1159	structure unknown	yes (slightly)	
P. syringae B728a	1123	εLys-OH Asp-Thr-(Thr-Ser-OH Asp-Ser)	yes (slightly)	
P. brassicacearum LMG 21623T	1150	Ser-AOHOrn-Ala-Gly-(Ser-Ser-OHAsp-Thr)	yes (slightly)	*P. fluorescens* PL9
P. putida KT2440	1072	Asp-Orn-(OHAsp-Dab)-Gly-Ser-cOHOrn	No	*P. putida* G4R and BIRD-1
P. putida L1	1349	Asp-εLys-OHAsp-Ser-aThr-Ala-Thr-Lys-cOHOrn	No	*P. putida* GB-1 and WCS358
P. fluorescens BTP2	1049	Ser-Val-OHAsp-Gly-Thr-Ser-cOHOrn	No	
P. fluorescens Pf0-1	1381	Ala-AcOHOrn-Orn-Ser-Ser-Ser-Arg-OHAsp-Thr	No	
P. entomophila L48	1314	Ala-xxx-OHHis-Asp-Gly-Gly-Ser-Thr-Ser-cOHOrn	No	
P. vancouverensis LMG 20222T		structure unknown	No	
P. asplenii LMG 21749T		structure unknown	No	
P. koreensis MFY71		structure unknown	No	

Attempts to determine the mass of the molecule or to determine its structure by ^{1}H or ^{13}C NMR were however unsuccessful.

D. Other gene clusters for putative secondary metabolites. Given the fact that there is no characteristic polyketide synthase gene cluster in the genome of W15Oct28, we decided to use the antiSMASH algorithm [16,17] to search for putative biosynthetic gene clusters containing genes corresponding to fatty acid desaturase and PKS tailoring enzymes, in an approach similar to the one used to solve the puzzle of the tunicamycin biosynthetic gene cluster [42]. One cluster of ten genes was selected by antiSMASH for its potential involvement in secondary metabolism, which is also present in the recently sequenced *Pseudomonas* sp. HYS [43]. This gene cluster presented in **Figure 3b** is preceded by a gene encoding an AraC transcription regulator, followed by genes coding for a NAD-dependent epimerase similar to the *mupV* gene from *P. fluorescens* NCIMB10586 strain involved in the synthesis of the polyketide antibiotic mupirocin [44], an acetylornithine aminotransferase that can catalyze the reversible transfer of an acetyl-group from a basic (ornithine) to an acidic (glutamate) amino acid [45], a PKS cyclase, an isobutylamine *N*-hydroxylase similar to VlmH involved in valanimycin biosynthesis [46], a FabG like short chain

dehydrogenase, a fatty acid desaturase, an hypothetical protein and three proteins forming an RND family efflux transporter (MFP subunit, EmrB/Qac drug resistant transporter, and outer membrane lipoprotein of RND efflux system). However, attempts to inactivate genes in this cluster have so far failed and its involvement in the production of a bioactive molecule remains hypothetical.

Safracins were reported to be produced by two strains of *P. fluorescens*, A2-2 and SC 12695 [47]. The biosynthesis of this antitumor compound drew attention due to the fact that safracin B can be used as an intermediate for the synthesis of ecteinascidin, which is a potent anti-tumor agent [48]. The biosynthesis gene cluster of safracin in *P. fluorescens* A2-2 contains three NRPSs (SacA, SacB, SacC), three precursor biosynthetic enzymes (SacD, SacF, SacG), two tailoring enzymes (SacI, SacJ), one resistance protein (SacH) and one protein with unknown function (SacE) [47]. In the genome of *P. putida* W15Oct28, we found three genes corresponding to the beginning of the cluster (*sacJ*, *sacI*, *sacA*) (**Figure 3C**). SacA is a NRPS, but in W15Oct28 it is truncated since it only contains one condensation domain. A second short ORF containing only one adenylation domain follows, which shows no similarity with any of the genes in the safracin cluster in

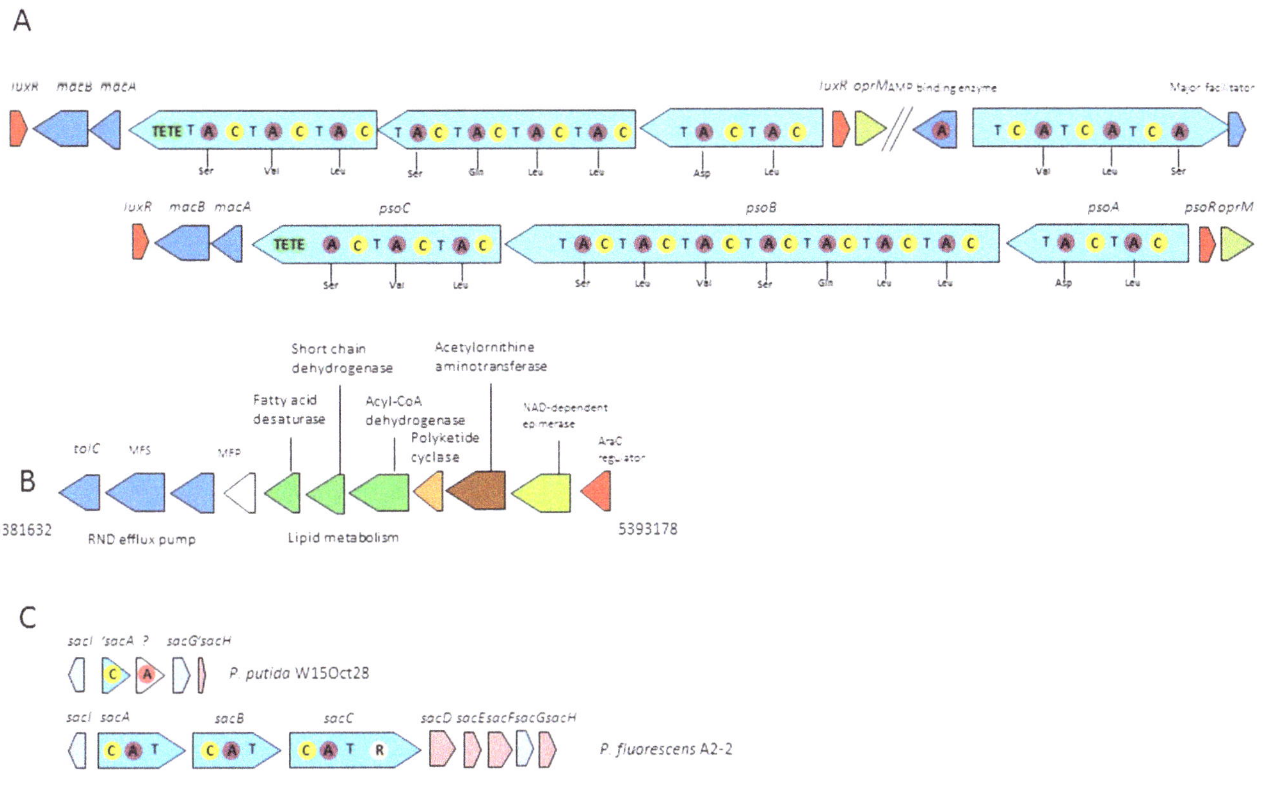

A

B

C

Figure 3. Genes involved in the production of secondary metabolites. A. Genes involved in the production of putisolvins in strain W15Oct28 (top) and strain *P. putida* PCL1145 (bottom): *psoA*, *psoB*, and *psoC* are the genes encoding NRPS enzymes with their condensation (C), adenylation (A) domains showing the predicted activated amino acid, and the T domain for the thioester attachment of the activated amino acid. The two thioesterase domains responsible for the detachment of the completed peptide at the end of *psoC* are also indicated (TE). The *macA* and *macB* genes correspond to a transporter, the *oprM* gene coding for an efflux system porin, and the two orphan *luxR* genes are shown in red. The amino acids predicted to be activated by the different A domains by the antiSMASH analysis are indicated without mentioning whether they are in the D- or L- form. B. The ten genes cluster possibly involved in the biosynthesis of a secondary metabolite. The cluster is preceded by a gene encoding an AraC regulator. See the text for details. C. The incomplete safracin gene cluster of W15Oct28 compared to the complete safracin gene cluster of *P. fluorescens* A2-2.

P. fluorescens A2-2. The partial cluster ends with *sacG* and *sacH* while the part from the thioesterase domain of *sacA* till the end of *sacF* is missing (**Figure 3C**).

Secretion systems: first evidence for a Type IV secretion system in a pseudomonad

P. putida W15Oct28 genome contains genes for different secretion systems, including a Fap amyloid fiber secretion system, a curli production system, one type I, one type II, one type II/IV secretion system, as well as one putative type IVb secretion system, three autotransporter genes (type V) and several genes for a type VI secretion system (**Figure 6**). There is however no system for type III secretion in this strain. Most of those secretion systems are conserved in other *P. putida* group species, with the exception of the Dot/Icm type IVb secretion system, although a highly similar gene cluster is present in the genome of *P. putida* BIRD-1 (PPUBIRD1_4462-4506) [49]. The new putative type IV secretion system is most similar to the T4SSb from the pathogenic Gram-negative bacterium *Legionella pneumophila* [50], which is characterized by the presence of several conjugative transfer proteins, such as TraA, IncP type conjugative transfer protein TrbN, and IncI/plasmid conjugation transfer protein TraN. Although this is the first description of the presence of genes encoding the components of Type IV secretion system in *Pseudomonas*, we do not know at this stage if it is functional.

Another interesting observation is the presence in the genome of five different VgrG type VI effector proteins.

Evidence for the presence of pyocins genes

Beside the production of antimicrobial compounds synthesized by secondary metabolism which are often characterized by a broad inhibitory spectrum, bacteria also produce antimicrobial peptides or proteins termed bacteriocins to kill their close relatives (usually within the same species) [51–53]. The best studied bacteriocins are those produced by *E. coli*, termed colicins, and the pyocins produced by *P. aeruginosa* and other pseudomonads [51,54]. Pyocins production genes are found in the genome of strain W15Oct28, including a CvpA colicin V production protein and the CreA colicin E2 immunity protein which are conserved in different *P. putida* genomes. There is also the presence of an S-type pyocin gene encoding a protein with translocation domain and killing domain similar to the one in strain KT2440 while the immunity gene is similar to the one in strain GB-1 (results not shown). The killing domain contains a HNH DNA endonuclease motif, which shares similarity with the S1/S2 type pyocins which are present in different *P. aeruginosa* strains [51]. Besides its presence in strain KT2440 as described by Parret and De Mot [54], this pyocin killing domain could also be found in *P. putida* strains ND6 and S13.12. The immunity gene adjacent to the gene

Table 3. Antagonistic activity of *P. putida* W15Oct28 against different microorganisms.

Bioassay indicator strain	Activity
Gram-positive bacteria	
Staphylococcus aureus NCTC 8325	+
Staphylococcus aureus ATCC 29213	+
Methicillin-resistant *S. aureus* B6	+
Methicillin-resistant *S. aureus* D6	+
Staphylococcus epidermidis RP62A	+
Micrococcus luteus ATCC 4698	+
Curtobacterium flaccumfaciens	+
Gram-negative bacteria	
Pseudomonas syringae pv. tomato DC3000	+
Pseudomonas syringae pv. syringae B301D	+
Pseudomonas syringae pv syringae B728a	+
Pseudomonas entomophila L48	+
Pseudomonas aeruginosa PAO1	+
Pseudomonas aeruginosa PA14	+
Xanthomonas translucens pv. cerealis LMG679	+
Pseudomonas aeruginosa PA7	−
Pseudomonas putida KT2440	−
Pseudomonas protegens Pf-5	−
Escherichia coli MG1655	−
Klebsiella pneumoniae 8401	−
Klebsiella pneumoniae 8410	−
Salmonella enterica sp. Typhimurium X3000	−
Salmonella enteritidis 76Sa88	−
Eukaryotic microorganisms	
Pythium ultimum	−
Saccharomyces cerevisiae	+

of killing domain shows similarity to the immunity gene of the *E. coli* E7/8 colicin.

Discussion

The *Pseudomonas* genus definition has a long history, which first started with phenotypic characterizations and metabolic pathway profiling, later followed by 16s rDNA phylogeny, which resulted in the exclusion of some species from the true pseudomonads which cluster in the rRNA group I [55,56]. Now a new chapter is opening based on whole genome phylogeny as exemplified in this and other studies [57,58]. Although the species *P. putida* and *P. fluorescens* represent a large group of *Pseudomonas* species adapted to diverse environmental niches, the number of sequenced genomes is still low compared to the large number of genomes from the well-studied pathogenic species *P. aeruginosa* [49]. Next generation sequencing methods which became available in the last few years promoted a surge in taxonomic study based on whole genome comparison. Recently, an increasing number of genomes of strains belonging to the *P. putida* group have been reported, providing an opportunity to better understand how this species can adapt to different environmental niches using a combination of bioinformatics and experimental approaches. By combining a multi locus sequencing approach (MLSA) and whole genome comparison, we could

confirm that our strain W15Oct28 is indeed a true *P. putida* since its closest relative is the *P. putida* type strain NBRC 14164[T] [28]. We believe that the strain W15Oct28 described in this study represents a good example of how diverse *P. putida* strains can be. Strain W15Oct28 was isolated from the river Woluwe in Brussels [11] and it attracted our attention because of its antagonism towards *Staphylococcus aureus* and *P. aeruginosa*. This strain produces a pyoverdine with a mass as 1,624 Da as the only siderophore [12]. The very low surface tension of cultures supernatants also suggested the production of biosurfactant(s). We could indeed identify that W15Oct28 produced two putisolvins by mass spectrometry, and the amino acid sequence of putisolvins deduced from the analysis of NRPS genes showed that they shared 80% identity with the *P. putida* PCL1445 putisolvins [36,37].

The most striking characteristic of W15Oct28 is its strong antagonistic activity against several bacterial pathogens, which is due to the production of a highly hydrophobic antimicrobial molecule, the structure of which could not yet be determined, due to the difficulty to ionize the molecule. An intriguing observation was that a majority of transposon mutants which were unable to produce the compound had insertions in genes involved in the biosynthesis of pyoverdine, despite the fact that pyoverdine itself showed no antimicrobial activity. One possible explanation is that

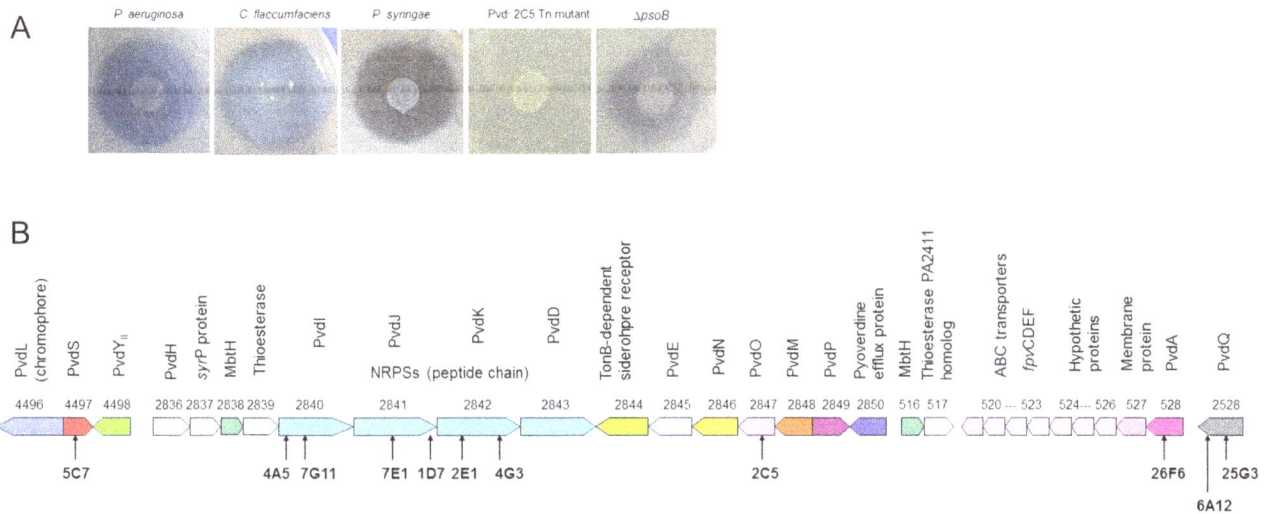

Figure 4. Antagonistic activity and role of pyoverdine in the antagonism. A. Antagonistic activity of the *P. putida* W15Oct28 strain against *Pseudomonas aeruginosa*, *Curtobacterium flaccumfaciens*, and *Pseudomonas syringae*. The pyoverdine-negative *pvdO* 2C5 transposon mutant has lost its antagonism against *P. aeruginosa* while the Δ*psoB* putisolvin-negative mutant keeps a reduced, but still visible level of antagonism. B. The pyoverdine genes clusters: the arrows indicate the places of transposon insertions causing the loss of pyoverdine production and of the antagonism.

the proteins involved in the production of the antibiotic compound share the same "siderosome" platform as the pyoverdine biosynthetic machinery recently described in *P. aeruginosa* [59–61]. Another surprising result was the discovery in the genome of an incomplete gene cluster for the biosynthesis of the antibiotic safracin, which is produced by a *P. fluorescens* strain [47]. We hypothesize that this entire gene cluster was originally present in the genome and became later on deactivated by deletion. Besides its particular secondary metabolism, W15Oct28 possesses other traits which are not shared by other *P. putida* group species. W15Oct28 was found to contain a putative type IVb secretion system, which could only be found in the genome of *Pseudomonas sp.* BIRD-1, but not in other known *Pseudomonas* genomes. Strain

W15Oct28 also contains a fatty acid biosynthesis gene cluster which is only present in two newly whole genome sequenced strains *P. putida* NBRC 14164[T] and H8234 [27,28], which are the closest relatives of W15Oct28. The W15Oct28 strain has also 56 TonB-dependent receptors and most of them have highly conserved homologs in other *P. putida* strains, especially GB-1 and NBRC 14164, which makes this strain one of the *Pseudomonas* having the largest repertoire of TonB-dependent receptors, comparable with *P. fluorescens* ATCC17400 which has 55 receptor genes [33]. Cross feeding results show that *P. putida* W15Oct28 is able to utilize pyoverdines produced by *P. aeruginosa* and *P. fluorescens* strains while it shows a limited capacity to use pyoverdines produced by other *P. putida*, in

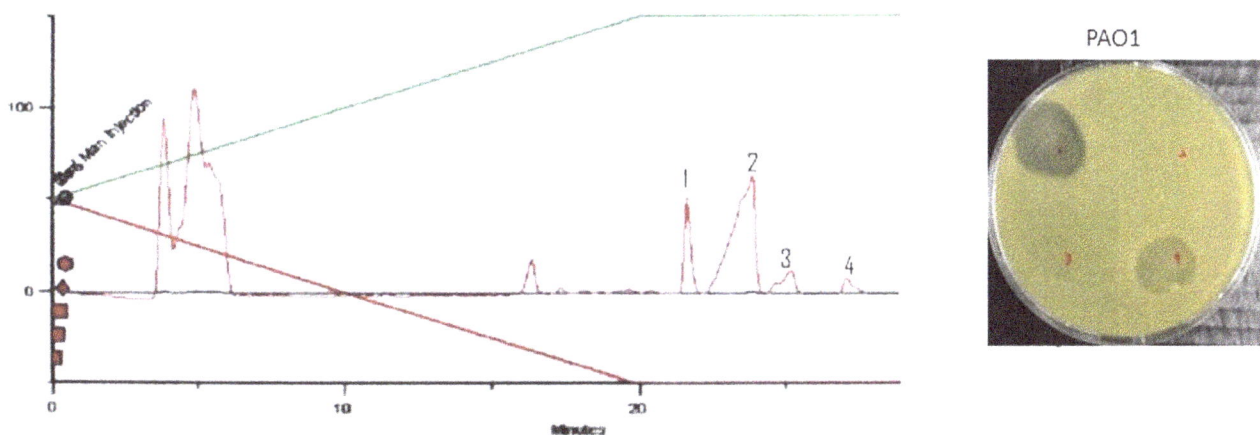

Figure 5. Fractionation of the antagonistic activity by HPLC. HPCL fractionation of a crude extract from a culture supernatant of *P. putida* W15Oct28 grown in M9-glucose medium: A, chromatography of crude extraction of W15Oct28 cultured in M9 minimal medium for 48 hours; B, chromatography of crude extraction of W15Oct28 cultured in M9 minimal medium for 48 hours plus 5 days stay at 6°C. The yield of fraction 1 and 4 were increased with longer time of cultivation. Fractions 1 to 4 were spotted on an agar plate inoculated with *P. aeruginosa* PAO1 (upper left, fraction 1, upper right, fraction 2, lower left, fraction 3, lower right, fraction 4). The green line represents the acetonitrile gradient.

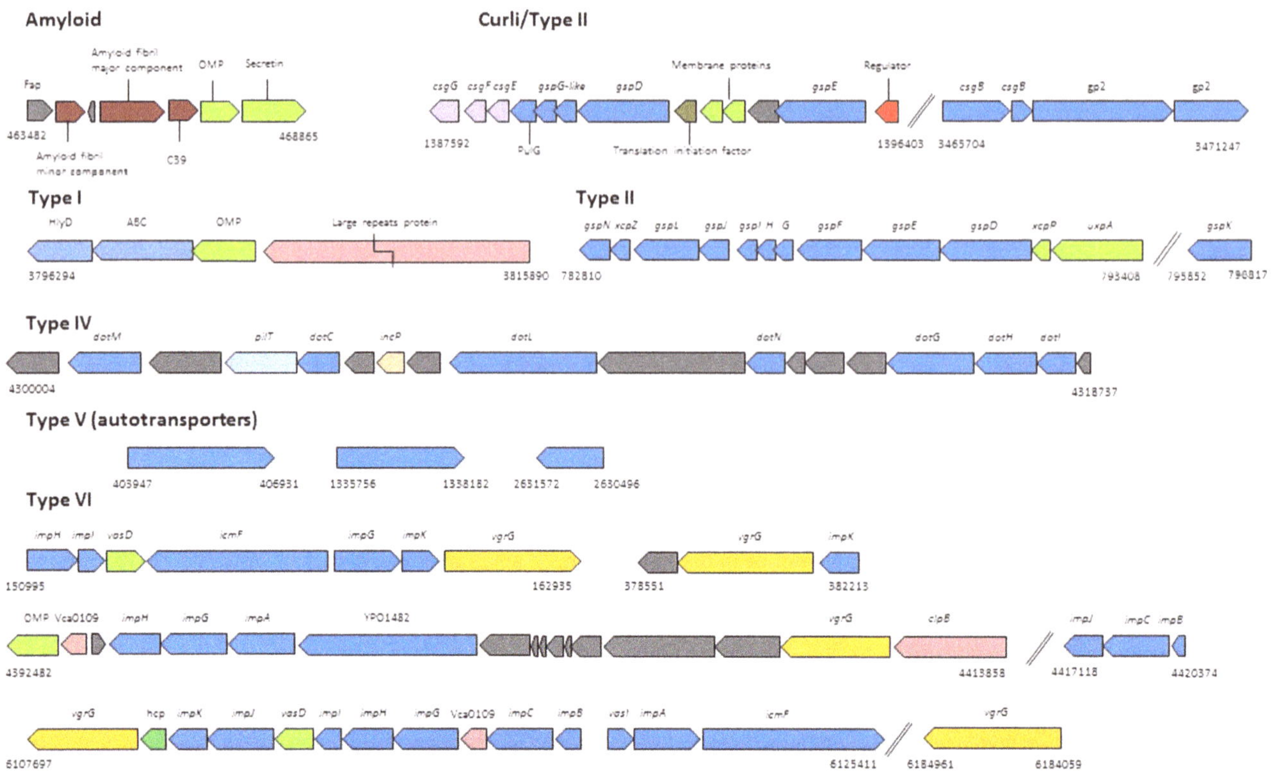

Figure 6. Secretion systems. Gene clusters coding for components of different secretion systems: amyloid, curli, type I, type II, type IV, type V (autotransporters) and type VI secretion systems. The genes in yellow for type VI secretion represent the different VgrG effector proteins.

agreement with the lack of known *P. putida* ferric-pyoverdine receptors in W15Oct28 genome (results not shown). On the other hand, we assume that the pyoverdine produced by W15Oct28 is not commonly found in *Pseudomonas* sp. strains since its corresponding receptor has no homolog with an identity above 50% in Genbank so far. Results of the antagonism tests show that this strain has a promising inhibitory activity against both Gram-positive and Gram-negative phytopathogens, especially against different pathovars of *P. syringae*. In contrast, strain W15Oct28 shows no antagonism against non-pathogenic *Pseudomonas* strains. It remains to be seen why the production of this antimicrobial compound(s) is dependent on the pyoverdine biosynthetic machinery, and to elucidate the structure of the bioactive compound(s) produced by this particular strain of *P. putida*.

Supporting Information

Figure S1 Circular representation of the *P. putida* W15Oct28 genome. The green and purple inner circle represents the GC skew while the black circle represents the GC content. The blue, green and purple circles represent the ORFs the products of which show a BLAST hit with *P. putida* NBRC 14164, BIRD-1 and GB1, respectively. The two external red circles represent the different ORFs in the bottom and the top strand, respectively. The genomic islands detected via the island finder (see text for details) are indicated by blue triangles. The yellow arrow shows the location in the genome where the ten genes cluster possibly involved in the biosynthesis and secretion of

the antimicrobial compound is found. The partial safracin gene cluster is indicated by a purple triangle and the other clusters for pyoverdine biosynthesis, fatty acid synthesis, and type IV secretion system are indicated by colored arcs.

Figure S2 LC/MS Mass spectra of extracted putisolvin I (A) and II (B) of *P. putida* W15Oct28. The arrows indicate the masses. See text in the results section for details.

File S1 Table S1. List of strains and plasmids used in this study. **Table S2.** List of TonB-dependent receptors.

Acknowledgments

The authors would like to thank Prof. Dr. J-P Hernalsteens from VUB for providing the *S. aureus*, *M. luteus*, *E. coli*, *K. pneumonia*, and *Salmonella* sp. strains for bioassay test, and Prof. R. De Mot from K.U. Leuven for providing the *Xanthomonas translucens* and *Pseudomonas savastanoi* pv. *savastanoi* LMG5484 strain. The authors are very grateful for the material and protocol of yeast recombineering method for deletion provided by Prof. M. Höfte and Dr. K. De Maeyer.

Author Contributions

Conceived and designed the experiments: LY PC. Performed the experiments: LY FH JD SB GL RB SM. Analyzed the data: LY FH JD GL RB SM PC. Contributed reagents/materials/analysis tools: PC FH JD SB GL SM RB. Wrote the paper: LY PC RB.

References

1. Nelson KE, Weinel C, Paulsen IT, Dodson RJ, Hilbert H, et al. (2002) Complete genome sequence and comparative analysis of the metabolically versatile *Pseudomonas putida* KT2440. Environ Microbiol 4: 799–808.

2. Rojas A, Duque E, Mosqueda G, Golden G, Hurtado A, et al. (2001) Three efflux pumps are required to provide efficient tolerance to toluene in *Pseudomonas putida* DOT-T1E. J Bacteriol 183: 3967–3973.

3. Roca A, Pizarro-Tobias P, Udaondo Z, Fernandez M, Matilla MA, et al. (2013) Analysis of the plant growth-promoting properties encoded by the genome of the rhizobacterium *Pseudomonas putida* BIRD-1. Environ Microbiol 15: 780–794.

4. Ramos JL, Duque E, Godoy P, Segura A (1998) Efflux pumps involved in toluene tolerance in *Pseudomonas putida* DOT-T1E. J Bacteriol 180: 3323–3329.

5. Matilla MA, Ramos JL, Bakker PA, Doornbos R, Badri DV, et al. (2010) *Pseudomonas putida* KT2440 causes induced systemic resistance and changes in Arabidopsis root exudation. Environ Microbiol Rep 2: 381–388.

6. Espinosa-Urgel M, Ramos JL (2004) Cell density-dependent gene contributes to efficient seed colonization by *Pseudomonas putida* KT2440. Appl Environ Microbiol 70: 5190–5198.

7. Espinosa-Urgel M, Kolter R, Ramos JL (2002) Root colonization by *Pseudomonas putida*: love at first sight. Microbiology 148: 341–343.

8. Gross H, Loper JE (2009) Genomics of secondary metabolite production by *Pseudomonas* spp. Nat Prop Rep 26: 1408–1446.

9. Loper JE, Henkels MD, Shaffer BT, Valeriote FA, Gross H (2008) Isolation and identification of rhizoxin analogs from *Pseudomonas fluorescens* Pf-5 by using a genomic mining strategy. Appl Environ Microbiol 74: 3085–3093.

10. Mulet M, Lalucat J, Garcia-Valdes E (2010) DNA sequence-based analysis of the *Pseudomonas* species. Environ Microbiol 12: 1513–1530.

11. Pirnay JP, Matthijs S, Colak H, Chablain P, Bilocq F, et al. (2005) Global *Pseudomonas aeruginosa* biodiversity as reflected in a Belgian river. Environ Microbiol 7: 969–980.

12. Ye L, Ballet S, Hildebrand F, Laus G, Guillemyn K, et al. (2013) A combinatorial approach to the structure elucidation of a pyoverdine siderophore produced by a *Pseudomonas putida* isolate and the use of pyoverdine as a taxonomic marker for typing *P. putida* subspecies. Biometals 26: 561–575.

13. Matthijs S, Laus G, Meyer JM, Abbaspour-Tehrani K, Schafer M, et al. (2009) Siderophore-mediated iron acquisition in the entomopathogenic bacterium *Pseudomonas entomophila* L48 and its close relative *Pseudomonas putida* KT2440. Biometals 22: 951–964.

14. Zerbino DR, Birney E (2008) Velvet: algorithms for de novo short read assembly using de Bruijn graphs. Genome Res 18: 821–829.

15. Aziz RK, Bartels D, Best AA, DeJongh M, Disz T, et al. (2008) The RAST Server: rapid annotations using subsystems technology. BMC Genomics 9: 75.

16. Medema MH, Blin K, Cimermancic P, de Jager V, Zakrzewski P, et al. (2011) antiSMASH: rapid identification, annotation and analysis of secondary metabolite biosynthesis gene clusters in bacterial and fungal genome sequences. Nucleic Acids Res 39: W339–346.

17. Blin K, Medema MH, Kazempour D, Fischbach MA, Breitling R, et al. (2013) antiSMASH 2.0—a versatile platform for genome mining of secondary metabolite producers. Nucleic Acids Res 41: W204–212.

18. Stachelhaus T, Mootz HD, Marahiel MA (1999) The specificity-conferring code of adenylation domains in nonribosomal peptide synthetases. Chem Biol 6: 493–505.

19. Grant JR, Stothard P (2008) The CGView Server: a comparative genomics tool for circular genomes. Nucleic Acids Res 36: W181–184.

20. Langille MG, Brinkman FS (2009) IslandViewer: an integrated interface for computational identification and visualization of genomic islands. Bioinformatics 25: 664–665.

21. Grissa I, Vergnaud G, Pourcel C (2007) CRISPRFinder: a web tool to identify clustered regularly interspaced short palindromic repeats. Nucleic Acids Res 35: W52–57.

22. Dennis JJ, Zylstra GJ (1998) Plasposons: modular self-cloning minitransposon derivatives for rapid genetic analysis of gram-negative bacterial genomes. Appl Environ Microbiol 64: 2710–2715.

23. Shanks RM, Kadouri DE, MacEachran DP, O'Toole GA (2009) New yeast recombineering tools for bacteria. Plasmid 62: 88–97.

24. Richter M, Rossello-Mora R (2009) Shifting the genomic gold standard for the prokaryotic species definition. Proc Natl Acad Sci USA 106: 19126–19131.

25. Hartney SL, Mazurier S, Girard MK, Mehnaz S, Davis EW, 2nd, et al. (2013) Ferric-pyoverdine recognition by Fpv outer membrane proteins of *Pseudomonas protegens* Pf-5. J Bacteriol 195: 765–776.

26. Mulet M, Garcia-Valdes E, Lalucat J (2013) Phylogenetic affiliation of *Pseudomonas putida* biovar A and B strains. Res Microbiol 164: 351–359.

27. Molina L, Bernal P, Udaondo Z, Segura A, Ramos JL (2013) Complete genome sequence of a *Pseudomonas putida* clinical isolate, strain H8234. Genome A1: 00496-13.

28. Ohji S, Yamazoe A, Hosoyama A, Tsuchikane K, Ezaki T, et al. (2014) The complete genome sequence of *Pseudomonas putida* NBRC 14164T confirms high intraspecies variation. Genome A 2: 00029–14.

29. Schalk IJ, Guillon L (2013) Fate of ferrisiderophores after import across bacterial outer membranes: different iron release strategies are observed in the cytoplasm or periplasm depending on the siderophore pathways. Amino Acids 44: 1267–1277.

30. Visca P, Imperi F, Lamont IL (2007) Pyoverdine siderophores: from biogenesis to biosignificance. Trends Microbiol 15: 22–30.

31. Ravel J, Cornelis P (2003) Genomics of pyoverdine-mediated iron uptake in pseudomonads. Trends Microbiol 11: 195–200.

32. Cornelis P (2010) Iron uptake and metabolism in pseudomonads. Appl Microbiol Biotechnol 86: 1637–1645.

33. Ye L, Matthijs S, Bodilis J, Hildebrand F, Raes J, et al. (2014) Analysis of the draft genome of *Pseudomonas fluorescens* ATCC17400 indicates a capacity to take up iron from a wide range of sources, including different exogenous pyoverdines. Biometals 27: 633–644.

34. Dubern JF, Bloemberg GV (2006) Influence of environmental conditions on putisolvins I and II production in *Pseudomonas putida* strain PCL1445. FEMS Microbiol Lett 263: 169–175.

35. Dubern JF, Lugtenberg BJ, Bloemberg GV (2006) The *ppuI-rsaL-ppuR* quorum-sensing system regulates biofilm formation of *Pseudomonas putida* PCL1445 by controlling biosynthesis of the cyclic lipopeptides putisolvins I and II. J Bacteriol 188: 2898–2906.

36. Dubern JF, Coppoolse ER, Stiekema WJ, Bloemberg GV (2008) Genetic and functional characterization of the gene cluster directing the biosynthesis of putisolvin I and II in *Pseudomonas putida* strain PCL1445. Microbiology 154: 2070–2083.

37. Kuiper I, Lagendijk EL, Pickford R, Derrick JP, Lamers GE, et al. (2004) Characterization of two *Pseudomonas putida* lipopeptide biosurfactants, putisolvin I and II, which inhibit biofilm formation and break down existing biofilms. Mol Microbiol 51: 97–113.

38. Li W, Rokni-Zadeh H, De Vleeschouwer M, Ghequire MG, Sinnaeve D, et al. (2013) The antimicrobial compound xantholysin defines a new group of *Pseudomonas* cyclic lipopeptides. PloS one 8: e62946.

39. Rokni-Zadeh H, Li W, Yilma E, Sanchez-Rodriguez A, De Mot R (2013) Distinct lipopeptide production systems for WLIP (white line-inducing principle) in *Pseudomonas fluorescens* and *Pseudomonas putida*. Environ Microbiol Rep 5: 160–169.

40. Mossialos D, Ochsner U, Baysse C, Chablain P, Pirnay JP, et al. (2002) Identification of new, conserved, non-ribosomal peptide synthetases from fluorescent pseudomonads involved in the biosynthesis of the siderophore pyoverdine. Mol Microbiol 45: 1673–1685.

41. Yeung AT, Bains M, Hancock RE (2011) The sensor kinase CbrA is a global regulator that modulates metabolism, virulence, and antibiotic resistance in *Pseudomonas aeruginosa*. J Bacteriol 193: 918–931.

42. Chen W, Qu D, Zhai L, Tao M, Wang Y, et al. (2010) Characterization of the tunicamycin gene cluster unveiling unique steps involved in its biosynthesis. Protein & cell 1: 1093–1105.

43. Gao J, Yu X, Xie Z (2012) Draft genome sequence of high-siderophore-yielding *Pseudomonas* sp. strain HYS. J Bacteriol 194: 4121.

44. El-Sayed AK, Hothersall J, Cooper SM, Stephens E, Simpson TJ, et al. (2003) Characterization of the mupirocin biosynthesis gene cluster from *Pseudomonas fluorescens* NCIMB 10586. Chem Biol 10: 419–430.

45. Iqbal A, Clifton IJ, Chowdhury R, Ivison D, Domene C, et al. (2011) Structural and biochemical analyses reveal how ornithine acetyl transferase binds acidic and basic amino acid substrates. Org Biomol Chem 9: 6219–6225.

46. Garg RP, Ma Y, Hoyt JC, Parry RJ (2002) Molecular characterization and analysis of the biosynthetic gene cluster for the azoxy antibiotic valanimycin. Mol Microbiol 46: 505–517.

47. Velasco A, Acebo P, Gomez A, Schleissner C, Rodriguez P, et al. (2005) Molecular characterization of the safracin biosynthetic pathway from *Pseudomonas fluorescens* A2-2: designing new cytotoxic compounds. Mol Microbiol 56: 144–154.

48. Jin W, Metobo S, Williams RM (2003) Synthetic studies on ecteinascidin-743: constructing a versatile pentacyclic intermediate for the synthesis of ecteinascidins and saframycins. Org Lett 5: 2095–2098.

49. Winsor GL, Lam DK, Fleming L, Lo R, Whiteside MD, et al. (2011) Pseudomonas Genome Database: improved comparative analysis and population genomics capability for *Pseudomonas* genomes. Nucleic Acids Res 39: D596–600.

50. Zink SD, Pedersen L, Cianciotto NP, Abu-Kwaik Y (2002) The Dot/Icm type IV secretion system of *Legionella pneumophila* is essential for the induction of apoptosis in human macrophages. Infect Immun 70: 1657–1663.

51. Michel-Briand Y, Baysse C (2002) The pyocins of *Pseudomonas aeruginosa*. Biochimie 84: 499–510.

52. Cascales E, Buchanan SK, Duche D, Kleanthous C, Lloubes R, et al. (2007) Colicin biology. Microbiol Mol Biol Rev 71: 158–229.

53. Penfold CN, Walker D, Kleanthous C (2012) How bugs kill bugs: progress and challenges in bacteriocin research. Biochem Soc Trans 40: 1433–1437.

54. Parret AH, De Mot R (2002) Bacteria killing their own kind: novel bacteriocins of *Pseudomonas* and other gamma-proteobacteria. Trends Microbiol 10: 107–112.

55. Palleroni NJ (2010) The Pseudomonas story. Environ Microbiol 12: 1377–1383.

56. Mulet M, Garcia-Valdes E, Lalucat J (2013) Phylogenetic affiliation of *Pseudomonas putida* biovar A and B strains. Res Microbiol 164: 351–359.

57. Loper JE, Hassan KA, Mavrodi DV, Davis EW 2nd, Lim CK, et al. (2012) Comparative genomics of plant-associated *Pseudomonas* spp: insights into diversity and inheritance of traits involved in multitrophic interactions. PLoS Genet 8: e1002784.

58. Kyrpides NC, Hugenholtz P, Eisen JA, Woyke T, Goker M, et al. (2014) Genomic encyclopedia of bacteria and archaea: sequencing a myriad of type strains. PLoS Biol 12: e1001920.

59. Guillon L, El Mecherki M, Altenburger S, Graumann PL, Schalk IJ (2012) High cellular organization of pyoverdine biosynthesis in *Pseudomonas aeruginosa*: clustering of PvdA at the old cell pole. Environ Microbiol 14: 1982–1994.

60. Schalk IJ, Guillon L (2013) Pyoverdine biosynthesis and secretion in *Pseudomonas aeruginosa*: implications for metal homeostasis. Environ Microbiol 15: 1661–1673.

61. Guillon L, Altenburger S, Graumann PL, Schalk IJ (2013) Deciphering protein dynamics of the siderophore pyoverdine pathway in *Pseudomonas aeruginosa*. PloS One 8: e79111.

14

Architecture and Assembly of the *Bacillus subtilis* Spore Coat

Marco Plomp[1], Alicia Monroe Carroll[2], Peter Setlow[2]*, Alexander J. Malkin[1]*

1 Biosciences and Biotechnology Division, Physical and Life Sciences Directorate, Lawrence Livermore National Laboratory, Livermore, California, United States of America,
2 Department of Molecular Biology and Biophysics, University of Connecticut Health Center, Farmington, Connecticut, United States of America

Abstract

Bacillus spores are encased in a multilayer, proteinaceous self-assembled coat structure that assists in protecting the bacterial genome from stresses and consists of at least 70 proteins. The elucidation of *Bacillus* spore coat assembly, architecture, and function is critical to determining mechanisms of spore pathogenesis, environmental resistance, immune response, and physicochemical properties. Recently, genetic, biochemical and microscopy methods have provided new insight into spore coat architecture, assembly, structure and function. However, detailed spore coat architecture and assembly, comprehensive understanding of the proteomic composition of coat layers, and specific roles of coat proteins in coat assembly and their precise localization within the coat remain in question. In this study, atomic force microscopy was used to probe the coat structure of *Bacillus subtilis* wild type and *cotA, cotB, safA, cotH, cotO, cotE, gerE,* and *cotE gerE* spores. This approach provided high-resolution visualization of the various spore coat structures, new insight into the function of specific coat proteins, and enabled the development of a detailed model of spore coat architecture. This model is consistent with a recently reported four-layer coat assembly and further adds several coat layers not reported previously. The coat is organized starting from the outside into an outermost amorphous (crust) layer, a rodlet layer, a honeycomb layer, a fibrous layer, a layer of "nanodot" particles, a multilayer assembly, and finally the undercoat/basement layer. We propose that the assembly of the previously unreported fibrous layer, which we link to the darkly stained outer coat seen by electron microscopy, and the nanodot layer are *cotH-* and *cotE-* dependent and *cotE-*specific respectively. We further propose that the inner coat multilayer structure is crystalline with its apparent two-dimensional (2D) nuclei being the first example of a non-mineral 2D nucleation crystallization pattern in a biological organism.

Editor: Etienne Dague, LAAS-CNRS, France

Funding: This work was supported by a grant from the National Institutes of Health (http://www.nih.gov/) (GM-19698) (PS), by a Department of Defense Multi-disciplinary University Research Initiative (http://www.arl.army.mil/www/default.cfm?page=472) through the United States Army Research Laboratory and the United States Army Research Office under contract number W911F-09-1-0286 (PS), and by the Lawrence Livermore National Laboratory (https://www.llnl.gov/) through Laboratory Directed Research and Development Grant 04-ERD-002 (AJM). Part of this work was performed under the auspices of the United States Department of Energy by the University of California, Lawrence Livermore National Laboratory under Contract W-7405-Eng-48. The funders had no role in study design, data collection and analysis, or preparation of the manuscript. This document was cleared by the Lawrence Livermore National Laboratory for publication.

* Email: setlow@nso2.uchc.edu (PS); malkin1@llnl.gov (AJM)

Introduction

Spores of bacteria of *Bacillus* species are formed in sporulation and are metabolically dormant and resistant to a large variety of environmental stress factors. While multiple factors contribute to spore resistance, one striking spore feature is the multilayer spore coat that provides protection against many toxic chemicals, as well as digestion by lytic enzymes and being eaten by several types of predatory eukaryotes [1–4]. The spore coat is assembled moderately late in sporulation from components synthesized in the mother cell compartment of the sporulating cell, and comprises the outer layers of spores of many *Bacillus* species, although spores of some species contain an outermost exosporium. Spore coat structure and assembly have been best studied in the model spore former *Bacillus subtilis* and ~70 spore specific proteins have been identified in the spore coat [2,3,5,6]. In addition, a number of these coat proteins undergo covalent modifications including proteolytic cleavage, cross-linking, and tyrosine peroxidation.

The spore coat of *B. subtilis* has drawn attention not only because of its role in spore resistance but also because some coat proteins play significant roles in spore germination. However, much recent work on the spore coat has focused on determining overall spore coat structure as well as the mechanisms involved in the assembly of this large multi-molecular structure. Work to date has indicated that there are at least four coat layers that can be distinguished by electron microscopy (EM) as well as other means – undercoat, inner coat, outer coat, and an outermost glycoprotein layer called the crust [2,4,7,8]. Several of these individual layers also have sublayers, as the inner and outer coats have multiple lamellae. Most of the proteins in these various layers do not have specific roles in spore properties with the exception of a few coat enzymes, and most importantly, proteins that are essential for coat morphogenesis. The morphogenetic proteins include coat proteins such as CotE, CotH, CotO, SafA, and SpoVID, loss of any of which have drastic effects on overall coat architecture, as these proteins direct the assembly of different subsets of proteins into the coat [2,4,9–11]. In addition, the SpoIIID, GerE and GerR proteins have major effects on the expression of genes encoding coat proteins that are transcribed during sporulation, and this in turn has significant effects on coat properties and morphology [4–6].

A variety of studies of the functional repertoire of coat proteins have focused on the determination of the locations of these proteins in the spore coat and their specific roles in spore coat morphogenesis [5,7,12–15]. These studies have been extended and complemented by studies of direct interactions between various coat proteins, both *in vitro* and *in vivo* [4,16–20]. All of this work has given a picture of the molecular interactions in the spore coat, as well as the dependencies of the assembly of specific proteins into the coat. However, this type of analysis has not yet been complemented by detailed analysis of the structures of the various spore layers. Atomic force microscopy (AFM) has been used to unravel high-resolution structures of the coats of dormant and germinating spores of various *Bacillus* [14,21–30] and *Clostridium* [31] species. However, this analysis has generally been conducted on wild-type spores, with AFM data on only a few mutants lacking specific coat layers. Consequently, in this work we have used high-resolution AFM to analyze the surface structure of spores of wild-type *B. subtilis* spores as well as spores of a variety of mutant strains in order to reveal the surface morphology of various layers of the spore coat. The results from these analyses have provided high-resolution visualization of the various spore coat structures as well as several coat layers not reported previously. This information has allowed the formulation of a model for coat structure and provided further insight into the assembly of the spore coat.

Materials and Methods

Strains used in this study

The *B. subtilis* strains used in this study (Table 1) except one are isogenic with the wild-type strain PS832, a prototrophic derivative of strain 168. Preparation of strains by transformation with chromosomal DNA was as described [32].

Spore preparation

B. subtilis strains were grown at 37°C in Luria-Bertani (LB) [33] medium supplemented with the appropriate antibiotics when necessary. Chloramphenicol was used at a final concentration of 5 mg/liter, kanamycin at a final concentration of 10 mg/liter, and tetracycline at a final concentration of 10 mg/liter.

For spore preparation, *B. subtilis* strains were grown for 3 h in LB medium and then spread on 2× Schaeffer's-glucose medium agar plates without antibiotics [34]. Spores were harvested after incubation at 37°C for 5 d followed by incubation at room temperature for 2 d, and purified as described [34] by brief sonication and repeated washing with distilled water. All spore preparations, except for strain PS3735 (Δ*spoVID::kan*) (see below) were free (>98%) of vegetative and sporulating cells and germinated spores as determined by phase-contrast microscopy.

Spores of strain PS3735 (Δ*spoVID::kan*) were generally significantly contaminated with germinated spores and these germinated spores were removed by centrifugation in a one-step Histodenz™ (Sigma, St. Louis, MO) gradient. Four samples, each containing ~3 mg (dry weight) crude spores were suspended in 100 μl of 20% Histodenz™ that was layered on top of 2 ml of 50% Histodenz™ in four Ultra-Clear™ (11×34 mm) centrifuge tubes (Beckman Instruments, Palo Alto, CA) and then centrifuged at 14,000 rpm for 45 min at 20°C in a TLS 55 rotor. After centrifugation, the germinated spores in the supernatant fluid were removed, the pellets containing the dormant spores washed 5 times with 500 μl water and the final pellets were suspended in 500 μl water and combined. These purified spores were free (>98%) from vegetative and sporulating cells as well as germinated spores as determined by phase contrast microscopy.

Chemical decoating of spores

Spores (~6 mg dry weight) were decoated as described previously [35,36]. Briefly, spores were incubated for 90 min at 37°C in 1 ml of 50 mM Tris-HCl (pH 8.0)-8 M urea-10 mM EDTA-1% sodium dodecyl sulfate (SDS)-50 mM dithiothreitol (DTT). After incubation, the spores were centrifuged and the pellets were washed with 1 ml of water 6–10 times.

Atomic force microscopy

Droplets of ~2.0 μm of spore suspensions (~3×10⁹ spores/ml) were deposited on plastic cover slips and incubated for 10 min at room temperature and the sample substrate was carefully rinsed

---

Table 1. *B. subtilis* strains used in this study.

Strain	Genotype	Phenotype[a]	Source or reference[b]
PS832	wild-type		Laboratory stock
PS3394	Δ*cotE::tet*	Kan[r] Tet[r]	[1]
PS3735	Δ*spoVID::kan*	Kan[r]	[1]
PS3736	Δ*cotH::cat*	Cm[r]	[1]
PS3738	Δ*safA::tet*	Tet[r]	[1]
PS4133	Δ*cotB::cat*	Cm[r]	DL067→PS832
PS4134	Δ*cotO::tet*	Tet[r]	PE250→PS832
DL063	Δ*cotA::cat*	Cm[r]	[71]
DL067	Δ*cotB::cat*	Cm[r]	[71]
PE250	Δ*cotO::tet*	Tet[r]	[87]

[a]Abbreviations: Cm[r], chloramphenicol resistant; Kan[r], kanamycin resistant; Tet[r], tetracycline resistant.
[b]DNA from the strain to the left of the arrow was used to transform the strain to the right of the arrow.

with double-distilled water and allowed to dry. Our prior work with spores of other *Bacillus spp.* [22–25] demonstrated that spore morphological and structural attributes were reproduced both for spores analyzed within the same spore preparation and when multiple spore preparations were analyzed. Thus, in this study for each spore strain a single spore batch was analyzed by AFM with ~50–75 spores being imaged for each spore strain. Detailed experimental procedures for AFM imaging of spores were as described previously [22,24]. Images were collected using a Nanoscope IV atomic force microscope (Bruker Corporation, Santa Barbara, CA) operated in tapping mode. For rapid low-resolution analysis of spore samples, fast scanning AFM probes (DMASP Micro-Actuated, Bruker Corporation, Santa Barbara, CA) with resonance frequencies of ~210 kHz were utilized. For high-resolution imaging, SuperSharpSilicon (SSS) AFM probes (NanoWorld Inc, Neuchâtel, Switzerland) with tip radii <2 nm and resonance frequencies of ~300 kHz were used. Nanoscope software 5.30r3sr3 was used for acquisition and subsequent processing of AFM images. In order to successfully assess both overall low-resolution and high-resolution spore features, raw AFM images typically need to be modified. In particular, the *contrast enhancement* command, which runs a statistical differencing filter on the current image, was typically utilized. This filter can bring all the features of an image to the same height and equalize the contrast among them. This allows all features of an image to be seen simultaneously, and thus a single spore or a group of spores can be imaged at relatively low resolution while visualizing spore coat attributes at high resolution. Heights of spore surface features (i.e. folds, coat layers, etc.) were measured from *height* images using the *section* command, which allows measurements of vertical distance (height), horizontal distance, and the angle between two or more points on the surface. Tapping amplitude, phase and height images were collected simultaneously. Height images allow quantitative height determinations, providing precise measurements of spore surface topography. Amplitude and phase images do not provide height information. While amplitude and phase images provide similar morphological and structural information as do height images, they can often display a greater amount of structural detail and contrast compared with height images, often making them a preferred choice for presentation purposes. The surface roughness of spore surfaces for wild type and *cotE gerE* spores was evaluated as the root mean square (RMS) value R_q using AFM height images. R_q is the standard deviation of the Z values (height) within the given area and is calculated as: $\sqrt{\Sigma(Z_i-Z_{ave})^2/N}$, where Z_{ave} is the average of Z values within the given area, Z_i is the current Z value, and N is the number of points within the given area. R_q was determined for each spore from 4 μm^2 height images (pixel number ~512^2) of multiple spores using a 400 nm^2 zoomed in area in the center of the spore. In order to eliminate tilt on the spore surface, prior to the measurement of the roughness, the image was flattened using the third flatten order in the *flatten* command. Step roughness levels were determined by manually digitizing steps' contour from AFM capture images with a plot digitizer (http://plotdigitizer.sourceforge.net/). Once the x and y coordinates of the step contours were obtained, the step perimeter length, S, was estimated from the sum of all segment lengths given by $S = \Sigma\sqrt{(\Delta x^2+\Delta y^2)}$, where the sum is carried over all digitized contour segments. The sinuosity index, which is a measure of step meandering/roughness, is then calculated by taking the ratio of the contour length S over the shortest path length between the two end points of the step (straight line). Note, that the value of the sinuosity index ranges from 1 (case of straight line) to infinity (case of a closed loop, where the shortest path length is zero).

Results

Surface architecture of wild-type and decoated spore surfaces

As seen previously by AFM [21,22], the prominent surface features of air-dried wild-type *B. subtilis* spores are surface ridges extending along the long axis of the spore (Fig. 1a,b; light blue arrows). The height of these surface ridges was generally 15–30 nm, occasionally exceeding 40 nm. Similar surface ridges have been observed on spores of *Bacillus anthracis* [29], *Bacillus cereus* [22,23], *Bacillus atrophaeus* [22,24], *Bacillus thuringiensis* [22,23], and *Clostridium novyi* NT [31]. This ridge formation appears to be due to coat folding caused by changes in spore size upon dehydration [22,24,30,37,38]. RMS roughness R_q of wild-type coat surfaces measured as described in the Methods section for 20 spores varied between 3.49 nm to 8.71 nm with an average R_q value of ~5.26 nm.

AFM studies of protozoal-digested coat-defective *B. subtilis* spores [27] showed that the *B. subtilis* spore's outer surface exhibits a thin layer without prominent structural features, which was defined as an amorphous layer (Fig. 1c,e; green arrows). EM of ruthenium red stained *B. subtilis* spores demonstrated the presence of an outermost glycoprotein layer, and it was suggested that this layer is an exosporium that is tightly attached to the coat layer [8]. Later, a combination of EM, fluorescence microscopy, and genetic analysis also demonstrated the existence of this outermost glycoprotein layer that was named the spore crust [7]. Thus the outermost layer revealed by AFM and the crust layer correspond to the same spore layer. The thickness of the outermost amorphous layer in *B. subtilis* spores as measured from AFM images (Fig. 1e) was not uniform and varied between 4–15 nm with an RMS roughness R_q of ~3 nm. Typically, the coverage of surfaces of *B. subtilis* spores with the amorphous layer was not complete, revealing an underlying rodlet layer, seen on all visualized wild-type spores, with a periodicity of ~7–8.5 nm (Fig. 1c–e; red arrows); note that these rodlets are also seen on the surfaces of the surface ridges. Rodlet structures similar to ones seen in Fig. 1 were previously described in freeze-etching EM [39–41] and AFM studies of both fungal [42,43] and bacterial (*B. atrophaeus, B. cereus* and *B. thuringiensis*) [22–25] spores, with rodlet structures on *B. atrophaeus*, and *B. cereus* spores exhibiting ~8 nm periodicity. Note, that depending on sporulation conditions for *B. thuringiensis*, rodlet structures were found either on the spore coat or as extrasporal structures that were present in spore preparations [30].

In order to remove spores' outer coat, *B. subtilis* spores were chemically decoated with urea-SDS at slightly alkaline pH as described in Methods. The great majority of the proteins removed by this type of treatment have been well characterized in work from a number of laboratories [44]. This treatment partially or completely removed the amorphous layer, and the outer surface of the decoated spores was now comprised primarily of the intact rodlet layer (Fig. 2b–f; red arrows), which was covered in some cases with remnants of the amorphous layer (Fig. 2a–c; green arrows). The 15–30 nm surface ridges were also seen on the air-dried decoated spores, similar to what was seen on intact spores, and again these ridges appear to contain rodlets (Fig. 2a–d; light blue arrows).

Surface architecture of spores lacking CotA, CotB and SafA

CotA and CotB are two outer coat proteins that are likely localized on or very near the spore's outer surface [5,13,14]. Loss of either of these proteins has no notable effect on spore resistance

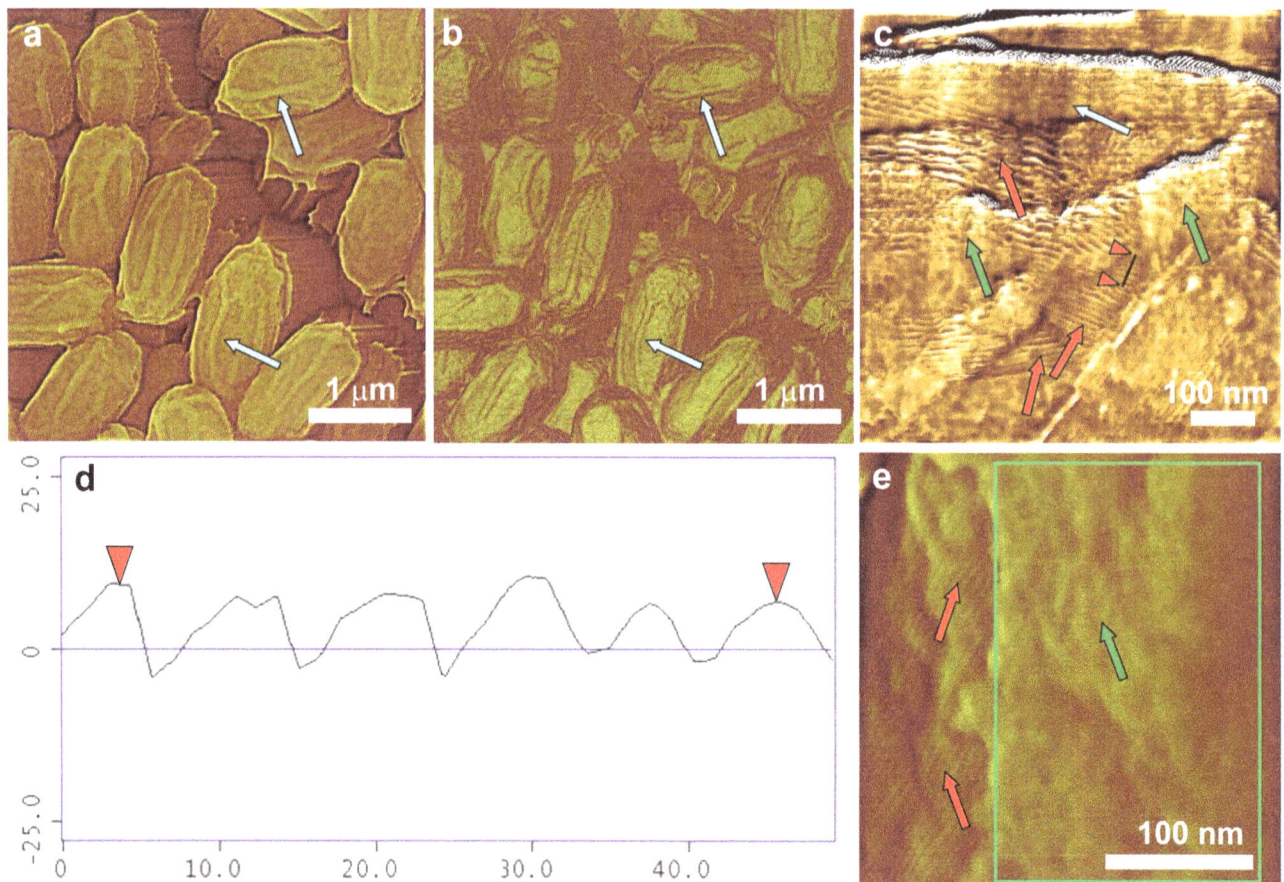

Figure 1. AFM images of *B. subtilis* **wild-type spores.** (a) Height and (b) phase images of spores with surface ridges (coincidental in both images) extending along the entire length of spores (several surface ridges noted by light blue arrows). (c) High-resolution height image of an area on a surface of a single spore showing surface ridges (light blue arrow), patches of an amorphous outermost layer (green arrows), and a rodlet layer (red arrows) seen beneath the amorphous layer. (d) A cross section line drawn perpendicular to rodlets (indicated with red arrows in (c)) showing a periodicity of ~8.2 nm. (e) High-resolution height image of an area on the surface of a single spore showing patches of an amorphous outermost layer (green arrow and green rectangle), and a rodlet layer (red arrows) seen beneath the amorphous layer.

properties or gross spore coat structure. We found that both *cotA* and *cotB* spore morphologies were indistinguishable from wild-type spores by AFM (Fig. 3), as all *cotA* and *cotB* spores were encased in the outermost amorphous and rodlet layers (Fig. 3c,d; green and red arrows, respectively) and exhibited 20–40 nm thick surface ridges (Fig. 3a–d; light blue arrows). The *cotA* and *cotB* spores also had an undulating surface topography from a subsurface layer (Fig. 3c,d; red circles) that was also seen in wild-type spores (data not shown).

In contrast to CotA and CotB, SafA plays a significant role in the assembly of at least some components of the spore's outer coat, and much of the coat in *safA* spores does not adhere tightly and can peel off [4,45]. We observed that the general surface morphology of *safA* spores as seen by AFM (Fig. 4) appears similar to that of wild-type, *cotA* and *cotB* spores, with amorphous and rodlet layers (Fig. 4c; green and red arrows, respectively) forming the outermost *safA* spores' coat layer. However, the degree of *safA* spore coat folding was different from that in wild-type spores. This resulted in the formation of surface ridges in *safA* spores (Fig. 4a–c; light blue arrows) that appeared shorter (e.g. not running along the whole spore surface as in Fig. 1a) and smaller (ridge heights of 10–20 nm) than in wild-type spores. Furthermore, some *safA* spores had no or minimal surface ridges (Fig. 4a; dark blue arrow), and ~25% of *safA* spores had an oversized spore coat

sacculus that appeared not to be firmly attached to the body of the spore itself (Fig. 4a,b; spores with adjacent green stars, and data not shown), consistent with previous work [44].

Surface architecture of spores lacking CotO and CotH

In addition to SafA, CotO and CotH also play significant roles in outer coat assembly, with perhaps some role in inner coat assembly as well [26,46]. As seen by AFM (Fig. 5), the outer surface of *cotO* spores was covered either completely or partially by a layer with a grainy appearance (Fig. 5b; brown arrow) and exhibited 15–40 nm thick ridges (Fig. 5a; light blue arrows). The thickness of the grainy layer was 8–20 nm as measured from the AFM images. High-resolution imaging of areas where the grainy structure density was low revealed that this layer actually has a fibrous structure, with the thickness of the thinnest fibers being ~2–4 nm (Fig. 5c; several fibers indicated with light yellow arrows). Thus, high densities of these fibrous structures appear to have assembled on the inner coat to form a layer that has a granular structure (Fig. 5b,c). Underneath the granular structure, multiple structural layers were observed (Fig. 5c; terraces of 3 consecutive layers numbered 1–3, and the edge of one terrace indicated by a purple arrow), and these terraces were decorated with "nanodot" particles (Fig. 5b,c; groups of nanodots indicated with black arrows, and a circle in 5c). While the heights of some

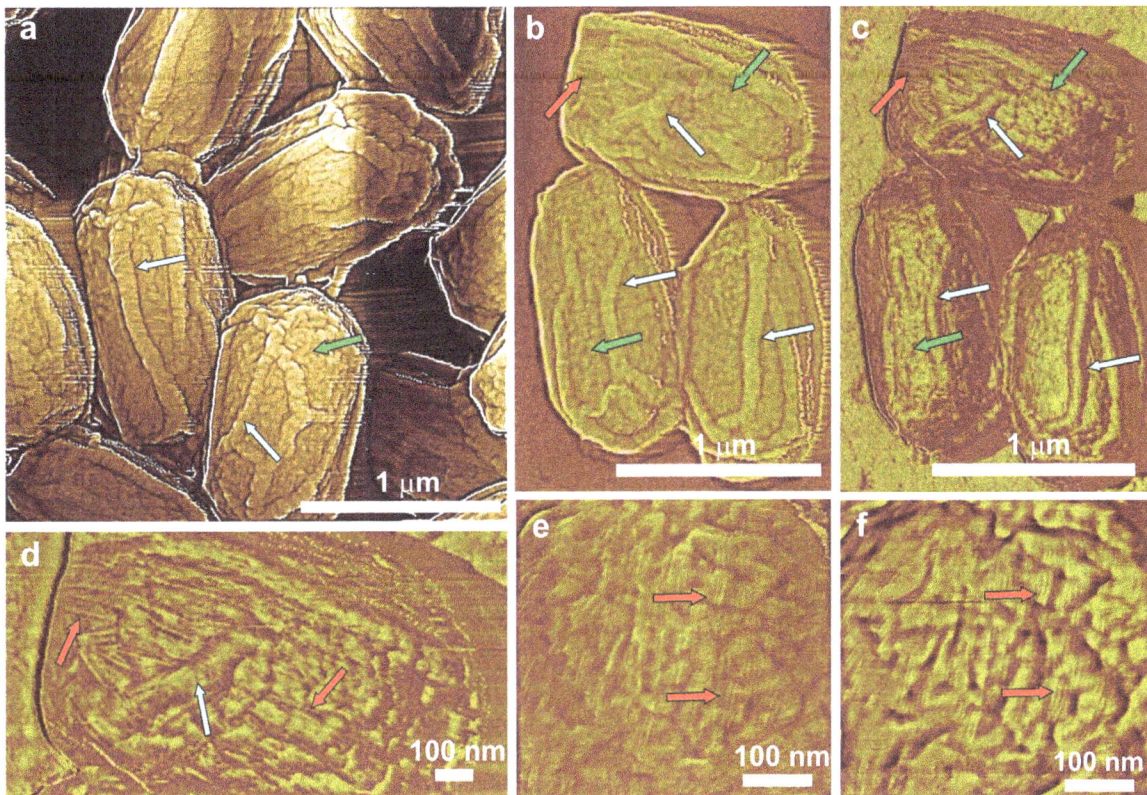

Figure 2. AFM images of decoated *B. subtilis* **wild-type spores.** Surface ridges extending along the entire length of spores are indicated with light blue arrows in height (a, b) and phase (c, d) images. Patches of rodlet structures are indicated with red arrows in (b–d). The green arrows in (a–c) indicate remnants of the amorphous outermost layer. High resolution height (e) and phase (f) images showing coincidental patches of rodlet structures denoted with red arrows.

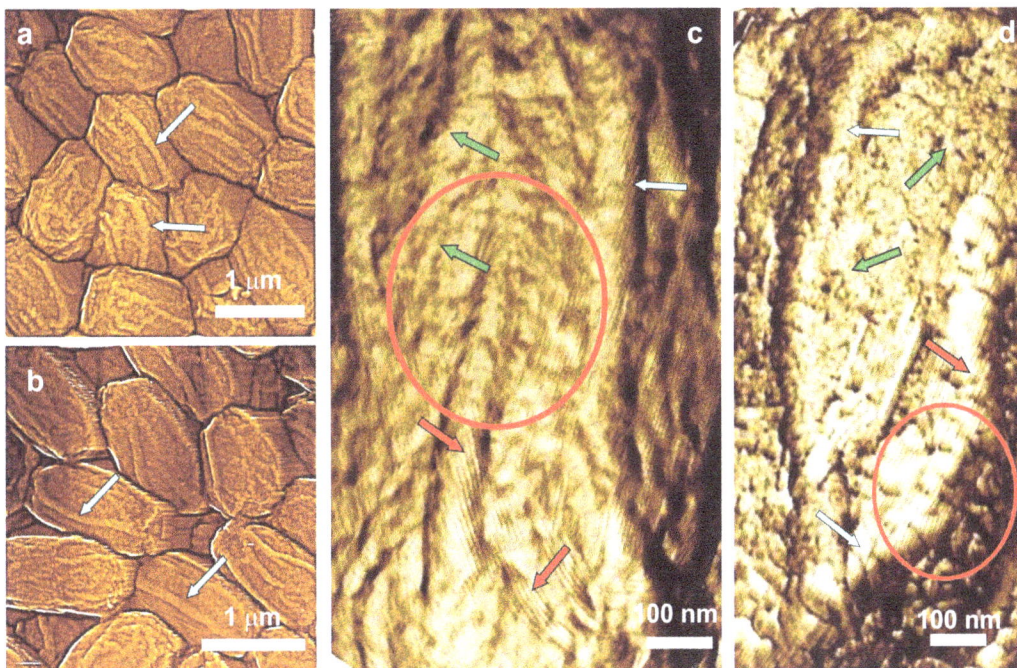

Figure 3. AFM images of *cotA* **and** *cotB* **spores.** Height images of *cotA* (a) and *cotB* (b) spores exhibit surface ridges similar to those in wild-type spores (light blue arrows). High-resolution phase images of single *cotA* (c) and *cotB* (d) spores show an irregular outermost amorphous layer (green arrows) as well as underlying rodlets (red arrows). In addition to the amorphous layer and rodlets seen on these spores' outermost surface, a strong undulating topography from a sub-surface layer is also present (red circles). Surface ridges in (c,d) are indicated with light blue arrows.

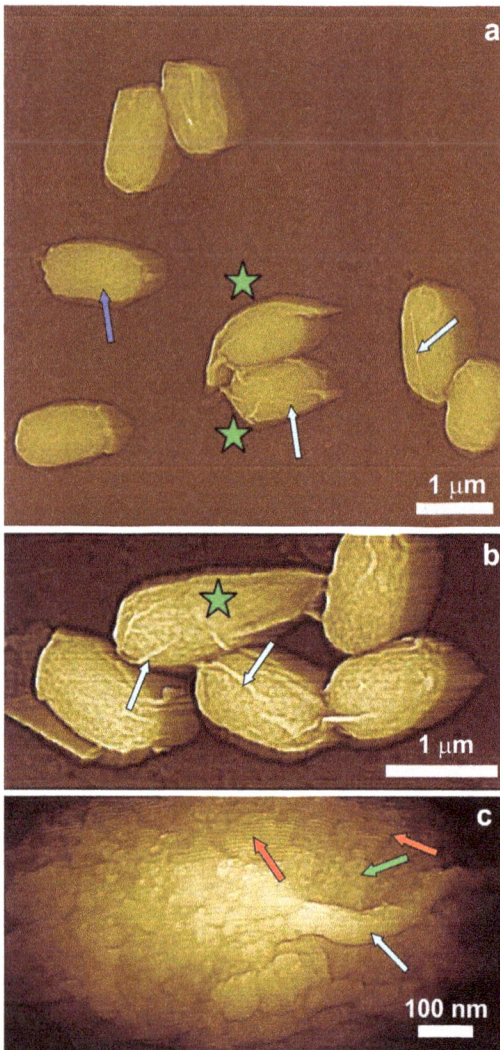

Figure 4. AFM height images of *safA* spores. (a–c) Several surface ridges are indicated with light blue arrows, and in (a) and (b) spores with an oversized sacculus are marked with adjacent green stars. A spore with at most minimal ridges is indicated with a dark blue arrow in (a). In panel (c), two patches of rodlet structure are indicated with red arrows, and a patch of an amorphous layer is indicated with a green arrow.

nanodots were as small as 3–4 nm, their typical height was 10–22 nm.

Significant numbers of *cotH* spores (Fig. 6) were also encased in the outermost amorphous and rodlet layers (Fig. 6b; green arrow and red arrows, respectively) and exhibited 15–40 nm thick surface ridges (Fig. 6a,b; light blue arrows). However, 10–15% of *cotH* spores were partially (Fig. 6b) or completely (Fig. 6c) devoid of outer spore coat layers. These *cotH* spores with a defective outer coat exhibited multilayer structures (Fig. 6c; two layers indicated with purple arrows) similar to ones observed on *cotO* spores (Fig. 5b,c). As seen in Fig. 6b,c, these layers again exhibited high densities of nanodots (Fig. 6b; one group of nanodots indicated with a black arrow) similar to ones seen on *cotO* spores (Fig. 5). Nanodot heights appeared smaller and more uniform on *cotH* spores compared with those on *cotO* spores, varying between 2.5–3.5 nm, and none of the *cotH* spores exhibited the fibrous/granular structural layer observed on *cotO* spores.

Surface architecture of *cotE*, *gerE* and *cotE gerE* spores

CotE is one of the major morphogenetic proteins in spore coat assembly, and *cotE* spores lack an outer coat and also have alterations in the inner coat layer [2]. AFM images (Fig. 7) revealed that the outermost surface of *cotE* spores is also a multilayer structure, composed of ~6 nm thick smooth layers (Fig. 7c; three consecutive layers marked as 1–3). These structures are identical to ones observed for *cotO* and *cotH* spores (Fig. 5,6). Note also that: i) the surface of *cotE* spores was devoid of nanodots; ii) the vast majority of *cotE* spores appeared to lack the outermost amorphous and rodlet layers; and iii) *cotE* spores exhibited no granular/fibrous surface structures.

In contrast to wild-type spores, 20–25% of *cotE* spores had no surfaces ridges (Fig. 7a; dark blue arrow) or shorter, thinner ridges (Fig. 7a; light blue arrows) that did not extend across the entire spore surface. The thickness of surface ridges that were seen were only 5–15 nm, less than for surface ridges on intact and decoated wild-type spores (Fig. 1,2). The other interesting morphological feature observed on many *cotE* spores was an oversized spore coat sacculus (Fig. 7a,b; green stars). This was also seen on some *safA* spores (Fig. 4a,b; adjacent green stars), while the wild-type spore coat was always tightly fitted (Fig. 1).

The multilayer outer structure of *cotE* spores (Fig. 7b,c) exhibited step growth patterns similar to those observed on surfaces of inorganic [47,48] and macromolecular [49–52] crystals. An example of similar structures observed on the surface of a growing trypsin crystal [53] is shown in Fig. 7d. Similar patterns were also observed for the inner coat of *C. novyi* NT [31] and *B. anthracis* spores (Plomp and Malkin, unpublished data). As seen in Fig. 7c, layers of structure forming the inner coat of *B. subtilis* spores are similar in morphology to the surface of trypsin crystals (Fig. 7d), with both showing rough steps with many kinks and a number of 5–10 nm (Fig. 7c) and 70–90 nm (Fig. 7d) wide holes (Fig. 7c,d; purple arrows and circles, respectively). The sinuosity index, which is a further measure of the step roughness (see Methods), was estimated for steps on surfaces of *cotE* spores (Fig. 7c) and trypsin crystals (Fig. 7d) as 3.84 and 1.49 respectively. Note that high-resolution AFM observations, which allow at least 1 nm resolution for macromolecular crystalline layers [50,54], do not result in molecular scale visualization of the molecular packing within the spore coat layers.

While most *cotE* spores are encased only in the multilayer coat structure, <5% spores were completely covered by a rodlet layer (Fig. 8a,b; red arrow), with a periodicity of ~7.2 nm (Fig. 8c; insert). Occasionally, as seen in Fig. 8a,b 4–10 nm thick patches of the outermost amorphous layer were observed atop the rodlet layer of <10% of *cotE* spores (Fig. 8a,b; green arrows). In addition, on <5% of *cotE* spores a honeycomb-like coat layer with a periodicity of ~8–9 nm was observed atop the inner coat layer (Fig. 9a,b, orange arrow). Note, that loose honeycomb layers with remnants of rodlet structures on top of a honeycomb layer (Fig. 9c, orange and red arrows respectively), were occasionally observed in spore preparations. While >75% of *cotE* spores lacked a complete rodlet layer, these spores still exhibited patches of rodlet structure of different sizes assembled atop the inner spore coat layer (Fig. 9a; red arrows).

In contrast to CotE and other proteins noted above, GerE is a transcription factor that modulates the expression of some coat protein genes late in sporulation, including genes that encode proteins in the insoluble fraction of the spore coat [6]. A *gerE* mutation has drastic effects on overall spore coat structure, as: i) much of the *gerE* spores' coat adheres poorly [55]; and ii) some coat component(s) responsible for the strong X-ray scattering by the spore coat is either absent or misassembled on *gerE* spores,

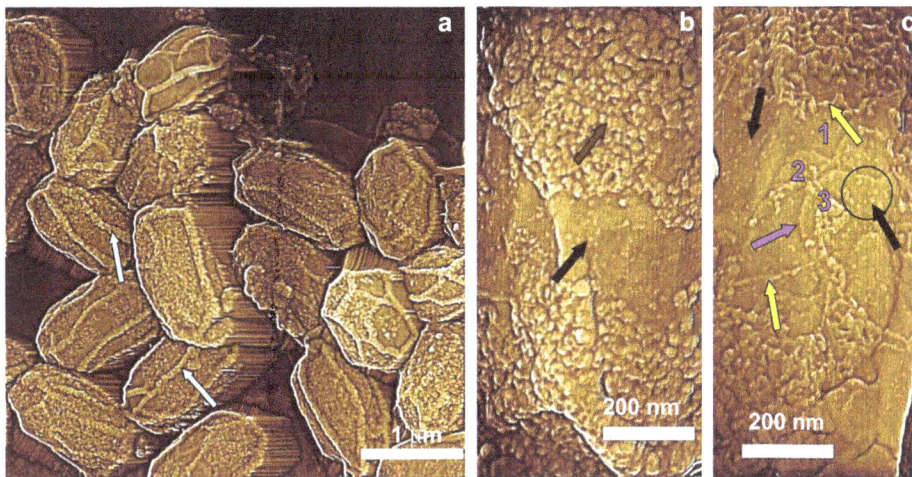

Figure 5. AFM images of *cotO* **spores.** (a) Height image of spores with surface ridges extending along the entire length of spores (light blue arrows). (b,c) High-resolution height images of areas on surfaces of single spores showing a dense fiber structure forming a granular structure (b; brown arrow) and individual fibers (c; light yellow arrows). In panel (c), three layers (terraces) of inner coat structure are numbered 1, 2, and 3 in purple. Step edges representing boundaries of each layer (one marked with a purple arrow) are visible. In panels (b) and (c), nanodots are marked with black arrows and one area with a high density of nanodots is circled in panel (c).

while this X-ray scattering is observed from *cotE* spores [56]. As seen by AFM (Fig. 10), *gerE* spores were devoid of the outer amorphous and rodlet layers, and fibrous structures. Most of these spores were only partially or completely covered by patches of irregular material (Fig. 10a; black stars; Fig. 10b), and 40–45% of *gerE* spores had only patches of this material (Fig. 10a; grey stars; Fig. 10c; grey arrow), with a thin layer of material covering the spore surface (Fig. 10b,c). The thickness of these patches of coat material was ~6 nm, a value similar to the thickness of the inner coat layers forming the multilayered coat structure (Fig. 5b, 7d).

The combination of *cotE* and *gerE* mutations has an even more drastic effect on spore coat structure than either mutation alone, as *cotE gerE* spores are almost completely devoid of a coat (Fig. 11),

except for a thin rind of insoluble material [28]. As reported previously [28], with the exception of small numbers of spores which have remnants of coat material (Fig. 11b; grey arrow), >90% of *cotE gerE* spores had none of the spore coat structures described above and their outer surface appeared rather smooth (Fig. 11a,b), although high-resolution imaging revealed a slightly bumpy textured outermost surface (Fig. 11c). The RMS roughness R_q of *cotE gerE* spore surfaces measured for 20 spores as described in the Methods section varied between 0.25 nm to 0.49 nm with average R_q value of ~0.38 nm. These severely coat-defective spores also appeared less rigid than intact spores, as *cotE gerE* spores within a closely packed monolayer were more deformed compared to ones with fewer near neighbors (Fig. 11a). Approx-

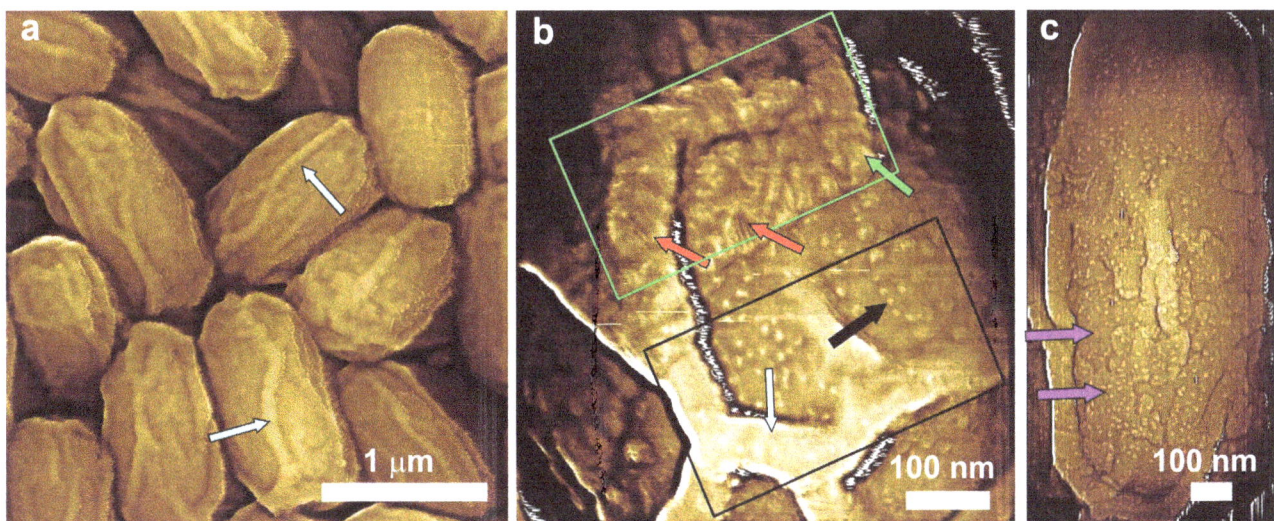

Figure 6. AFM images of *cotH* **spores.** (a) Height image of spores with surface ridges extending along the entire length of spores (light blue arrows). (b) High-resolution height image of a spore surface area showing the upper surface area (green rectangle) covered with an amorphous layer (green arrow) and rodlets (red arrows). The lower part of the outermost layer-free area (black rectangle) is covered with nanodots (black arrow). One of the surface ridges in (b) is indicated with a light blue arrow. In panel (c), a two–layer inner coat structure (two purple arrows noting the two layers) decorated with nanodots can be seen.

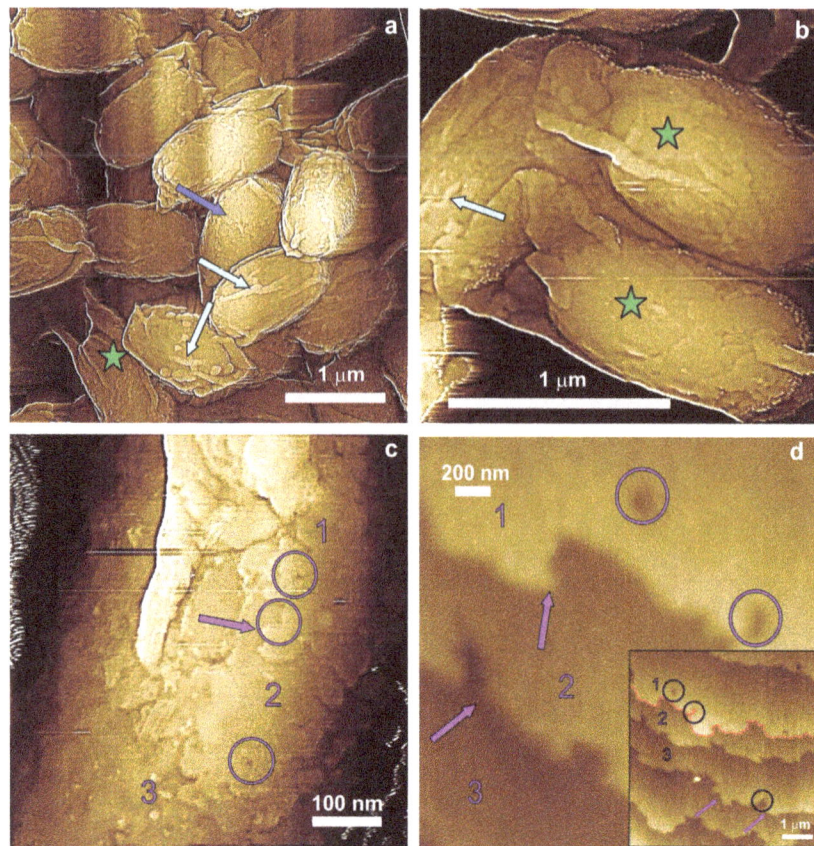

Figure 7. AFM images of *cotE* spores. (a,b) Height images of spores that exhibit surface ridges (light blue arrows), and several spores with an oversize sacculus are labeled with green stars. In (a) a spore with no apparent ridges is indicated with a dark blue arrow. (c) Height image of a multilayer inner coat structure. Three layers are indicated with numbers, and a kink on a step edge is marked with a purple arrow. Several holes in the layered structure are also indicated with purple circles. The hole in the middle circle corresponds to a pinning point on the step. (d) Height image of a multilayer layer structure similar to ones seen in Figs. 5b,c, 6c, and 7c, as seen on the surface of a trypsin crystal. Similar to the spore coat layers in (c), three layers, kinks and several holes are indicated with purple numbers, arrows and circles, respectively. The insert in (d) is a larger area of the crystal surface seen in (d). The same holes and three layers seen in (d) are indicated in the insert. The red line in (d) denotes the step contour, which was utilized for the measurement of the sinuosity index. Panel (d) is reprinted with permission from Plomp M, McPherson A, Larson SB, Malkin AJ (2001). Growth mechanisms and kinetics of trypsin crystallization. J Phys Chem B 105: 542–551. [52]. © (2001) American Chemical Society.

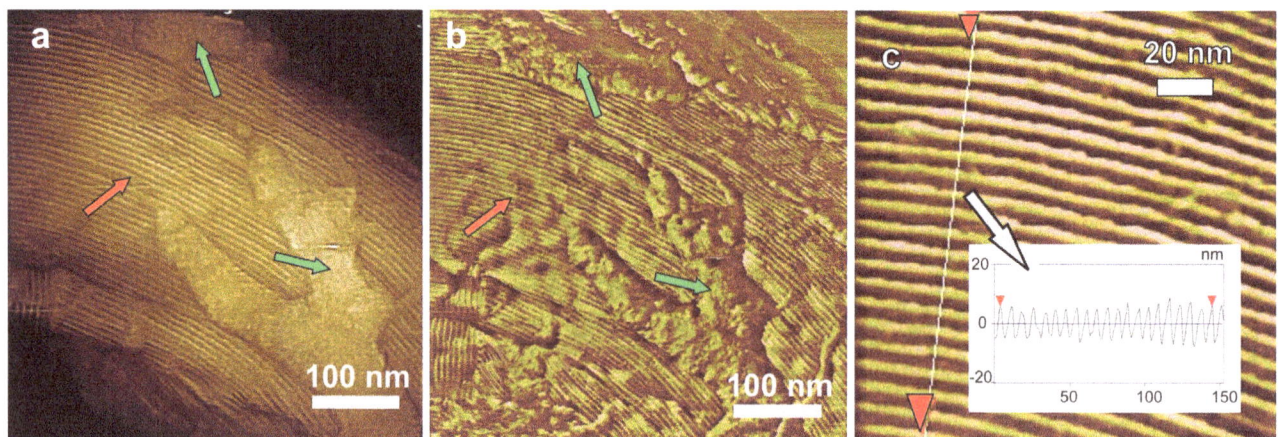

Figure 8. AFM images of *cotE* spores. (a) High-resolution height (a) and phase (b) images of the spore surface showing (coincidental in both images) a rodlet structure (red arrows) covered with patches of an amorphous layer (green arrows). (c) High-resolution height image of the spore surface with an insert with a cross section line drawn perpendicular to rodlets showing the periodicity of ~7.2 nm.

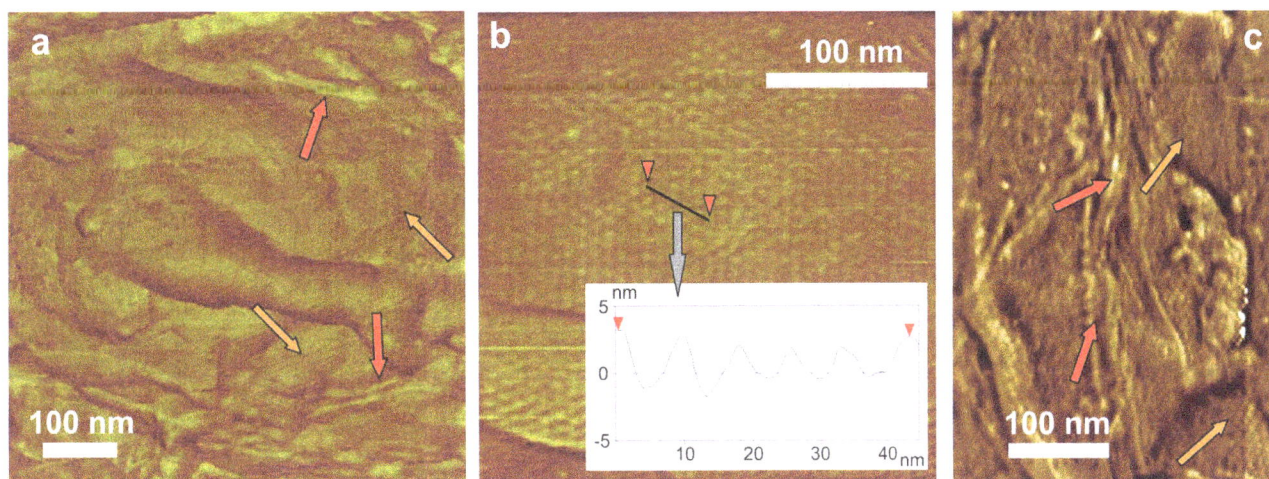

Figure 9. AFM images of *cotE* spores. (a,b) High-resolution height images of the spore surface showing a honeycomb structure (orange arrows in (a)) and patches of rodlets on top seen in (a) (red arrows). The insert in (b) is a cross section line along a honeycomb structure (indicated with a black line and red arrows in (b)) showing a periodicity of ~8.5 nm. (c) A portion of a loose honeycomb layer (orange arrows) with remnants of rodlet structures (red arrows), which were seen in *cotE* spore preparations.

imately 7% of *cotE gerE* spores also exhibited 80–100 nm wide and 30–40 nm deep depressions (Fig. 11a; black circles), which were also observed on some *gerE* spores (data not shown). Note, that neither *gerE* nor *cotE gerE* spores exhibited surface ridges (Fig. 10,11).

Surface architecture of *spoVID* spores

SpoVID is another major morphogenetic protein in spore coat assembly. This protein is essential for the adherence and assembly of the coat, and while the peptidoglycan cortex forms relatively normally in *spoVID* spores, the coat largely assembles as swirls in the cytoplasm, giving rise to spores with little coat material [2,10]. Consequently, the surface architecture of *spoVID* spores is drastically different from that of wild-type spores, as a number of *spoVID* spores were again encased in only loosely fitted coat sacculi (Fig. 12a; green stars). Indeed, for a number of *spoVID* spores, the coat sacculi were partially (Fig. 12 c,d; grey stars) or completely (Fig. 12b,d; white stars) sloughed off, releasing empty

sacculi (Fig. 12a, insert; dark blue star) and leaving spores encased in what appeared at lower resolution to be a rather smooth structure (Fig. 12a). Note that the shape of a number of the coatless *spoVID* spores was altered significantly compared either to other mutant spores described above or to *spoVID* spores still encased in coat sacculi. The shape of the coatless *spoVID* spores also varied significantly (Fig. 12b; spores with white stars), sometimes having a shape resembling a bowling pin.

The outer and internal surface structures of the coat sacculi released from *spoVID* spores were similar to the outermost surface structure of wild-type spores, as seen in a high-resolution image of a *spoVID* spore sacculus (Fig. 13a), and consisted of rodlet layers (Fig. 13a; red arrows) covered with amorphous material (Fig. 13a; green arrows). As illustrated in Fig. 13b, high-resolution images of the surfaces of the coatless *spoVID* spores revealed a 2–6 nm thick amorphous layer (Fig. 13b; grey arrow) and an underlying pitted surface structure (pink arrow).

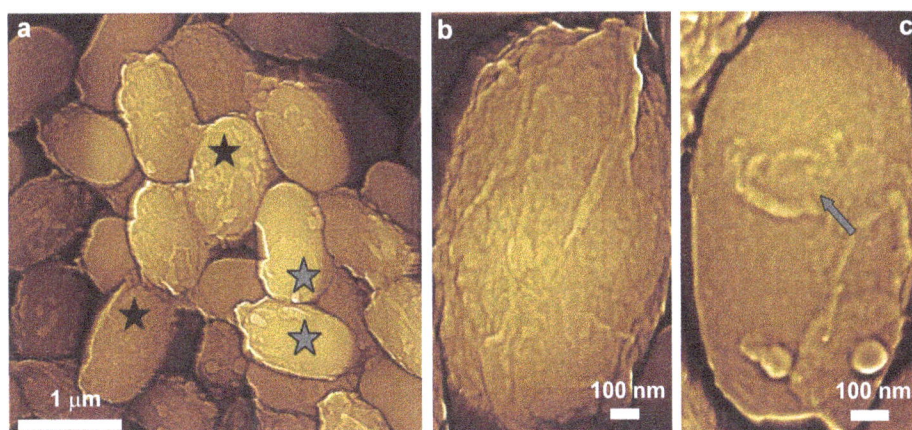

Figure 10. AFM height images of *gerE* spores. (a) *gerE* spores are either completely (black stars) or partially (grey stars) covered with coat material. (b) A spore that is completely encased in the coat material, and (c) a spore with patches of coat material (grey arrow).

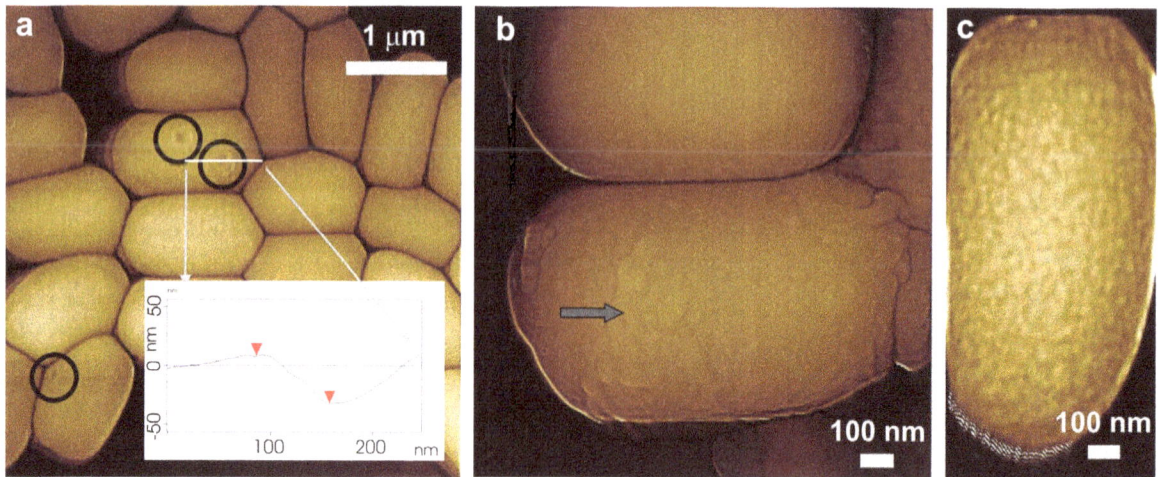

Figure 11. AFM height images of *cotE gerE* spores. (a) Spores which appeared to be devoid of spore coat material. Closely packed spores are more deformed than ones that are not surrounded by other spores. Some spores exhibit 80–100 nm wide and 30–40 nm deep depressions (black circles). The insert in (a) is a cross section line (indicated with a white line) drawn across the ~100 nm wide depression showing a depth of ~40 nm. (b) Image showing small patches of coat material (grey arrow) on the spore surface. (c) High-resolution image of a spore devoid of any obvious coat material, and showing a textured outermost surface.

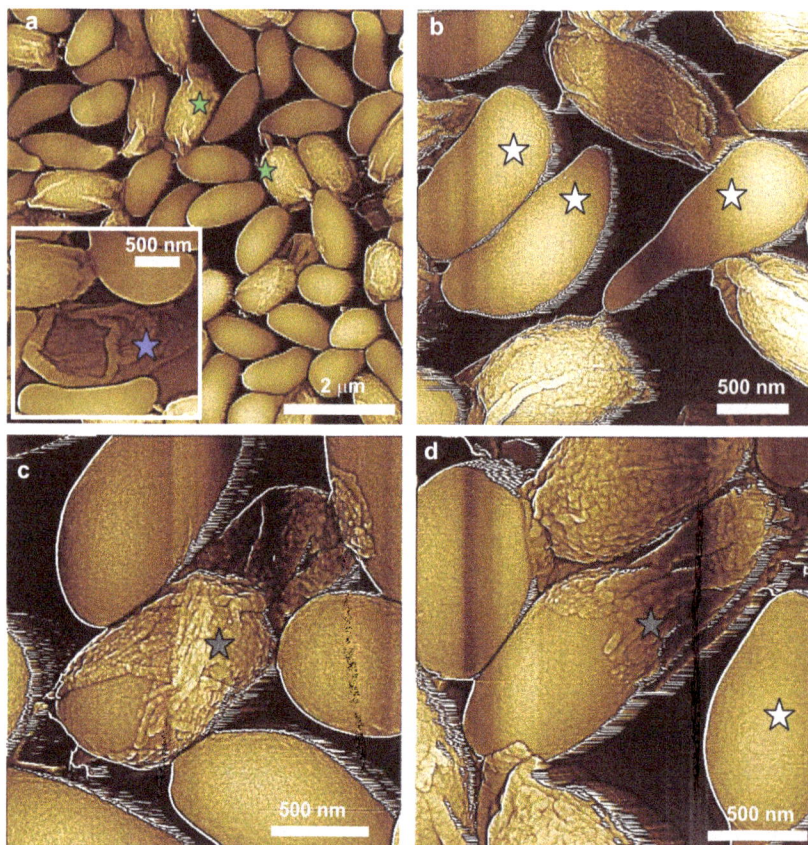

Figure 12. AFM height images of *sspoVID* spores. (a) Many of the *spoVID* spores are devoid of obvious spore coat material, although some *spoVID* spores are encased in loosely fitting coat sacculi (green stars); insert: an empty intact sacculus (blue star) present in a spore preparation. (b,d) Severely deformed spores without any visible coat material are indicated with white stars. (c, d) Spores with partially sloughed off coat sacculii are indicated with grey stars.

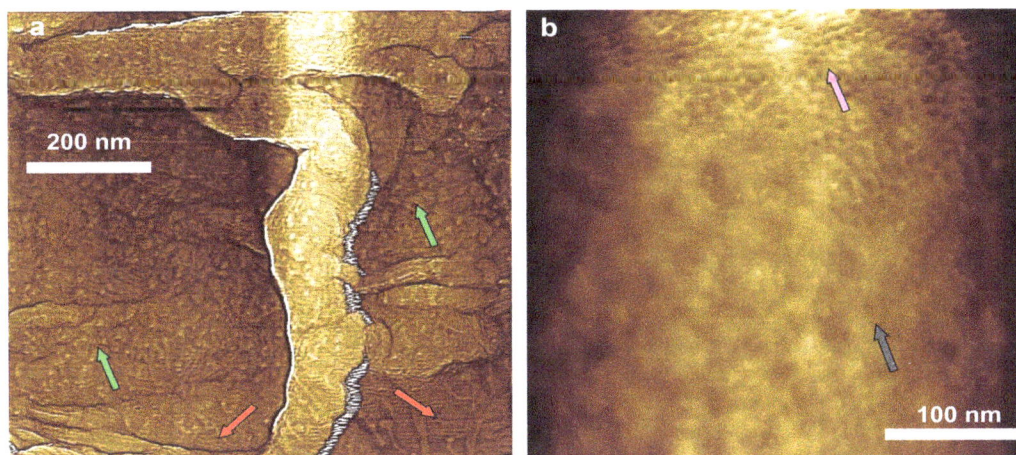

Figure 13. High-resolution AFM height images of *spoVID* spores. (a) External and internal surfaces of the empty coat sacculus in Fig. 12a, insert exhibit morphology similar to that of the outermost wild-type spore layer seen in Fig. 1b. The surface is comprised of rodlets (red arrows) and patches of amorphous material (green arrows). In (b) a pitted layer (pink arrow) is seen beneath a layer of coat material (grey arrow).

Discussion

Topography of the outer spore surface

The ridges on the surfaces of air-dried *B. subtilis* spores (Fig. 1a) have been seen previously in EM [39,40] and AFM [22–24,29,30] studies of spores of various *Bacillus* species. These ridges have been proposed to form due to the folding of the coat in response to dehydration, likely as a consequence of decreases in spores' internal volume [22,37,38]. Indeed, AFM measurements of the morphology of fully hydrated and air-dried spores demonstrate that surface ridges on dehydrated spores mostly disappear or decrease in size upon hydration [22]. Thus, spore coat flexibility can compensate for decreases in spore surface area upon drying by surface folding and ridge formation [22]. Current work demonstrated that this surface folding takes place in the spore coat, since dry *gerE*, *cotE gerE* and *spoVID* spores lacking much of the spore coat exhibit no surface ridges (Fig. 10–12). Note, that while *gerE* and *cotE gerE* spores as well as coat rinds produced by protozoal

digestion of spores exhibit some coat material [1,28] (Fig. 9,10), the amount of this material is either not sufficient or its proteomic composition is not appropriate to form surface ridges. In contrast, the presence of surface ridges on *cotO*, *cotH* and *cotE* spores lacking the amorphous and rodlet layers (Fig. 5–7) indicates again that surface ridge formation takes place within the multilayer spore coat structure.

The spore coats of *B. subtilis* mutants lacking morphogenetic coat proteins as well as chemically decoated wild-type spores have different thickness and composition, and these variations could affect the coat's elastic properties and thus change its folding and surface ridges. However, affecting the outer spore coat architecture by mutation or chemical decoating gave no large changes in spore surface ridge parameters or patterns. These results suggest that formation of spore surface ridges originates within multilayer coat structures, which are relatively unaffected by loss of some coat proteins, with the amorphous, rodlet and fibrous layers only following the ridge-associated topography. Our data showing pronounced changes in the surface folding of *safA* spores (Fig. 4) may indicate that these spores' multilayer coat structure is either thinner or more flexible than in wild-type, decoated wild-type, *cotA*, *cotB*, *cotH*, and *cotO* spores (Fig. 1–3,5,6). However, *cotE* spore coats with lower levels of surface folding typically have the same number of layers as do *cotH* and *cotO* spore coats. Perhaps the decreased surface folding of *cotE* spore coats is due to changes in the elastic properties of inner coat layers because one or more inner coat proteins are not assembled in *cotE* spores. Note, that the wide range of surface ridge parameters and folding patterns observed with spores of different species [22–24] and isogenic strains (this work) makes it problematic to assign these parameters as spore species-specific structural attributes.

Spore coat architecture

In addition to providing information on spore surface topography, AFM images allowed construction of a detailed model of coat architecture (Fig. 14). In this model starting from the outside, the coat consists of an amorphous layer/crust, a rodlet layer, a honeycomb layer, a fibrous layer, a nanodot particle layer, the multilayer assembly, and the undercoat/basement layer just above the pitted surface, which we tentatively assign as the spore cortex.

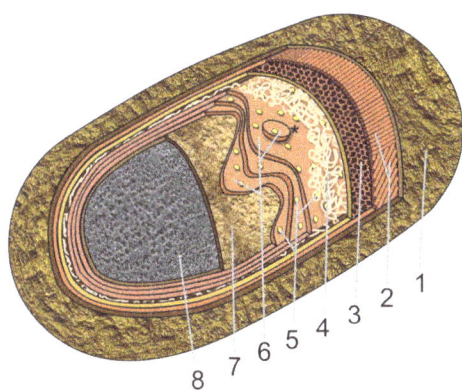

Figure 14. Model of the spore coat architecture of a single *B. subtilis* spore. The layers of the spore coat and the cortex are depicted as: (1) an outermost amorphous layer (the crust); (2) the rodlet layer; (3) the honeycomb layer; (4) the fibrous/granular layer, (5) the nanodot layer on top of a multilayer structure (6) ((with a 2D nucleus (indicated with *) seen on the upper layer)); and the basement layer (7), which is on the top of the cortex's outer pitted surface (8). Structural features of spore coat layers are not shown to scale.

The existence of an outermost tightly fitting spore layer was initially reported in thin section EM images of *B. subtilis* spores treated with a reducing agent [57]. It was suggested that this layer is an exosporium-like structure, which in EM images of untreated spores is usually indistinguishable from the darkly stained outer spore coat. While the amorphous spore coat layer reported here (Fig. 1) could correspond to this outer coat structure, it does not resemble an exosporium, as this outer layer has no paracrystalline basal layer typical of the exosporium of spores of the *B. cereus* group [41,58]. Rather the amorphous layer likely corresponds to the outer crust layer of *B. subtilis* spores that stains with ruthenium red and is glycoprotein-rich [7,8]. Patches of an outermost amorphous layer are also observed in AFM studies of *B. atrophaeus* spores (Plomp and Malkin, unpublished data).

Rodlet structures, similar to ones seen in Fig. 1b,c, were previously described on the outer surface of a diverse set of microorganisms (see [59]), including Gram-negative bacteria and various fungi. The fungal rodlet layers were resistant to treatment by detergents, organic solvents, enzymes, alkali and mild acids [60,61], and the structural proteins hydrophobins [62,63] and chaplins [64] were integral components of fungal rodlet structures. In several cases, these rodlets had a cross β-structure similar to that in the amyloid fibrils [64] associated with several neurodegenerative diseases [65]. The amyloid-like rodlet fibrils forming microbial outer surface layers appear to play important roles in attachment, dispersal and pathogenesis [59].

Rodlet structures were also reported in EM [38–40] and AFM studies [22–24,30] of spores of several *Bacillus* species, although the proteins that form these rodlet structures are not known. The structural similarities between *B. atrophaeus* rodlets seen during their germination-induced disassembly [25] and amyloid rodlets found on surfaces of fungi and bacteria suggest that *B. subtilis* coat rodlets may also be amyloids. However, full understanding of the function of rodlet structures in spores awaits elucidation. Interestingly, micro-etch pits form in the rodlet layer early in spore germination [25], and these could facilitate access of degradative enzymes to their targets in an otherwise tightly packed coat. Characterization of the strength and mechanical stiffness of individual amyloid fibrils of insulin reveals that these parameters are similar to those of steel and silk [66]. Thus, the spore coat' rodlet layer could play a role in protecting spores from mechanical stress, and a combination of rodlet and amorphous structures could provide spores with a wide range of physicochemical properties. Indeed, the existence of both hydrophobic (rodlets) and hydrophilic (glycoproteins) structures on the outermost layer might enable spores' successful dissemination as both as air-born and fully hydrated particles.

Current results indicate that the assembly of the outermost coat layer does not require CotA and CotB, two of the most abundant outer coat proteins [2,14,67–71]. While interactions between CotB and CotG are critical in guiding assembly of the outer coat layer, no coat assembly defect has been observed in *cotA* or *cotB* mutants [69,70]. In addition, *cotA* and *cotB* mutations have no effects on spore lysozyme resistance, germination [2] or surface appearance (Fig. 3), and CotB is absent from *cotH* spores [72] yet the outermost coat structure of CotH spores is similar to that of wild-type spores (Fig. 6), including both the rodlet and amorphous layers. Thus, neither CotA nor CotB appear to play important roles in directing the assembly of spores' outer layers. The similar surface ridges on *cotA* and *cotB* spores further suggests that loss of these proteins does not significantly alter the elastic properties of the spore coat.

Loss of SafA also does not affect the high-resolution architecture of spores' amorphous and rodlet layers (Fig. 4), consistent with

safA spores' lysozyme resistance [2,45]. However, SafA plays an important role in coat assembly, as in many *safA* spores the coat is loosely attached to the spore (Fig. 4). SafA is localized in the spore cortex near the inner coat, and SafA may help associate the spore cortex and coat [2,73]. The absence of surface ridges on a large portion of *safA* spores, along with relatively thinner existing surface ridges, also suggest that *safA* spores' coat is thinner and/or more flexible. This is consistent with EM analyses that indicate the *safA* spore coat often has 1–2 layers instead of the typical 3–5 layers [45].

Inner coat structures

In *B. cereus* [22] and *B. atrophaeus* spores [25] the coat's rodlet layer is underlain by a honeycomb structure also observed in *B. subtilis* spores (Fig. 9). Since disordered microporous inorganic substrates can effectively initiate three-dimensional protein crystallization [74], perhaps the spore coat's honeycomb structure represents a biological example of a microporous matrix that facilitates the ordered self-assembly of the coat's rodlet structure. Note that the 8–9 nm periodicity of the *B. subtilis* coat honeycomb layer is similar to periodicities of honeycomb structures for *B. cereus*, *B. thuringiensis* [22] and *C. novyi* NT [31] spores. This indicates that molecular dimensions of proteins forming these honeycomb structures are similar for different bacterial species, and the molecular composition of the honeycomb layer in different bacterial species may thus be similar.

Studies of *cotO*, *cotH* and *cotE* spores revealed consecutive structural layers of granular/fibrous material (*cotO* spores; Fig. 5), nanodots (*cotO* and *cotH* spores; Fig. 5, 6), and multilayer structures (*cotO*, *cotH* and *cotE* spores; Fig. 5–7). While spores of these mutants had the multilayer structure, only *cotO* spores retained the granular/fibrous structure and *cotE* spores lacked the nanodot layer. We propose that the granular/fibrous layer represents an outer spore coat layer that appears as a darkly stained irregular layer in EM images [69]. The thickness of this outer coat layer varies significantly in EM images on both the same spore and between spores, consistent with the range of granular/fibrous layer thickness observed on *cotO* spores.

On wild-type spores and spores of some mutants lacking specific coat proteins (i.e. *cotA* and *cotB*), the grainy/fibrous outer coat layer was largely obscured by the rodlet and amorphous layers. However, the force exerted by the AFM probe tip on the outermost spore layer allows visualization of underlying structures, as in the AFM visualization of a cytoskeleton beneath a cellular plasma membrane [75]. AFM phase imaging can probe micromechanical properties of sample materials (e.g. viscoelasticity) [76] and map surface inhomogeneity of these properties. Furthermore, when mechanical properties of two layers are significantly different, phase imaging can provide structural information on layers beneath the topmost layer [77]. Thus an irregular grainy layer can often be seen in AFM phase images beneath the outer rodlet structure (Fig. 3c,d), and we suggest that this underlying layer corresponds to a grainy/fibrous outer coat layer (Fig. 5, 414). Note, that an undulating surface morphology similar to that seen on *cotA* and *cotB* spores (Fig. 3c,d) was also observed on the surface of wild-type spores (data not shown).

Typically, multilayer structures on *cotO*, *cotH* and *cotE* spores contained 3–5 layers, consistent with the appearance of the lightly staining lamellar inner coat of *B. subtilis* spores seen by EM [68], and thus these multilayer structures may correspond to the *B. subtilis* spores' inner coat. We further suggest that the nanodots between the outer and inner coat layers but absent on *cotE* spores, might be CotE molecules that facilitate the assembly of the grainy/granular outer coat layer. The height of the smallest nanodots seen on *cotH* spores was ~3 nm, consistent with CotE's mol wt of 20.9 kDa [2], and this suggestion is consistent with current models of *B. subtilis* spore coat assembly that have CotE positioned between the inner and outer coat layers [2,26]. However, these nanodots could also be small coat protein aggregates, and further experiments, perhaps using AFM-based immunolabeling techniques (29), will be needed to identify the protein(s) forming the nanodots.

The *cotO* spores have no amorphous or rodlet layers, which could explain the partial lysozyme sensitivity of *cotO* spores [26]. However, the presence of these outer layers on the majority of *cotH* spores (Fig. 6) is consistent with their relatively normal lysozyme resistance [1]. The outer coat of *cotO* spores often appears disorganized and missing in EM thin sections [26] and is generally indistinguishable from that of *cotH* spores. CotO and CotH are suggested to be localized below the coat surface [13,26] and to participate in a late phase of coat assembly. However, our AFM analyses showed pronounced differences between *cotO* and *cotH* spore coats. In particular, CotO plays a critical role in the assembly of the amorphous and rodlet layers, while assembly of the fibrous outer coat requires CotH and CotE. AFM studies also indicated that these proteins play a role in assembly of the coat's amorphous and rodlet layers, consistent with biochemical, genetic and EM studies [26,46,72]. It has been suggested [26] that CotO and CotH also play an important role in inhibiting the tendency of outer coat protein layers to stack up resulting in the polymerization of the coat layers into closed shells. However, AFM demonstrates that *cotO* and *cotH* spore coats self-assemble to form contiguous shells rather than disorganized coats. At the same time, many *cotE* spores exhibited only a loose coat sacculus (Fig. 7), indicating that CotE plays an important role in the assembly of the inner coat and/or its attachment to the cortex as noted above.

The crucial role for CotE and GerE in proper coat assembly was further highlighted by the AFM of *gerE* and *cotE gerE* spores. First, loss of *gerE* prevented formation of the outer coat, rodlet, and amorphous layers. Second, while most *gerE* spores are encased in a loose structure formed by what appeared to be patches of the inner coat (Fig. 10), these structures do not resemble the inner coat multilayer structures described above. The *cotE gerE* spores were devoid of the amorphous and rodlet layers, and both complete inner and outer coats, and these spores' surface exhibited some roughness (Fig. 11). This surface likely corresponds to the basement/undercoat layer [4]. Thus, both CotE and GerE are crucial in proper assembly of the inner coat. Note, also that *cotE gerE* spores are less rigid than *cotE* or *gerE* spores. This increased deformability is due either to the loss of the inner coat or a role for CotE in the assembly and elastic properties of the basement layer. The nature of the 80–100 nm wide and 30–40 nm deep depressions seen in Fig. 11a is unclear, but we speculate that these holes may facilitate germinant access to the spore inner membrane, and are perhaps associated with the GerP proteins important in germinant movement through spores' outer layers [78].

Another coat protein important for proper spore coat assembly and attachment to the cortex is SpoVID, as a large percentage of *spoVID* spores lacked obvious coat structures, with some encased in a misassembled sacculus composed of amorphous and rodlet structures (Fig. 12). The thickness of the sacculi walls varied between 15–30 nm, indicating that the sacculi could contain coat material in addition to the rodlet and amorphous layers (Fig. 13a). Note that none of the *spoVID* spores visualized in this study exhibited the multilayer inner coat structures seen on *cotO*, *cotH* and *cotE* spores indicating that the inner coat is absent on *spoVID* spores. Most *spoVID* sacculi were only loosely attached to the

spore body and were partially sloughed off, exposing a relatively smooth spore surface (Fig. 12). These AFM data are consistent with observations of swirls of spore coat in *spoVID* mother cells [10] (Fig. 12a, insert) and that SpoVID is required for the stable attachment of the coat. High-resolution imaging of the surface of *spoVID* spores indicated the existence of two prominent layers (Fig. 13b). One layer (13b; square) could correspond to the basement layer [4] and a pitted layer (Fig. 13b; black arrow) could correspond to either a subbasement coat layer or the cortex. Note, that *spoVID* spores lacking sacculi exhibit very high deformability (Fig. 12).

During wild-type *B. subtilis* sporulation, proteins forming honeycomb and rodlet coat layers self-assemble on the outer spore coat layer. Based on AFM results with *cotE* spores, the complete outer coat layer is not essential for formation of patches of the honeycomb and rodlet coat layers. Thus, the underlying integument is not crucial for assembly of the rodlet and honeycomb layers. Proteins that form honeycomb and rodlet spore coat structures must therefore be present during *cotE* spore formation, and self-assemble on the outer spore surface producing amorphous and rodlet layers (Fig. 8). Indeed, during *B. thuringiensis* sporulation rodlet proteins can self-assemble on the underlying spore coat, or in either the mother cell cytoplasm or the sporulation medium [23]. Hydrophobins, which form fungal rodlet layers, also self-assemble into rodlet fibrils *in vitro* (for review see [59]).

The multilayer structure forming the inner coat of *B. subtilis* spores exhibits patterns similar to ones described for the inner coats of spores of *C. novyi* NT [31] and *B. anthracis* (Plomp and Malkin, unpublished data). These patterns are also similar to those observed on surfaces of inorganic and macromolecular crystals. In addition to growth steps, these patterns include two-dimensional (2D) nuclei and screw dislocations that are major growth sources of inorganic, organic, and macromolecular crystals [79]. The presence of these growth patterns plus the smooth appearances of coat layers strongly point to a crystalline nature [79] of *B. subtilis* inner coat layers. While no screw dislocation sources similar to ones observed on the *C. novyi* NT inner spore coat [31] were seen on *B. subtilis* spores, on some spores with a low density of the grainy/fibrous outer layer, circular 2D nuclei were observed on the inner coat (Fig. 15a; dark blue arrows). This indicates that *B. subtilis* spores could represent the first case of non-mineral 2D nucleation growth patterns in a biological organism.

The observations above strongly suggest that assembly of inner spore coat layers proceeds by formation of 2D nuclei and their subsequent growth, similar to the birth-and-spread growth mechanism of conventional and macromolecular crystals [51,79]. In this model, 2D crystal growth takes place by generation and subsequent spread of 2D nuclei that provide a new crystalline layer on crystalline surfaces. Subsequent formation and growth of new 2D nuclei on this layer result in the formation of a new crystalline layer. An example of such growth, showing 2D nuclei on the surface of a crystal of satellite tobacco mosaic virus that are similar to ones seen in Fig. 15a, is presented in Fig. 15b (dark blue arrows). Typically, 2D growth takes place at high supersaturation (e.g. protein and precipitant concentrations used in macromolecular crystallization) [51,52,79], suggesting that relatively high concentrations of inner coat protein(s) are present during *B. subtilis* sporulation.

Step edges seen on the inner coat of *B. subtilis* spores showed significant roughness with many kinks (Fig. 7b), suggesting that formation of the inner coat was strongly affected by impurities. Similar patterns have been described for a wide range of crystalline surfaces (illustrated in Fig. 7d), where adsorption of

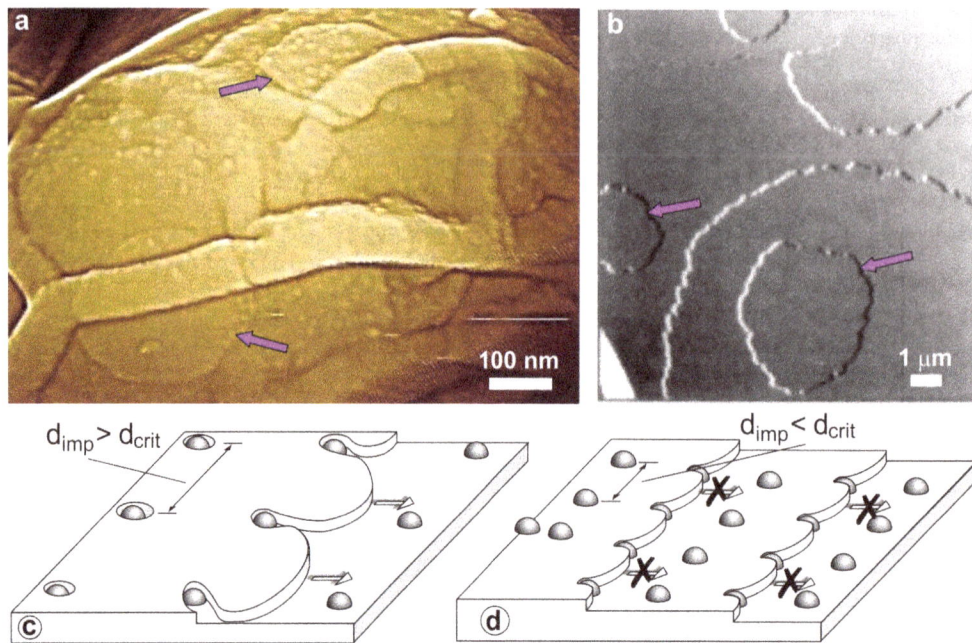

Figure 15. 2D nucleation and growth of inner spore coat layers. Panel (a) shows two putative 2D nuclei (purple arrows) on the inner coat surface of a *cotO* spore. Panel (b) shows 2D nuclei (purple arrows) on the surface of a satellite tobacco mosaic virus (STMV) crystal. This illustration is reproduced with permission from Malkin AJ, Kuznetsov YuG, Land TA, DeYoreo JJ, McPherson A (1995) Mechanisms of growth for protein and virus crystals. Nature Struct Biol. 2: 956–959 [48]. © (1995) Nature Publishing Group. (c) At a relatively small impurity (indicated as small balls) density, the average impurity distance d_{imp} is larger than d_{crit} and steps are able to advance. (d) At higher impurity densities, $d_{imp} < d_{crit}$, the curvature of step segments between impurities increases and steps are halted. Panels (c) and (d) are reproduced, with permission from Plomp M, McPherson A, Malkin AJ (2003). Repair of impurity-poisoned protein crystal surfaces. Proteins: Struct, Function, Bioinform 50: 486–495 [82]. © (2003) John Wiley and Sons.

impurities (ones present in solution), but not forming a layer at the step terraces and edges results in step roughening and cessation of growth [80–82]. Indeed the roughness and sinuosity of step edges on the inner coat of *B. subtilis* spores (Fig. 7c) are higher than observed for step edges on the surface of the trypsin crystal (Fig. 7d). This may indicate [80–83] that higher levels of impurities are adsorbed on the inner coat surface of *B. subtilis* spores compared to ones on the surface of trypsin crystals. Growth steps stop at sites of contact with impurity particles (indicated as small balls in Fig. 15c,d) that are adsorbed to the surface. However, portions of steps between neighboring impurity particles continue to grow, resulting in pinning of growth steps (Fig. 15c) as seen in Fig. 7c,d. Step advancement ceases (Fig. 15d) when at increased impurities' concentration, the distance between impurities/pinning points d_{imp} becomes smaller than the diameter of critical nuclei d_c necessary for step advancement [80]. One interesting feature of the inner coat is a number of ~5–10 nm holes (Fig. 7c), which may indicate locations of clusters of impurities [53,83]. Note, that in general the size of such holes is a function of the size of impurities or their clusters adsorbed on the surface. As described for a number of systems [80–83], these clusters of impurities may be responsible for pinning the advancement and cessation of growth of spore coat layers observed in Fig. 7c. Alternatively, such holes that were also observed on inner coat layers of *C. novyi* NT spores [31] and *B. anthracis* spores (Plomp and Malkin, unpublished data) could be an intrinsic feature of spore inner coat layers having a particular function. These results, combined with prior observation of screw dislocations on the inner coat of *C. novyi* NT spores [31], strongly suggest that inner spore coat assembly is governed by two crystallization mechanisms – growth on dislocations and 2D nucleation. These observations suggest that while spore coat proteins are produced

enzymatically [84], the assembly of these proteins into coat layers may be a self-assembly process similar to crystallization, and may be influenced by the sporulation conditions (protein and salt concentrations, pH, temperature, impurities) when these proteins assemble.

The lack of high-resolution crystalline lattice structures of the *B. subtilis* inner coat layers is similar to prior observations of *C. novyi* NT [31] and *B. anthracis* inner spore coat layers (Plomp and Malkin, unpublished data). It was suggested that proteins forming the *C. novyi* NT inner coat layers [31] are not globular, but rather peptides 'standing upright' in the layers, similar to peptide arrangements found in several organic crystals [85,86]. This hypothesis was based on the fact that for globular proteins, the ~6 nm height of the inner spore coat layers would not be considerably different in either perpendicular or lateral unit cell parameters, with the latter being amenable for AFM visualization [54]. Based on the lack of molecular scale AFM resolution of the crystalline lattice forming the *B. subtilis* inner coat layer, it is reasonable to suggest that proteins forming the inner coat might be also "standing upright" peptides [31,85,86].

In conclusion, the results presented in this communication provide further understanding of the structure and assembly of the *B. subtilis* spore coat. Furthermore, morphological and structural attributes of *B. subtilis* spores described here could thus serve as a baseline for future studies of effects of sporulation conditions on these structures. In addition, the similarities of some of the new findings with *B. subtilis* spores to findings with spores of *C. novyi* NT and other *Bacillus* species, suggest that the coat structure proposed in this work may generally be similar for spores of all of these species. While there is extensive knowledge of the individual proteins in the spore coat, as well as their location and assembly, there is much less knowledge of precise coat structure. In particular, the new high-resolution AFM studies have identified a number of new coat structural features, including the nanodots, the fibrous layer, and the terraced multilayer inner spore coat. Based on these results, we propose that the amorphous/crust layer and rodlets form the outermost spore structure, the fibrous layer and multilayer structure correspond to the outer coat and the inner coat respectively, with honeycomb and nanodot structures sandwiched between the outermost layer and the inner coat and the inner and the outer coats respectively.

Note, that high-resolution studies of fully hydrated *B. atrophaeus* [22,25] and *Clostridium novyi* NT spores [31] demonstrated that rodlet, honeycomb, and inner coat layer structures, similar to ones described here for *B. subtilis*, maintained the same patterns, lattice periodicities, and step heights as seen on air-dried spores.

Finally, the striking similarity between the appearance of the terraces and likely 2D nuclei in the multilayer inner coat and in inorganic and macromolecular crystals suggest that at least this part of the coat may assemble by crystallization mechanisms. A consequence of a crystallization spore coat assembly mechanism is that coat structure will be influenced by conditions during which these proteins self-assemble. In particular, variations in rates of 2D nucleation on spores could change the growth rate and hence the thickness of the spore coat, and this could influence spore properties such as their resistance and germination. The challenge now will be to correlate spore coat features identified in this work with specific coat proteins, and to understand how individual proteins contribute to these coat features, in particular, by using AFM-based immunolabeling techniques [29].

Acknowledgments

The authors are grateful to Patrick Eichenberger and Selim Elhadj for strain PE250 (Δ*cotO*::*tet*) and assisting in data analysis, respectively.

Author Contributions

Conceived and designed the experiments: AJM PS. Performed the experiments: MP AMC PS AJM. Analyzed the data: MP AJM AMC PS. Contributed reagents/materials/analysis tools: MP AMC PS AJM. Wrote the paper: MP PS AJM.

References

1. Klobutcher LA, Ragkousi K, Setlow P (2006) The *Bacillus subtilis* spore coat provides "eat resistance" during phagocytic predation by the protozoan *Tetrahymena thermophila*. Proc Natl Acad Sci USA 103: 165–70.
2. Henriques AO, Moran CP Jr (2007) Structure, assembly, and function of the spore surface layers. Annu Rev Microbiol 61: 555–588.
3. Laaberki MH, Dworkin J (2008) Role of spore coat proteins in the resistance of *Bacillus subtilis* spores to *Caenorhabditis elegans* predation. J Bacteriol 190: 6197–6203.
4. McKenney PT, Driks A, Eichenberger P (2013) The *Bacillus subtilis* endospore: assembly and functions of the multilayered coat. Nature Rev Microbiol 11: 33–44.
5. McKenney PT, Eichenberger P (2012) Dynamics of spore coat morphogenesis in *Bacillus subtilis*. Mol Microbiol 83: 245–260.
6. de Hoon MJL, Eichenberger P, Vitkup D (2010) Hierarchical evolution of the bacterial sporulation network. Curr Biol 20: 735–745
7. McKenney PT, Driks A, Eskandarian HA, Grabowski P, Guberman J, et al. (2010) A distance-weighted interaction map reveals a previously uncharacterized layer of the *Bacillus subtilis* spore coat. Curr Biol 20: 934–938.
8. Waller LN, Fox N, Fox KF, Fox A, Price RL (2004) Ruthenium red staining for ultrastructural visualization of a glycoprotein layer surrounding the spore of *Bacillus anthracis* and *Bacillus subtilis*. J Microbiol Meth 58: 23–30.
9. Wang KH, Isidro AL, Domingues L, Eskandarian HA, McKenney PT, et al. (2009) The coat morphogenetic protein SpoVID is necessary for spore encasement in *Bacillus subtilis*. Mol Microbiol 74: 634–649.
10. Beall BA, Driks A, Losick R, Moran CP Jr (1993) Cloning and characterization of a gene required for assembly of the *Bacillus subtilis* spore coat. J Bacteriol 175: 1705–1716.
11. Isticato R, Sirec T, Giglio R, Baccigalupi L, Pesce G, et al. (2013) Flexibility of the programme of spore coat formation in *Bacillus subtilis*: bypass of CotE requirement by overproduction of CotH. PLoS One 8: e74949.
12. Imamura D, Kuwana R, Takamatsu H, Watabe K (2010) Localization of proteins to different layers and regions of *Bacillus subtilis* spore coats. J Bacteriol 192: 518–524.
13. Imamura D, Kuwana R, Takamatsu H, Watabe K (2011) Proteins involved in formation of the outermost layer of *Bacillus subtilis* spores. J Bacteriol 193: 4075–4080.
14. Tang J, Krajcikova D, Zhu R, Ebner A, Cutting S, et al. (2007) Atomic force microscopy imaging and single molecule recognition force spectroscopy of coat proteins on the surface of *Bacillus subtilis* spore. J Mol Recog 20: 483–489.
15. Abhyankar W, Ter Beek A, Dekker H, Kort R, Brul S, et al. (2011) Gel-free proteomic identification of the *Bacillus subtilis* insoluble coat protein fraction. Proteomics 11: 4541–4550.
16. De Francesco M, Jacobs JZ, Nunes F, Serrano M, McKenney PT, et al. (2012) Physical interactions between coat morphogenetic proteins SpoVID and CotE is necessary for spore encasement in *Bacillus subtilis*. J Bacteriol 194: 4941–4950
17. Kim H, Hahn M, Grabowski P, McPherson DC, Otte MM, et al. (2006) The *Bacillus subtilis* spore coat protein interaction network. Mol Microbiol 59: 487–502.
18. Krajcikova D, Lukacova M, Mullerova D, Cutting SM, Barak I (2009) Searching for protein-protein interactions within the *Bacillus subtilis* spore coat. J Bacteriol 191: 3212–3219.
19. Mullerova D, Krajcikova D, Barak I (2009) Interactions between *Bacillus subtilis* early spore coat morphogenetic proteins. FEMS Microbiol Lett 299: 74–85.
20. Qiao H, Krajcikova D, Xing C, Lu B, Hao J, et al. (2013) Study of the interactions between the key spore coat morphogenetic proteins CotE and SpoVID. J Struct Biol 181: 128–135.
21. Chada VGR, Sanstad EA, Wang R, Driks A (2003) Morphogenesis of *Bacillus* spore surfaces. J Bacteriol 185: 6255–6261.
22. Plomp M, Leighton TJ, Wheeler KE, Malkin AJ (2005) The high-resolution architecture and structural dynamics of *Bacillus* spores. Biophys J 88: 603–608.
23. Plomp M, Leighton TJ, Wheeler KE, Malkin AJ (2005) Architecture and high-resolution structure of *Bacillus thuringiensis* and *Bacillus cereus* spore surfaces. Langmuir 21: 7892–7898.
24. Plomp M., Leighton TJ, Wheeler KE, Pitesky ME, Malkin AJ (2005) *Bacillus atrophaeus* outer spore coat assembly and ultrastructure. Langmuir 21: 10710–10716.
25. Plomp M., Leighton TJ, Wheeler KE, Hill HD, Malkin AJ (2007) *In vitro* high-resolution structural dynamics of single germinating bacterial spores. Proc Nac Acad Sci USA 104: 9644–9649.
26. McPherson DC, Kim H, Hahn M, Wang R, Grabowski P, et al. (2005) Characterization of the *Bacillus subtilis* spore morphogenetic coat protein CotO. J Bacteriol 187: 8278–8290.
27. Carroll AM, Plomp M, Malkin AJ, Setlow P (2008) Protozoal digestion of coat-defective *Bacillus subtilis* spores produces "rinds" composed of insoluble coat protein. Appl Environ Microbiol 74: 5875–5881.
28. Ghosh S, Setlow B, Wahome PG, Cowan AE, Plomp M, et al. (2008) Characterization of spores of *Bacillus subtilis* that lack most coat layers. J Bacteriol 190: 6741–6748.
29. Plomp M, Malkin AJ (2009) Mapping of proteomic composition on the surfaces of *Bacillus* spores by atomic force microscopy. Langmuir 25: 403–409.
30. Malkin AJ, Plomp M (2010) High-resolution architecture and structural dynamics of microbial and cellular system: Insights from high-resolution *in vitro* atomic force microscopy. In: Kalinin SV, Gruverman A, editors. Scanning probe microscopy of functional materials: nanoscale imaging and spectroscopy. New York: Springer. pp. 39–68.
31. Plomp M, McCaffery JM, Cheong I, Huang X, Bettegowda C, et al. (2007) Spore coat architecture of *Clostridium novyi* NT spores. J Bacteriol 189: 6457–6468.
32. Anagnostopoulos C, Spizizen J (1961) Requirements for transformation in *Bacillus subtilis*. J Bacteriol 81: 741–746.
33. Maniatis T, Fritsch EF, Sambrook J (1982) Molecular cloning: A laboratory manual. Cold Spring Harbor, NY: Cold Spring Harbor Laboratory. 545 p.
34. Nicholson WL, Setlow P (1990) Sporulation, germination and outgrowth. In: Harwood CR, Cutting SM, editors. Molecular biological methods for *Bacillus*. Chichester, UK: John Wiley & Sons Ltd. pp. 391–450.
35. Ragkousi K, Setlow P (2004) Transglutaminase-mediated cross-linking of GerQ in the coats of *Bacillus subtilis* spores. J Bacteriol 186: 5567–75.
36. Monroe A, Setlow P (2006) Localization of the transglutaminase cross-linking sites in the *Bacillus subtilis* spore coat protein GerQ. J Bacteriol 188: 7609–16.
37. Westphal AJ, Price PB, Leighton TJ, Wheeler KE (2003) Kinetics of size changes of individual *Bacillus thuringiensis* spores in response to changes in relative humidity. Proc Natl Acad Sci USA 100: 3461–3466.
38. Driks A (2003) The dynamic spore. Proc Natl Acad Sci USA 100: 3007–3009.
39. Aronson AI, Fitz-James P (1976) Structure and morphogenesis of the bacterial spore coat. Bact Rev 40: 360–402.
40. Holt SC, Leadbetter ER (1969) Comparative ultrastructure of selected aerobic spore-forming bacteria: a freeze etching study. Bacteriol Rev 33: 346–378.

41. Wehrli E, Scherrer P, Kübler O (1980) The crystalline layers in spores of *Bacillus cereus* and *Bacillus thuringiensis* studied by freeze-etching and high resolution electron microscopy. Eur J Cell Biol 20: 283–289.

42. Dufrène YF, Boonaert CJP, Gerin PA, Asther M, Rouxhet PG (1999) Direct probing of the surface ultrastructure and molecular interactions of dormant and germinating spores of *Phanerochaete chrysosporium*. J Bacteriol 181: 5350–5354.

43. Dufrène YF (2004) Using nanotechnologies to explore microbial surfaces. Nature Rev Microbiol 2: 451–458.

44. Henriques AO, Moran CP Jr (2000) Structure and assembly of the bacterial endospore coat. Methods 20: 95–110.

45. Takamatsu H, Kodama T, Nakayama T, Watabe K (1999) Characterization of the *yrbA* gene of *Bacillus subtilis*, involved in resistance and germination of spores. J Bacteriol 181: 4986–4994.

46. Zilhão R, Naclerio G, Baccigalupi L, Henriques AO, Moran CP Jr, et al. (1999) Assembly requirements and role of CotH during spore coat formation in *Bacillus subtilis*. J Bacteriol 181: 2631–2633.

47. Maiwa K, Plomp M, van Enckevort WJP, Bennema P (1998) AFM observation of barium nitrate {111} and {100} faces: spiral growth and two-dimensional nucleation growth. J Cryst Growth 186: 214–223.

48. Rogilo DI, Fedina LI, Kosolobov SS, Ranguelov BS, Latyshev AV (2013) Critical terrace width for two-dimensional nucleation during Si growth on Si (111)-(7×7) surface. Phys Rev Lett 111: 036105.

49. Malkin AJ, Kuznetsov YuG, Land TA, DeYoreo JJ, McPherson A (1995) Mechanisms of growth for protein and virus crystals. Nature Struct Biol 2: 956–959.

50. Malkin AJ, Kuznetsov YuG, McPherson A (1999) *In situ* atomic force microscopy studies of surface morphology, growth kinetics, defect structure and dissolution in macromolecular crystallization. J Cryst Growth 196: 471–488.

51. Malkin AJ, McPherson A (2004) Probing of crystal interfaces and the structures and dynamic properties of large macromolecular ensembles with in situ atomic force microscopy. In: Lin XY, DeYoreo JJ, editors. From solid-liquid interface to nanostructure engineering, vol. 2. New York: Plenum/Kluwer Academic Publisher. pp. 201–208.

52. DeYoreo JJ, Vekilov PG (2003) Principles of crystal nucleation and growth. In: Dove PM, DeYoreo JJ, Weiner S, editors. Biomineralization. Washington, DC: Mineral Society of America. pp. 57–93.

53. Plomp M, McPherson A, Larson SB, Malkin AJ (2001). Growth mechanisms and kinetics of trypsin crystallization. J Phys Chem B 105: 542–551.

54. Kuznetsov YuG, Malkin AJ, Land TA, DeYoreo JJ, Barba AP, et al. (1997) Molecular resolution imaging of macromolecular crystals by atomic force microscopy. Biophys J 72: 2357–2364.

55. Moir A (1981) Germination properties of a spore coat-defective mutant of *Bacillus subtilis*. J. Bacteriol 146: 1106–1116.

56. Qiu X, Setlow P (2009) Structural and genetic analysis of X-ray scattering by spores of *Bacillus subtilis*. J Bacteriol 191: 7620–7622.

57. Sousa JCF, Silva MT, Balassa G (1976) Exosporium-like outer layer in *Bacillus subtilis* spores. Nature 263: 53–54.

58. Kaillas L, Terry C, Abbott N, Taylor R, Mullin N, et al. (2011) Surface architecture of endospores of the *Bacillus cereus/anthracis/thuringiensis* family at the subnanometer scale. Proc Natl Acad Sci USA 108: 16014–16019.

59. Gebbink MF, Claessen D, Bouma B, Dijkhuizen L, Wösten HA (2005) Amyloids – a functional coat for microorganisms. Nature Rev Microbiol 3: 333–341.

60. HashimotoT, Wu-Yuan CD, Blumenthal HJ (1976) Isolation and characterization of the rodlet layer of *Trichophyton mentagrophytes* microconidial wall. J Bacteriol 127: 1543–1549.

61. Beever RE, Redgewell RJ, Dempsey G (1979) Purification and chemical characterization of the rodlet layer of *Neurospora crassa* conidia. J Bacteriol 140: 1063–1070.

62. Wessels JGH (1998) Hydrophobins: Proteins that change the nature of the fungal surface. Adv Microbial Physiol 38: 1–45.

63. Wösten HAB, de Vocht ML (2000) Hydrophobins, the fungal coat unravelled. Biochim Biophys Acta 1469: 79–86.

64. Claessen D, Stokroos I, Deelstra HJ, Penninga, NA Bormann C, et al. (2004) The formation of the rodlet layer of streptomycetes is the result of the interplay between rodlins and chaplins. Mol Microbiol 53: 433–443

65. Dobson CM (2003) Protein folding and misfolding. Nature 426: 884–890.

66. Smith JF, Knowles TPJ, Dobson CM, MacPhee CE, Welland ME (2006) Characterization of the nanoscale properties of individual amyloid fibrils. Proc Natl Acad Sci USA 103: 15806–15811.

67. Isticato R, Cangiano G, Tran HT, Ciabattini A, Medaglini D, et al. (2001) Surface display of recombinant proteins on *Bacillus subtilis* spores. J Bacteriol 183: 6294–6301.

68. Driks A (1999) *Bacillus subtilis* spore coat. Microbiol Mol Bio Rev 63: 1–20.

69. Zheng LB, Donovan WP, Fitz-James PC, Losick R (1988) Gene encoding a morphogenic protein required in the assembly of the outer coat of the *Bacillus subtilis* endospore. Genes Dev 2: 1047–1054.

70. Zilhão R, Serrano M, Isticato R, Ricca E, Moran CP Jr, et al. (2004) Interactions among CotB, CotG, and CotH during assembly of the *Bacillus subtilis* spore coat. J Bacteriol 186: 1110–1119.

71. Donovan W, Zheng LB, Sandman K, Losick R (1987) Genes encoding spore coat polypeptides from *Bacillus subtilis*. J Mol Biol 196: 1–10.

72. Naclerio G, Baccigalupi L, Zilhao R, de Felice M, Ricca E (1996) *Bacillus subtilis* spore coat assembly requires *cotH* gene expression. J Bacteriol 178: 4375–4380

73. Ozin AJ, Henriques AO, Yi H, Moran CP Jr (2000) Morphogenetic proteins SpoVID and SafA form a complex during assembly of the *Bacillus subtilis* spore coat. J Bacteriol: 1828–1833.

74. Frenkel D (2006) Physical chemistry - Seeds of phase change. Nature 443: 641–641.

75. Kuznetsov YuG, Malkin AJ, McPherson A (1997). Atomic force microscopy studies of living cells: Visualization of motility, division, aggregation, transformation and apoptosis. J Struct Biol 120: 180–191.

76. Magonov SN, Elings V, Whangbo MH (1997) Phase imaging and stiffness in tapping-mode atomic force microscopy. Surf Sci 375: L385–L391.

77. Magonov SN, Cleveland J, Elings V, Denley D, Whangbo M-H (1997) Tapping-mode atomic force microscopy study of the near-surface composition of a styrene-butadiene-styrene triblock copolymer film. Surf Sci 389: 201–211.

78. Butzin XY, Troiano AJ, Coleman WH, Griffiths KK, Doona CJ, et al. (2012) Analysis of the effects of a *gerP* mutation on the germination of spores of *Bacillus subtilis*. J Bacteriol 194: 5749–5758.

79. Chernov AA (1984) Modern crystallography. III. Crystal growth. Berlin: Springer-Verlag. 517 p.

80. Cabrera N, Vermilyea DA (1958) The growth of crystals from solution. In: Doremus RH, Roberts BW, Turnbul D, editors. Growth and perfection of crystals. New York: Wiley. pp. 393–410.

81. van Enckevort WJP, van der Berg ACJF, Kreuwel KBG, Derksen AJ, Couto MS (1996) Impurity blocking of growth steps: experiments and theory. J Cryst Growth 166: 156–161.

82. Land TA, Martin TL, Potapenko S, Palmore GT, DeYoreo JJ. (1999) Recovery of surfaces from impurity poisoning during crystal growth. Nature 399: 442–445.

83. Plomp M, McPherson A, Malkin AJ (2003). Repair of impurity-poisoned protein crystal surfaces. Proteins: Struct, Function, Bioinform 50: 486–495

84. Driks A (2002) Maximum shields: the assembly and function of the bacterial spore coat. Trends Microbiol 10: 251–254.

85. Hollander FFA, Plomp M, van de Streek CJ, van Enckevort WJP (2001) A two-dimensional Hartman-Perdok analysis of polymorphic fat surfaces observed with atomic force microscopy. Surf Sci 471: 101–113.

86. Plomp M, van Enckevort MJV, van Hoof PJCM, van de Streek CJ (2003) Morphology and dislocation movement in n-$C_{40}H_{82}$ paraffin crystals grown from solution. J Cryst Growth 249: 600–613.

87. Eichenberger P, Jensen ST, Conlon EM, van Ooij C, Silvaggi J, et al. (2003) The σ^E regulon and the identification of additional sporulation genes in *Bacillus subtilis*. J Mol Biol 327: 945–972.

15

Loss of Cln3 Function in the Social Amoeba *Dictyostelium discoideum* Causes Pleiotropic Effects That Are Rescued by Human CLN3

Robert J. Huber*, Michael A. Myre⁹, Susan L. Cotman⁹

Center for Human Genetic Research, Massachusetts General Hospital, Harvard Medical School, Boston, Massachusetts, United States of America

Abstract

The neuronal ceroid lipofuscinoses (NCL) are a group of inherited, severe neurodegenerative disorders also known as Batten disease. Juvenile NCL (JNCL) is caused by recessive loss-of-function mutations in *CLN3*, which encodes a transmembrane protein that regulates endocytic pathway trafficking, though its primary function is not yet known. The social amoeba *Dictyostelium discoideum* is increasingly utilized for neurological disease research and is particularly suited for investigation of protein function in trafficking. Therefore, here we establish new overexpression and knockout *Dictyostelium* cell lines for JNCL research. *Dictyostelium* Cln3 fused to GFP localized to the contractile vacuole system and to compartments of the endocytic pathway. *cln3⁻* cells displayed increased rates of proliferation and an associated reduction in the extracellular levels and cleavage of the autocrine proliferation repressor, AprA. Mid- and late development of *cln3⁻* cells was precocious and *cln3⁻* slugs displayed increased migration. Expression of either *Dictyostelium* Cln3 or human CLN3 in *cln3⁻* cells suppressed the precocious development and aberrant slug migration, which were also suppressed by calcium chelation. Taken together, our results show that Cln3 is a pleiotropic protein that negatively regulates proliferation and development in *Dictyostelium*. This new model system, which allows for the study of Cln3 function in both single cells and a multicellular organism, together with the observation that expression of human CLN3 restores abnormalities in *Dictyostelium cln3⁻* cells, strongly supports the use of this new model for JNCL research.

Editor: Thierry Soldati, Université de Genève, Switzerland

Funding: This work was supported by a Postdoctoral Fellowship from the Canadian Institutes of Health Research (289813 to R.J.H.) and the National Institutes of Health: National Institute of Neurological Disorders & Stroke (R01NS073813 to S.L.C.). The funders had no role in study design, data collection and analysis, decision to publish, or preparation of the manuscript.

Competing Interests: The authors have declared that no competing interests exist.

* Email: rhuber@mgh.harvard.edu

⁹ These authors contributed equally to this work.

Introduction

The neuronal ceroid lipofuscinoses (NCL) are a group of inherited, severe neurodegenerative disorders also known as Batten disease [1]. At the cellular level, NCL disorders characteristically display aberrant lysosomal function and an excessive accumulation of lipofuscin in neurons and other cell types [2,3]. Clinical manifestations include vision loss, seizures, the progressive loss of motor function and psychological ability, and a reduced lifespan [4]. Recent evidence also points to pathology outside of the central nervous system, more specifically the cardiac and immune systems [5–8]. North American and Northern European populations have the highest rates of incidence, however the NCL disorders have a worldwide distribution with varying incidence rates depending on the region (1:14000 to 1:100000) [9]. Currently there are no effective treatments or cure for NCL disorders.

Juvenile NCL (JNCL), the most common subtype of NCL, occurs due to recessive mutations in the *CLN3* gene with the majority of JNCL patients carrying a ~1-kb genomic deletion spanning exons 7 and 8 [10]. Indel, missense, nonsense, and splice site mutations have also been documented in JNCL patients [11,12]. In mammals, *CLN3* encodes a 438 amino acid multi-pass transmembrane protein (CLN3/battenin; ceroid-lipofuscinosis, neuronal 3) that is primarily found in endosomes and lysosomes with evidence that it may also traffic to other subcellular membranes [3,13,14]. In neurons, CLN3 may be important for events localized at the synapse [15]. Evidence from yeast and mouse models independently suggests that CLN3 may function in lysosomal pH homeostasis, endocytic trafficking, and autophagy [16–20]. Despite substantial research efforts using a variety of systems, the precise function of CLN3 remains unclear [21].

A new, unexplored approach to studying CLN3 function involves the use of the social amoeba *Dictyostelium discoideum*, which has been selected by the National Institutes of Health as a model organism for biomedical and human disease research. This genetically tractable model eukaryote is being used successfully to study the function of genes linked to neurodegenerative disorders and is particularly suited to modeling human lysosomal and trafficking diseases [22–29]. *Dictyostelium* is a soil microbe that undergoes an asexual life cycle composed of a growth phase in which single cells grow and divide mitotically as they feed on bacteria and a multicellular developmental stage that is induced upon starvation. During the early stages of *Dictyostelium* development, the starving population of cells secretes cAMP in a pulsatile manner, which serves to attract individual cells chemotactically to form a multicellular aggregate also referred to as a mound. After a series of morphological changes, the mound develops into a slug-like structure that is capable of both photo- and thermotaxis. When conditions are suitable, the slug, composed of predominantly two cell types (i.e., pre-stalk and pre-spore), completes the life cycle by forming a fruiting body comprised of a mass of spores supported by a stalk of dead cells. When a food source becomes available, the spores germinate allowing the amoeba to re-start the life cycle. Thus, *Dictyostelium* serves as a valuable system for studying a variety of cell and developmental processes [30–32].

Understanding the normal function of CLN3 is a key step in designing targeted therapies for JNCL. Therefore, in this study, we have established new tools for research into CLN3 function by generating a Cln3-deficient *Dictyostelium* mutant by targeted homologous recombination and introducing GFP-tagged *Dictyostelium* Cln3 and human CLN3 into *Dictyostelium* cells. Assessment of the knockout and overexpression cells during growth and development strongly indicates that the function of CLN3 is conserved from *Dictyostelium* to human. Furthermore, our results strongly support a key role for CLN3 in regulating the endocytic pathway and calcium-dependent developmental events.

Materials and Methods

Cells and chemicals

AX3 and *cln3⁻* cells were grown and maintained at room temperature on SM agar with *Klebsiella aerogenes* and in HL5 medium supplemented with ampicillin (100 µg/ml) and streptomycin sulfate (300 µg/ml). *cln3⁻* cells also required blasticidin S hydrochloride (10 µg/ml), while strains carrying the extrachromosomal vector pTX-GFP required G418 (10 µg/ml) [33]. HL5, FM minimal medium, and low fluorescence HL5 were purchased from ForMedium (Hunstanton, Norfolk, UK). The QIAquick PCR Purification Kit, QIAquick Gel Extraction Kit, and QIAprep Spin Miniprep Kit were used for all PCR purifications, gel extractions, and plasmid isolations, respectively, and were all purchased from Qiagen Incorporated (Valencia, CA, USA). Restriction enzymes were purchased from New England BioLabs Incorporated (Ipswich, MA). All primers were purchased from Integrated DNA Technologies Incorporated (Coralville, IA, USA). EGTA and FITC-dextran were purchased from Sigma-Aldrich (St. Louis, MO, USA). Mouse monoclonal anti-p80 was purchased from the Developmental Studies Hybridoma Bank (University of Iowa, Iowa City, IA, USA).

Axenic growth and pinocytosis

Cells in the mid-log phase of growth ($1–5\times10^6$ cells/ml) were diluted to $1–2\times10^5$ cells/ml in fresh HL5 or FM and incubated at 22°C and 150 rpm. Cell concentrations were measured every 24 hours over a 120- or 144-hour growth period with a hemocytometer. Pinocytosis assays were performed as previously described [34]. Briefly, AX3 and *cln3⁻* cells (5×10^6 cells/ml) were grown in HL5. FITC-dextran (70,000 M_r, 100 µl of a 20 mg/ml solution) was added to a 5-ml cell suspension, which was then incubated for 90 minutes at room temperature and 150 rpm. Equal volumes of cells (500 µl) were harvested at the indicated times, washed 2 times with ice-cold Sorenson's buffer (2 mM Na_2HPO_4, 14.6 mM KH_2PO_4, pH 6.0), and then lysed with 1 ml of buffer containing

50 mM Na_2HPO_4 pH 9.3 and 0.2% Triton-X. Lysates were placed in black 96-well plates and fluorescence was measured with a Molecular Devices SpectraMax M2 Multi-Mode Microplate Reader (excitation 470, emission 515). For axenic growth and pinocytosis assays, statistical significance was assessed in GraphPad Prism 5 (GraphPad Software Incorporated, La Jolla, CA, USA) using two-way ANOVA followed by Bonferroni post-hoc analysis. A p-value<0.05 was considered significant (i.e., n = # of independent cell cultures; see relevant Figure legends for additional details). For experiments assessing the effect of *cln3* knockout on the intra- and extracellular levels of AprA and CfaD, AX3 and *cln3⁻* cells grown axenically in HL5 (as described above) were harvested and lysed after 48 and 72 hours of growth. At each of these time points, cells from 15 ml of culture were also spun down and conditioned media was collected and filtered through a 0.45 µm filter unit. Samples were standardized by loading volumes of conditioned media according to cell number (i.e., media from 100000 cells). Whole cell lysates and samples of conditioned media were separated by SDS-PAGE and analyzed by western blotting.

Development

Development assays were performed as previously described [35]. Briefly, cells grown in HL5 were harvested in the mid-log phase of growth ($1–5\times10^6$ cells/ml) and washed two times with ice-cold KK2 phosphate buffer (2.2 g/L KH_2PO_4, 0.7 g/L K_2HPO_4, pH 6.5). Washed cells (3×10^7 cells/ml) were deposited in four individual cell droplets (25 µl each droplet) on black, gridded, cellulose filters (0.45 mm pore size) (EMD Millipore Corporation, Billerica, MA, USA) overlaid on four Whatman #3 cellulose filters (EMD Millipore Corporation, Billerica, MA, USA) pre-soaked in KK2 buffer. Cells were maintained in the dark in a humidity chamber at room temperature. Structures were viewed and photographed at the indicated times with a Nikon SMZ800 microscope (Nikon Instruments Incorporated, Melville, NY, USA) equipped with a SPOT Insight color camera 3.2.0 (Diagnostic Instruments Incorporated, Sterling Heights, MI, USA). Images were captured with SPOT for Windows (Diagnostic Instruments Incorporated, Sterling Heights, MI, USA). For each independent experiment, developmental phenotypes were scored for each cell droplet (i.e., 4 total) and then averaged to obtain a mean value for that experiment (i.e., n = # of independent experiments; see relevant Figure legends for additional details). Statistical significance was assessed in GraphPad Prism 5 (GraphPad Software Incorporated, La Jolla, CA, USA). Data that satisfied parametric requirements were analyzed using one-way ANOVA followed by the Bonferroni multiple comparison test. Non-parametric data were analyzed using the Kruskal-Wallis test followed by the Dunn multiple comparison test. A p-value<0.05 was considered significant. See relevant Figure legends for additional details.

Live cell imaging, fxation, and immunolocalization

Cells were viewed live in 6-well dishes containing water or low fluorescence HL5. Fixation in ultra-cold methanol (for cells probed with anti-VatM or anti-Rh50) or 4% paraformaldehyde (for cells probed with anti-p80) followed by immunolocalization, were performed as previously described [36,37]. Prior to fixation, cells were grown overnight on coverslips in low fluorescence HL5. The following primary and secondary antibodies were used; rabbit polyclonal anti-GFP (1:1000) (Life Technologies Incorporated, Carlsbad, CA, USA), mouse monoclonal anti-GFP (1:50) (Santa Cruz Biotechnology Inc., Santa Cruz, CA, USA), mouse monoclonal anti-VatM (1:10–1:25) [38], rabbit polyclonal anti-Rh50 (1:1500–1:2000) [39], mouse monoclonal anti-p80 (1:50) [40], donkey anti-rabbit Alexa Fluor 488, donkey anti-rabbit

Alexa Fluor 555, donkey anti-mouse Alexa Fluor 488, and donkey anti-mouse Alexa Fluor 555 (1:50–1:100) (Life Technologies Incorporated, Carlsbad, CA, USA). Coverslips were mounted on slides with Prolong Gold anti-fade reagent with DAPI (Life Technologies Incorporated, Carlsbad, CA, USA) and sealed with nail polish. Live cells were viewed with a Nikon Eclipse TE2000-U microscope equipped with Nikon Digital Sight DS-Qi1Mc and Nikon Digital Sight DS-Fi1 digital cameras (Nikon Instruments Incorporated, Melville, NY, USA). Fixed cells were imaged either with a Leica SP5 AOBS scanning laser confocal microscope (Leica Microsystems, Buffalo Grove, IL, USA) or a Zeiss Axioskop2 mot plus epifluorescence microscope equipped with a Zeiss AxioCam MRm digital camera (Carl Zeiss Microscopy LLC, Thornwood, NY, USA). For confocal analysis, the separate channels were imaged using sequential scanning mode and z-sections were taken with a pinhole setting of 1 airy unit (AU). Separate channel and overlay (i.e., merge) images were exported from the Leica imaging software (LAS AF), or from the Zeiss AxioVision imaging software (version 4.6.3), as .tif files and opened into Adobe Photoshop CS5 for compilation of figures. For epifluorescence images, the merge of the separate channel images was produced using ImageJ/Fiji software. If minor brightness and contrast adjustments were necessary, these were made in Photoshop uniformly for each set of images of a given co-stain combination.

SDS-PAGE and western blotting

Cells were lysed with a buffer containing 50 mM Tris–HCl pH 8.0, 150 mM sodium chloride, 0.5% NP-40, 5 mM EDTA, 10 mM sodium fluoride, 1 mM sodium orthovanadate, and a protease inhibitor cocktail tablet (Roche Diagnostics Corporation, Indianapolis, IN, USA). Proteins were separated by SDS-PAGE and analyzed by western blotting with mouse monoclonal anti-tubulin (1:1000) (12G10, Developmental Studies Hybridoma Bank, The University of Iowa, IA, USA), mouse monoclonal anti-actin (1:1000), mouse monoclonal anti-GFP (1:1000) (Santa Cruz Biotechnology Inc., Santa Cruz, CA, USA), rabbit polyclonal anti-AprA (1:1000) [41], and rabbit polyclonal anti-CfaD (1:1000) [42]. Immunoblots were digitally scanned using a GS800 Calibrated Densitometer scanner and Quantity One software (Bio-Rad Laboratories Incorporated, Hercules, CA, USA). Identified bands were quantified with ImageJ/Fiji and levels were normalized to ß-actin levels. Results were pooled from four independent experiments, each with at least two technical replicates. Statistical significance was determined using a one-sample t-test (mean, 100; two-tailed). A p-value<0.05 was considered significant.

Bioinformatic and phylogenetic analysis

Sequence alignments between *Dictyostelium* Cln3 and human CLN3 were performed using the dictyBase BLAST server (http://www.dictybase.org/tools/blast). For phylogenetic analyses, the amino acid sequence of *Dictyostelium* Cln3 was inputted into the NCBI BLASTp server. Amino acid sequences for significant hits corresponding to CLN3 orthologs from 20 different organisms (i.e., mammals and NIH model systems) were obtained and aligned using ClustalX version 1.83. Neighbor-Joining trees were created using ClustalX version 1.83 and PAUP version 4.0 (Sinauer Associates Incorporated Publishers, Sunderland, MA, USA) and viewed using TreeView version 1.6.6.

Gene knockout and validation

Targeted disruption of the *cln3* gene in *Dictyostelium discoideum* was accomplished using an approach that has been previously described [26]. Targeting arms were amplified by PCR

using the Expand High-Fidelity PCR System (Roche Diagnostics Corporation, Indianapolis, IN, USA) and cloned into vector pLPBLP, which knocked out the gene of interest by homologous recombination and introduced a blasticidin resistance (*bsr*) cassette [43]. The 5′ targeting arm was amplified using the following primers, which incorporated *Kpn*I and *Hin*dIII sites (underlined) to facilitate directional cloning into pLPBLP; 5′-GGTACCTCTTTATACTATATATTATACCTCCTTCTC-3′ (forward) and 5′-AAGCTTCATCTTGAAACTAAAC-CAAATGCAATATTTGC-3′ (reverse). The 3′ targeting arm was amplified using the following primers, which incorporated *Pst*I and *Spe*I (underlined) to facilitate directional cloning into pLPBLP; 5′-CTGCAGAAAACAAAGATATATTCGTTGTG-CACG-3′ (forward) and 5′-ACTAGTATGAAGAAT-CAGTTTTTGGAACCTCAGAG-3′ (reverse). AX3 cells were electroporated with 10 μg of linearized gene-targeting DNA. 96 colonies resistant to blasticidin S hydrochloride (10 μg/ml) were collected and replica-plated into a 96-well plate. Genomic DNA was extracted using the DNeasy Blood and Tissue Kit (Qiagen Incorporated, Valencia, CA, USA) and targeted gene disruption was validated by nine PCR reactions using a combination of primers (File S1, Table S1). PCR analysis identified eight positive clones that all showed a similar growth phenotype (discussed in Results). Two of these clones were further analyzed by Southern blotting. Genomic DNA from each clone was isolated and digested overnight with *Hin*dIII at 37°C, separated by agarose gel electrophoresis, and transferred to positively charged nylon membranes by capillary transfer. Blots were hybridized with a DIG-labelled probe corresponding to the entire sequence of the *bsr* gene using the PCR DIG Probe Synthesis Kit and the DIG High Prime DNA Labeling and Detection Starter Kit II according to the manufacturer's instructions (Roche Diagnostics Corporation, Indianapolis, IN, USA). The *bsr* gene was amplified from pLPBLP using the following primers; 5′-ATGGATCAATTTAA-CATTTCTCAAC-3′ (forward) and 5′-TTAATTTCGGGTA-TATTTGAGTGG-3′ (reverse). Based on the position of *Hin*dIII cut sites in the *Dictyostelium* genome, a single 2746 bp fragment was expected on Southern blots probed with the DIG-labelled *bsr* probe (www.dictybase.org). A ~2750 bp fragment was detected in both clones however one of the clones also contained an unexpected ~6600 bp fragment. Since this implied an unintended and possibly complex integration event, we chose to work with the clone containing the single ~2750 bp fragment. We designated this clone as the *cln3* knockout strain and used these cells in all subsequent analyses.

Construction of GFP expression constructs and cell lines

Vector pTX-GFP, which incorporates an N-terminal GFP tag, was used to generate all GFP-fusion protein constructs [33]. Full-length *Dictyostelium cln3* was amplified from cDNA using the following primers, which incorporated *Sac*I and *Xho*I sites (underlined) to facilitate directional cloning into pTX-GFP; 5′-GAGCTCATGGGAAAGGATTATACATT-3′ (forward) and 5′-CTCGAGTTATGTTGAGGATGAAGAAT-3′ (reverse). Full-length human *CLN3* was amplified from cDNA using the following primers, which also incorporated *Sac*I and *Xho*I sites (underlined); 5′-GAACTTGAGCTCATGGGAGGCTGTG-3′ (forward) and 5′-TAATCCCTCGAGTCAGGAGAGCTGGC-3′ (reverse). To facilitate the expression of *Dictyostelium* GFP-Cln3 and human GFP-CLN3 at close to endogenous levels, the *act15* promoter and the first 11 codons of the GFP open reading frame, which contained the initiation methionine and an amino-terminal 8x histidine tag, was removed from pTX-GFP by digesting the plasmid with *Sal*I and *Kpn*I. Three fragments containing DNA from the non-coding region directly upstream of *cln3* were amplified from AX3 gDNA using primers cln3_up_elem_F1, cln3_up_elem_F2, cln3_up_elem_F3, and cln3_up_elem_R1 (File S1, Table S1). Forward primers incorporated *Sal*I restriction sites and reverse primers incorporated *Kpn*I restriction sites to facilitate directional cloning into pTX-GFP. The longest fragment (i.e., upstream element 1) spanned the entire region upstream of the *cln3* start site up to the end of the preceding gene (File S1, Fig. S1). The other two fragments (i.e., upstream elements 2 and 3) spanned regions within upstream element 1 up to the *cln3* start site. The three upstream elements, which also included the first 36 base pairs (12 codons) of the *cln3* open reading frame, were then separately cloned into pTX-GFP upstream and in-frame with the GFP open reading frame. All constructs were validated by agarose gel electrophoresis and DNA sequencing (CHGR Genotyping Resource, Genomics Core Facility, Massachusetts General Hospital, Boston, MA, USA). The ability of each *cln3* upstream element to drive GFP expression in AX3 cells was verified by western blotting (File S1, Fig. S1). Since upstream element 1 was the strongest driver of gene expression (File S1, Fig. S1), we used this fragment of DNA, hereafter referred to as 'cln3 upstream element', to drive gene expression in our modified version of pTX-GFP (i.e., *act15* promoter removed).

Results

Sequence analysis of *Dictyostelium* Cln3

The 438 amino acid sequence of human CLN3 was inputted into the dictyBase BLASTp server (http://www.dictybase.org/tools/blast). The highest match was a 421 amino acid protein (Cln3; DDB_G0291157). There were 117 exact matches (27% identical) and 197 positive matches (46% similar) within a 429 amino acid region of similarity (Fig. 1A). In comparison, the CLN3 homolog in *Saccharomyces cerevisiae*, Btn1p, is 38% identical and 49% similar to the human protein, while the *Schizosaccharomyces pombe* homolog is 32% identical and 47% similar. However, the CLN3 homologs in yeast are comparatively smaller than *Dictyostelium* Cln3 (408 aa and 396 aa vs. 421 aa). Residues that are myristoylated or glycosylated in human CLN3 are conserved in *Dictyostelium* Cln3 and a putative prenylation motif near the C-terminus of the protein (i.e., 398-CFIL-401) is present, although it does not precisely align with the prenylation motif in the human protein, which is found at the end of the protein (i.e., 435-CQLS-438) (Fig. 1A). Importantly, point mutations (missense and nonsense) documented from JNCL patients are highly conserved in the *Dictyostelium* ortholog (Fig. 1A). Together, these similarities indicate that the function of the protein is likely conserved from *Dictyostelium* to human. A phylogenetic tree showing the relationship of *Dictyostelium* Cln3 to CLN3 orthologs from 20 different organisms of interest (i.e., NIH model systems and mammals) firmly places *Dictyostelium* Cln3 within the CLN3 family of proteins (Fig. 1B).

Dictyostelium Cln3 fused to GFP localizes to the contractile vacuole network and to vesicles of the endocytic pathway

To gain insight into the function of the CLN3 ortholog in *Dictyostelium*, we transformed AX3 cells with a vector that expressed *Dictyostelium* Cln3 fused to GFP. We chose to place the GFP tag on the N-terminus since a previous study has reported the mis-localization of CLN3 tagged with C-terminal GFP, presumably due to the masking of the prenylation motif [44]. Protein expression was verified by western blotting and a thorough discussion and analysis of the banding pattern is provided in the supporting information (File S1, Fig. S2). In live AX3 cells incubated in water, *Dictyostelium* GFP-Cln3 localized to the membranes of vacuolar-shaped structures and small cytoplasmic vesicles, to tubular-like structures within the cytoplasm, and as punctate distributions within the cytoplasm (Fig. 2A). Time-lapse video microscopy of these cells showed multiple vacuoles undergoing dynamic events of expansion and contraction (File S1, Fig. S3). In free-living amoebae and protozoa, the contractile vacuole (CV) system acts as an osmoregulatory organelle that controls the intracellular water balance by collecting and expelling excess water out of the cell. In *Dictyostelium*, the CV system consists of tubules and vacuoles that function to collect and expel excess water, respectively [45]. Based on our initial observations of *Dictyostelium* GFP-Cln3 localization in AX3 cells, we next fixed and probed cells expressing GFP-Cln3 with antibodies directed against two established *Dictyostelium* CV system markers, the V-ATPase membrane subunit (VatM) and the rhesus-like glycoprotein Rh50 [38,39]. VatM generates an acidic environment in several intracellular compartments and is found in both the CV and endosomal systems, however it is enriched ~10-fold in the CV system, while Rh50 is more specific to the CV system [38,39,46,47]. GFP-Cln3 was found to strongly localize to both VatM- and Rh50-positive compartments (Fig. 2B). Interestingly, much like VatM, GFP-Cln3 localized to both small cytoplasmic vesicles and at distinct punctate distributions within the cytoplasm (Fig. 2B). GFP-Cln3 was also observed to localize as punctate clusters on the vacuolar membrane (Fig. 2B). Since localization of GFP-Cln3 was observed on the smaller, VatM-positive punctate distributions, we also assessed localization of GFP-Cln3 to p80-positive compartments. The p80 protein localizes to late endosomes during *Dictyostelium* growth [40]. Although GFP-Cln3 localized primarily to the vacuoles of the CV system (Fig. 2A,B), which were unstained by the p80 antibody, we did observe GFP-Cln3 localization on the membranes of a subset of small cytoplasmic vesicles that were also stained by the p80 antibody (Fig. 2B).

To further support the localization of *Dictyostelium* GFP-Cln3 to VatM-, Rh50-, and p80-positive subcellular compartments, we analyzed the localization of GFP-Cln3 using immunofluorescence and confocal microscopy. Across multiple z-sections of the amoeboid *Dictyostelium* cells, GFP-Cln3 localized to VatM-positive vesicles and punctate distributions, Rh50-positive tubules and vacuolar-shaped structures, and a subset of p80-positive vesicles (Fig. 3). Taken together, our data strongly suggest that Cln3 localizes to both the CV and endocytic systems in *Dictyostelium*.

Cln3⁻ cells show enhanced rates of proliferation and increased intracellular accumulation of FITC-dextran

To further study the function of Cln3 in *Dictyostelium*, a *cln3* knockout mutant was generated by targeted homologous recombination, which deleted the entire region spanning amino acids 61–421 (Fig. 4A–C). RNA-Seq data shows that expression of *cln3* mRNA decreases by ~30% during the first 4 hours of development, but then increases dramatically during the next 8 hours (i.e., ~8-fold increase), with expression peaking after 12 hours of development [48]. Expression decreases slightly between 12 and 20 hours (~15% decrease), but overall remains high during the mid- to late stages of *Dictyostelium* development.

Since growth is a major phase of the *Dictyostelium* life cycle, we first assessed the effect of Cln3 deficiency on the rate of cell proliferation in axenic media. In HL5, *cln3⁻* cells proliferated at a significantly enhanced rate compared to parental AX3 cells (genotype effect, two-way ANOVA, p<0.001) (Fig. 5A). However,

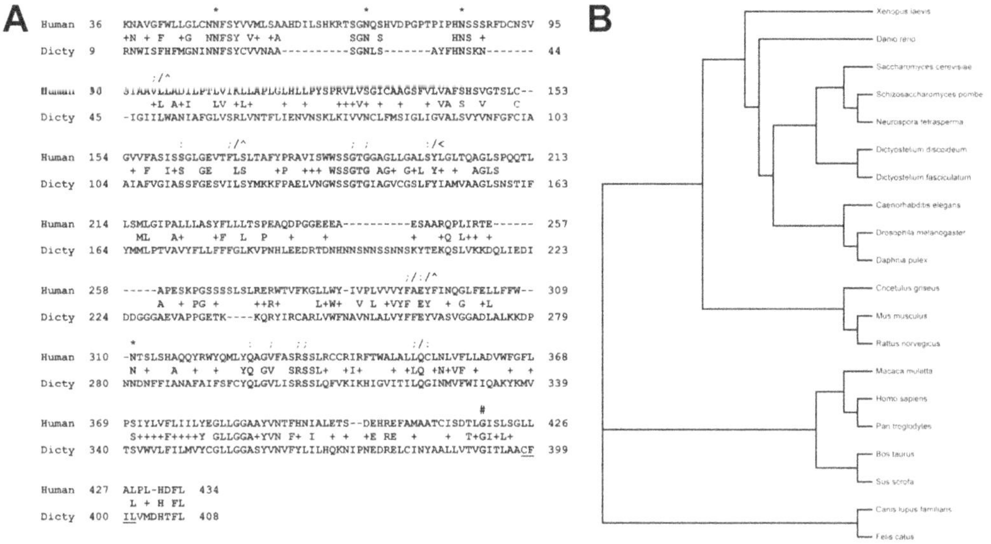

Figure 1. Bioinformatic analysis of _Dictyostelium_ Cln3. (A) Alignment of human CLN3 and the _Dictyostelium_ ortholog. The following residues are conserved; N-linked glycosylation sites (*), sites of missense point mutations (;), sites of nonsense point mutations (:), target for myristoylation (#), sites that when mutated cause a slower disease progression in compound heterozygosity with the common 1.02 kb deletion mutation (^), sites that when mutated cause a slower disease progression in homozygosity (<) [10,13,85–87]. _Dictyostelium_ Cln3 also contains a putative prenylation motif (i.e., CFIL; underlined). (B) Phylogenetic tree showing the relationship of _Dictyostelium_ Cln3 to CLN3 orthologs from 20 different organisms (i.e., mammals and NIH model systems).

no significant difference was observed between the highest densities attained by both strains after 120 hours of growth (Fig. 5A). Since we were able to successfully overexpress _Dictyostelium_ GFP-Cln3 in AX3 and $cln3^-$ cells, we next assessed the ability of GFP-Cln3 to alter the enhanced rate of proliferation of $cln3^-$ cells and the effect of GFP-Cln3 overexpression on AX3 cell proliferation. GFP-Cln3 overexpression significantly suppressed the enhanced proliferation of $cln3^-$ cells to levels observed in AX3 cells (Fig. 5A). Overexpression of GFP-Cln3 in AX3 cells had no significant effect on cell proliferation however these cells reached a significantly lower final density after 120 hours when compared to all other strains (Fig. 5A). Based on these results, we then assessed the growth of $cln3^-$ cells in FM minimal media to determine whether limiting available nutrients would suppress the enhanced growth rate. When grown in FM, cells of both strains proliferated at a reduced rate compared to growth in HL5 (Fig. 5A, B). We did not detect any significant differences in the growth rates of AX3 and $cln3^-$ cells during the first 96 hours of growth in FM (Fig. 5B). However, at the 120- and 144-hour time points, $cln3^-$ cells were at a significantly higher density than AX3 cells, and the genotype was found to have a significant effect on the overall growth curve, as determined by two-way ANOVA (p< 0.01) (Fig. 5B).

Since pinocytosis is required for the growth of _Dictyostelium_ cells in liquid media, we used a well-established assay to assess whether this process was dysregulated in $cln3^-$ cells. AX3 and $cln3^-$ cells were incubated with FITC-dextran, and the amount of intracellular fluorescence was measured at specific time intervals over a 90-minute incubation period. At the 40-minute time point, the intracellular fluorescence was relatively higher ($\sim$50%) in $cln3^-$ cells compared to AX3 cells (Fig. 5C). However, two-way ANOVA analysis of the pinocytic uptake of FITC-dextran over the entire 90-minute incubation period did not indicate a statistically significant genotype effect (p>0.05) (Fig. 5C). Although one of the pathological hallmarks of JNCL is the accumulation of lysosomal storage material in neurons and other

cell types [2,3], we were unable to observe any autofluorescent material in $cln3^-$ cells during growth (unpublished data).

Cln3 deficiency negatively affects the secretion and cleavage of autocrine proliferation repressor a during growth

In an attempt to gain further insight into the possible mechanisms by which Cln3 deficiency leads to enhanced proliferation, we next investigated two secreted proteins that modulate growth in _Dictyostelium_ by repressing cell proliferation: autocrine proliferation repressor A (AprA) and counting factor-associated protein D (CfaD) [41,42]. Whole cell lysates (i.e., intracellular) and conditioned growth media (i.e., extracellular) from AX3 and $cln3^-$ cells were analyzed for the levels of AprA and CfaD present in each sample. In whole cell lysates, anti-AprA strongly detected a 60-kDa protein and weakly detected a 55-kDa protein (Fig. 6A), consistent with the banding pattern observed in another parental strain of _Dictyostelium_, AX2 [41]. After 48 and 72 hours of axenic growth, the amount of the 55-kDa protein in $cln3^-$ whole cell lysates was significantly greater than the amount in AX3 cells (Fig. 6A). In contrast, there were no significant differences in levels of the 60-kDa protein (Fig. 6A). In samples of conditioned growth media, anti-AprA detected the 60-kDa and 55 kDa proteins as well as a 37-kDa protein, which had not been observed in whole cell lysates from either AX3 or $cln3^-$ cells (Fig. 6A). After 72 hours of growth, the amount of 60-kDa protein in $cln3^-$ conditioned media, was significantly reduced compared to the amount present in AX3 conditioned media (Fig. 6A). After 48 and 72 hours of growth, the amount of 37-kDa protein in conditioned media from $cln3^-$ cells was also significantly reduced compared to amounts present in AX3 conditioned media (Fig. 6A). In contrast, the 55-kDa protein was present in significantly greater amounts at each time point in $cln3^-$ conditioned media (Fig. 6A).

In whole cell lysates and samples of conditioned growth media, anti-CfaD detected two proteins of molecular weights 65-kDa and

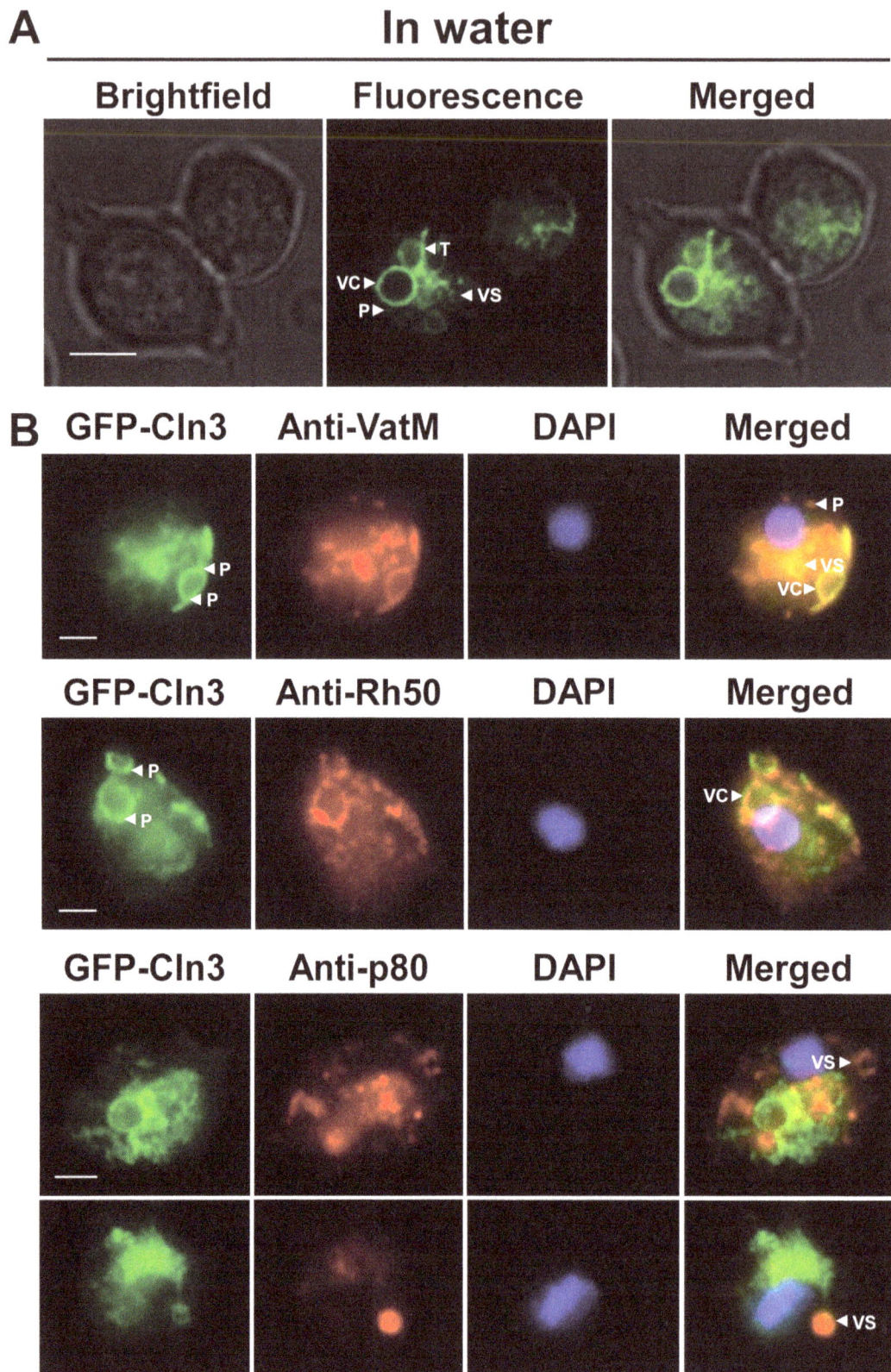

Figure 2. Intracellular localization of *Dictyostelium* **GFP-Cln3 using epifluorescence microscopy.** (A) AX3 cells overexpressing GFP-Cln3 imaged live in water. Scale bar = 5 µm. (B) AX3 cells overexpressing GFP-Cln3 were fixed in either ultra-cold methanol (for VatM and Rh50 immunostaining) or 4% paraformaldehyde (for p80 immunostaining) and then probed with anti-VatM, anti-Rh50, or anti-p80, followed by the appropriate secondary antibody linked to Alexa 555. Cells were stained with DAPI to reveal nuclei (blue). Images were merged with ImageJ/Fiji. VC, vacuolar-shaped structures; VS, cytoplasmic vesicles; T, tubular-like structures within the cytoplasm; P, punctate distributions within the cytoplasm. Scale bars (B, C) = 2.5 µm.

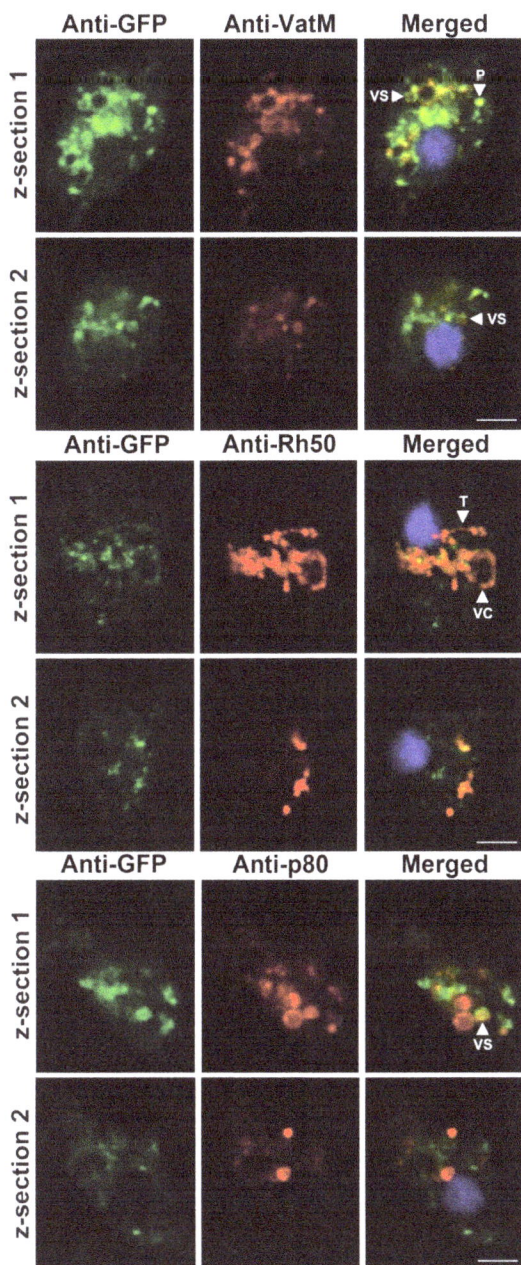

Figure 3. Intracellular localization of *Dictyostelium* **GFP-Cln3 using confocal microscopy.** AX3 cells overexpressing GFP-Cln3 were fixed in either ultra-cold methanol (for VatM and Rh50 immunostaining) or 4% paraformaldehyde (for p80 immunostaining) and then probed with anti-GFP (rabbit polyclonal anti-GFP for anti-VatM and anti-p80 co-staining and mouse monoclonal anti-GFP for anti-Rh50 co-staining) followed by anti-rabbit or anti-mouse Alexa 488. Cells were then probed with one of anti-VatM, anti-Rh50, or anti-p80 followed by the appropriate secondary antibody linked to Alexa 555. Two z-sections are shown for each cell. Z-sections 1 and 2 are approximately 1 µm and 3 µm, respectively, from the bottom of each cell. VC, vacuolar-shaped structures; VS, cytoplasmic vesicles; T, tubular-like structures within the cytoplasm; P, punctate distributions within the cytoplasm. Scale bars = 2.5 µm.

27-kDa, consistent with the predicted molecular weights of full-length CfaD and its putative cleavage product (Fig. 6B) [42]. After 48 hours of growth, there was significantly more CfaD (i.e., both 65-kDa and 27-kDa proteins) in $cln3^-$ whole cell lysates compared to AX3 lysates (Fig. 6B). However, there was no significant difference between strains in the intracellular level of either protein after 72 hours of growth (Fig. 6B). There was no significant effect resulting from Cln3 deficiency on the amounts of full-length CfaD or its cleavage product in conditioned media after 48 and 72 hours of growth (Fig. 6B). The absence of actin and tubulin from samples of conditioned growth media verified that the samples were not contaminated with intracellular proteins (Fig. 6C). Together, these data suggest that Cln3 deficiency in *Dictyostelium* leads to an enhanced rate of cell proliferation that is concomitant with alterations in secretory proteins that regulate extracellular proliferation signaling.

Cln3 deficiency accelerates the formation of tipped mounds and slugs during mid-development

Given the dramatic increase in $cln3$ expression upon entering developmental phases of the *Dictyostelium* life cycle, we next sought to extend our analysis of Cln3 function to developmental processes. After 12 hours of development, $33\pm5\%$ of $cln3^-$ structures had progressed to the tipped mound stage of development, compared to only $3\pm1\%$ of AX3 structures (Fig. 7A,B). By 15 hours, $83\pm3\%$ of $cln3^-$ multicellular structures had developed into either fingers or slugs compared to only $19\pm3\%$ of AX3 structures (Fig. 7A, C). Overexpression of *Dictyostelium* GFP-Cln3, or expression of *Dictyostelium* GFP-Cln3 or human GFP-CLN3 under the control of the $cln3$ upstream element in $cln3^-$ cells, suppressed the precocious development of $cln3^-$ cells at both the 12- and 15-hour time points to levels that were not significantly different from AX3 (Fig. 7A–C). Thus, Cln3 deficiency leads to precocious mid-stage development of *Dictyostelium* and this acceleration can be returned to near-normal levels by re-introducing *Dictyostelium* Cln3 or human CLN3 in an N-terminal fusion with GFP.

Cln3 deficiency increases slug migration and accelerates fruiting body formation during late development

During the later stages of *Dictyostelium* development, a larger number of $cln3^-$ slugs were observed to migrate outside the spot of deposition compared to AX3 slugs (Fig. 8A). After 18 hours, $41\pm2\%$ of $cln3^-$ slugs migrated out of the spot of deposition compared to only $16\pm2\%$ of AX3 slugs (Fig. 8B). Notably, this could not be accounted for by the overall accelerated rate of development observed in $cln3^-$ cells, since a significantly higher percentage of $cln3^-$ slugs also migrated out of the spot after 21 hours compared to AX3 slugs (Fig. 8A, unpublished data). Overexpression of *Dictyostelium* GFP-Cln3, or expression of *Dictyostelium* GFP-Cln3 or human GFP-CLN3 under the control of the $cln3$ upstream element in $cln3^-$ cells, significantly suppressed this slug migration phenotype to levels observed for AX3 slugs (Fig. 8A,B). Interestingly, the slug migration phenotype could not be explained by a defect in phototaxis, since we observed no obvious effect of $cln3$ knockout on slug migration in a phototaxis assay (unpublished data).

Finally, Cln3 deficiency significantly accelerated fruiting body formation for those structures that remained in the deposition spot. After 18–21 hours of development, $86\pm3\%$ of $cln3^-$ structures had developed into fruiting bodies compared to only $55\pm6\%$ of AX3 structures (Fig. 8A, C). As it did for the slug migration stage, overexpression of *Dictyostelium* GFP-Cln3 or expression of *Dictyostelium* GFP-Cln3 or human GFP-CLN3 under the control of the $cln3$ upstream element, in $cln3^-$ cells, significantly

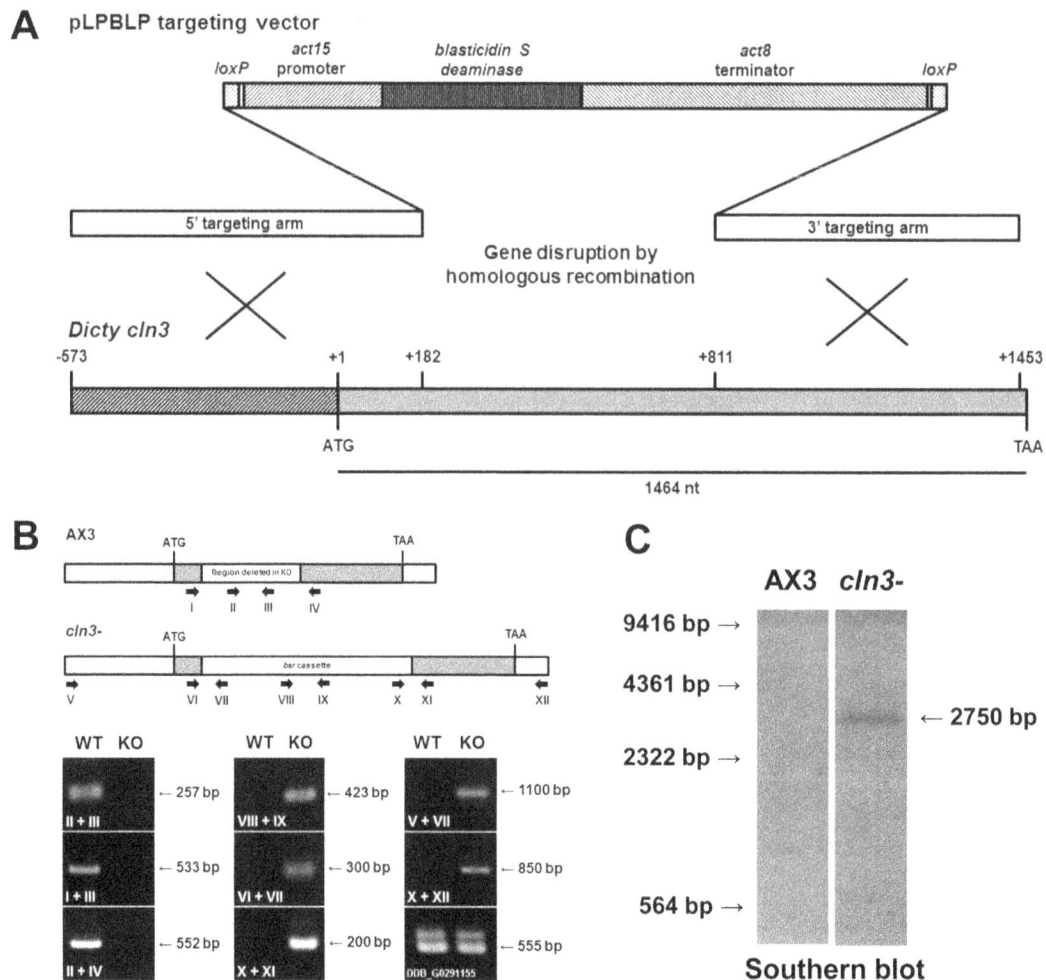

Figure 4. Generation of a *Dictyostelium cln3* knockout mutant. (A) Creation of a *Dictyostelium cln3* knockout mutant by homologous recombination. The pLPBLP targeting vector and sites of recombination are shown. (B) Validation of *cln3* knockout by PCR analysis. Primers are denoted by Roman numerals and arrows. The *Dictyostelium* gene denoted DDB_G0291155 lies downstream of *cln3* and was amplified to confirm that the insertion of the *bsr* cassette did not affect this gene. (C) Validation of *cln3* knockout by Southern blotting. DNA ladder (in bp) is shown to the left of the blot.

suppressed the accelerated fruiting body formation to levels that were not significantly different from AX3 (Fig. 8C).

Taken together, these data strongly indicate that Cln3 deficiency causes an overall accelerated rate of development in *Dictyostelium*, but that development nevertheless proceeds to the fruiting body stage (Fig. 8D). The ability to rescue the precocious development of *cln3⁻* cells by introducing human CLN3 strongly supports the notion that these steps require a function that is conserved between *Dictyostelium* and humans.

Calcium chelation restores the timing of *cln3⁻* slug formation and suppresses the abnormal migration of *cln3⁻* slugs

Since calcium signaling has been shown to be involved in regulating a number of developmental processes in *Dictyostelium* [49–52], the effect of calcium chelation on the substantial acceleration of mid-developmental events in *cln3⁻* cells was assessed. AX3 and *cln3⁻* cells were deposited on filters soaked in EGTA at concentrations that have previously been shown to be effective at chelating calcium during *Dictyostelium* development [51,52]. The timing of slug formation and the extent of slug

migration were then assessed. Interestingly, EGTA (1 mM and 2 mM) suppressed the accelerated formation of *cln3⁻* slugs and fingers after 15 hours of development, and suppressed the enhanced migration of *cln3⁻* slugs at the 18-hour time point to levels that were not significantly different from AX3 (Fig. 9A–D). EGTA had no significant effect on the accelerated formation of *cln3⁻* fruiting bodies (unpublished data).

Discussion

In this study, we have shown that *Dictyostelium* contains an ortholog of CLN3, for which loss-of-function mutations in humans causes the childhood onset neurodegenerative disorder JNCL. We generated a *Dictyostelium cln3* knockout mutant that was validated by PCR and Southern blotting and have provided evidence that links Cln3 function to axenic growth and multicellular development. *Dictyostelium* GFP-Cln3 localizes primarily to the CV system, and to a lesser extent, to compartments of the endocytic pathway. Expression of *Dictyostelium* GFP-Cln3 or human GFP-CLN3 in *cln3⁻* cells suppresses the aberrant proliferation, precocious development, and slug migration phenotypes observed in knockout cells. Together, our data strongly suggest that Cln3 is

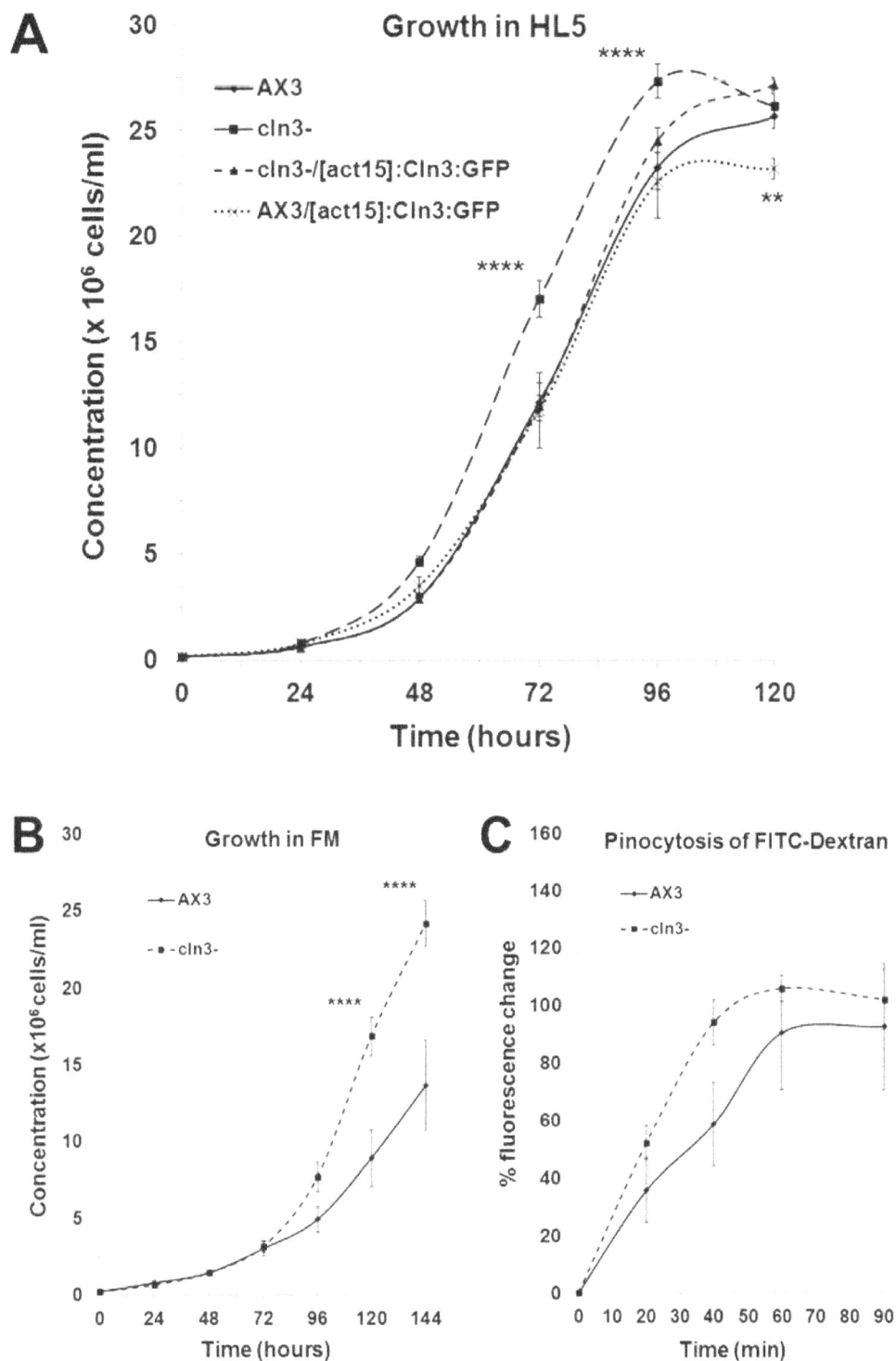

Figure 5. Effect of *cln3* knockout on cell proliferation and pinocytosis. (A) Axenic growth of AX3 *cln3⁻*, *cln3⁻*/[*act15*]:Cln3:GFP, and AX3/[*act15*]:Cln3:GFP cells in HL5 medium. Data presented as mean concentration (×10⁶ cells/ml) ± s.e.m (n = 10–20). (B) Axenic growth of AX3 and *cln3⁻* cells in FM medium. Data presented as mean concentration (×10⁶ cells/ml) ± s.e.m (n = 8). (C) Effect of *cln3* knockout on the intracellular accumulation of FITC-dextran. Data presented as mean % fluorescence change ± s.e.m (n = 10). Statistical significance was assessed using two-way ANOVA followed by Bonferroni post-hoc analysis. Two-way ANOVA revealed a significant effect of genotype on the growth curves shown in panels A and B (p<0.001 and p<0.01, respectively). **p-value<0.01 and ****p-value<0.0001 vs. AX3 as determined from Bonferroni post-hoc analysis at the indicated time points.

A AprA

B CfaD

C

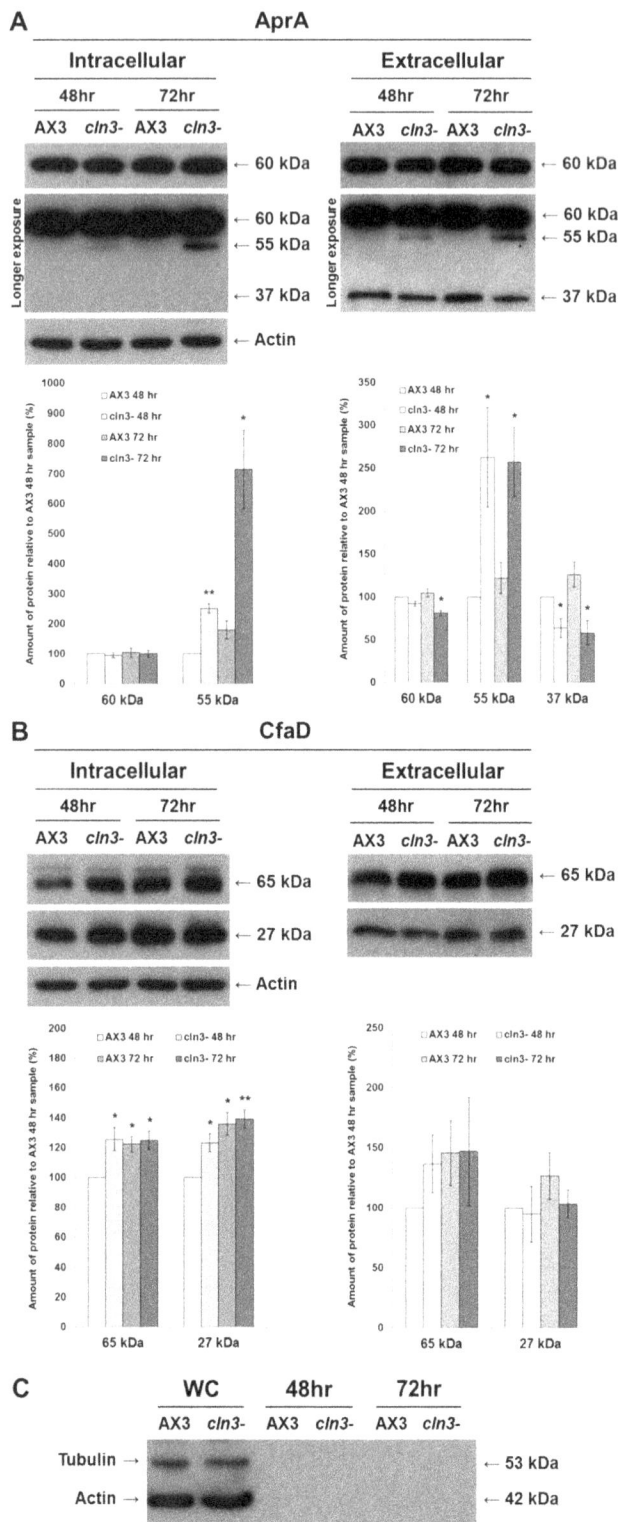

Figure 6. Effect of *cln3* knockout on the intra- and extracellular levels of AprA and CfaD. AX3 and *cln3⁻* cells grown axenically in HL5 were harvested and lysed after 48 and 72 hours of growth. Whole cell lysates (20 µg) (i.e., intracellular) and samples of conditioned growth media (i.e., extracellular) were separated by SDS-PAGE and analyzed by western blotting with anti-AprA, anti-CfaD, anti-tubulin, and anti-actin. Molecular weight markers (in kDa) are shown to the right of each blot. (A) Intra- and extracellular protein levels of AprA. Immunoblots that were exposed for a longer period of time (i.e., longer exposure) are included to show the 55-kDa and 37-kDa protein

bands detected by anti-AprA. Note that the 37-kDa protein was detected in samples of conditioned growth media, but not in whole cell lysates. (B) Intra- and extracellular protein levels of CfaD. Data in all plots presented as mean amount of protein relative to AX3 48 hour sample (%) ± s.e.m (n = 4 independent experimental means, from 2 replicates in each experiment). Statistical significance was determined using a one-sample t-test (mean, 100; two-tailed) vs. the AX3 48 hour sample. *p-value<0.05. **p-value<0.01. (C) Detection of tubulin and actin in whole cell lysates (WC; lanes 1–2), but not in samples of conditioned growth media (lanes 3–6).

a negative regulator of proliferation and development in *Dictyostelium*. Finally, we have provided evidence linking AprA secretion and cleavage to Cln3 function during growth, and calcium signaling to Cln3 function during multicellular development.

The enhanced proliferation of *cln3⁻* cells, coupled with the observation that *Dictyostelium* GFP-Cln3 overexpression in AX3 cells significantly reduces the final density of stationary phase cultures, strongly support the notion that Cln3 negatively regulates this cellular process in *Dictyostelium*. In *Dictyostelium*, extracellular liquid is ingested by macropinocytosis. [53]. An increased rate of pinocytosis would conceivably allow cells to ingest nutrients required for growth at an enhanced rate. Moreover, Journet et al. [54] identified Cln3 in an analysis of the macropinocytic proteome of *Dictyostelium* amoeba. Our pinocytosis analysis of *cln3⁻* cells during axenic growth only revealed minor differences suggesting further work is needed to fully elucidate the mechanisms by which Cln3 deficiency affects cell proliferation in *Dictyostelium*. In other systems, CLN3 has also been reported to localize to the endocytic pathway and its deficiency impairs endocytosis in those systems [19,55–59]. Together, our results, coupled with those reported by others, indicate that further research is required to determine the precise function of CLN3 in the endocytic pathway, which may be organism or cell-type dependent.

Based on our observations of the intra- and extracellular amounts of AprA and the fact that AprA negatively regulates cell proliferation in *Dictyostelium* [41], it would appear that the enhanced proliferation of *cln3⁻* cells could be at least partially explained by the lack of full-length AprA and its putative 37-kDa cleavage product in conditioned media. Since the intracellular amounts of 60-kDa AprA were not significantly different between AX3 and *cln3⁻* cells, thus excluding the possibility that *aprA* gene expression or translation were affected by Cln3 deficiency, our results suggest that Cln3 facilitates the secretion of AprA during growth. The detection of a 37-kDa protein by the highly specific anti-AprA antibody in conditioned media, but not whole cell lysates, suggests that AprA is cleaved extracellularly during growth. Since the amount of the 37-kDa protein was significantly reduced in *cln3⁻* cells, these results suggest that Cln3 deficiency also negatively affects the secretion of a protease required for AprA cleavage. This is supported by previous studies that have reported the proteolytic cleavage of extracellular proteins during growth and development [60–63]. Furthermore, a study describing the secreted proteome profile of growing and developing *Dictyostelium* cells also reports the detection of a large number of extracellular proteases in conditioned media [64]. Like AprA, CfaD is part of a ~150 kDa complex that functions extracellularly to repress cell proliferation in *Dictyostelium*, and chromatography and pull-down assays suggest that CfaD interacts with AprA [42]. Since increased levels of intracellular CfaD were observed in *cln3⁻* cells during the early stages of axenic growth, our results suggest that altered CfaD secretion could also explain the enhanced proliferation of *cln3⁻* cells. However, we observed no correlated decrease in the extracellular levels of CfaD over the same time period. Neverthe-

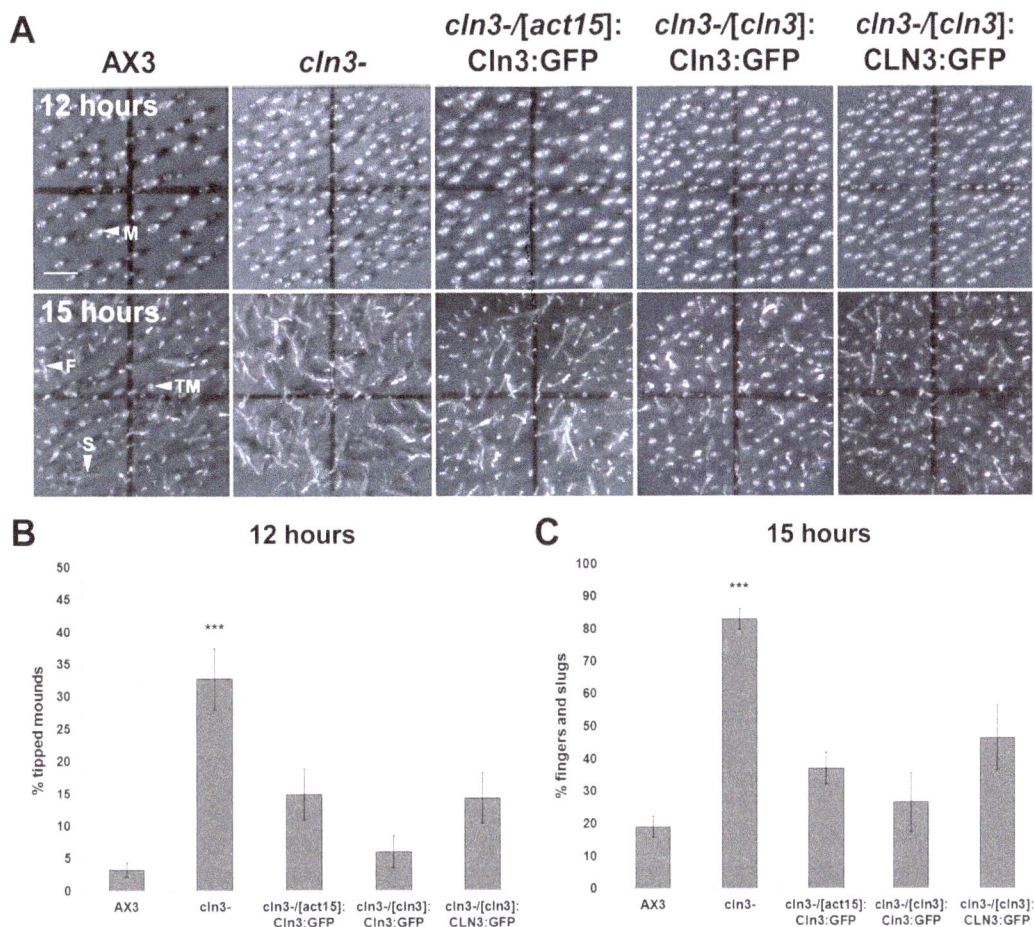

Figure 7. Effect of *cln3* knockout on the formation of tipped mounds and slugs. (A) AX3, *cln3⁻*, or *cln3⁻* cells overexpressing GFP-Cln3 or expressing GFP-Cln3 or GFP-CLN3 under the control of the *cln3* upstream element imaged after 12 and 15 hours of development. Images are a top-view of developing cells. (B) Quantification of the number of tipped mounds observed after 12 hours of development. Data presented as mean % tipped mounds ± s.e.m (n = 10–19). (C) Quantification of the number of fingers and slugs observed after 15 hours of development. Data presented as mean % fingers and slugs ± s.e.m (n = 10–33). Statistical significance was assessed using the Kruskal-Wallis test followed by the Dunn multiple comparison test (***p-value<0.001 vs. AX3). Scale bars = 1 mm. M, mound; TM, tipped-mound; F, finger; S, slug.

less, our data indicate that Cln3 facilitates the secretion of AprA, and may to a lesser extent, also facilitate CfaD secretion. Taken together, the altered secretion of these extracellular signaling proteins could explain the enhanced proliferation of *cln3⁻* cells.

During growth, *Dictyostelium* GFP-Cln3 localized primarily to the CV system in live and fixed cells, and to a lesser extent to the endocytic system. In *Dictyostelium*, the CV system is dynamic and functions in a number of cellular processes including osmoregulation, calcium storage, protein transport to the plasma membrane, and secretion [45,65,66]. Although *Dictyostelium* GFP-Cln3 was observed to localize to the CV system, we observed no obvious sensitivity of *cln3⁻* cells to hypo-osmotic conditions during growth in HL5 (25% HL5, 75% double-distilled water) or during starvation in double-distilled water (unpublished data). However, further analysis is required to determine if there are subtle effects of Cln3-deficiency on osmoregulation during *Dictyostelium* growth. In *Dictyostelium*, the CV and endosomal systems appear to be physically separated from each other. However, some experimental evidence also indicates that controlled intracellular transport might occur between these two systems [67–69]. The observation that GFP-Cln3 localizes to both the CV and endocytic systems in *Dictyostelium* is consistent with the localization of mammalian CLN3 to multiple subcellular compartments

including the endocytic and lysosomal systems [13]. Notably, endogenous Cln3 has been reported within fractions of the macropinocytic pathway in *Dictyostelium*, consistent with our localization data presented here [3,46,47,53]. Finally, since *Dictyostelium* GFP-Cln3 is able to rescue growth and developmental phenotypes, we are confident that we have correctly identified the subcellular localization of Cln3 in *Dictyostelium*.

Phenotypes were observed in *cln3⁻* cells during mid- and late *Dictyostelium* development that further support Cln3 as a negative regulator in *Dictyostelium*. Consistent with the relatively higher expression of *cln3* mRNA during mid- and late development, loss of *cln3* by gene knockout significantly accelerated the formation of mid- and late developmental structures. Precocious development has been observed in a number of *Dictyostelium* knockout mutants. Specifically, early tipped mound formation has been reported in strains overexpressing cyclin C, cyclin-dependent kinase 8, or the G-protein alpha 5 subunit, and in knockout mutants of histidine kinase C, a metabotropic glutamate receptor-like protein, protein inhibitor of STAT, and SCAR/WAVE [70–75]. Several knockout mutants that display increased slug migration have been described, including mutants for genes important for oxysterol binding, the assembly of mitochondrial complex I, and the targeting of proteins for degradation via proteasomes [76–78]. This phenotype has also

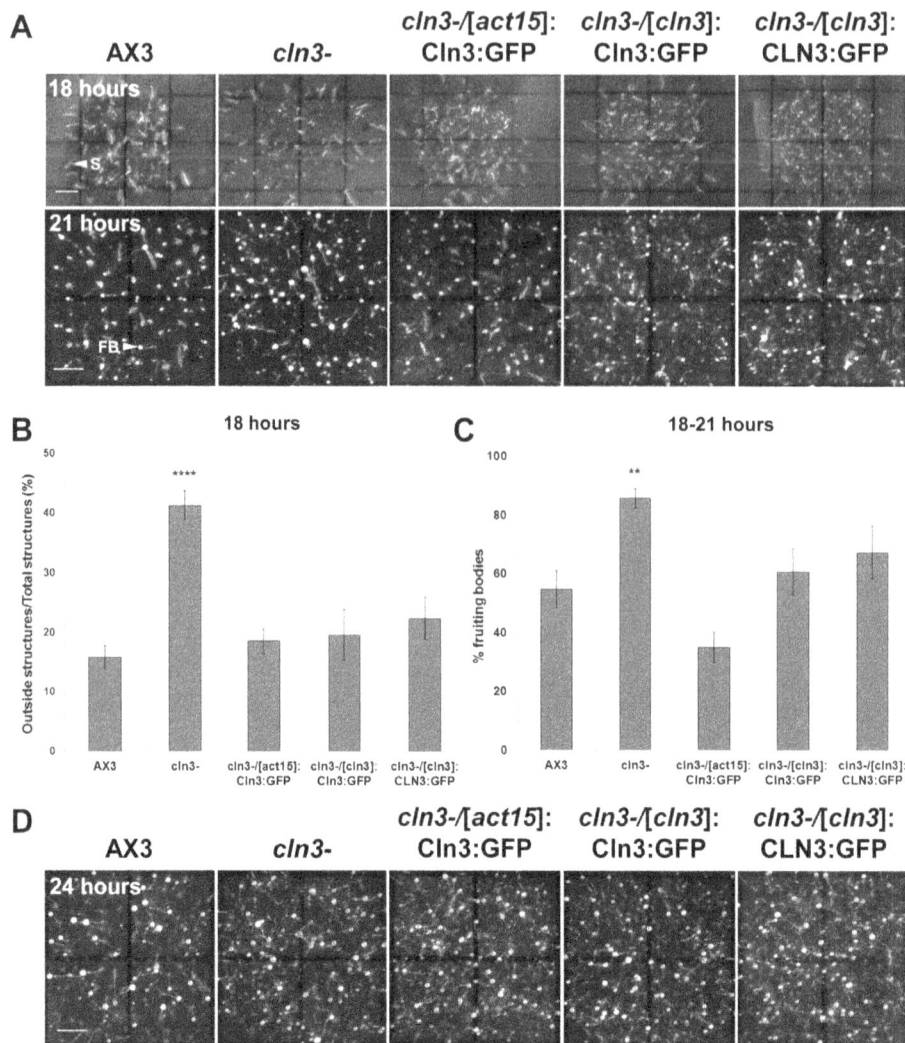

Figure 8. Effect of *cln3* knockout on slug migration and fruiting body formation. (A) AX3, *cln3⁻*, or *cln3⁻* cells overexpressing GFP-Cln3 or expressing GFP-Cln3 or GFP-CLN3 under control of the *cln3* upstream element imaged after 18 and 21 hours of development. (B) Quantification of the number of slugs that migrated outside the spot of deposition after 18 hours. Data presented as mean outside structures/total structures (%) ± s.e.m (n = 10–28). (C) Quantification of the number of fruiting bodies observed after 18–21 hours of development. Data presented as mean % fruiting bodies ± s.e.m (n = 10–32). (D) Fruiting bodies formed after 24 hours of development. Images in A and D are a top-view of developing cells. Statistical significance in B was assessed using one-way ANOVA (p<0.0001) followed by the Bonferroni multiple comparison test (****p-value<0.0001 vs. AX3). Statistical significance in C was assessed using the Kruskal-Wallis test followed by the Dunn multiple comparison test (**p-value<0.01 vs. AX3). Scale bars = 1 mm. S, slug; FB, fruiting body.

been observed in cells overexpressing histidine kinase C or in cells where calcium-binding protein 3 expression has been knocked down by RNAi [72,79]. The diversity of functions associated with these proteins as well as those discussed above for the other developmental phenotypes in *cln3⁻* cells, highlight the importance of elucidating the signal transduction pathways underlying the function of Cln3 during *Dictyostelium* development.

The ability to completely restore the timing of *cln3⁻* slug formation and the enhanced slug migration through the chelation of calcium provides some mechanistic insight into the signaling pathways affected by Cln3 deficiency during these stages of the life cycle. These results are interesting given that *Dictyostelium* GFP-Cln3 localizes predominantly to the CV system, which has been shown to be a highly efficient store of intracellular calcium, and to be required for cAMP-induced calcium influx [65]. In addition, the primary sensor of intracellular calcium, calmodulin, is found predominantly on the membranes of the CV system [80,81]. Our

results are consistent with studies in mammalian systems that have reported altered calcium homeostasis in the absence of functional CLN3, which may lead to synaptic dysfunction and neuronal apoptosis [82–83]. Furthermore, CLN3 has been shown to bind to the neuronal calcium-binding protein, calsenilin, in a calcium-dependent manner [84].

Taken together, our data strongly supports Cln3 as a negative regulator of proliferation and development in *Dictyostelium*. Furthermore, our study indicates that *cln3* knockout in *Dictyostelium* compromises the cell's ability to respond to extracellular and/or environmental cues. This first report of a *Dictyostelium* model to study NCL should spur further research using this important model organism. In addition to *CLN3*, *Dictyostelium* also possesses homologs to most of the other known NCL genes (e.g., *CLN1-5, CLN7, CLN10-14*) indicating that the NCL biological pathway is likely to be conserved in this model system. The cellular processes and signaling pathways that regulate the

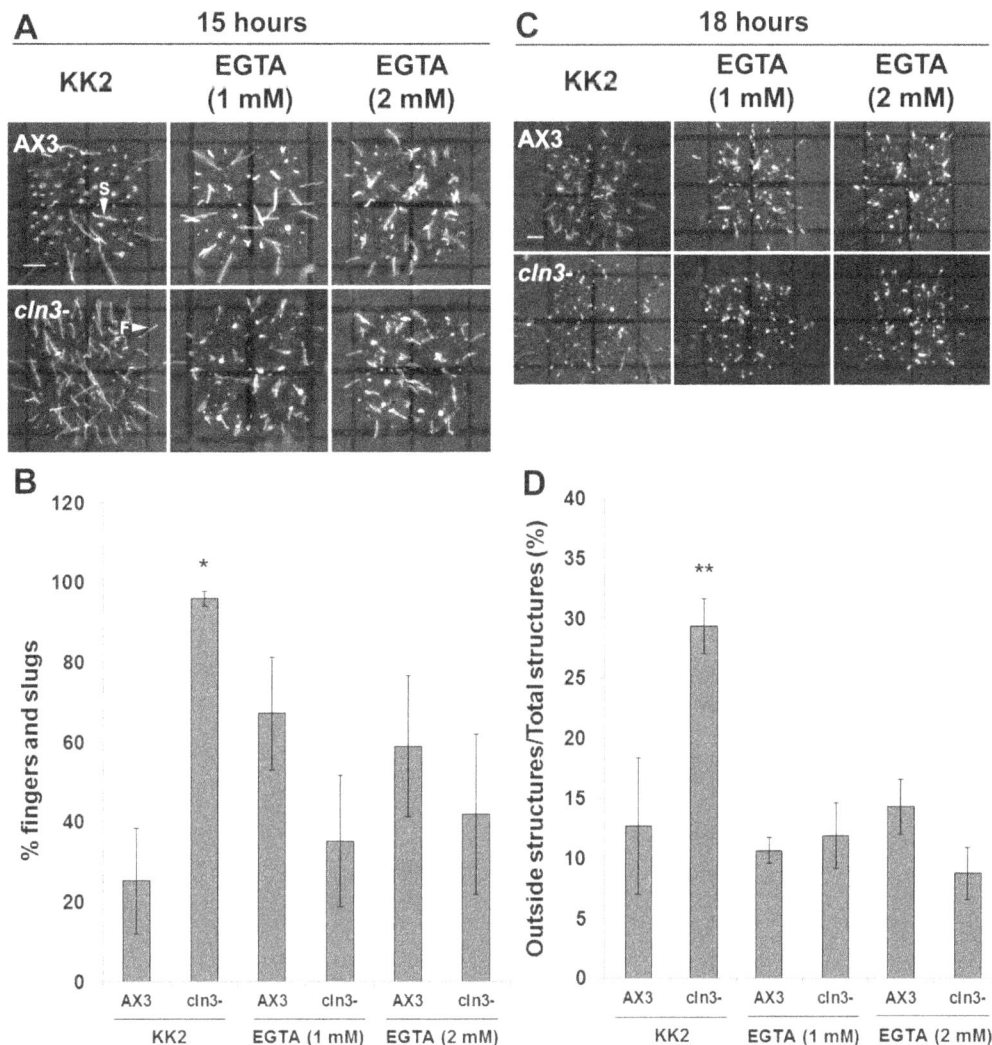

Figure 9. Effect of calcium chelation on AX3 and *cln3⁻* slug formation and migration. (A) AX3 and *cln3⁻* cells developed in the presence of KK2± EGTA and imaged after 15 hours of development. Scale bar = 1 mm. (B) Quantification of the number of fingers and slugs observed after 15 hours of development. Data presented as mean % fingers and slugs ± s.e.m (n≥4). (C) AX3 and *cln3⁻* cells developed in the presence of KK2± EGTA and imaged after 18 hours of development. Scale bar = 1 mm. (D) Quantification of the number of slugs that migrated outside the spot of deposition after 18 hours. Data presented as mean outside structures/total structures (%) ± s.e.m (n≥5). Images in A and C are a top-view of developing cells. Statistical significance in B was assessed using the Kruskal-Wallis test followed by the Dunn multiple comparison test (*p-value<0.05 vs. AX3). Statistical significance in D was assessed using one-way ANOVA (p<0.0001) followed by the Bonferroni multiple comparison test (**p-value<0.01 vs. AX3). F, finger; S, slug.

behavior of *Dictyostelium* cells are remarkably similar to those observed in human cells, strengthening the argument that investigation of NCL gene function in this model organism offers something unique to the study of this devastating group of inherited neurodegenerative disorders.

Supporting Information

Figure S1 Analysis of gene expression driven by endogenous *cln3* upstream elements. AX3 cells were transformed with the appropriate construct (pTX-GFP; *act15* promoter replaced with *cln3* upstream element 1, 2, or 3) and grown in HL5. Cells were harvested and lysed. Proteins (20 µg) were separated by SDS-PAGE and analyzed by western blotting with anti-GFP, anti-tubulin (loading control), or anti-actin (loading

control). Molecular weight markers (in kDa) are shown to the right of each blot.

Figure S2 Western blot analysis of *Dictyostelium* strains expressing *Dictyostelium* GFP-Cln3 or human GFP-CLN3 under the control of the *act15* promoter or *cln3* upstream element 1. (A–C) AX3 and *cln3⁻* cells were transformed with the appropriate construct (gene expression driven by the *act15* promoter) and grown in HL5. Cells were lysed and sample loading buffer was added to whole cell lysates which were either loaded directly into polyacrylamide gels or heated for 5 minutes at 95°C prior to loading into gels. Proteins (20 µg) were separated by SDS-PAGE and analyzed by western blotting with anti-GFP, anti-tubulin (loading control), or anti-actin (loading control). (D) AX3 and *cln3⁻* cells were transformed with the appropriate construct (gene expression driven by *cln3* upstream element 1) and grown in HL5. Cells were lysed and samples were prepared and analyzed as described above. Molecular weight markers (in kDa) are shown to the left of each blot.

Figure S3   Video of *Dictyostelium* GFP-Cln3 localization in AX3 cells incubated in water. AX3 cells expressing *Dictyostelium* GFP-Cln3 were grown overnight in low-fluorescence HL5. Cells were washed two times with double distilled water and then resuspended in double distilled water.

Table S1   List of primers used for *cln3* knockout validation and amplification of *cln3* upstream elements. The following primers were designed to amplify gDNA from AX3 and *cln3*⁻ cells to validate the knockout of the *cln3* gene in the *bsr* resistant clone and to amplify fragments upstream of the *cln3* start site. The *Dictyostelium* gene denoted DDB_G0291155 lies

downstream of *cln3* and was amplified to confirm that the insertion of the *bsr* cassette did not affect gene DDB_G0291155.

File S1   Results, Discussion, and References specific to the Supplemental Table and Figures.

Acknowledgments

The authors would like to thank Dr. Danton H. O'Day for critically reviewing the manuscript prior to submission. The anti-AprA and anti-CfaD antibodies were kindly provided as a gift by Dr. Richard H. Gomer.

Author Contributions

Conceived and designed the experiments: RJH MAM SLC. Performed the experiments: RJH MAM. Analyzed the data: RJH MAM SLC. Contributed reagents/materials/analysis tools: MAM SLC. Contributed to the writing of the manuscript: RJH MAM SLC.

References

1. Santavuori P (1988) Neuronal ceroid lipofuscinosis in childhood. Brain Dev 10: 80–83.
2. Anderson GW, Goebel HH, Simonati A (2013) Human pathology in NCL. Biochim Biophys Acta 1832: 1807–1826.
3. Kollmann K, Uusi-Rauva K, Scifo E, Tyynelä J, Jalanko A, et al. (2013) Cell biology and function of neuronal ceroid lipofuscinosis-related proteins. Biochim Biophys Acta 1832: 1866–1881.
4. Schulz A, Kohlschütter A, Mink J, Simonati A, Williams R (2013) NCL diseases - clinical perspectives. Biochim Biophys Acta 1832: 1801–1806.
5. Hofman I, van der Wal A, Dingemans K, Becker A (2001) Cardiac pathology in neuronal ceroid lipofuscinoses – a clinicopathologic correlation in three patients. Eur J Paediatr Neurol 5(Suppl A): 213–217.
6. Ostergaard J, Rasmussen T, Molgaard H (2011) Cardiac involvement in juvenile neuronal ceroid lipofuscinosis (Batten disease). Neurology 76: 1245–1251.
7. Chattopadhyay S, Ito M, Cooper JD, Brooks AI, Curran TM, et al. (2002) An autoantibody inhibitory to glutamic acid decarboxylase in the neurodegenerative disorder Batten disease. Hum Mol Genet 11: 1421–1431.
8. Castaneda J, Pearce D (2008) Identification of alpha-fetoprotein as an autoantigen in juvenile Batten disease. Neurobiol Dis 29: 92–102.
9. Haltia M, Goebel HH (2013) The neuronal ceroid-lipofuscinoses: A historical introduction. Biochim Biophys Acta 1832: 1795–1800.
10. The International Batten Disease Consortium (1995) Isolation of a novel gene underlying Batten disease, *CLN3*. Cell 82: 949–957.
11. Mole S (2012) NCL Mutation and Patient Database. NCL Resource – A Gateway for Batten Disease. Available: http://www.ucl.ac.uk/ncl/mutation.shtml.
12. Warrier V, Vieira M, Mole SE (2013) Genetic basis and phenotypic correlations of the neuronal ceroid lipofuscinoses. Biochim Biophys Acta 1832: 1827–1830.
13. Cotman SL, Staropoli JF (2012) The juvenile Batten disease protein, CLN3, and its role in regulating anterograde and retrograde post-Golgi trafficking. Clin Lipidol 7: 79–91.
14. Uusi-Rauva K, Kyttälä A, van der Kant R, Vesa J, Tanhuanpää K, et al. (2012) Neuronal ceroid lipofuscinosis protein CLN3 interacts with motor proteins and modifies location of late endosomal compartments. Cell Mol Life Sci 69: 2075–2089.
15. Luiro K, Kopra O, Lehtovirta M, Jalanko A (2001) CLN3 protein is targeted to neuronal synapses but excluded from synaptic vesicles: new clues to Batten disease. Hum Mol Genet 10: 2123–2131.
16. Pearce DA, Nosel SA, Sherman F (1999) Studies of pH regulation by Btn1p, the yeast homolog of human Cln3p. Mol Genet Metab 66: 320–323.
17. Gachet Y, Codlin S, Hyams JS, Mole SE (2005) btn1, the *Schizosaccharomyces pombe* homologue of the human Batten disease gene CLN3, regulates vacuole homeostasis. J Cell Sci 118: 5525–5536.
18. Holopainen JM, Saarikoski J, Kinnunen PK, Järvelä I (2001) Elevated lysosomal pH in neuronal ceroid lipofuscinoses (NCLs). Eur J Biochem 268: 5851–5856.
19. Luiro K, Yliannala K, Ahtiainen L, Maunu H, Järvelä I, et al. (2004) Interconnections of CLN3, Hook1 and Rab proteins link Batten disease to defects in the endocytic pathway. Hum Mol Genet 13: 3017–3027.
20. Cao Y, Espinola JA, Fossale E, Massey AC, Cuervo AM, et al. (2006) Autophagy is disrupted in a knock-in mouse model of juvenile neuronal ceroid lipofuscinosis. J Biol Chem 281: 20483–20493.
21. Getty AL, Pearce DA (2011) Interactions of the proteins of neuronal ceroid lipofuscinosis: clues to function. Cell Mol Life Sci 68: 453–474.
22. van Egmond WN, Kortholt A, Plak K, Bosgraaf L, Bosgraaf S, et al. (2008) Intramolecular activation mechanism of the *Dictyostelium* LRRK2 homolog Roco protein GbpC. J Biol Chem 283: 30412–30420.
23. Gilsbach BK, Ho FY, Vetter IR, van Haastert PJ, Wittinghofer A, et al. (2012) Roco kinase structures give insights into the mechanism of Parkinson disease-related leucine-rich-repeat kinase 2 mutations. Proc Natl Acad Sci USA 109: 10322–10327.
24. Meyer I, Kuhnert O, Gräf R (2011) Functional analyses of lissencephaly-related proteins in *Dictyostelium*. Semin Cell Dev Biol 22: 89–96.
25. McMains VC, Myre M, Kreppel L, Kimmel AR (2010) *Dictyostelium* possesses highly diverged presenilin/gamma-secretase that regulates growth and cell-fate specification and can accurately process human APP: a system for functional studies of the presenilin/gamma-secretase complex. Dis Model Mech 3: 581–594.
26. Myre MA, Lumsden AL, Thompson MN, Wasco W, MacDonald ME, et al. (2011) Deficiency of huntingtin has pleiotropic effects in the social amoeba *Dictyostelium discoideum*. PLoS Genet 7: e1002052.
27. Lo Sardo V, Zuccato C, Gaudenzi G, Vitali B, Ramos C, et al. (2012) An evolutionary recent neuroepithelial cell adhesion function of huntingtin implicates ADAM10-Ncadherin. Nat Neurosci 15: 713–721.
28. Myre MA (2012) Clues to γ-secretase, huntingtin and Hirano body normal function using the model organism *Dictyostelium discoideum*. J Biomed Sci 10: 19–41.
29. Maniak M (2011) *Dictyostelium* as a model for human lysosomal and trafficking diseases. Semin. Cell Dev Biol 22: 114–119.
30. Huber RJ, O'Day DH (2012) A matricellular protein and EGF-like repeat signalling in the social amoebozoan *Dictyostelium discoideum*. Cell Mol Life Sci 69: 3989–3997.
31. Muller-Taubenberger A, Kortholt A, Eichinger L (2013) Simple system - substantial share: The use of *Dictyostelium* in cell biology and molecular medicine. Eur J Cell Biol 92: 45–53.
32. Huber RJ (2014) The cyclin-dependent kinase family in the social amoebozoan *Dictyostelium discoideum*. Cell Mol Life Sci 71: 629–639.
33. Levi S, Polyakov M, Egelhoff TT (2000) Green fluorescent protein and epitope tag fusion vectors for *Dictyostelium discoideum*. Plasmid 44: 231–238.
34. Rivero F, Maniak M (2006) Quantitative and microscopic methods for studying the endocytic pathway. In: Eichinger L, Rivero F, editors. Methods in Molecular Biology 346: Dictyostelium discoideum Protocols. New Jersey: Humana Press/Totowa. 423–438.
35. Huber RJ, O'Day DH (2012) The cyclin-dependent kinase inhibitor roscovitine inhibits kinase activity, cell proliferation, multicellular development, and Cdk5 nuclear translocation in *Dictyostelium discoideum*. J Cell Biochem 113: 868–876.
36. Huber RJ, O'Day DH (2011) Nucleocytoplasmic transfer of cyclin dependent kinase 5 and its binding to puromycin-sensitive aminopeptidase in *Dictyostelium discoideum*. Histochem Cell Biol 136: 177–189.
37. Charette SJ, Cosson P (2006) Exocytosis of late endosomes does not directly contribute membrane to the formation of phagocytic cups or pseudopods in *Dictyostelium*. FEBS Lett 580: 4923–4928.
38. Fok AK, Clarke M, Ma L, Allen RD (1993) Vacuolar H+-ATPase of *Dictyostelium discoideum*. A monoclonal antibody study. J Cell Sci 106: 1103–1113.
39. Benghezal M, Gotthardt D, Cornillon S, Cosson P (2001) Localization of the Rh50-like protein to the contractile vacuole in *Dictyostelium*. Immunogenetics 52: 284–288.
40. Ravanel K, de Chassey B, Cornillon S, Benghezal M, Zulianello L, et al. (2001) Membrane sorting in the endocytic and phagocytic pathway of *Dictyostelium discoideum*. Eur J Cell Biol 80: 754–764.
41. Brock DA, Gomer RH (2005) A secreted factor represses cell proliferation in *Dictyostelium*. Development 132: 4553–4562.

42. Bakthavatsalam D, Brock DA, Nikravan NN, Houston KD, Hatton RD, et al. (2008) The secreted Dictyostelium protein CfaD is a chalone. J Cell Sci 121: 2473–2480.

43. Faix J, Kreppel L, Shaulsky G, Schleicher M, Kimmel AR. (2004) A rapid and efficient method to generate multiple gene disruptions in Dictyostelium discoideum using a single selectable marker and the Cre-loxP system. Nucleic Acids Res 32: e143.

44. Haskell RE, Derksen TA, Davidson BL (1999) Intracellular trafficking of the JNCL protein CLN3. Mol Genet Metab 66: 253–260.

45. Gerisch G, Heuser J, Clarke M (2002) Tubular-vesicular transformation in the contractile vacuole system of Dictyostelium. Cell Biol Int 26: 845–852.

46. Rodriguez-Paris JM, Nolta KV, Steck TL (1993) Characterization of lysosomes isolated from Dictyostelium discoideum by magnetic fractionation. J Biol Chem 268: 9110–9116.

47. Temesvari L, Rodriguez-Paris J, Bush J, Steck TL, Cardelli J (1994) Characterization of lysosomal membrane proteins of Dictyostelium discoideum. A complex population of acidic integral membrane glycoproteins, Rab GTP-binding proteins and vacuolar ATPase subunits. J Biol Chem 269: 25719–25727.

48. Rot G, Parikh A, Curk T, Kuspa A, Shaulsky G, et al. (2009) dictyExpress: A Dictyostelium discoideum gene expression database with an explorative data analysis web-based interface. BMC Bioinformatics 10: 256.

49. Sakamoto H, Nishio K, Tomisako M, Kuwayama H, Tanaka Y, et al. (2003) Identification and characterization of novel calcium-binding proteins of Dictyostelium and their spatial expression patterns during development. Dev Growth Differ 45: 507–514.

50. Scherer A, Kuhl S, Wessels D, Lusche DF, Raisley B, et al. (2010) Ca²⁺ chemotaxis in Dictyostelium discoideum. J Cell Sci 123: 3756–3767.

51. Poloz Y, O'Day DH (2012) Ca²⁺ signaling regulates ecmB expression, cell differentiation and slug regeneration in Dictyostelium. Differentiation 84: 163–175.

52. Poloz Y, O'Day DH (2012) Colchicine affects cell motility, pattern formation and stalk cell differentiation in Dictyostelium by altering calcium signaling. Differentiation 83: 185–199.

53. Maniak M (2003) Fusion and fission events in the endocytic pathway of Dictyostelium. Traffic 4: 1–5.

54. Journet A, Klein G, Brugière S, Vandenbrouck Y, Chapel A, et al. (2012) Investigating the macropinocytic proteome of Dictyostelium amoebae by high-resolution mass spectrometry. Proteomics 12: 241–245.

55. Fossale E, Wolf P, Espinola JA, Lubicz-Nawrocka T, Teed AM, et al. (2004) Membrane trafficking and mitochondrial abnormalities precede subunit c deposition in a cerebellar cell model of juvenile neuronal ceroid lipofuscinosis. BMC Neurosci 5: 57.

56. Codlin S, Haines RL, Mole SE (2008) btn1 affects endocytosis, polarization of sterol-rich membrane domains and polarized growth in Schizosaccharomyces pombe. Traffic 9: 936–950.

57. Cao Y, Staropoli JF, Biswas S, Espinola JA, MacDonald ME, et al. (2011) Distinct early molecular responses to mutations causing vLINCL and JNCL presage ATP synthase subunit C accumulation in cerebellar cells. PLoS One 6: e17118.

58. Tecedor L, Stein CS, Schultz ML, Farwanah H, Sandhoff K, et al. (2013) CLN3 loss disturbs membrane microdomain properties and protein transport in brain endothelial cells. J Neurosci 33: 18065–18079.

59. Vidal-Donet JM, Cárcel-Trullols J, Casanova B, Aguado C, Knecht E (2013) Alterations in ROS activity and lysosomal pH account for distinct patterns of macroautophagy in LINCL and JNCL fibroblasts. PLoS One 8: e55526.

60. Suarez A, Huber RJ, Myre MA, O'Day DH (2011) An extracellular matrix, calmodulin-binding protein from Dictyostelium with EGF-like repeats that enhance cell motility. Cell Signal 23: 1197–1206.

61. Huber RJ, Suarez A, O'Day DH (2012) CyrA, a matricellular protein that modulates cell motility in Dictyostelium discoideum. Matrix Biol 31: 271–280.

62. Brock DA, Gomer RH (1999) A cell-counting factor regulating structure size in Dictyostelium. Genes Dev 13: 1960–1969.

63. Brock DA, Hatton RD, Giurgiutiu DV, Scott B, Ammann R, et al. (2002) The different components of a multisubunit cell number-counting factor have both unique and overlapping functions. Development 129: 3657–3668.

64. Bakthavatsalam D, Gomer RH (2010) The secreted proteome profile of developing Dictyostelium discoideum cells. Proteomics 10: 2556–2559.

65. Malchow D, Lusche DF, Schlatterer C, De Lozanne A, Müller-Taubenberger A (2006) The contractile vacuole in Ca²⁺-regulation in Dictyostelium: its essential function for cAMP-induced Ca²⁺-influx. BMC Dev Biol 6: 31.

66. Sriskanthadevan S, Lee T, Liu Z, Yang D, Siu CH (2009) Cell adhesion molecule DdCAD-1 is imported into contractile vacuoles by membrane invagination in a Ca²⁺- and conformation-dependent manner. J Biol Chem 284: 36377–36386.

67. Hacker U, Albrecht R, Maniak M (1997) Fluid-phase uptake by macropinocytosis in Dictyostelium. J Cell Sci 110: 105–112.

68. Gabriel D, Hacker U, Köhler J, Müller-Taubenberger A, Schwartz JM, et al. (1999) The contractile vacuole network of Dictyostelium as a distinct organelle: its dynamics visualized by a GFP marker protein. J Cell Sci 112: 3995–4005.

69. Mercanti V, Charette SJ, Bennett N, Ryckewaert JJ, Letourneur F, et al. (2006) Selective membrane exclusion in phagocytic and macropinocytic cups. J Cell Sci 119: 4079–4087.

70. Greene DM, Hsu DW, Pears CJ (2010) Control of cyclin C levels during development of Dictyostelium. PLoS One 5: e10543.

71. Hadwiger JA, Natarajan K, Firtel RA (1996) Mutations in the Dictyostelium heterotrimeric G protein alpha subunit G alpha5 alter the kinetics of tip morphogenesis. Development 122: 1215–1224.

72. Singleton CK, Zinda MJ, Mykytka B, Yang P (1998) The histidine kinase dhkC regulates the choice between migrating slugs and terminal differentiation in Dictyostelium discoideum. Dev Biol 203: 345–357.

73. Prabhu Y, Müller R, Anjard C, Noegel AA (2007) GrlJ, a Dictyostelium GABAB-like receptor with roles in post-aggregation development. BMC Dev Biol 7: 44.

74. Kawata T, Hirano T, Ogasawara S, Aoshima R, Yachi A (2011) Evidence for a functional link between Dd-STATa and Dd-PIAS, a Dictyostelium PIAS homologue. Dev Growth Differ 53: 897–909.

75. Bear JE, Rawls JF, Saxe III CL (1998) SCAR, a WASP-related protein, isolated as a suppressor of receptor defects in late Dictyostelium development. J Cell Biol 142: 1325–1335.

76. Fukuzawa M, Williams JG (2002) OSBPa, a predicted oxysterol binding protein of Dictyostelium, is required for regulated entry into culmination. FEBS Lett 527: 37–42.

77. Torija P, Vicente JJ, Rodrigues TB, Robles A, Cerdán S, et al. (2006) Functional genomics in Dictyostelium: MidA, a new conserved protein, is required for mitochondrial function and development. J Cell Sci 119: 1154–1164.

78. Nelson MK, Clark A, Abe T, Nomura A, Yadava N, et al. (2000) An F-Box/WD40 repeat-containing protein important for Dictyostelium cell-type proportioning, slug behaviour, and culmination. Dev Biol 224: 42–59.

79. Lee CH, Jeong SY, Kim BJ, Choi CH, Kim JS, et al. (2005) Dictyostelium CBP3 associates with actin cytoskeleton and is related to slug migration. Biochim Biophys Acta 1743: 281–290.

80. Zhu Q, Clarke M (1992) Association of calmodulin and an unconventional myosin with the contractile vacuole complex of Dictyostelium discoideum. J Cell Biol 118: 347–358.

81. Sriskanthadevan S, Brar SK, Manoharan K, Siu CH (2013) Ca²⁺-calmodulin interacts with DdCAD-1 and promotes DdCAD-1 transport by contractile vacuoles in Dictyostelium cells. FEBS J 280: 1795–1806.

82. An Haack K, Narayan SB, Li H, Warnock A, Tan L, et al. (2011) Screening for calcium channel modulators in CLN3 siRNA knock down SH-SY5Y neuroblastoma cells reveals a significant decrease of intracellular calcium levels by selected L-type calcium channel blockers. Biochim Biophys Acta 1810: 186–191.

83. Warnock A, Tan L, Li C, An Haack K, Narayan SB, et al. (2013) Amlodipine prevents apoptotic cell death by correction of elevated intracellular calcium in a primary neuronal model of Batten disease (CLN3 disease). Biochem Biophys Res Commun 436: 645–649.

84. Chang JW, Choi H, Kim HJ, Jo DG, Jeon YJ, et al. (2007) Neuronal vulnerability of CLN3 deletion to calcium-induced cytotoxicity is mediated by calsenilin. Hum Mol Genet 16: 317–326.

85. Bause E (1983) Structural requirements of N-glycosylation of proteins. Biochem J 209: 331–336.

86. Munroe PB, Mitchison HM, O'Rawe AM, Anderson JW, Boustany RM, et al. (1997) Spectrum of mutations in the Batten disease gene, CLN3. Am J Hum Genet 61: 310–316.

87. Haskell RE, Carr CJ, Pearce DA, Bennett MJ, Davidson BL (2000) Batten disease: evaluation of CLN3 mutations on protein localization and function. Hum Mol Genet 9: 735–744.

The Mucoid Switch in *Pseudomonas aeruginosa* Represses Quorum Sensing Systems and Leads to Complex Changes to Stationary Phase Virulence Factor Regulation

Ben Ryall[1], Marta Carrara[1], James E. A. Zlosnik[1¤b], Volker Behrends[1,2], Xiaoyun Lee[1¤a], Zhen Wong[1], Kathryn E. Lougheed[1], Huw D. Williams[1]*

1 Department of Life Sciences, Faculty of Natural Sciences, Imperial College London, Sir Alexander Fleming Building, London, United Kingdom, **2** Department of Surgery and Cancer, Faculty of Medicine, Imperial College London, Sir Alexander Fleming Building, London, United Kingdom

Abstract

The opportunistic pathogen *Pseudomonas aeruginosa* chronically infects the airways of Cystic Fibrosis (CF) patients during which it adapts and undergoes clonal expansion within the lung. It commonly acquires inactivating mutations of the anti-sigma factor MucA leading to a mucoid phenotype, caused by excessive production of the extracellular polysaccharide alginate that is associated with a decline in lung function. Alginate production is believed to be the key benefit of *mucA* mutations to the bacterium in the CF lung. A phenotypic and gene expression characterisation of the stationary phase physiology of *mucA22* mutants demonstrated complex and subtle changes in virulence factor production, including cyanide and pyocyanin, that results in their down-regulation upon entry into stationary phase but, (and in contrast to wildtype strains) continued production in prolonged stationary phase. These findings may have consequences for chronic infection if mucoid *P. aeruginosa* were to continue to make virulence factors under non-growing conditions during infection. These changes resulted in part from a severe down-regulation of both AHL-and AQ (PQS)-dependent quorum sensing systems. *In trans* expression of the cAMP-dependent transcription factor Vfr restored both quorum sensing defects and virulence factor production in early stationary phase. Our findings have implications for understanding the evolution of *P. aeruginosa* during CF lung infection and it demonstrates that *mucA22* mutation provides a second mechanism, in addition to the commonly occurring *lasR* mutations, of down-regulating quorum sensing during chronic infection this may provide a selection pressure for the mucoid switch in the CF lung.

Editor: Martin Pavelka, University of Rochester, United States of America

Funding: BBSRC PhD studentship. The funders had no role in study design, data collection and analysis, decision to publish, or preparation of the manuscript.

Competing Interests: The authors have declared that no competing interests exist.

* E-mail: h.d.williams@imperial.ac.uk

¤a Current address: Departement de Microbiologie Fondamentale, Universite de Lausanne, Lausanne, Switzerland
¤b Current address: Centre for Understanding and Preventing Infection in Children and Department of Pediatrics/Faculty of Medicine, University of British Columbia, Vancouver, British Columbia, Canada

Introduction

Pseudomonas aeruginosa forms intractable, chronic infections in around 80% of Cystic Fibrosis (CF) sufferers and is associated with decreased lung function and an increased risk of respiratory failure and death [1,2]. It's success as a CF lung pathogen is aided by the expression of a range of virulence factors including the potent toxin hydrogen cyanide, which is regulated by quorum sensing and expressed maximally under low oxygen tension, a condition that *P. aeruginosa* is believed to experience in the thickened mucus of the CF lung [3,4]. *P. aeruginosa* cyanide production is the mediating factor in the paralytic killing of *Caenorhabditis elegans* [5], is toxic to *Drosophila melanogaster* [6] and cyanide has been detected in burn wound infections caused by *P. aeruginosa* [7]. The potential clinical significance of cyanide in CF lung infections was recently demonstrated by the detection of cyanide in sputum from cystic fibrosis patients infected with *P. aeruginosa* and its presence is

associated with a decline in lung function [8,9]. However, the clinical significance of cyanogenesis and its consequences for the host remain unclear [10,11]. Cyanide is volatile and its usefulness as a surrogate marker for the diagnosis of *P. aeruginosa* and *Bcc* infection in children who cannot expectorate sputum is currently being investigated [12,13]. Cyanide has the potential to inhibit aerobic and anaerobic respiration [16]. *P. aeruginosa* avoids the toxic effects of cyanide in part by synthesising a respiratory chain terminated by a cyanide insensitive terminal oxidase [14,15] and by the action of detoxification mechanisms [16,17].

A major event in the course of chronic *P. aeruginosa* lung infection is a switch of infecting strains to an alginate over-producing, mucoid phenotype that is often associated with a poor prognosis for the patient [2,18]. Understanding the consequences of the switch to alginate overproduction is important to an understanding of how *P. aeruginosa* is able to achieve such persistent and problematic CF infections. The mucoid switch most commonly results from loss of regulation of the alternate sigma factor AlgU via mutation in the anti-sigma factor *mucA* [2,19]. The *P. aeruginosa* MucA protein along with MucB sequesters AlgU to the cytoplasmic membrane. Loss of function mutations of *mucA* result in AlgU being constitutively free to interact with core RNA polymerase and direct transcription from its target genes, which include the alginate biosynthetic genes located in the *algD* operon as well as AlgU's own operon [20,21], leading to alginate over production and elevated levels of AlgU. One of the most common mutations in *mucA* in CF clinical isolates is the *mucA22* mutation, which results in the truncation of MucA [20,21].

It is believed that alginate overproduction *per se* provides a major benefit to *P. aeruginosa* of the mucoid switch in the CF lung, due to it providing the bacterium with protection from the host immune response by inhibiting phagocytosis, protecting it against the oxidative burst and providing an immunomodulatory role [22,23]. However, AlgU regulates genes in addition to the alginate biosynthetic genes, including other transcriptional regulators, such as *algR* and the heat shock sigma factor *rpoH* [24,25]. Previous studies have shown an increase in type III secretion systems, a decrease in flagellum production and the decrease in C4-HSL-dependent quorum sensing in biofilms are associated with *mucA* mutation [26–28]. Recently it was shown that the *mucA22* mutation also affects the osmotic stress sensitivity of *P. aeruginosa* but only in stationary phase, suggesting that *mucA22*, in addition to affecting alginate production during growth, affects cellular physiology in non-growing cells, a factor that could be significant during chronic infections [29,30].

Given the evidence for the presence of cyanide in the CF lung together with its stationary phase synthesis and the chronic nature of *P. aeruginosa* CF lung infection, our aim was to investigate the effect of the switch to mucoidy on cyanide production by *P. aeruginosa*. We show here that *mucA* mutation causes a more complex and subtle change in cyanide and pyocyanin production than previously recognised, resulting in down-regulation upon entry into early stationary phase and, in contrast to wildtype strains, continued production in prolonged stationary phase. We further show that *mucA* mutation is associated with a profound down regulation of the three major quorum sensing systems of *P. aeruginosa*. We argue that these findings have significant implications in terms of chronic CF lung infection.

Materials and Methods

Bacterial strains and growth conditions

The bacterial strains used are listed in Table S1. Overnight starter cultures were prepared by inoculating several colonies into 5 ml LB broth in a 25 ml sterile universal bottle at 37°C with shaking at 200 rpm. For all assays and RNA extractions strains were grown by inoculating 1 ml of overnight culture into 50 ml LB broth in 250 ml conical flasks incubated in an orbital shaker at 200 rpm and 37°C.

Construction of *mucA22* mutants

Unmarked *mucA22* mutants (loss of a single base) were made in two *P. aeruginosa* wild type backgrounds: PAO1 and PAO381 by gene replacement of the wildtype allele as described previously [29], with the presence of the correct mutation checked by sequencing of PCR amplicons of the putative *mucA22* mutants. Non-mucoid suppressor strains were isolated from *mucA22* strains by growing on LBA plates for several days at 37°C, after which time non-mucoid outgrowths appeared from the edge of the mucoid colony. Suppressor strains are denoted by adding an S to the end of the name of the *mucA22* mutant they were isolated from.

Cyanide, pyocyanin and elastase assays

Cyanide measurement was performed using a cyanide ion-selective micro-electrode (Lazar Research Laboratories, L.A., CA, USA) as described in [31] and elastase was assayed using the elastin-congo red method of Ohman et al [32]. Pyocyanin was assayed as described by Essar et al [33]. Briefly, the culture supernatant was extracted into chloroform, which was then acidified with 0.2M HCl and the pyocyanin-containing acidic, aqueous layer was removed and it's absorbance determining at 520 nm. The A520 reading was then normalised by dividing by the final OD600 reading of the culture.

Quorum sensing signal molecule bioluminescent assay

lux-reporter strains were used to assay the levels of quorum sensing signal molecules in culture supernatants. Plasmid pSB401 in *E. coli* S17.1 was used for assaying 2-oxo-C12-HSL, plasmid pSB536 in *E. coli* S17.1 was used for assaying C4-HSL and PAO1Δ*pqsA:pqsA::CTXlux* [34,35] was used for assaying PQS. Luminescence and OD was measured in a FLUOstar Omega plate Reader (BMG Labtech). The luminescence output for each well was blank corrected and normalised to OD. Concentrations of 2-oxo-C12HSL and C4HSL were calculated with reference to an AHL standard curve. PQS levels were calculated relative to PAO1 8 hour supernatant levels.

qRT-PCR

RNA was extracted using RNeasy Protect Bacteria Mini Kit (Qiagen) using enzymatic lysis and proteinase K digestion with on column DNAase treatment. qRT-PCR was carried out using Rotor-Gene SYBR Green PCR Kit (Qiagen) in a Rotor-Gene 3000 quantitative PCR machine (Corbett). Expression of the target genes was normalised to the expression of 16S RNA.

Results

mucA22 mutants and isolation and use of spontaneous suppressor strains

Isogenic *mucA22* mutants were constructed in PAO1 and PAO381 backgrounds and spontaneous mucoid suppressor strains were isolated from each of the *mucA22* mutants as described in Materials and Methods. We used these as control strains in this study as we wanted to know whether suppression of mucoidy affected key phenotypes associated with *mucA* mutation, particularly as suppressors are likely to accumulate in the CF lung [36]. We did not complement or repair the *mucA22* mutations as a control, because the high frequency of spontaneous mucoid to non-mucoid conversion in our hands made it difficult to be confident that a change to non-mucoidy did not result from the acquisition of a secondary mutation. By gene sequencing we confirmed that the suppressor strains, for which data is reported here, did not have a *mucA22* reverting mutation and neither did they have mutations in *algU*, *algR*, *algD2*, or *lasR*.

P. aeruginosa mucA22 strains are deficient in cyanide, pyocyanin and elastase production in early stationary phase

As an initial test cyanide levels were measured in 13 independently isolated *mucA22* strains 2 hours after entry of the cultures into stationary phase. Each of the *mucA22* strains had markedly reduced culture cyanide levels compared to their isogenic wildtype strain equating to an approximate 75% reduction in peak cyanide levels (comparing mean cyanide concentration for each *mucA22* strain to their respective wild type, t-test, all p values<0.001; Fig. S1A). All spontaneous non-mucoid suppressor strains isolated from *mucA22* strains showed a return to wildtype cyanide production. PAO579, a mucoid PAO381 derivative that has made the switch to mucoidy by an *algU/mucA22* independent mechanism [37], produces ~3 times the cyanide levels of the PAO381 *mucA22* strains (data not shown), strongly suggesting that the phenotype observed is a consequence of *mucA22* mutation rather than alginate over production *per se*. At 2 h into stationary phase the *mucA22* strains produced significantly less pyocyanin than their wild type parents (t-test p<0.001 for each pair; Fig. S1B). In contrast, the suppressor strains showed wild type pyocyanin production. In stationary phase *mucA22* mutants were also defective in elastase production (p<0.01 for each pair) and this phenotype was partially reversed in suppressor strain (Fig. S1C).

P. aeruginosa mucA22 strains maintain cyanide and pyocyanin production in late stationary phase

Investigating the kinetics of cyanide production throughout growth indicated marked differences between wildtype and *mucA22* mutants (Fig. 1A; Fig. S2). For the wild type strain PAO1, cyanide concentrations in culture media increased rapidly at the onset of stationary phase and reached a maximum of ~300 μM after 8–9 hours in culture, followed by a steady decrease in levels until cyanide was no longer detectable after 20 hours. In contrast, the *mucA22* mutant (M04) had the expected lower peak-level of ~100 μM cyanide, but these levels were maintained into late stationary phase. The suppressor strains behave similarly to PAO1 and a similar pattern of cyanide production was found in *mucA22* mutants isolated in PAO381 background (Fig. S2). Culture pH was similar for all strains throughout growth, indicating that cyanide gas formation, was not responsible for the differential levels of solution cyanide. We conclude that the wildtype strains switch off cyanide production a few hours into stationary phase and any remaining cyanide is lost from the culture, presumably by blowing off or being metabolised [15]. The kinetics of pyocyanin production was affected in a similar way in *mucA22* mutants, with a decrease relative to wildtype and suppressor in early stationary phase followed by continued production leading to 3 –fold higher levels than the wildtype and suppressor strain in late stationary phase cultures (Fig. 1B). Turning to elastase activity, levels were negligible in mid-log phase but throughout stationary phase the *mucA22* strains were deficient in elastase production compared to their parental strain (p<0.01 for each pair) (Fig. 1C), this being reversed in most of the suppressor strains. These findings indicate that *mucA22* mutation is associated with a global alteration of virulence factor regulation in stationary phase, although, the regulation of elastase production appears to differ from that of cyanide and pyocyanin.

The mucA22 mutation alters the expression of cyanide, pyocyanin and elastase biosynthetic genes

qRT-PCR was used to determine whether the phenotypic changes observed in *mucA22* mutants resulted from changes in gene expression. PAO1 and the suppressor strain M04S (Fig. 2A) showed similar *hcnA* expression levels to each other, with highest levels at 2 hours (mid-log phase) and 6 hours (early stationary phase) and decreased levels in late stationary phase (24 hours). It is interesting that significant *hcnA* expression is observed in the mid-log phase cultures at a time when no cyanide was detectable in the culture medium (Fig. 1), indicating that *hcnA* expression is induced in exponential phase well ahead of cyanide production in stationary phase. Expression in the *mucA22* mutant M04 was

significantly lower at 2 and 6 hours and markedly increased at 24 hours. Overall, the pattern of *hcnA* expression is consistent with the cyanide production data presented in Fig. 1 (and Figs. S1 and S2) suggesting that the low-level cyanide production in early stationary phase followed by continued production in later stationary phase seen in *mucA22* strains results from differences in the expression of the cyanide synthase genes. *hcnA* transcripts were still evident in PAO1 at 24 hours even though cyanide levels in culture supernatant had dropped to zero (Fig. 2), suggesting that either post-transcriptional regulation is affecting cyanide production or it is being lost from culture as fast as it is being made.

The expression of *phzM* and *phzS* genes was very low in mid-log phase cultures, although lowest in *mucA22* strains (Fig. 2B and C), but was expressed at similar levels in early stationary phase cultures, suggesting that deceased pyocyanin production in *mucA22* strains may be a consequence of a delay in induction of pyocyanin gene expression (but this is not as marked as with the *hcnA* gene) or it may not be wholly explained by changes in gene expression. However, the *mucA22* strain M04 showed significantly increased expression of *phzM* and *phzS* compared to the wild type and suppressor strains at 24 hours, indicating that increased pyocyanin levels in late stationary phase cultures of mucoid strains correlated with increased pyocyanin gene expression.

lasB expression was detected only at very low levels in exponential phase cultures (2 h), but there was significantly increased *lasB* expression in all strains at 6 h and a significant difference between the wild type and *mucA22* strain (M04). At 24 h, all strains showed similar *lasB* expression.

The qRT-PCR gene expression data are broadly consistent with the phenotypic data and in particular show markedly increased levels of the *hcnA* and *phz* genes that are consistent with maintenance of increased cyanide and pyocyanin levels in late stationary phase.

The mucA22 mutation results in suppression of quorum sensing signal molecule production throughout growth

As cyanide, pyocyanin and elastase are quorum sensing-(QS)-regulated [38], we tested whether *mucA22* mutation led to a defect in the two N-acyl homoserine lactone (AHL)-dependent QS systems of *P. aeruginosa*; the *lasRI* and *rhlRI* systems that synthesise 3-oxo-C12-HSL and C4–HSL respectively [38]. The *mucA22* mutant M04 was greatly impaired in the production of both these AHL molecules (Fig. 3A and B; as were other independently isolated *mucA22* mutants of PAO1 and PAO381, Fig. S3). Wild type levels of both AHLs were observed in the non-mucoid suppressor strain.

The third quorum sensing system in *P. aeruginosa* is the 2-alkyl-4-quinolone (AQ or PQS)-dependent quorum sensing system (mediated by PQS; 2-heptyl-3-hydroxy-4-quinolone and HHQ; 2-heptyl-4-quinolone) and we found that the *mucA22* mutant M04 was also impaired in production of PQS (Fig. 3C). This was most apparent in early to mid-stationary phase (8 hours, t-test, p<0.05), whereas in late stationary phase (24 hours), the differences between wildtype and *mucA22* and the suppressor strain are less marked but still significant (t-test, p<0.05). These data are consistent with a quorum sensing defect being responsible for the changes in cyanide, pyocyanin and elastase production in early stationary phase cultures of mucoid strains.

We used qRT-PCR to determine the expression of *lasR*, *lasI* (2-oxo-C12-HSL system), *rhlR*, *rhlI* (C4-HSL system) and the *pqsA*, *pqsH* and *pqsR* genes (PQS system) (Fig. 4). In early stationary phase (6 h), M04 had a significant reduction in expression of AHL and PQS quorum sensing genes compared to the wild-type (Fig. 4), which suggested that reduced AHL and PQS levels in early

A. Cyanide profile

B. Pyocyanin profile

C. Elastase profile

■ PA01 wt non-mucoid   ▨ M04 mucA22 mucoid   ☐ M04S suppressor non-mucoid

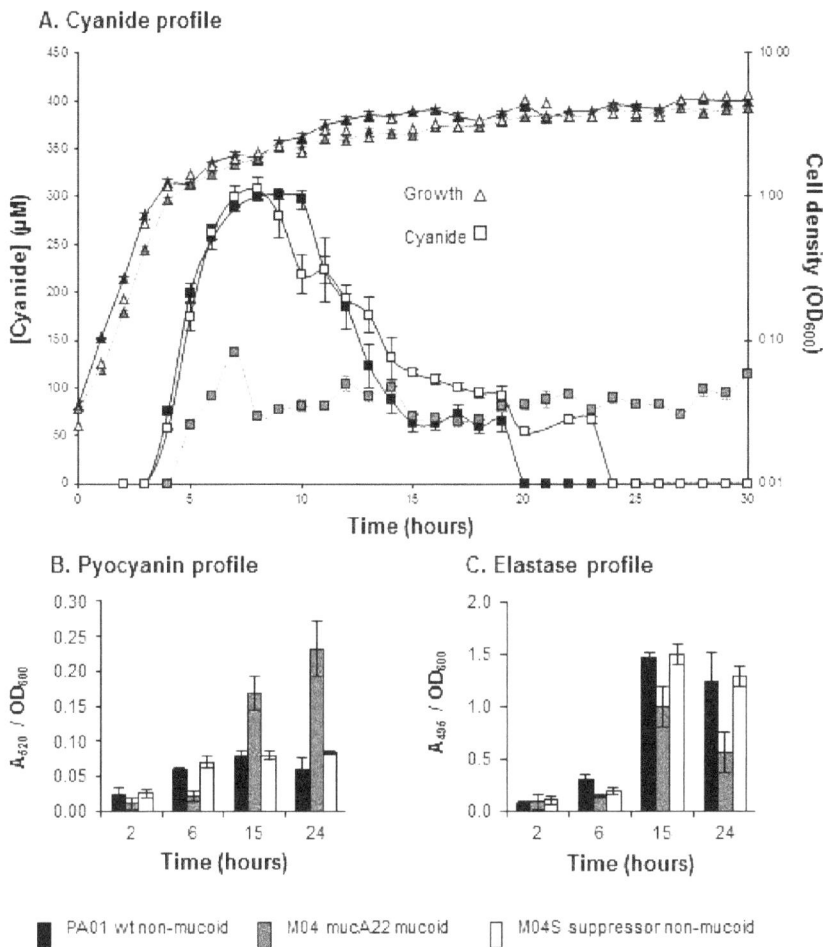

Figure 1. *mucA22*, mucoid strains show differential regulation of virulence factor production throughout grow. A, Change in cyanide concentration in liquid culture supernatant (primary y axis, square markers) throughout growth (secondary y axis, triangle markers) for PAO1 (wild type, non-mucoid) (black); M04 (*mucA22*, mucoid) (dark grey); and M04S (non-mucoid suppressor isolated from M04) (light grey). **B** and **C** respectively, pyocyanin and elastase levels in culture supernatant normalised to OD_{600} of culture at 2 hours (mid-exponential phase), 6 hours (early stationary phase),15 hours (mid stationary phase) and 24 hours (late stationary phase). Values are means of 3 independent replicates with SE error bars. Culture conditions and cyanide, pyocyanin and elastase assays were as described in Figure 1.

stationary phase (Fig. 3) result from gene expression changes. However, quorum-sensing gene expression levels in the suppressor strain M04S was similar to M04 and did not reflect the levels of AHL and PQS molecules detected (Fig. 3); the reason for this is unclear. The correlation between the levels of quorum sensing gene expression and the AHL and PQS levels is less clear at 24 hours (Fig. 4).

We also tested whether adding back C4-HSL or 3-oxo-C12-HSL was able to recover the early stationary phase cyanide-production phenotype of *mucA22* strains. However, supplementing the culture medium with 100 µM C4-HSL or 3-oxo-C12 HSL did not recover early stationary phase (6–8 h) cyanide production in *mucA22* mutants (data not shown). However, as *rhlR* and *lasR* expression is repressed in *mucA22* strains (Fig. 4), it is likely that the levels of the RhlR and LasR proteins are too low to respond effectively to the added AHLs. We also considered the possibility that our *mucA22* mutants had acquired a *lasR* mutation, but PCR amplification and sequencing of the *lasR* gene from mutants showed none to be present.

These experiments demonstrate that the switch to mucoidy has a profound effect on the multiple quorum sensing systems of *P. aeruginosa*.

mucA22 mutation does not act through disruption of the *rsmY/rsmZ* regulatory network

Next we explored further how a defect in *mucA* could lead to a loss of quorum sensing and affect cyanide production. We reasoned that a defect in *mucA*, leading to increased expression of *algU* might act through disruption of *rsmY* and *rsmZ* regulatory systems, which through sequestration of the negative regulator of quorum sensing RsmA activate quorum sensing and cyanide production [39–43] and that this might occur through competition with an unidentified sigma factor that is a proposed component of the HptB -mediated region of this regulatory pathway [44]. To test this we introduced *rsmY*-and *rsmZ-lacZ* fusions into wild type (PAO1), *mucA22* mutant (M04) and suppressor strain (M04S) and measured β-galactosidase activity of reporter strains through growth into stationary phase. No significant differences in *rsmY*- and *rsmZ-lacZ* fusion expression were observed between wild type and *mucA22* strains at any stage during growth, indicating that expression *rsmY* and *rsmZ* is not compromised by *mucA22* mutation (Fig. S4).

A. *hcnA* (cyanide)

B. *phzM* (pyocyanin)

C. *phzS* (pyocyanin)

D. *lasB* (elastase)

■ PAO1 (wt. non-mucoid)

■ M04 (*mucA22*. mucoid)

☐ M04S (suppressor. non-mucoid)

*	= p<0.07
**	= p<0.05
***	= p<0.01
****	= p<0.001

(ANOVA, Turkeys Multi Comparison Test)

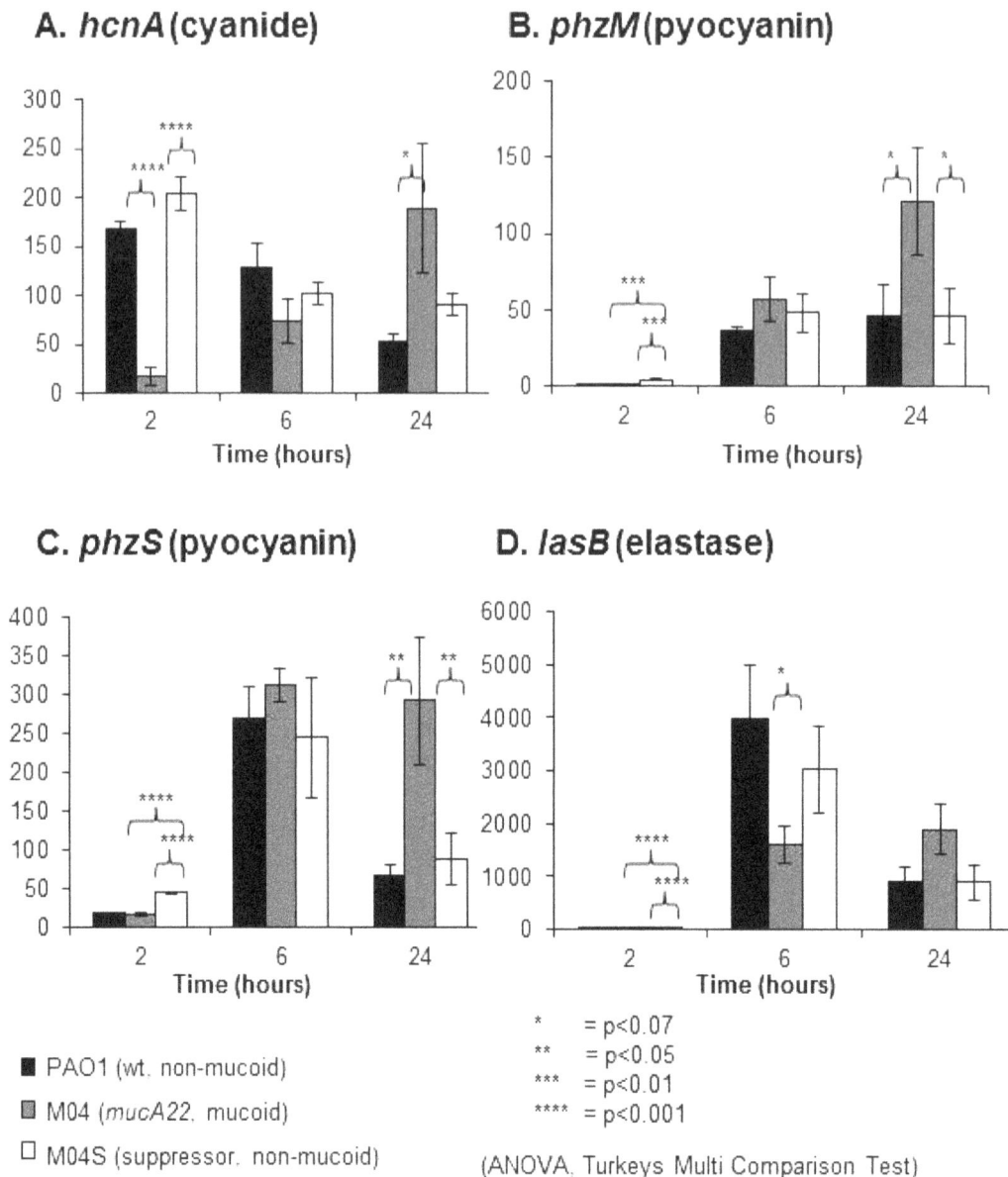

Figure 2. *mucA* mutation alters the expression of cyanide, pyocyanin and elastase genes. Expression profiles of *hcnA* (A), *phzM* (B), *phzS* (C) and *lasB*, (D) for PAO1 (wild type, non-mucoid) (black); M04 (*mucA22*, mucoid) (dark grey); and M04S (non-mucoid suppressor isolated from M04) (light grey). Values are averages of 3 biological replicates and the value for each replicate was the average of duplicate qRT-PCR technical replicates. Error bars are ±SEM. Data were analysed by ANOVA and post-hoc by Tukey's Multiple Comparison Test.

Plasmid expression of the cAMP-dependent transcription factor Vfr restores the quorum sensing defect and early stationary phase cyanide production in the *mucA22* mutant

AlgU activates the expression of a number of genes, including the gene encoding the AlgR response regulator and together they activate the transcription of the alginate biosynthetic genes [45,46]. The cAMP-dependent Vfr signalling system has been shown to be defective in *mucA* mutants at the level of *vfr* expression via a mechanism involves AlgU and AlgR [47]. Given that plasmid encoded Vfr restores cAMP-dependent Vfr signalling in a *mucA22* mutant [47], we investigated whether introduction of pPa-*vfr* into our isogenic strains resulted in the restoration of cyanide production. It is clear that cyanide production in early stationary phase is restored to wildtype levels in the *mucA22* mutant by pPa-*vfr* (Fig. 5A), suggesting that loss of Vfr contributes to the reduced peak cyanide production in *mucA22* mutants. The presence of pPa-*vfr* in wildtype and suppressor strains leads to them maintaining the production of cyanide into late stationary phase (24 h) in a way similar to M04, rather than the complete loss of cyanide production associated with the wildtype in the absence of pPa-*vfr* (Fig. 5B, Fig. 1). Furthermore, we found that plasmid borne expression of Vfr was able to restore 3-oxo-C12-HSL and C4–HSL and PQS levels to wildtype in 6 h stationary phase cultures of M04 (Fig. 5C–E).

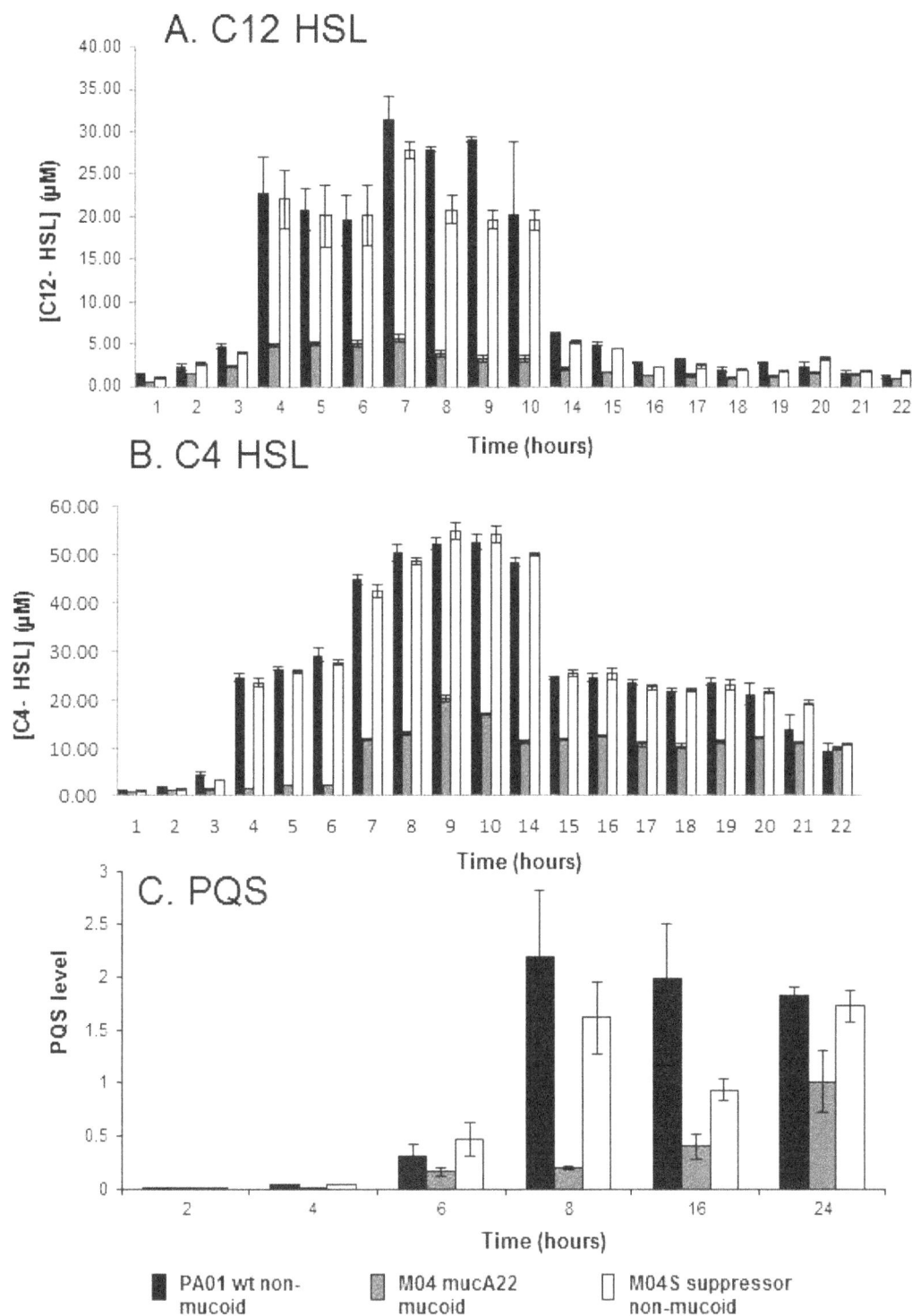

Figure 3. *mucA* mutation leads to suppression of quorum sensing signal molecule production throughout growth. Activities of the LasIR (A) RhlIR (B) and PQS (C) systems was compared in PAO1 (wild type, non-mucoid) (black), M04 (*mucA22*, mucoid) (dark grey) and M04S (non-mucoid suppressor isolated from MucA04) (light grey) using bioluminescent reporter strains to assay 3-oxo-C_{12} (lasIR) C_4-HSL (RhlIR) and PQS signal molecule levels in supernatant. Bioluminescence was measured using a luminometer. Culture conditions were as described in Figure 1. Values are means of 3 independent replicates and error bars are ±SEM.

Discussion

The *mucA22* mutation is the predominant mutation leading to mucoidy in *P. aeruginosa* during chronic CF lung infections. In this present work we report that the switch to a mucoid phenotype via a *mucA22* mutation has a profound effect on stationary phase physiology. An important finding is that a persistent low-level production of cyanide and pyocyanin occurs in mucoid strains in

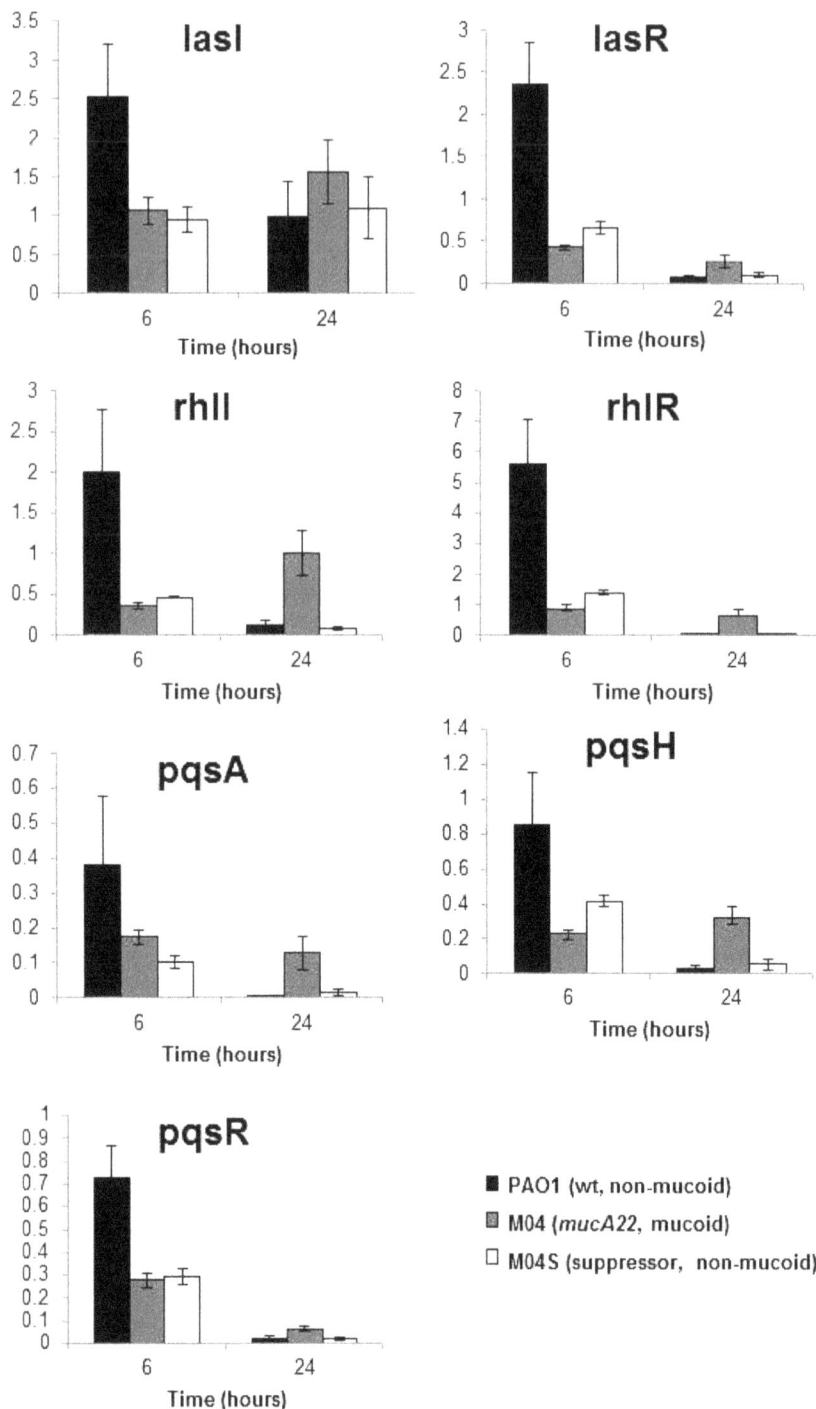

Figure 4. *mucA* **mutation leads to suppression of quorum sensing gene expression.** qRT-PCR expression profiles of quorum sensing regulatory genes for: PAO1 (wild type, non-mucoid) (black); M04 (*mucA22*, mucoid) (dark grey); and M04S (non-mucoid suppressor isolated from MucA04) (light grey). Values are averages of 3 biological replicates and the value for each replicate was the average of duplicate qRT-PCR technical replicates Error bars are ±SE and data were analysed by ANOVA and post-hoc by Tukey's Multiple Comparison Test.

late stationary phase compared to non-mucoid strains. We used spontaneous non-mucoid revertants of mucoid strains in this study due to the problematic nature of complementing or repairing *mucA22* mutations in the context of the unstable nature of the mucoid phenotype. The mucoid phenotype is frequently reported as unstable and non-mucoid variants emerge through suppressor mutations in culture and in the CF lung and where both phenotypes can coexist [36,48–51]. Suppressor mutations in the *algU* gene have been described [51], but we did not find *algU* mutations in any of the suppressor stains used in this study and neither did they have identifiable mutations in known regulars of mucoidy including; *mucA*, *algD*, *algR*, and *lasR*.

Cody et al (2009) have reported the regulation of *hcnABC* genes by the AlgZ-AlgR two component regulatory systems and

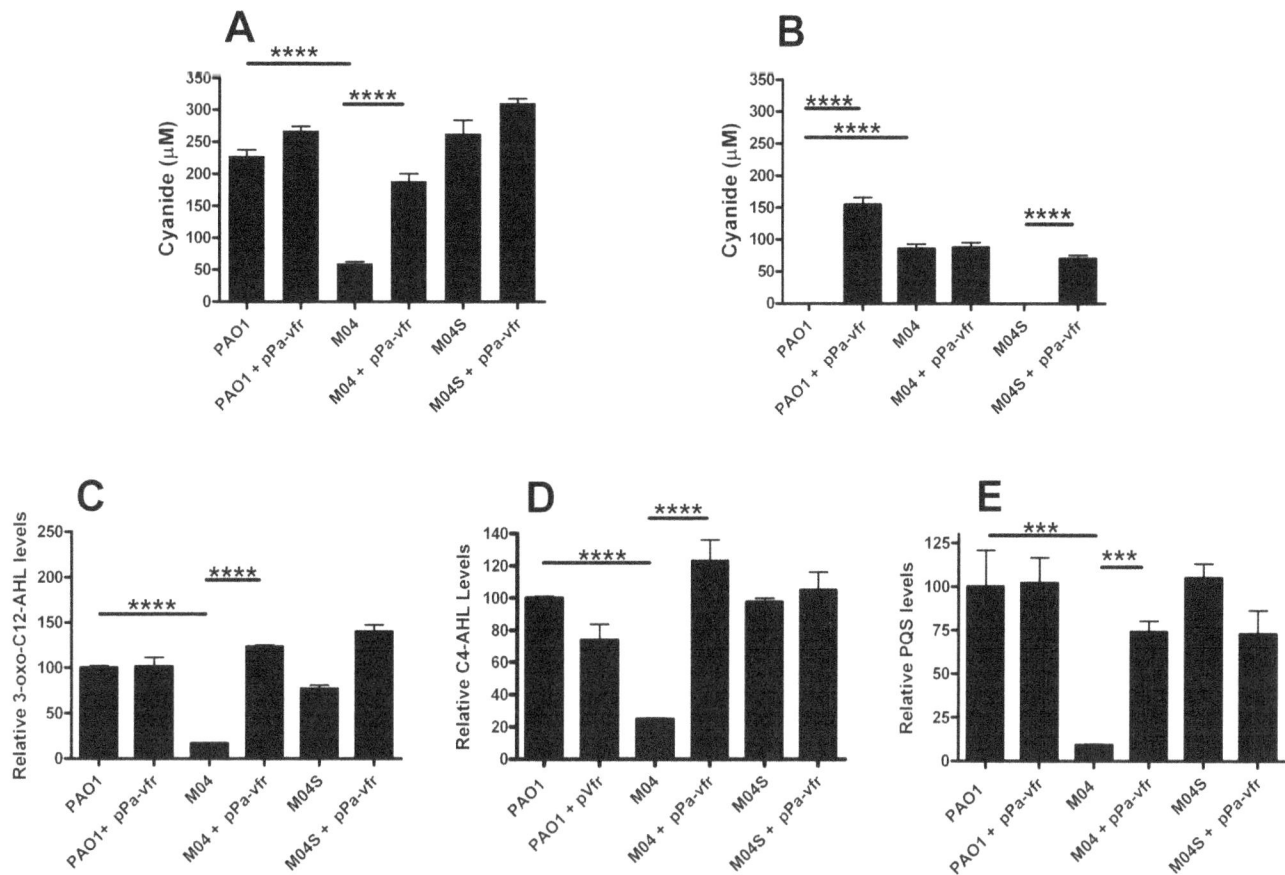

Figure 5. Plasmid expression of the cAMP-dependent transcription factor Vfr restores the early stationary phase cyanide production and the quorum sensing defect in the *mucA22* mutant. Cyanide concentration and QS molecule levels were measured in culture supernatants of wildtype (PAO1), *muc22* mutant (M04) and its suppressor strain (M04S) with or without the plasmid pPa-vfr. Cyanide levels after (A) 6 hours growth (early stationary phase) and (B) 24 hours growth (late stationary phase). (C) Relative 3-oxo-C12 AHL levels; (D) relative C4-AHL levels; (E) relative PQS levels after 8 hours growth (early stationary phase). Values are relative to the PAO1 8 hour value. All experiments are averages of three independent replicates error bars are ±SEM and data were analysed by ANOVA and post-hoc by Tukey's Multiple Comparison Test. **** $p < 0.001$; *** $p < 0.01$.

demonstrated the presence of HCN in cultures after 36 hours of growth [52]. This contrasts with our findings, which were highly reproducible in our hands. The only time we have seen cyanide in late stationary phase was using sealed flasks that prevent loss of cyanide gas by blowing off [15].Previously, mucoid CF isolates have been reported to be deficient in motility (due to inhibition of flagella production), pyocyanin production, elastase and total protease production, and decreased exotoxin A and exoenzyme S production as well as having reduced expression of type III secretion systems and type IV pili [26,53–56]. Here we demonstrate that *mucA22* mutation and the consequent increased availability of the sigma factor AlgU leads to a more complex and subtle change in the stationary phase regulation of virulence factor production than previously recognised (Figs. 1, 2 and S1). We show that *mucA22* mutation results in repression of AHL- and AQ-dependent quorum sensing systems (Figs. 3 and 4) leading to decreased production of the virulence factors cyanide, pyocyanin and elastase in early stationary phase, but also changes the regulation of cyanide and pyocyanin in late stationary phase leading to their continued production in non-growing cultures which is in stark contrast to non-growing strains (Fig. 1). qRT-PCR data indicates that *mucA22* mutation leads to changes in the regulation of expression of the HCN synthase genes (*hcnABC*), the

phz genes involved in pyocyanin synthesis and the elastase encoding *lasB* gene (Fig. 2). The effects of *mucA* mutation on stationary phase properties that we observe fit with the recent finding that MucA modulates osmotic stress tolerance in stationary phase, but not exponential phase, cultures of *P. aeruginosa* [29].

Previous publications provide contradictory findings regarding cyanide production in mucoid and non-mucoid strains. Carterson et al found that mucoid strains produced seven times **more** cyanide than non-mucoid strains [57]. This is in contrast with a study [58] that found a **decrease** in expression of the *hcnABC* genes in mucoid *P. aeruginosa*, and a recent survey of hydrogen cyanide (HCN) released into the gas phase by 96 genotyped *P. aeruginosa* strains that found reduced cyanide accumulation by mucoid strains [59]. The apparent incongruous nature of these reports is explained by our data. While Carterson *et al* [57] measured cyanide given off from plate grown cultures (most cells of which will have reached stationary phase), Rau *et al* [58] performed a transcriptomic study on mid-exponential phase cultures. Our data shows that in exponential phase cultures the *hcnA* gene is expressed in non-mucoid strains (in advance of cyanide being detectable in the growth medium), but there is a 10-fold lower level of expression in *mucA22* mucoid strains, which is consistent with data of Rau *et al* [58]. This differential gene expression is maintained into early stationary phase when cyanide levels peak at a 2.5-fold lower level in *mucA* strains (Fig. 1 and 2). There is complete loss of cyanide production in non-mucoid strains in late stationary phase (Fig. 1), whereas mucoid strains maintain cyanide production, which explains, we suggest, the high levels of cyanide trapped from the plate grown cultures by Carterson *et al* [57]. Indeed when we assayed cyanide in our strains by the same plate assay method we obtained a similar result to Carterson *et al* (2004) (data not shown). In an earlier study, Firoved and Deretic [25] reported an increase in expression of *lasB*, *hcnA* and *hcnC* in a *mucA* strain, grown to mid-exponential phase, compared to its non-mucoid parent, but cyanide levels were not measured. These data are different to both our findings and those of Rau et al [58], and are not easily explained other than by possible growth condition and strain differences.

Elastase peak levels were also reduced in *mucA22* mutants, but *lasB* showed different regulation in late stationary phase to cyanide and pyocyanin, indicating operation of a different regulatory mechanism.

As cyanide, pyocyanin and elastase are all positively regulated by quorum sensing we tested whether the change in their regulation in a *mucA22* mutant might be through an effect on quorum sensing. We found that while *mucA22* strains showed induction of LasR/3-oxo-C12-HSL, RhlR/C4-HSL and AQ-dependent quorum sensing systems their levels were reduced to approximately 17%, 35%, and 10% respectively of wild type levels at the time of peak production in early stationary phase (Fig. 3), which can explain the decreased cyanide, pyocyanin and *lasB* expression in early stationary phase in *mucA* strains. In late stationary phase (22 to 24 hours) *mucA* strains have comparable or lower levels of QS molecules compared to wild type and suppressor strains, and so the maintenance of cyanide and pyocyanin production at these time points must be dependent on other regulatory factors. Morici et al have previously demonstrated a link between the AlgR response regulator (a positive regulator of *hcnABC*) and the RhlR/C4-HSL quorum sensing system [28]. They showed by both transcriptomic and gene fusion studies that AlgR partly suppressed the RhlR/C4-HSL quorum sensing system, but in a biofilm specific manner; it did not affect planktonic cultures. Rau et al [58] reported an effect of *mucA* mutation on the RhlR/C4-AHL quorum sensing system, showing decreased *rhlR* and *rhlI* expression at a single time point in mid-exponential phase cultures and also decreased C4-HSL production when cultures reached an OD$_{600}$ of 1.0. Our study has confirmed and extended these findings through the growth cycle from early exponential phase to late stationary phase and to effects on the *lasR* and PQS quorum sensing systems.

We next addressed how the *mucA22* mutation could lead to such marked effects on quorum sensing? We found no evidence for a role for the regulatory small RNAs *rsmY* and *rsmZ* in this regulation, but we did find data supporting a role for Vfr, the cAMP-dependent positive regulator of the Las quorum sensing system (Fig. 5). Vfr has recently been shown to be down regulated in *mucA* strains via an AlgR dependant mechanism [47] and we found that plasmid expression of Vfr restored the quorum sensing defect in *mucA22* mutants and consequently restored cyanide production. So while at this stage a complete mechanism to explain our findings is not deducible, a testable model building on these findings is shown in Fig. 6.

The switch to a mucoid phenotype is one of the most frequently encountered and clinically significant changes undergone by *P. aeruginosa* and occurs during chronic infection of the cystic fibrosis lung. It occurs in most CF lung infections and is concurrent with a decrease in the clinical condition of the patient [2,18]. Evidence

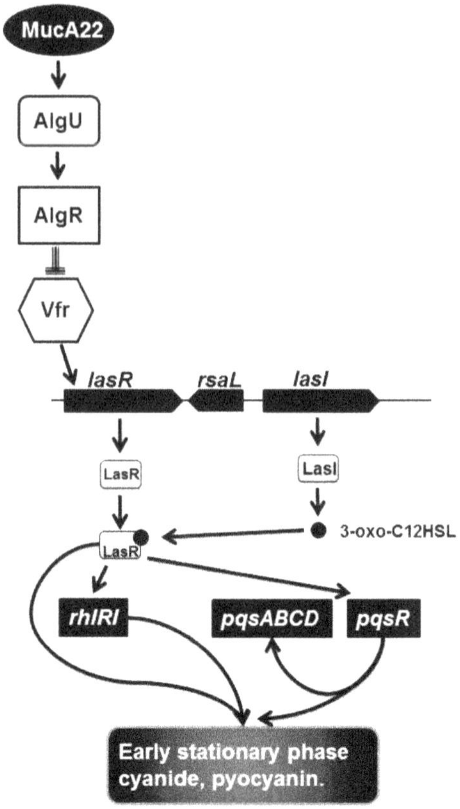

Figure 6. Model to explain early stationary phase down regulation of cyanide, pyocyanin and elastase production by mucA22 mutation of *P. aeruginosa*. This figure is an adaptation and extension of Figure 7 from [47]. Solid arrows indicate positive regulation and T-bar negative regulation. *MucA22* mutation results in release of AlgU, which is then free to associate with core RNA polymerase and direct expression from its target genes, which include the gene for the transcriptional regulator AlgR. AlgR is a repressor of the cAMP-dependent transcriptional regulator, VfR, which is a positive regulator of the LasRI quorum sensing system. AlgR inhibition VfR transcription will result in down regulation of the LasRI quorum sensing system, which since it is the dominant QS system will act to down regulate both the RhlRI and AQ (PQS) QS systems, resulting in suppression of the entire QS network and decreased expression of virulence factors in early stationary phase.

indicates that alginate provides *P. aeruginosa* with protection from the host immune response by inhibiting phagocytosis, protecting against the oxidative burst and by providing an immunomodulatory role and so provides a selective advantage in the lung [60,61]. However the pleiotropic effects of *mucA22* mutation on quorum sensing and virulence factor regulation may also provide a powerful selection pressure for the mucoid switch. Genetic analysis of *P. aeruginosa* isolates from the airways of multiple CF patients has revealed that one of the most common targets of mutation in CF isolates is the *lasR* gene [62], suggesting a strong selection for the loss of *lasR* function and subsequent down-regulation of quorum sensing pathways in the CF airways. Our data indicates that the switch to mucoidy provides a second route for down regulation of quorum sensing pathways, which could provide a selection pressure for the switch to mucoidy during CF lung infections. Whether these findings are of significance *in vivo* requires further study in appropriate model systems and of clinical strains.

If this mechanism was reproduced in vivo during CF lung infection it would lead to continued production and cyanide and pyocyanin in non-growing populations of *P. aeruginosa*, which could be significant as these compounds can damage lung tissue and provide a source of nutrients for growth and survival. While cyanide is only synthesised aerobically HCN made by P. aeruginosa in oxic regions of the lung that diffuse into anoxic areas would potentially inhibit anaerobic respiration in the absence of protective mechanism against HCN [16,63].

Supporting Information

Figure S1 *mucA22*, mucoid strains are deficient in cyanide, pyocyanin and elastase production in early stationary phase. Cyanide (**A**), Pyocyanin (**B**) and Elastase (**C**) in supernatant of liquid cultures of non-mucoid wild type (black), mucoid *mucA22* (dark grey) and non-mucoid suppressor (light grey) strains at 2 hours post stationary phase initiation. Cultures were incubated in 50 ml LB in 250 ml conical flasks in orbital shaker at 200 rpm and 37°C and grown until 2 hour post stationary phase initiation (6 hours in all cases); growth was followed by determining OD_{600} every hour. Cyanide concentrations were measured with an ion selective micro electrode. Pyocyanin levels were determined by chloroform/HCl extraction followed by absorbance measurement 520 nm. Elastase levels were determined by assaying enzymatic breakdown of Elastin-Congo red then measurement of liberated Congo red in a spectrophotometer at A_{495}. Values are means of 3 independent replicates and error bars are ±SEM.

Figure S2 Growth and Cyanide production for PA0381 and PAO578 (*mucA22*). Optical density on primary y axis, (solid markers) cyanide concentration on secondary y axis, (empty markers) for (A) PAO381 (wild type, non-mucoid) (black); and PAO578 (*mucA22*, mucoid derivative of PAO381) (dark grey).

Figure S3 *mucA* mutation leads to suppression of quorum sensing signal molecule production in independently isolated *mucA22* mutants isolated in PAO1 and PAO381 backgrounds. Activities of the C4-AHL (A) and 3-oxo-C12-AHL (B) was compared in wild type, non-mucoid strains (WT), *mucA22* (MUCOID) and suppressor strains after 2 hours (mid-log) and 6 hours (early stationary phase) of growth in LB medium. Values are means of 3 independent replicates and error bars are ±SEM.

Figure S4 *mucA22* mutation does not act through disruption of the *rsmY/rsmZ* regulatory network. The wildtype strain PAO1, the *mucA22* mutant M04 and its suppressor strain M04S carrying the *rsmY-lacZ* or *rsmZ-lacZ* gene fusions [44] were grown in LB and growth followed (A) and samples assayed for β-galactosidase (B), the values plotted being ±SEM (n = 3 biological replicates).

Acknowledgments

We thank Alain Filloux, John Govan, Paul Williams and Matthew Wolfgang for kindly providing strains and plasmids.

Author Contributions

Conceived and designed the experiments: BR JEAZ HDW. Performed the experiments: BR MC JEAZ VB XL ZW KEL. Analyzed the data: BR MC JEAZ VB XL ZW KEL HDW. Wrote the paper: BR HDW.

References

1. Schaedel C, de Monestrol I, Hjelte L, Johannesson M, Kornfalt R, et al. (2002) Predictors of deterioration of lung function in cystic fibrosis. Pediatr Pulmonol 33: 483–491.
2. Govan JR, Deretic V (1996) Microbial pathogenesis in cystic fibrosis: mucoid *Pseudomonas aeruginosa* and *Burkholderia cepacia*. Microbiol Rev 60: 539–574.
3. Castric PA (1983) Hydrogen cyanide production by Pseudomonas aeruginosa at reduced oxygen levels. Can J Microbiol 29: 1344–1349.
4. Worlitzsch D, Tarran R, Ulrich M, Schwab U, Cekici A, et al. (2002) Effects of reduced mucus oxygen concentration in airway *Pseudomonas* infections of cystic fibrosis patients. J Clin Invest 109: 317–325.
5. Gallagher LA, Manoil C (2001) *Pseudomonas aeruginosa* PAO1 kills *Caenorhabditis elegans* by cyanide poisoning. J Bacteriol 183: 6207–6214.
6. Broderick KE, Chan A, Balasubramanian M, Feala J, Reed SL, et al. (2008) Cyanide produced by human isolates of *Pseudomonas aeruginosa* contributes to lethality in *Drosophila melanogaster*. J Infect Dis 197: 457–464.
7. Goldfarb WB, Margraf H (1967) Cyanide production by *Pseudomonas aeruginosa*. Ann Surg 165: 104–110.
8. Ryall B, Davies JC, Wilson R, Shoemark A, Williams HD (2008) *Pseudomonas aeruginosa*, cyanide accumulation and lung function in CF and non-CF bronchiectasis patients. Eur Respir J 32: 740–747.
9. Sanderson K, Wescombe L, Kirov SM, Champion A, Reid DW (2008) Bacterial cyanogenesis occurs in the cystic fibrosis lung. Eur Respir J 32: 329–333.
10. Anderson RD, Roddam LF, Bettiol S, Sanderson K, Reid DW (2010) Biosignificance of bacterial cyanogenesis in the CF lung. J Cyst Fibros 9: 158–164.
11. Lenney W, Gilchrist FJ (2011) *Pseudomonas aeruginosa* and cyanide production. Eur Respir J 37: 482–483.
12. Enderby B, Smith D, Carroll W, Lenney W (2009) Hydrogen cyanide as a biomarker for *Pseudomonas aeruginosa* in the breath of children with cystic fibrosis. Pediatr Pulmonol 44: 142–147.
13. Stutz MD, Gangell CL, Berry LJ, Garratt LW, Sheil B, et al. (2011) Cyanide in bronchoalveolar lavage is not diagnostic for *Pseudomonas aeruginosa* in children with cystic fibrosis. Eur Respir J 37: 553–558.
14. Cunningham L, Pitt M, Williams HD (1997) The *cioAB* genes from *Pseudomonas aeruginosa* code for a novel cyanide-insensitive terminal oxidase related to the cytochrome *bd* quinol oxidases. Molecular Microbiology 24: 579–591.
15. Zlosnik JE, Tavankar GR, Bundy JG, Mossialos D, O'Toole R, et al. (2006) Investigation of the physiological relationship between the cyanide-insensitive oxidase and cyanide production in *Pseudomonas aeruginosa*. Microbiology 152: 1407–1415.
16. Williams HD, Ziosnik JEA, Ryall B (2007) Oxygen, cyanide and energy generation in the cystic fibrosis pathogen *Pseudomonas aeruginosa*. Advances in Microbial Physiology, Vol 52 52: 1–71.
17. Cipollone R, Ascenzi P, Tomao P, Imperi F, Visca P (2008) Enzymatic detoxification of cyanide: clues from *Pseudomonas aeruginosa* Rhodanese. J Mol Microbiol Biotechnol 15: 199–211.
18. Lyczak JB, Cannon CL, Pier GB (2002) Lung infections associated with cystic fibrosis. Clin Microbiol Rev 15: 194–222.
19. Martin DW, Schurr MJ, Mudd MH, Govan JRW, Holloway BW, et al. (1993) Mechanism of conversion to mucoidy in *Pseudomonas-aeruginosa* infecting cystic fibrosis patients. Proceedings of the National Academy of Sciences of the United States of America 90: 8377–8381.
20. Rowen DW, Deretic V (2000) Membrane-to-cytosol redistribution of ECF sigma factor AlgU and conversion to mucoidy in *Pseudomonas aeruginosa* isolates from cystic fibrosis patients. Mol Microbiol 36: 314–327.
21. Martin DW, Schurr MJ, Mudd MH, Govan JR, Holloway BW, et al. (1993) Mechanism of conversion to mucoidy in *Pseudomonas aeruginosa* infecting cystic fibrosis patients. Proc Natl Acad Sci U S A 90: 8377–8381.
22. Mathee K, Ciofu O, Sternberg C, Lindum PW, Campbell JIA, et al. (1999) Mucoid conversion of *Pseudomonas aeruginosa* by hydrogen peroxide: a mechanism for virulence activation in the cystic fibrosis lung. Microbiology-Sgm 145: 1349–1357.
23. Pedersen SS, Moller H, Espersen F, Sorensen CH, Jensen T, et al. (1992) Mucosal immunity to *Pseudomonas aeruginosa* alginate in cystic fibrosis. APMIS 100: 326–334.
24. Firoved AM, Boucher JC, Deretic V (2002) Global genomic analysis of AlgU (sigma(E)-dependent promoters (sigmulon) in *Pseudomonas aeruginosa* and implications for inflammatory processes in cystic fibrosis. J Bacteriol 184: 1057–1064.
25. Firoved AM, Deretic V (2003) Microarray analysis of global gene expression in mucoid *Pseudomonas aeruginosa*. J Bacteriol 185: 1071–1081.

26. Wu WH, Badrane H, Arora S, Baker HV, Jin SG (2004) MucA-mediated coordination of type III secretion and alginate synthesis in *Pseudomonas aeruginosa*. Journal of Bacteriology 186: 7575–7585.

27. Tart AH, Blanks MJ, Wozniak DJ (2006) The AlgT-dependent transcriptional regulator AmrZ (AlgZ) inhibits flagellum biosynthesis in mucoid, nonmotile *Pseudomonas aeruginosa* cystic fibrosis isolates. J Bacteriol 188: 6483–6489.

28. Morici LA, Carterson AJ, Wagner VE, Frisk A, Schurr JR, et al. (2007) *Pseudomonas aeruginosa* AlgR represses the Rhl quorum-sensing system in a biofilm-specific manner. J Bacteriol 189: 7752–7764.

29. Behrends V, Ryall B, Wang X, Bundy JG, Williams HD (2010) Metabolic profiling of *Pseudomonas aeruginosa* demonstrates that the anti-sigma factor MucA modulates osmotic stress tolerance. Mol Biosyst 6: 562–569.

30. Williams HD, Behrends V, Bundy JG, Ryall B, Zlosnik JE (2010) Hypertonic saline therapy in cystic fibrosis: do population shifts caused by the osmotic sensitivity of infecting bacteria explain the effectiveness of this treatment? Front Microbiol 1: 120.

31. Zlosnik JE, Williams HD (2004) Methods for assaying cyanide in bacterial culture supernatant. Lett Appl Microbiol 38: 360–365.

32. Ohman DE, Cryz SJ, Iglewski BH (1980) Isolation and characterization of *Pseudomonas aeruginosa* PAO mutant that produces altered elastase. J Bacteriol 142: 836–842.

33. Essar DW, Eberly L, Hadero A, Crawford IP (1990) Identification and characterization of genes for a second anthranilate synthase in *Pseudomonas aeruginosa*: interchangeability of the two anthranilate synthases and evolutionary implications. J Bacteriol 172: 884–900.

34. Winson MK, Swift S, Fish L, Throup JP, Jorgensen F, et al. (1998) Construction and analysis of *luxCDABE*-based plasmid sensors for investigating N-acyl homoserine lactone-mediated quorum sensing. FEMS Microbiol Lett 163: 185–192.

35. Fletcher MP, Diggle SP, Crusz SA, Chhabra SR, Camara M, et al. (2007) A dual biosensor for 2-alkyl-4-quinolone quorum-sensing signal molecules. Environ Microbiol 9: 2683–2693.

36. Hauser AR, Jain M, Bar-Meir M, McColley SA (2011) Clinical significance of microbial infection and adaptation in cystic fibrosis. Clin Microbiol Rev 24: 29–70.

37. Boucher JC, Schurr MJ, Deretic V (2000) Dual regulation of mucoidy in *Pseudomonas aeruginosa* and sigma factor antagonism. Mol Microbiol 36: 341–351.

38. Williams P, Camara M (2009) Quorum sensing and environmental adaptation in *Pseudomonas aeruginosa*: a tale of regulatory networks and multifunctional signal molecules. Curr Opin Microbiol 12: 182–191.

39. Pessi G, Haas D (2000) Transcriptional control of the hydrogen cyanide biosynthetic genes *hcnABC* by the anaerobic regulator ANR and the quorum-sensing regulators LasR and RhlR in *Pseudomonas aeruginosa*. J Bacteriol 182: 6940–6949.

40. Heurlier K, Williams F, Heeb S, Dormond C, Pessi G, et al. (2004) Positive control of swarming, rhamnolipid synthesis, and lipase production by the posttranscriptional RsmA/RsmZ system in *Pseudomonas aeruginosa* PAO1. J Bacteriol 186: 2936–2945.

41. Burrowes E, Baysse C, Adams C, O'Gara F (2006) Influence of the regulatory protein RsmA on cellular functions in *Pseudomonas aeruginosa* PAO1, as revealed by transcriptome analysis. Microbiology 152: 405–418.

42. Kay E, Humair B, Denervaud V, Riedel K, Spahr S, et al. (2006) Two GacA-dependent small RNAs modulate the quorum-sensing response in *Pseudomonas aeruginosa*. J Bacteriol 188: 6026–6033.

43. Brencic A, McFarland KA, McManus HR, Castang S, Mogno I, et al. (2009) The GacS/GacA signal transduction system of *Pseudomonas aeruginosa* acts exclusively through its control over the transcription of the RsmY and RsmZ regulatory small RNAs. Mol Microbiol 73: 434–445.

44. Bordi C, Lamy MC, Ventre I, Termine E, Hachani A, et al. (2010) Regulatory RNAs and the HptB/RetS signalling pathways fine-tune *Pseudomonas aeruginosa* pathogenesis. Mol Microbiol 76: 1427–1443.

45. Martin DW, Schurr MJ, Yu H, Deretic V (1994) Analysis of promoters controlled by the putative sigma factor AlgU regulating conversion to mucoidy in *Pseudomonas aeruginosa*: relationship to sigma E and stress response. J Bacteriol 176: 6688–6696.

46. Mohr CD, Leveau JH, Krieg DP, Hibler NS, Deretic V (1992) AlgR-binding sites within the *algD* promoter make up a set of inverted repeats separated by a large intervening segment of DNA. J Bacteriol 174: 6624–6633.

47. Jones AK, Fulcher NB, Balzer GJ, Urbanowski ML, Pritchett CL, et al. (2010) Activation of the *Pseudomonas aeruginosa* AlgU regulon through *mucA* mutation inhibits cyclic AMP/Vfr signaling. J Bacteriol 192: 5709–5717.

48. Schurr MJ, Martin DW, Mudd MH, Deretic V (1994) Gene cluster controlling conversion to alginate-overproducing phenotype in *Pseudomonas aeruginosa*: functional analysis in a heterologous host and role in the instability of mucoidy. J Bacteriol 176: 3375–3382.

49. Ohman DE, Goldberg JB, Flynn JL (1990) Molecular analysis of the genetic switch activating alginate production In: Silver S, Chakrabarty AM, Iglewski B, Kaplan S, editors. *Pseudomonas*: biotransformations, pathogenesis, and evolving biotechnology. Washington D.C.: American Society for Microbiology.

50. Govan JRW (1988) Alginate biosynthesis and other unusual characterisatics associated with the pathogenesis of *Pseudomonas aeruginosa* in cystic fibrosis. In: Griffiths E, Donachie W, Stephen J, editors. Bacterial infections od respiratory and gastrointestinal mucosae. Oxford: IRL Press.

51. Sautter R, Ramos D, Schneper L, Ciofu O, Wassermann T, et al. (2012) A complex multilevel attack on Pseudomonas aeruginosa algT/U expression and algT/U activity results in the loss of alginate production. Gene 498: 242–253.

52. Cody WL, Pritchett CL, Jones AK, Carterson AJ, Jackson D, et al. (2009) *Pseudomonas aeruginosa* AlgR controls cyanide production in an AlgZ-dependent manner. J Bacteriol 191: 2993–3002.

53. Elston HR, Hoffman KC (1967) Increasing incidence of encapsulated Pseudomonas aeruginosa strains. Am J Clin Pathol 48: 519–523.

54. Garrett ES, Perlegas D, Wozniak DJ (1999) Negative control of flagellum synthesis in *Pseudomonas aeruginosa* is modulated by the alternative sigma factor AlgT (AlgU). J Bacteriol 181: 7401–7404.

55. Mohr CD, Rust L, Albus AM, Iglewski BH, Deretic V (1990) Expression patterns of genes encoding elastase and controlling mucoidy: co-ordinate regulation of two virulence factors in *Pseudomonas aeruginosa* isolates from cystic fibrosis. Mol Microbiol 4: 2103–2110.

56. Woods DE, Sokol PA, Bryan LE, Storey DG, Mattingly SJ, et al. (1991) In vivo regulation of virulence in *Pseudomonas aeruginosa* associated with genetic rearrangement. J Infect Dis 163: 143–149.

57. Carterson AJ, Morici LA, Jackson DW, Frisk A, Lizewski SE, et al. (2004) The transcriptional regulator AlgR controls cyanide production in *Pseudomonas aeruginosa*. J Bacteriol 186: 6837–6844.

58. Rau MH, Hansen SK, Johansen HK, Thomsen LE, Workman CT, et al. (2010) Early adaptive developments of *Pseudomonas aeruginosa* after the transition from life in the environment to persistent colonization in the airways of human cystic fibrosis hosts. Environ Microbiol 12: 1643–1658.

59. Gilchrist FJ, Alcock A, Belcher J, Brady M, Jones A, et al. (2011) Variation in hydrogen cyanide production between different strains of *Pseudomonas aeruginosa*. Eur Respir J 38: 409–414.

60. Pedersen SS (1992) Lung infection with alginate-producing, mucoid *Pseudomonas aeruginosa* in cystic fibrosis. APMIS Suppl 28: 1–79.

61. Simpson JA, Smith SE, Dean RT (1989) Scavenging by alginate of free radicals released by macrophages. Free Radic Biol Med 6: 347–353.

62. Smith EE, Buckley DG, Wu Z, Saenphimmachak C, Hoffman LR, et al. (2006) Genetic adaptation by *Pseudomonas aeruginosa* to the airways of cystic fibrosis patients. Proc Natl Acad Sci U S A 103: 8487–8492.

63. Filiatrault MJ, Tombline G, Wagner VE, Van Alst N, Rumbaugh K, et al. (2013) *Pseudomonas aeruginosa* PA1006, which plays a role in molybdenum homeostasis, is required for nitrate utilization, biofilm formation, and virulence. PLoS One 8: e55594.

17

Magneto-Chemotaxis in Sediment: First Insights

Xuegang Mao[1,3], Ramon Egli[2]*, Nikolai Petersen[3], Marianne Hanzlik[4], Xiuming Liu[1,5]

1 College of Geographical Sciences, Fujian Normal University, Fuzhou, China, **2** Central institute for Meteorology and Geodynamics, Vienna, Austria, **3** Department of Earth and Environmental Sciences, Ludwig-Maximilians University, Munich, Germany, **4** Chemistry Department, Munich Technical University, Munich, Germany, **5** Department of Environment and Geography, Macquarie University, Sydney, Australia

Abstract

Magnetotactic bacteria (MTB) use passive alignment with the Earth magnetic field as a mean to increase their navigation efficiency in horizontally stratified environments through what is known as magneto-aerotaxis (M-A). Current M-A models have been derived from MTB observations in aqueous environments, where a >80% alignment with inclined magnetic field lines produces a one-dimensional search for optimal living conditions. However, the mean magnetic alignment of MTB in their most widespread living environment, i.e. sediment, has been recently found to be <1%, greatly reducing or even eliminating the magnetotactic advantage deduced for the case of MTB in water. In order to understand the role of magnetotaxis for MTB populations living in sediment, we performed first M-A observations with lake sediment microcosms. Microcosm experiments were based on different combinations of (1) MTB position with respect to their preferred living depth (i.e. above, at, and below), and (2) magnetic field configurations (i.e. correctly and incorrectly polarized vertical fields, horizontal fields, and zero fields). Results suggest that polar magnetotaxis is more complex than implied by previous experiments, and revealed unexpected differences between two types of MTB living in the same sediment. Our main findings are: (1) all investigated MTB benefit of a clear magnetotactic advantage when they need to migrate over macroscopic distances for reaching their optimal living depth, (2) magnetotaxis is not used by all MTB under stationary, undisturbed conditions, (3) some MTB can rely only on chemotaxis for macroscopic vertical displacements in sediment while other cannot, and (4) some MTB use a fixed polar M-A mechanisms, while other can switch their M-A polarity, performing what can be considered as a mixed polar-axial M-A. These observations demonstrate that sedimentary M-A is controlled by complex mechanical, chemical, and temporal factors that are poorly reproduced in aqueous environments.

Editor: James P. Brody, University of California, Irvine, United States of America

Funding: EG 294/1-1 and EG 294/2-1, German Research Foundation (http://www.dfg.de/en/); XM was funded by the China Scholarship Council (en.csc.edu.cn/). Natural Science Foundation of China (NSF grant 41210002). The funders had no role in study design, data collection and analysis, decision to publish, or preparation of the manuscript.

Competing Interests: The authors have declared that no competing interests exist.

* Email: ramon.egli@zamg.ac.at

Introduction

Magnetotactic bacteria (MTB) are a polyphyletic group of bacteria living in chemically stratified freshwater and marine environments within the so-called oxic-anoxic interface (OAI) [1,2]. MTB contain magnetite (Fe_3O_4) or greigite (Fe_3S_4) single-domain crystals, usually arranged in chains, whose magnetic moment ensures a passive alignment of the whole cell with the Earth magnetic field. Alignments >80% make MTB cells swim straight along magnetic field lines when observed under the optical microscope: this phenomenon is called magnetotaxis [1,3]. The swimming direction (parallel or antiparallel to the magnetic field vector **B**) is determined by the sense of flagellar rotation, which is controlled mainly by chemical factors, such as oxygen concentration and redox potential [5–7], and, to a certain extent, by physical stimuli such as light [8,9] and physical contact [10]. The MTB response to oxygen gradients in a magnetic field has been called magneto-aerotaxis (M-A); here we use this term in a broad sense to indicate the combination of passive magnetic alignment and chemical control of flagellar rotation.

The confinement of magneto-aerotaxis to displacements along inclined magnetic field lines that intersect horizontally stratified environments reduces a 3D search for optimal living conditions to a more efficient 1D search [1,11], providing MTB with a so-called magnetotactic advantage [12]. Two different magneto-aerotaxis mechanisms, called axial and polar magnetotaxis, have been observed in MTB cultures. Axial magnetotaxis is based on a temporal sensory mechanism that determines the swimming direction according to the sensed chemical gradient (e.g. *Magnetospirillum magnetotacticum*). Polar magnetotaxis, on the other hand, appears to be controlled by a threshold mechanism, so that cells swim consistently along a direction determined by the oxygen concentration, regardless of existing gradients (e.g. *Magnetococcus marinus* MC-1) [5]. In hanging drop assays with MTB grown in the Northern hemisphere, polar magnetotaxis is characterized by most cells swimming consistently parallel to the magnetic field, i.e. towards the magnetic North (N-seeking, or NS cells) and only few cells (<0.1%) swimming along the opposite direction (S-seeking, or SS cells). Because **B** points downwards in the Northern hemisphere, this behaviour is consistent with a downward displacement that enables cells to escape high oxygen concentrations in normally stratified environments. Axial magnetotaxis, on the other hand, appears as a random sequence of swimming direction reversals in hanging drop assays, due to the lack of a defined chemical gradient.

In most freshwater and marine environments, the OAI is located within the upper 3–10 cm of sediment, where living MTB populations have been systematically found (e.g [13,14]). M-A in sedimentary environments might be affected significantly by (1) mechanical interactions with sediment particles acting as additional randomizing forces against magnetic alignment, and (2) solid interfaces adding local perturbations to macroscopic chemical stratifications. Unfortunately, while realistic living conditions for natural MTB populations can be reproduced in the laboratory with sediment microcosms [15,16], M-A has never been studied outside liquid MTB cultures, due to the impossibility of performing direct MTB observations in sediment. Recently, the mean alignment of *Candidatus Magnetobacterium bavaricum* with Earth-like magnetic fields has been indirectly measured in sediment microcosms [17], where it does not exceed 1%. Accordingly, MTB displacement in sediment is described by a slightly biased random walk, rather than straight paths, with magnetotaxis overcoming the random diffusive component of the swimming path only over macroscopic distances estimated to exceed 1 mm [17]. Microscopic displacements, required for instance to overcome natural sediment mixing and avoid passive diffusion away from the preferred living depth, are therefore not expected to be assisted by the Earth magnetic field. Within this context, the role of magnetotaxis in sediment becomes less obvious than suggested by the 1D displacement model in water. Furthermore, simple M-A tests with wild-type MTB, such as hanging drop assays performed under strictly anoxic conditions and long-term observations of natural MTB populations in magnetically shielded microcosms, do not appear to support the current understanding of M-A.

In order to clarify the role of M-A in sediment, we have undertaken a systematic series of experiments aimed at investigating vertical migration and resilience of natural MTB populations in sediment subjected to selected favourable, unfavourable, or indifferent magnetic field configurations. Our results clarify the circumstances under which M-A supports navigation in sediment and reveal important differences between two MTB populations characterized by polar magnetotaxis.

Materials and Methods

1. Microcosms

All experiments have been performed on microcosms prepared with surface sediment from Lake Chiemsee, southern Germany [14,18]. Sediment has been collected with permission of the local authorities (Prien am Chiemsee municipality) using a bottom grab sampler on a boat of the local fire brigade, and transported to the laboratory, where it was transferred into several 30×20×20 cm glass aquaria (Figure 1a). Aquaria were allowed to stabilize for few weeks under constant laboratory conditions. Sediment was covered by 3–5 cm water and evaporation was steadily compensated by adding small amounts of distilled water. Microcosms have been prepared by transferring sediment slurry from aquaria into 1 L glass beakers (Figure 1b). The slurry was allowed to rest for at least 7 days in order to rebuild a stationary chemical stratification, characterized by an OAI located ~5 mm below the sediment-water interface (Figure 1c). Chemical stratification was controlled prior to every microcosm experiment by measuring profiles of dissolved oxygen with a motorized microprobe (OX50 from Unisense).

2. MTB characterization

Two relatively stable MTB populations occur in Chiemsee sediment after laboratory storage: *M. bavaricum* (Figure 1d), a large (~10 µm long), rod-shaped bacterium swimming at ~40 µm/s [16,19,20], and unspecified cocci (~1 µm diameter, Figure 1e) with up to 1 mm/s swimming velocities [21]. The identification of *M. bavaricum* (MBV) and magnetotactic cocci (MCC) under the optical microscope is based exclusively on their size and swimming behaviour (video S1). The two groups of bacteria appear to be homogeneous with respect to all experimental results presented here; although MCC might be phylogenetically heterogeneous [22]. Both groups perform polar magnetotaxis, as seen from consistently NS swimming directions in hanging drop assays. Observed NS:SS proportions were systematically >1000:1, and the existence of few incorrectly polarized cells is believed to be essential for allowing MTB to survive geomagnetic field reversals [23]. Magnetotactic spirilla and vibrios could be observed occasionally, but have not been considered for this study due to their erratic occurrence.

Vertical distributions of the two MTB populations in microcosm sediment represent the fundamental characterization tool for most experiments presented in this paper, and have been obtained as follows. Mini-cores of the uppermost ~3 cm of unconsolidated sediment have been sampled with a drinking straw (diameter 5 mm), and sliced in 1 mm increments. Each slice was diluted with distilled water (200 µL) and prepared for a hanging drop assay with a special optical microscope equipped with Helmholtz coils for producing controlled magnetic fields in the objective plane ([17]; text S1). In order to overcome natural MTB population heterogeneities, results from at least 7 minicores have been averaged to obtain a single profile. Mean MBV and MCC vertical distributions in undisturbed microcosms exposed to the Earth magnetic field are shown in Figure 1c. The MCC distribution is unimodal with ~90% cell counts comprised between 3 and 15 mm, while MBV extends to greater depths (>25 mm), according to a bimodal distribution. Similar MBV distributions have been reported for other microcosms prepared with the same type of sediment [16].

3. Hanging drop assays under anoxic conditions

In order to test the capability of MBV and MCC to switch swimming direction under strictly anoxic conditions, as expected from the M-A model, a sediment microcosm was placed inside a glove box, together with the microscope apparatus used for the hanging drop assay (text S1). The glove box was steadily flushed with nitrogen gas bubbled through a $FeCl_2$ solution for residual oxygen removal. After 4 weeks, oxygen dropped below detection threshold in sediment as well as the water column above it. Hanging drop assays have been performed directly inside the glove box with sediment samples taken at 2–25 mm below the sediment-water interface, using oxygen-free water from the microcosm. In order to avoid possible phototactic interferences, MTB were allowed to accumulate at the edges of the hanging drop in complete darkness for 20 minutes.

4. Resilience experiments

Long-term resilience experiments have been performed with a sediment-filled aquarium placed at the centre of three orthogonal ~1×1 m Helmholtz coil pairs connected to precision power supplies for generating controlled, homogeneous fields over the aquarium volume. The experiment, which lasted for >2 years, has been performed with the following field settings. During the first 92 days, the aquarium was exposed to the laboratory field (~44 µT with 71° downward inclination), with no current flowing

(a)

(b)

(c)

(d)

(e)

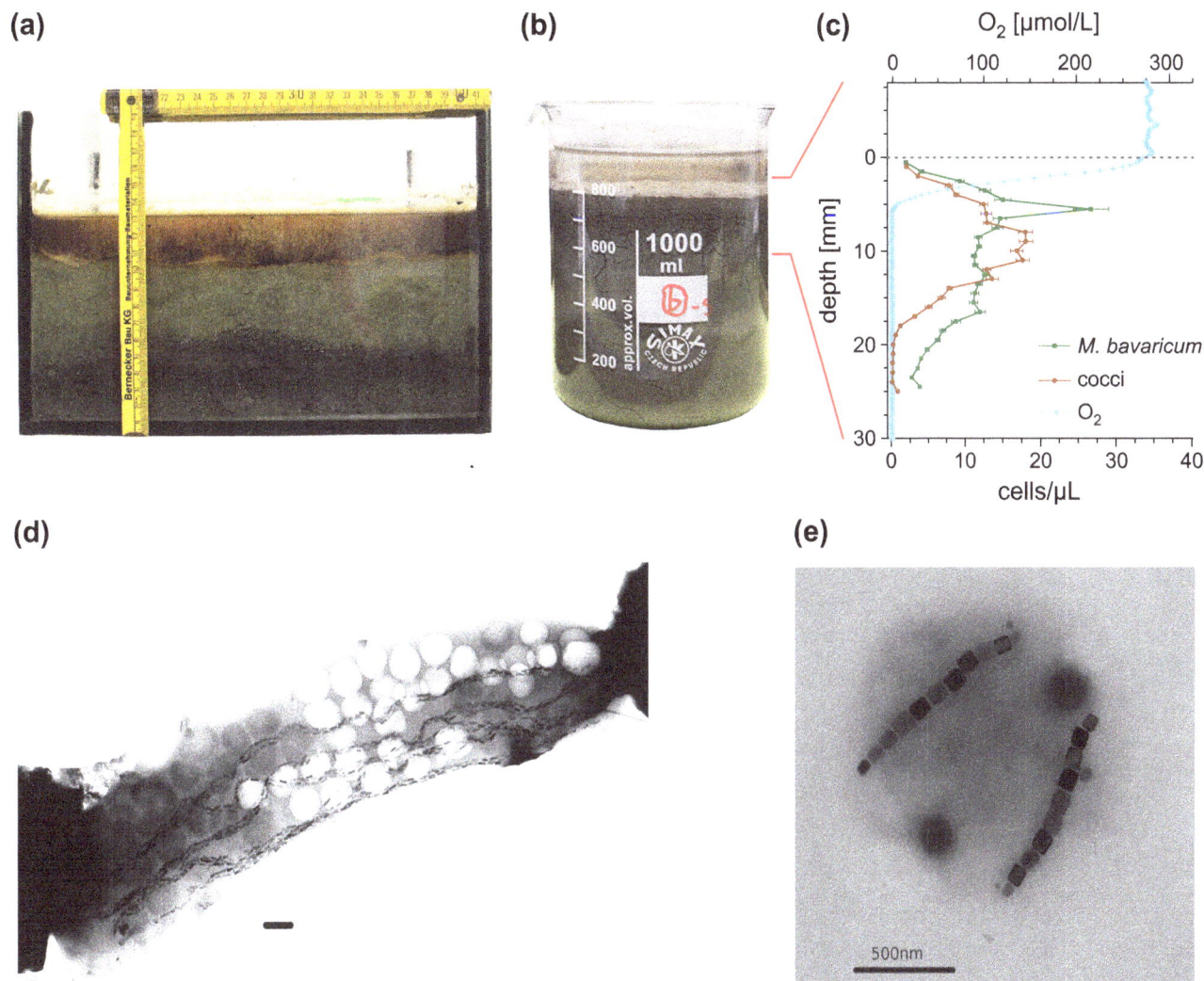

Figure 1. Microcosms and MTB characterization. (A) 30×20×20 cm glass aquarium filled with sediment. (B) 1 L sediment microcosm. (C) Oxygen profile measured after 7 days stabilization, and mean MTB profiles obtained from the average of $n = 88$ minicores. Error bars correspond to $\pm 2\varepsilon$, where $\varepsilon = \sigma/\sqrt{n}$ is the standard error estimated from the standard deviation σ of n individual minicore profiles. (D) Transmission electron micrograph (TEM) image of *M. bavaricum* containing five bundles of tooth-shaped magnetosomes and several empty (light) and filled (dark) sulfur inclusions. The horizontal bar corresponds to 0.5 μm. (E) TEM image of a coccus containing two chains of prismatic magnetosomes. The horizontal bar corresponds to 0.5 μm.

through the Helmholtz coils. The laboratory field was then carefully compensated with the Helmholtz coils, in order to obtain nearly zero-field conditions for the next 194 days. Natural fluctuations of the residual field were monitored at regular time intervals and never exceeded ±1.5 μT. A preferred magnetic direction was not available to MTB during this time, because residual field fluctuations are effectively unbiased (text S1). The natural field was re-established at the end of this experiment for 138 days. Finally, during the last 98 days, the aquarium was exposed to a ~120 μT vertical field generated by the Helmholtz coils. The field direction was reversed every 24 hours with an electronic commutator, so that MTB were exposed to upwards and downwards pointing fields every second day. This experimental setup aimed at testing the resilience of MTB population against field reversals, without exposing MTB to a consistent field direction favouring a particular magnetotaxis polarity. MTB profiles have been measured at regular time intervals during the whole experiment duration.

5. Observing vertical migration in sediment

Vertical MTB profiles represent the dynamic equilibrium eventually reached by motile cells in a stationary environment: as such, they do not provide information about cell migration. Because direct MTB observations are not possible in sediment, an indirect method has been devised for tracking macroscopic vertical displacements, which is based on the overwhelming NS polarity of cells observed in hanging drop assays. This polarity is switched by applying a short (~50 μs) magnetic pulse whose intensity (~100 mT) is sufficient for reversing the magnetic moment of magnetosome chains before reorientation of the whole cell can take place [17]. After imparting a magnetic pulse against the steady field used for cell alignment, MTB rotate by 180° because of magnetic moment reconfiguration, thereby reversing their swimming direction. Accordingly, former NS cells become SS after a magnetic pulse (Figure 2a–d). Because flagellar rotation is unaffected by magnetic pulses, pulsed cells can be considered identical to the normal ones, except for the reversed magnetotactic

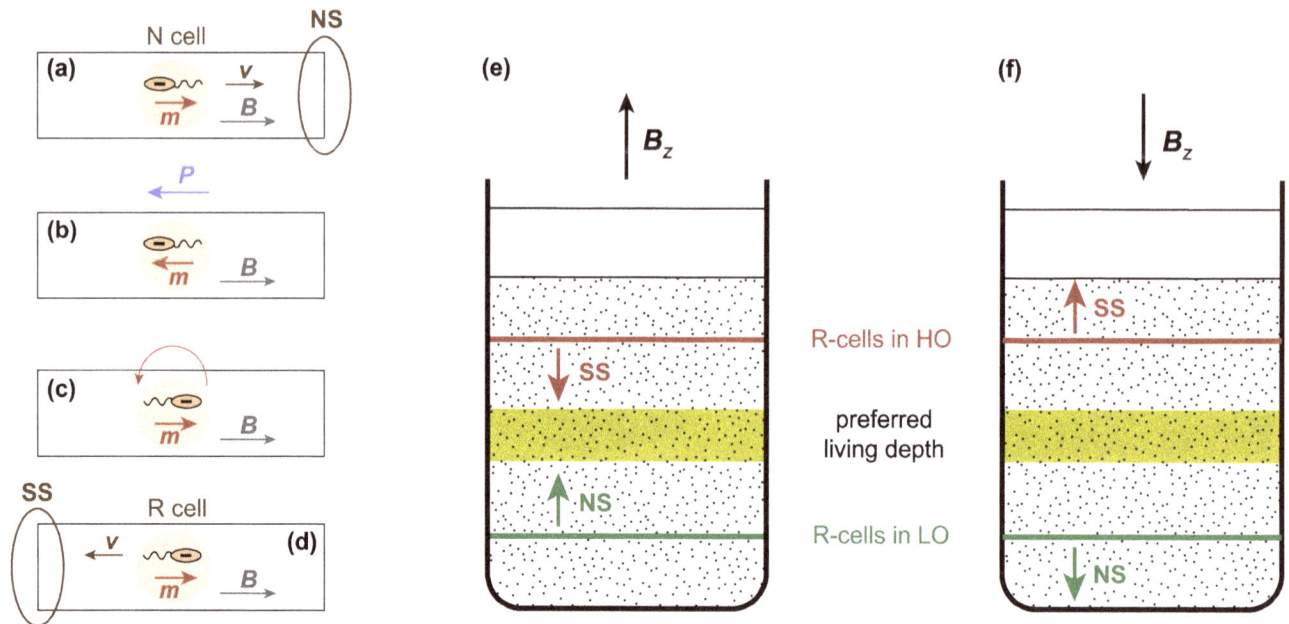

Figure 2. Production of 'reversed' MTB cells and vertical displacement detection in sediment. (A) 'Normal' MTB cell with polar magnetotaxis in the hanging drop assay. **B** is the magnetic field, **m** the cell magnetic moment, **v** the cell velocity. In oxygen-rich water, the cell is NS. (B) A short and strong magnetic pulse **P** antiparallel to **B** switches the cell magnetic moment before cell rotation takes place. (C) The switched cell rotates 180° so that **m** is again parallel to **B**. (D) Because flagellar rotation has not changed, the cell with reversed magnetic moment (R-cell) becomes SS. (E) Microcosm containing two thin R-cell layers above and below the preferred living depth. If **B** is pointing upward, R-cells performing polar M-A are expected to move toward the preferred living depth ('correct magnetotaxis'). (F) Same as (E) after reversing the direction of **B**. R-cells are expected to move away from the preferred living depth ('incorrect magnetotaxis').

polarity. In order to avoid confusion with the 'natural' SS behaviour controlled by flagellar rotation, we call 'normal cells' (N-cells) all polar MTB that have not been subjected to a magnetic pulse. These MTB behave as NS cells in the hanging drop assay, up to a negligible proportion of 'wrongly polarized' SS cells. On the other hand, polar MTB whose magnetic moments have been switched by a magnetic pulse behave as SS cells in the hanging drop assay and we call them 'reversed cells' (R-cells). Nearly equal amounts of NS and SS cells can be observed with hanging drop assays of pulsed sediment samples, instead of almost exclusively NS cells associated with untreated sediment. This result is expected if MTB living in sediment are almost randomly oriented, as reported in [17], because only cells whose magnetic moment from an angle of >90° to the pulsed field can be reversed.

N- and R-cells are always distinguishable in the hanging drop assay according to their NS and SS behaviours, respectively, regardless of the actual swimming polarity in sediment, which is controlled by chemical conditions. Therefore, if R-cells are created within a thin sediment layer by application of a magnetic pulse, the vertical migration of such cells can be monitored on sediment profiles by counting SS cells with the hanging drop assay. Because of the extremely low proportion of N-cells with SS behaviour, practically all SS cells retrieved at a certain depth in sediment must be R-cells coming from the thin layer where they have been created. Finding SS cells above (below) this layer means that R-cells have migrated upward (downward) in sediment. The migration direction is expected to depend on the position of R-cells relative to the preferred living depth, and on the vertical component of the local magnetic field (i.e. downward, as in the Northern hemisphere, or upward, as in the Southern hemisphere). The magnetic polarity of R-cells defines the following magnetotactic configurations: (1) 'correct magnetotaxis', if the field points

upwards, (2) 'incorrect magnetotaxis', if the field points downwards, (3) 'indifferent magnetotaxis', if the field is horizontal, and (4) 'no magnetotaxis' in zero fields.

The role of sediment chemistry in determining the direction of vertical displacements is tested with two limit situations of the polar M-A model, i.e. (1) cells are located above their preferred living depth and therefore in a 'high-oxygen' state (HO), and (2) cells are located below their preferred living depth and therefore in a 'low-oxygen' state (LO). HO cells have the same swimming polarity observed with the hanging drop assay (i.e. NS for N cells and SS for R cells), while the opposite is expected for cells in a LO state. HO and LO states might depend on the whole sediment chemistry (e.g. O_2, Eh, S^{2-}), rather than oxygen alone; however, the situation of undisturbed sediment is analogous to the stratification of MTB cultures where magneto-aerotaxis has been observed [5]. Therefore, M-A observations of cultured MTB can be replicated in sediment microcosms, using R-cells as 'tracer' for macroscopic (>1 mm) vertical displacements. The influence of magnetotaxis on such displacements can be tested by comparing results obtained with identical microcosms subjected to different field configurations.

Natural situations for R-cells are reproduced with 'correct magnetotaxis' configurations where the local magnetic field points upwards (Figure 2e): HO R-cells are SS and move down in sediment, eventually switching to a LO state after crossing their preferred living depth, in which case they become NS and move upwards. This mechanism keeps MTB within their preferred living depths. The opposite situation, similar to a geomagnetic field reversal ('incorrect magnetotaxis'), is created with a downward-pointing field (Figure 2f). In this case, R cells controlled by magnetotaxis move away from their preferred depth, and a population decline is expected over time due to sub-optimal living

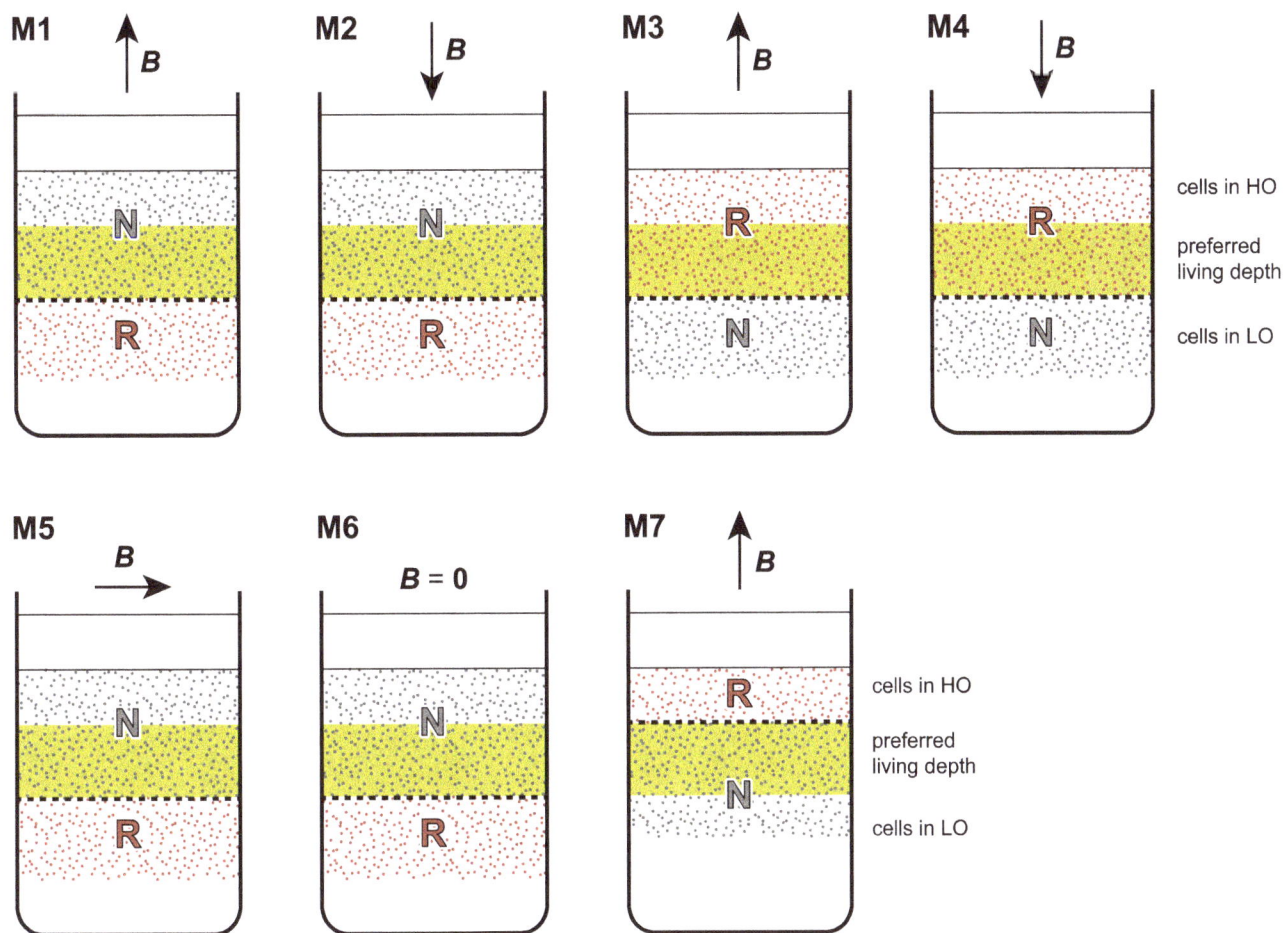

Figure 3. Sediment microcosms (M1–M7) used for vertical migration experiments. The microcosms consist of two sediment layers separated by a sharp boundary (dashed line) located above or below the preferred living depth (shaded area). Initially, only one layer contains R-cells, while N-cells occur in both layers. Vertical migration in a controlled field **B** is tracked by the appearance of R-cells in the other layer.

conditions. Finally, MTB must rely only on chemotaxis for vertical displacements in local magnetic field with zero vertical component ('indifferent magnetotaxis' or 'no magnetotaxis'). The combination of R-cells in HO and LO states with different magnetic field configurations can therefore be used to assess the role of magnetotaxis in sediment without the need of direct observations.

A necessary condition for the correct interpretation of vertical migration observations performed with the abovementioned method is that R-cells should not form spontaneously over the experimental time when field polarity is favourable to them. Because the appearance of R-cells in sediment originally containing N-cells has been documented [1,4], each experiment included an identical control microcosm without pulse-generated R-cells. All control microcosms displayed the usual NS:SS ratios > 1000 over the whole experiment time (not shown), therefore excluding the spontaneous generation of R-cells.

6. Vertical migration experiments

Seven vertical migration experiments have been prepared according to the principles explained in section 5. In all experiments, microcosms consisted of two layers of identical sediment material, with only one layer (the top or the bottom one)

containing R-cells. Microcosms have been prepared with the following procedure:

Microcosms with R-cells in bottom layer (M1–2 and M5–6 in Figure 3). Stabilized microcosms have been repeatedly exposed to pulsed fields generated by a small pair of coils placed in proximity of the sediment surface. The coil pair was moved all over the sediment surface to create a homogeneous layer containing ~50% R-cells. R-cells are concentrated within the uppermost 25 mm, where the pulse field is sufficiently strong for switching favourably oriented magnetic moments. Sediment containing exclusively N-cells has been subsequently deposited on the top of the existing microcosm sediment. In order to keep the interface between the two layers as sharp as possible, the second layer was formed by letting sediment material drop slowly from a sieve and deposit through the water column, in analogy with natural sedimentation (text S1).

Microcosms with R-cells in top layer (M3–4 and M7 in Figure 3). These microcosms have been produced with the procedure described above, except for the fact magnetic pulses have been applied to the sediment material used for depositing the top layer, instead of the bottom layer. Consequently, R-cells are initially contained only in the top layer, where their concentration is ~50%.

day 0　　　　　　　　　　**day 19**　　　　　　　　　　**day 38**

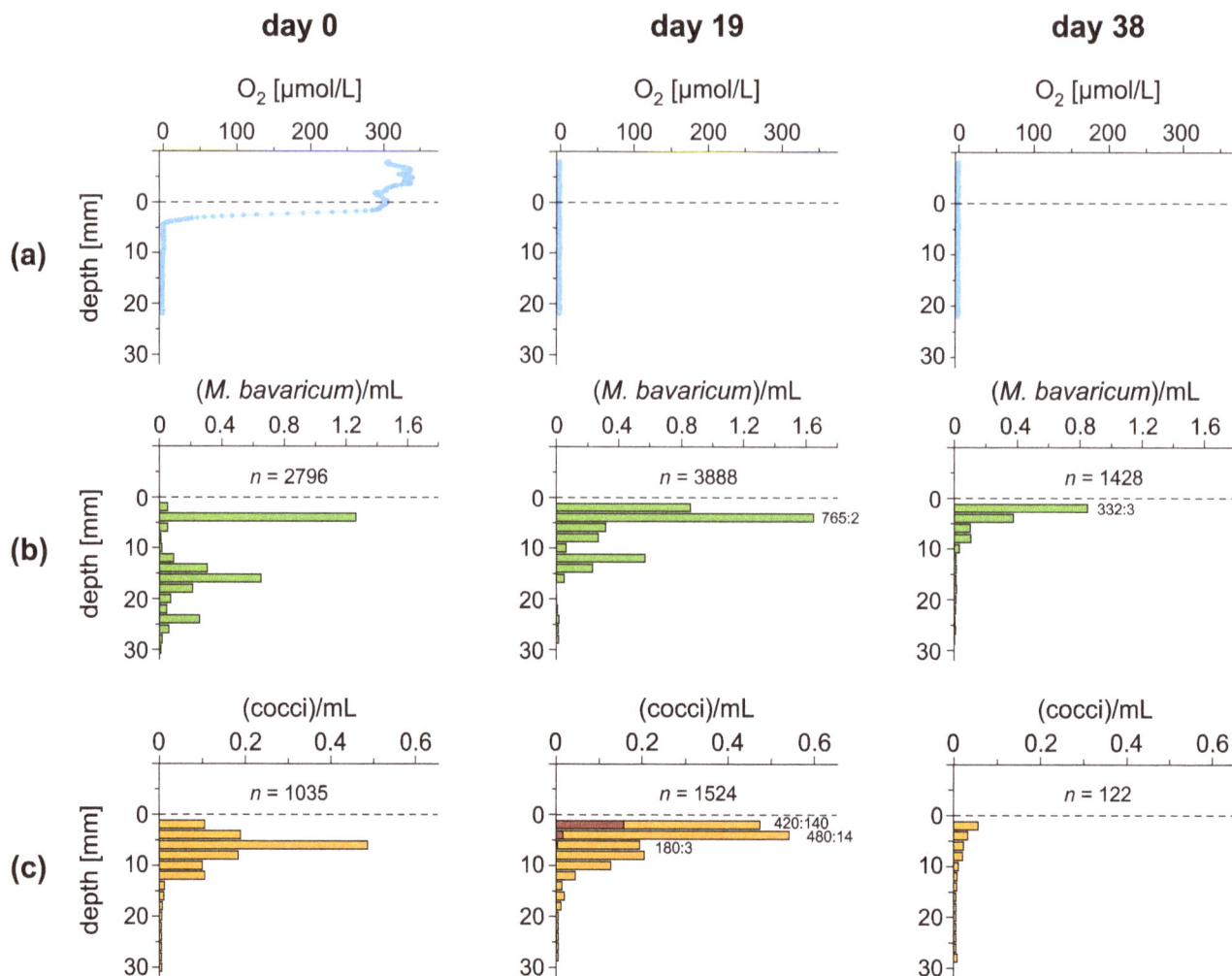

Figure 4. Evolution of oxygen concentration and cell counts in a microcosm under anoxic atmos-phere. (A) Oxygen profiles in water and sediment (0 = sediment/water interface), measured 0, 19, and 38 days after exposure to an oxygen-free atmosphere. (B–C) MBV and MCC cell concentrations as a function of sediment depth. Total MTB populations n (in 10^6 cells/m^2) and NS:SS proportions are given in each plot. Significant concentrations of SS cocci at day 19 are plotted as red bars.

Immediately after their preparation, microcosms have been placed in a Helmholtz coil system pro-ducing a homogeneous ~50 µT field directed vertically (experiments with 'correct' and 'incorrect' magnetotaxis), or horizontally (experiments with 'indifferent' magnetotaxis). Zero-field experiments have been performed by regulating the Helmholtz coils so, that the local Earth magnetic field was compensated up to a ~0.5 µT residual field that changed in an unsystematic manner with time. The chemical stratification of freshly prepared microcosms is obviously disturbed, because both the top layer and part of the bottom layer are initially saturated with oxygen. As excess oxygen is consumed in sediment, the usual chemical stratification, with the OAI occurring ~5 mm below the sediment-water interface, is re-established in 3–4 days. Therefore, all microcosms have been allowed to reach a new chemical equilibrium during ~7 days. Oxygen profiles have been measured to ensure that conditions similar to the case of undisturbed microcosms were established before taking MTB profiles. Finally, MTB profiles have been measured with the procedure described in section 2, obtaining NS cell counts (N-cells) as well as SS cell counts (R-cells). Detection of SS cells outside the pulsed layer must be attributed to vertical migration, if the same phenomenon is not observed in control microcosms containing only N cells.

Results

1. Hanging drop assays under anoxic conditions

Immediately after an oxygen-free atmosphere was established inside the glove box, the microcosm OAI moved upwards, reaching the sediment-water interface in 19 days and disappearing completely from the water column in 38 days (Figure 4a). MTB profiles moved up as well, following the OAI trend (Figure 4b–c), with maximum cell concentrations occurring at the sediment-water interface. On the other hand, MTB have never been found in the water column, while total cell counts decreased over time inside the sediment. Few SS cells have been observed in hanging drop assays prepared with sediment taken from the topmost 4 mm immediately after the OAI reached the water column, with up to 25% SS MCC-cells in a single profile. SS cell counts were extremely variable and decreased to <1% after 30 days.

A net predominance of SS cells in anoxic hanging drop assays, as expected for polar magnetotactic MTB in their LO state, could

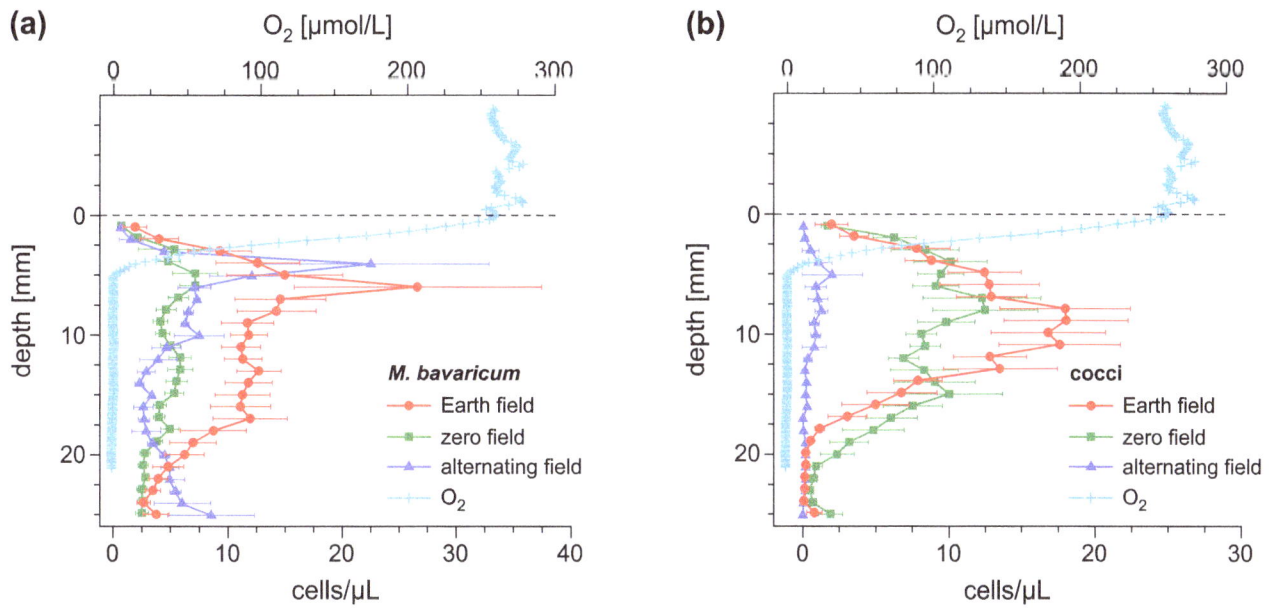

Figure 5. Mean MTB concentration profiles obtained during long-term exposures to different magnetic field configurations. Profiles refer to (A) *M. bavaricum*, and (B) cocci, during resilience experiments in the Earth magnetic field, zero field, and a daily switched vertical field. Error bars correspond to the standard deviation of individual minicors.

not be observed with this experimental setup. The transient increase of SS cell counts while the OAI was crossing the sediment-water interface, along with the fact that MTB did not follow the OAI into the water column, suggests that flagellar rotation might be controlled by additional factors besides oxygen concentration. Furthermore, strong mechanical stimulations and sudden chemical changes during preparation of the hanging drop assay can be considered analogous to sediment resuspension produced by a bioturbation event, and might therefore trigger a NS response useful to regain the preferred depth in sediment. Overall, results of hanging drop assays are inconclusive with respect to the characterization of sedimentary M-A.

2. Resilience experiments

Resilience experiments have been used to monitor the reaction of undisturbed MTB populations to different magnetic field conditions corresponding to: (1) normal, downward-pointing laboratory field, (2) zero field, and (3) vertical field whose polarity is switched daily. The long experiment duration enabled collection of a large number of sediment profiles, which define the mean MBV and MCC concentration profiles shown in Figure 5. Integration of these profiles over sediment depth gives the mean number of cells per unit area, which can be considered as a

measure for the total MTB population for the three above-mentioned experimental conditions (Table 1).

Zero-field conditions produced a ~50% decrease of the total MBV population by unchanged depth distribution. On the other hand, the total MCC population remained stable, but extended over a slightly larger depth range. Alternating field conditions produced the almost complete extinction of MCC, especially at > 12 mm depths, and a ~50% decrease of the total MBV population, whose depth distribution became markedly bimodal with two clearly distinct peaks at 4 mm and >25 mm. Population decrease in zero and alternating fields cannot be attributed to a natural declining trend, because complete recover was observed when the Earth magnetic field was re-established between the two time intervals. These results are difficult to reconcile with the M-A model, especially for the case of unchanged MCC concentrations in zero field (Table 1). At least in this case, it seems that magnetotaxis does not provide an advantage to MTB population living in undisturbed sediment.

3. Microcosm M3: R-cells at preferred depth, correct magnetotaxis

The sediment interface of this microcosm is located at 25 mm depth, R cells in the upper layer are within their preferred living range, and **B** points upwards (Figure 6a). This experiment

Table 1. Total MTB populations in the Earth magnetic field, zero field and vertically alternating field.

Field setting	Earth field (*n*=88)	Zero field (*n*=94)	Alternating field (*n*=47)
M. bavaricum	398 ± 17	166 ± 4 (rejected)	188 ± 5 (rejected)
cocci	279 ± 9	262 ± 11 (not rejected)	13 ± 1 (rejected)

Total MTB populations (in 10^6 cells/m^2 $\pm$ standard error) are obtained by integrating cell counts of *n* individual profiles sampled during exposure to the Earth magnetic field, zero field, and a vertical field whose direction was switched every day. Results of a Kolmogorov-Smirnov test with 95% confidence level are reported in parenthesis. The null hypothesis that zero or alternating fields did not affect MTB populations has been tested against the hypothesis of a population increase or decrease. In all cases, except for cocci in zero field, the population decreased significantly in zero or alternating fields.

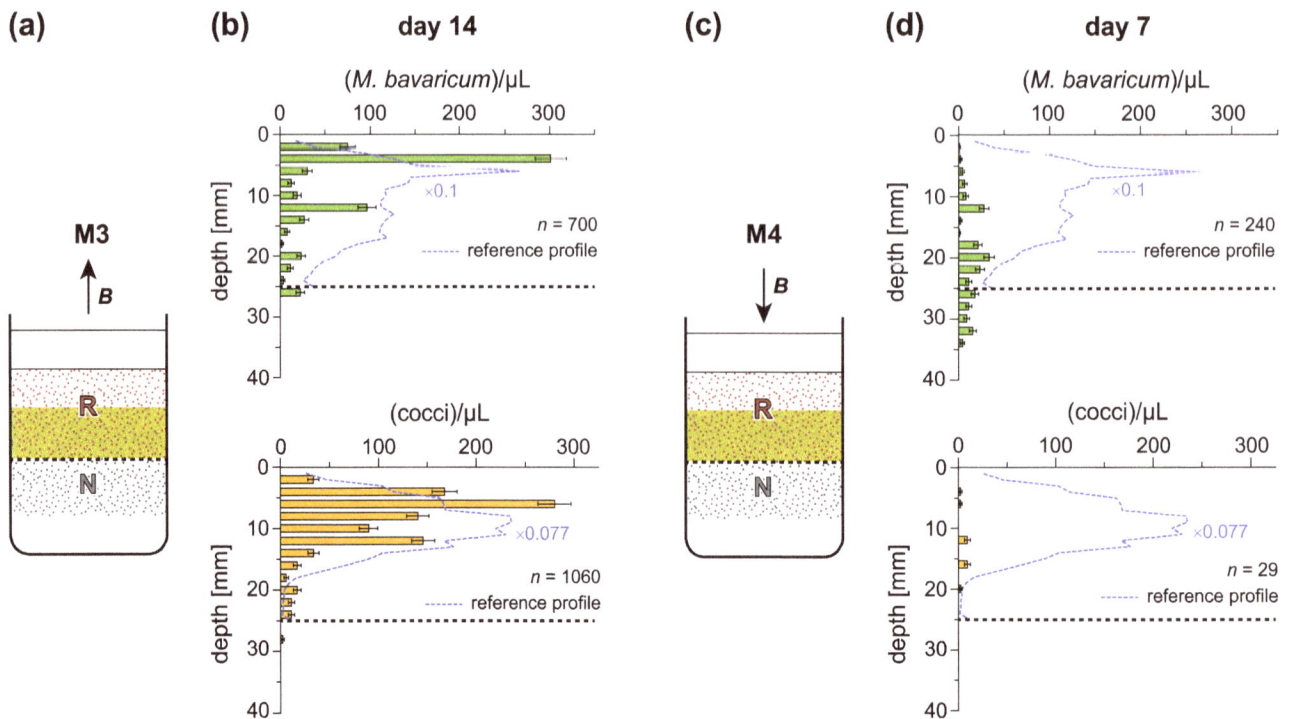

Figure 6. Results of microcosm experiments M3 and M4. R-cells were initially located in their preferred living depth range with (A) upward and (C) downward pointing vertical fields. (B) Cell concentration profiles after 14 days in upward field (correct magnetotaxis). (D) Same as (B), after 7 days in downward field (incorrect magnetotaxis). Long-term mean profiles in the Earth magnetic field obtained from a different microcosm (Figure 5) are shown for comparison (dashed curves), after rescaling cell counts to match concentrations observed in this experiment during correct magnetotaxis. Total MTB populations n, obtained by integrating cell counts in each depth, are expressed in 10^6 cells/m^2.

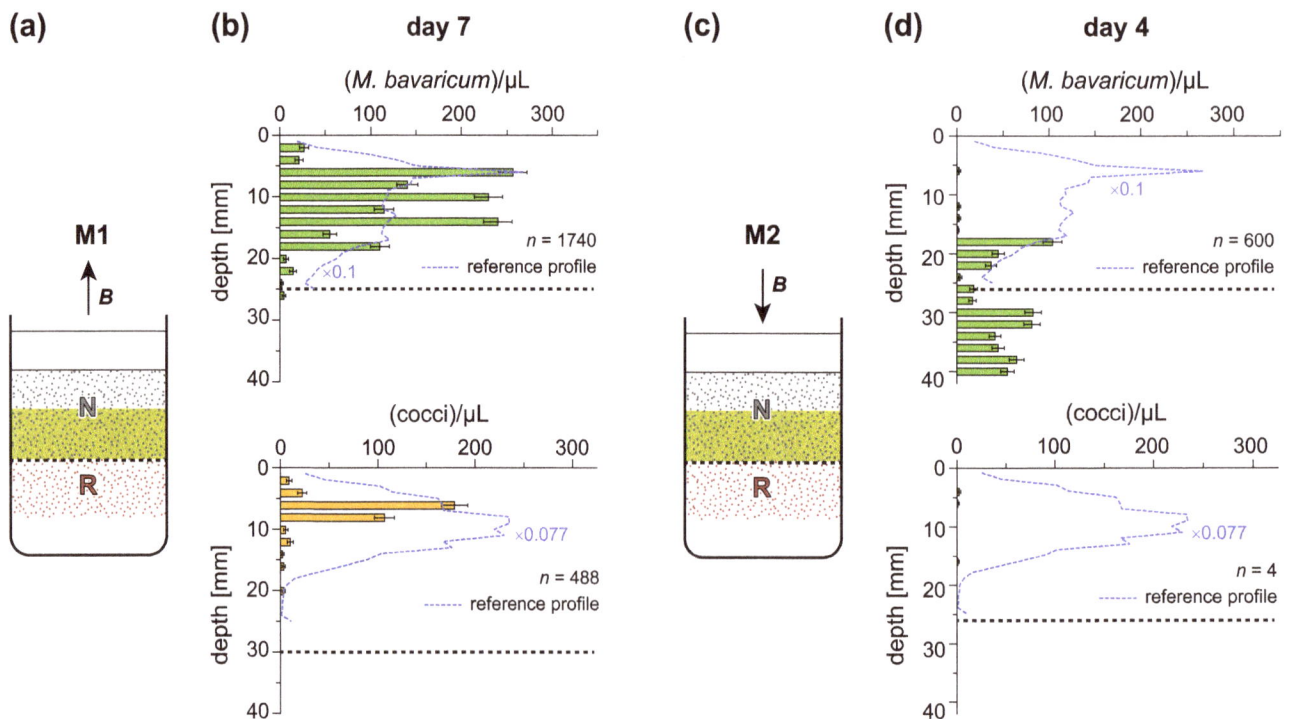

Figure 7. Results of microcosm experiments M1 and M2. R-cells were initially located below their preferred living depth range with (A) upward and (C) downward pointing vertical fields. (B) Cell concentration profiles after 7 days in upward field (correct magnetotaxis). (D) Same as (B), after 4 days in downward field (incorrect magnetotaxis). Long-term mean profiles in the Earth magnetic field (Figure 5) are shown for comparison (dashed curves), after rescaling cell counts as in Figure 6. Total MTB populations n, obtained by integrating cell counts in each depth, are expressed in 10^6 cells/m^2.

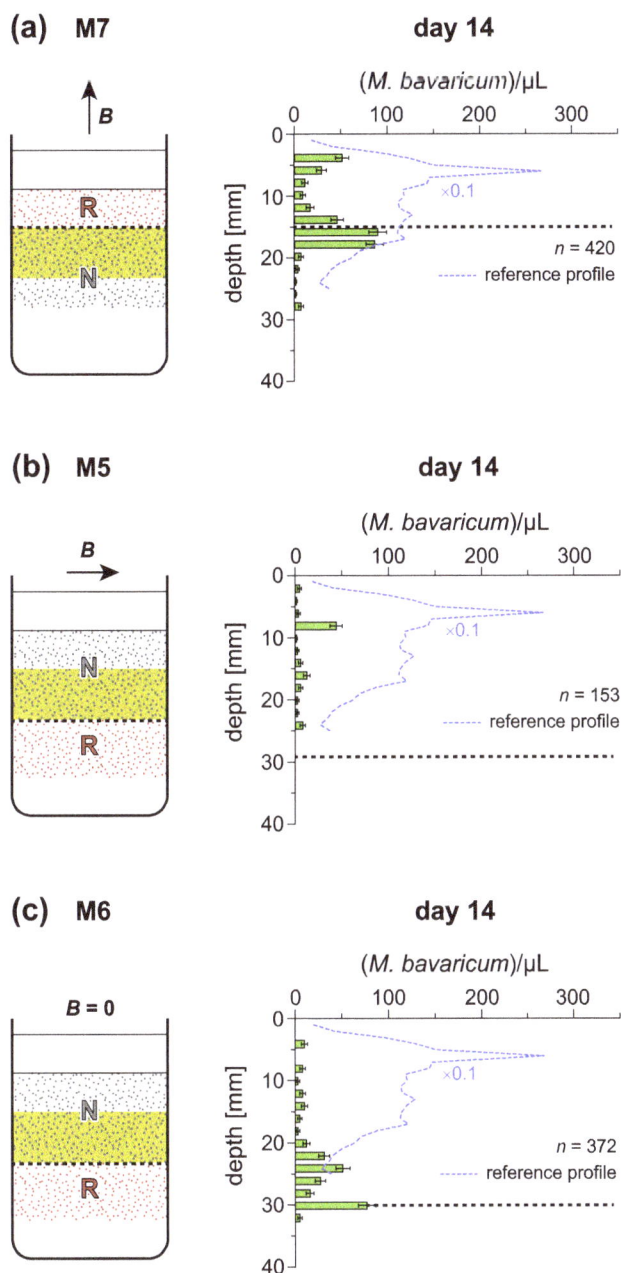

Figure 8. Results of microcosm experiments M7, M5 and M6. (A) Microcosm M7: R-cells of MBV were initially located above their preferred living depth, (<15 mm), in an upward pointing field (correct magnetotaxis of HO cells). (B) Microcosm M5: MBV R-cells were initially located below the preferred living depth (>30 mm), in a horizontal field (indifferent magnetotaxis of LO cells) (C) Microcosm M6: MBV R-cells were initially located below the preferred living depth (>25 mm), in zero field (LO cells with no magnetotaxis). The cell concentration profiles are shown after 14 days from experiment begin, together with long-term mean profiles in the Earth magnetic field (dashed curves), after rescaling cell counts as in Figure 6. Total MTB populations n, obtained by integrating cell counts in each depth, are expressed in 10^6 cells/m².

reproduces the normal conditions existing in the Southern hemisphere, except for the fact that R-cells have been created by switching the magnetic moment of N-cells. Vertical migration across the sediment boundary is not expected. At the same time,

magnetotaxis is incorrect for N-cells, whose number is expected to decline with time. The vertical distribution of R-cells 14 days after microcosm preparation did not change significantly (Figure 6b), and only few R-cells have been found immediately below the interface between the two sediment layers. On the other hand, N-cells of both MBV and MCC were hardly detectable after 14 days (not shown). Experiment results are compatible with the polar M-A model. Furthermore, R-cells created by magnetic moment switching appear to behave exactly like SS cells naturally occurring in the Southern hemisphere, thereby excluding any chemical sensing polarity.

4. Microcosm M4: R-cells at preferred depth, incorrect magnetotaxis

This is the same case as M3, up to a downward pointing magnetic field in which R-cells possess the wrong magnetotaxis polarity. In fact, microcosm M3 has been used for this experiment, after reversing the direction of the magnetic field. The initial condition of M4 is therefore given by Figure 6b. After exposure to a downward pointing field for 7 days, R-cell counts decreased markedly for MBV, while MCC became practically extinct (Figure 6d). The MBV population decrease is particularly evident in the topmost 10 mm, where HO cells would migrate upwards because of the incorrect magnetotaxis configuration. On the other hand, R-cells located near the sediment interface, which can be assumed to be in a LO state, are expected to migrate downwards, as seen by their appearance in the bottom sediment layer. Overall, these results can be explained by polar M-A. MBV appears to be more resilient than MCC to incorrectly polarized magnetotaxis.

5. Microcosm M1: R-cells in LO state, correct magnetotaxis

In this experiment, R-cells are initially confined in the bottom sediment layer, at >25 mm depths, and exposed to an upward-pointing field (Figure 7a). The initial depth range is below the preferred one, so that R-cells are expected to migrate upwards, assisted by correctly polarized magnetotaxis. This experiment aimed at testing the swimming direction of polar magnetotaxis under LO conditions, which was not unequivocally determined by anoxic hanging drop assays (section 1). The depth distribution of R-cells 7 days after microcosm preparation confirms that MTB were able to migrate back to their preferred living depth without losses (Figure 7b). Few R-cells appeared in the top layer already after 2 days and almost no R-cells remained in the bottom layer after 7 days. This result confirms the fact that polar MTB switch their swimming direction when exposed to LO conditions. The ~25 mm migration distance deduced from depth distributions is much larger than the ~1 mm minimum distance over which magnetotaxis is expected to be the dominant displacement mechanism in sediment [17]. Therefore, a mean vertical migration >3.6 mm/day can be deduced from the fact that ~25 mm were covered in <7 days. For comparison, migration velocities of the order of 10 mm/day have been predicted with a biased random walk model and 1% magnetic alignment [17].

6. Microcosm M2: R-cells in LO state, incorrect magnetotaxis

Microcosm M2 was prepared from M1 after adding a new 25 mm-thick sediment layer containing only N-cells (Figure 7c). This experiment reproduces the same situation of M1, but $\boldsymbol{B}$ now points downwards, providing an incorrect magnetotactic polarity to R-cells. With this configuration, R-cells in LO state are expected to migrate further down and decrease in number, while no R-cells are expected to appear in the top layer.

MTB profiles taken after 4 days from microcosm preparation (Figure 7d) are consistent with M-A model predictions with respect to the total population, whose decrease is more evident for MCC. On the other hand, some R-cells of MBV migrated upwards into the top layer, against the direction imposed by magnetotaxis on cells in LO state. The ~12 mm migration distance is sufficiently long for being dominated by magnetotaxis, and corresponds to a velocity of ~3 mm/day, which is similar to the estimate obtained with microcosm M1. Upward migration of MBV R-cells cannot be explained by current polar M-A models. If R-cells were in a LO state, as expected from the sediment interface depth, a spontaneous conversion of R-cells into N-cells must have occurred by switching the flagellar rotation senses corresponding to LO and HO states, respectively. On the other hand, the existence of HO cells ~25 mm below the sediment-water interface would be difficult to explain, because the starting point of this experiment was a stable MTB population in its preferred living depth, which soon evolved to LO conditions after addition of the top sediment layer and consequent upward migration of the OAI.

7. Microcosm M7: R-cells in HO, correct magnetotaxis

This experiment is in principle identical to the one performed with microcosm M3, with the only difference that the R-cell-containing top layer has been made as thin as possible (15 mm, Figure 8a), in order to monitor the migration of MBV cells in their HO state, assisted by a correct magnetotaxis polarity. After 14 days, high concentrations of R-cells were detected in vertical profiles down to 20 mm, and few cells have been found down to 27 mm (profile in Figure 8a). This experiment demonstrates the vertical migration of HO cells, as predicted by the M-A model. The concentration of N-cells, on the other hand, decreased significantly (data not shown), due to incorrectly polarized magnetotaxis.

8. Microcosm M5: R-cells in LO, horizontal field

This experiment aims at testing M-A in horizontal fields, as it would naturally occur at the geo-magnetic Equator. Microcosm preparation was identical to M1, except that R-cells in LO state inside the bottom layer are exposed to a horizontal field (Figure 8b). In this case, magnetotaxis does not provide any preferred direction for vertical migration (indifferent magnetotax-

is). The M-A model, combined with low magnetic alignments, predicts that any vertical migration of LO cells relies only on chemotaxis. N- and R-cells of MCC, though being abundant at the beginning of the experiments, disappeared completely after 14 days (data not shown). R-cells of MBV, on the other hand, migrated into the top layer (profile in Figure 8b) at cost of a significant population decrease. For comparison, vertical migration of LO cells assisted by a correctly oriented vertical field (microcosm M1) occurred without significant cell count decreases.

9. Microcosm M6: R-cells in LO, no magnetotaxis

This experiment explores the vertical migration capability of LO cells when magnetotaxis is elimi-nated by cancellation of the local magnetic field. In this case, MTB cells rely only on chemotaxis. R- and N-cells of MCC disappeared after 14 days, while R-cells of MBV migrated into the top layer without a significant population decrease (Figure 8c). The disappearance of MCC apparently contradicts the long-term resilience experiment (section 2) where the same type of bacteria was practically unaffected during >6 months in zero field. However, the resilience experiment has been performed with undisturbed sediment in which cells where not displaced from their preferred depth, as it is the case here.

Discussion

Experiments discussed in the results chapter are conveniently grouped according to (1) the initial MTB distribution with respect to the preferred living depth (i.e. at, above, and below) and (2) the magnetotactic configuration (i.e. 'correct' and 'incorrect' magnetotaxis in vertical fields, 'indifferent' magnetotaxis in horizontal fields, and no magnetotaxis in zero field). Results (Table 2) are expressed in terms of detectable vertical displacements (i.e. R-cells appearing in a sediment layer where they were not originally present) and evolution of total populations over time. Because cell counts are subjected to large uncertainties due to microcosm heterogeneity (i.e. lateral profile variations) and temporal fluctu-ations, population changes are considered significant only if they exceed a given factor that depends on the type of experiment. In case of resilience experiments (Table 1) the large number of measured profiles enabled the application of a Kolmogorov-Smirnov test for two variables with unknown probability distributions (i.e. total MBV and MCC populations subjected to normal and special field configurations, respectively). In all vertical

Table 2. Summary of microcosm experiment results.

Condition	Correct MT	Incorrect MT	Indifferent MT	No MT	Alternating MT
Above PD	M7	–	–	–	–
Migration	↓ MBV				
Population	↔ MBV				
At PD	M3	M4	–	R0	RA
Migration	↔ all	↓ MBV (partly)		↔ all	↔ all
Population	↔ all	↓ all		↓ MBV, ↔ MCC	↓ MBV, 0 MCC
Below PD	M1	M2	M5	M6	–
Migration	↑ all	↑ MBV	↑ MBV	↑ MBV	
Population	↔ all	↓ MBV, 0 MCC	↓ MBV, 0 MCC	↔ MBV, 0 MCC	

Abbreviations: PD = preferred depth, ↑ = upward migration, ↓ = downward migration or significant population decline, ↔ = no significant population decline or no vertical migration, 0 = disappearance (down to 0–1 counts), R0 = resilience in zero field, RA = resilience in alternating field.

Table 3. Summary of observed MTB behaviours under different magnetic field configurations.

MTB ability to:	Key experiments	*M. bavaricum*	Cocci
Use correct magnetotaxis when undisturbed	R0	Yes	No
Use correct magnetotaxis when displaced	M1 compared to M2, M5, and M6	Yes	Yes
Use chemotaxis when displaced in zero field	M6	Yes	No
Override incorrect magnetotaxis when displaced	M2 and RA	Yes	No

migration experiments (microcosms M1–M7), population changes are considered significant if they exceed a factor of 2. This factor has been chosen because it corresponds to the maximum cell count standard deviation of 88 individual sediment profiles taken under identical, stationary conditions (Figure 1c). Furthermore, a MTB population is considered nearly extinct if only 1–2 cells were counted in each depth interval.

Vertical displacements give information about how magneto-taxis is combined with chemical sensing, while a significant population decrease can be interpreted as the consequence of unfavourable envi-ronmental conditions that cannot be efficiently escaped. Experiment results appear contradictory with respect to the existence of a magnetotactic advantage in sediment: for example, resilience experiments in zero fields clearly demonstrate this advantage for MBV, but not for MCC, while a correct field polarity is essential for enabling displaced MTB cells of both types to reach their preferred depth. Furthermore, some experiments (e.g. M2) revealed the ability of MBV to migrate against the direction imposed by magnetotaxis. Overall, MBV and MCC appear to use magnetotaxis in a different manner. In order to gain some systematic understanding of our experiments, we summa-rized results according to the capability of MBV and MCC to cope with specific situations summarized in Table 3, which are discussed in the following.

1. Can MTB rely only on chemotaxis?

Zero-field experiments with microcosm M6 prove the capability of MBV to rely only on chemotaxis for reaching the optimal living depth in sediment. Similar results have also been obtained in a horizontal field (microcosm M5), where the absence of a vertical field component is equivalent to zero-field conditions as far as vertical migration is concerned. The capability of chemical reception to modify the MTB swimming behaviour (specifically by switching flagellar rotation) has been clearly demonstrated by M-A experiments with cultured cells [5]. However, polar M-A is not expected to support macroscopic displacements in zero field, because HO/LO cells would swim at random directions without run-and-tumble, run-reverse, or run-reverse-flick behaviours typical of non-magnetic bacteria performing chemotaxis [24,25]. The common feature of these chemotactic swimming behaviours is that nearly straight paths are interrupted by sharp directional changes, whose frequency is determined by the chemical environment (i.e. more frequent changes in presence of repellents). Directional changes are essential for introducing a chemically controlled, systematic bias to an otherwise fully random displace-ment.

Systematic observations of >1000 swimming trajectories of MBV and MCC with the hanging drop assay in precisely controlled zero fields did not reveal a single case where the continuity of (random) swimming paths was interrupted by tumbling or reverse-flick events [17]. On the other hand, MBV sometimes displays a tactile response when colliding with sediment

particles in hanging drop assays prepared with diluted sediment (video S2). Tactile responses consist in brief stops once a sediment particle is hit, as well as swimming direction reversals for variable amounts of time ranging from 1 s to the longest observation time (~10 s). Such tactile responses do not occur systematically and might therefore depend on some internal state of the cell. On the other hand, tactile responses could never be observed with MCC under identical conditions. Hanging drop observations have been performed in a magnetic field, which is required for 'extracting' MTB from opaque sediment, and then make cells swim back into sediment by reversing the field direction. Under such conditions, 'backward' swimming of MBV is characterized by a pronounced wiggling and ~30% speed reduction. The asymmetry between forward and backward swimming comes from the fact that, during forward swimming, flagella are arranged into a single, coherent bundle forming a helix that spirals around the body, minimizing viscous drag [26]. Reversing the sense of flagellar rotation make individual flagella sprout outwards, providing incoherent contri-butions to a less efficient propulsion. If an aligning field is absent, reversed flagellar rotation produces sudden cell reorientation by random angles, similar to those observed during chemotaxis (video S3). The fact that only MBV is able to rely exclusively on chemotaxis for macroscopic vertical displacements might be related to its ability to briefly reverse flagellar rotation even when being in a definite HO/LO state. This mechanism can be understood as a dominantly polar M-A with an axial component triggered by specific stimuli.

2. Is magnetotaxis used for overcoming vertical displacement from the optimal living depth?

The answer to this question seems obvious when considering that this should be the primary function of magnetotaxis. On the other hand, the poor (<1%) magnetic alignment of MTB in sediment [17] might nullify any magnetotactic advantage in this environment, leaving magnetotaxis as a guide for guiding the bacteria, when dislodged, back to the sediment, as postulated in [4]. We have investigated sedimentary magnetotaxis by monitor-ing the vertical migration of displaced R-cells under the following three field configurations: (1) 'correct' ($B_z > 0$), (2) 'incorrect' ($B_z < 0$), and (3) 'none' ($\mathbf{B} = 0$) or 'indifferent' ($B_z = 0$). Vertical migration occurs with no cell count decrease in case of correct confi-gurations for both MBV and MCC. On the other hand, MCC are intolerant to incorrect field orientations, while MBV appears to be able to migrate towards the preferred living depth even against the direction dictated by magnetotaxis, albeit at cost of a significant population decrease. This is only possible if an axial magnetotaxis mechanism somehow overrides polar M-A when necessary. The same capability is seen with resilience experiments, where the daily field polarity change was fatal to MCC but not to MBV. Finally, vertical migration of MBV in correct field configurations (25 mm in ≤7 days) appears to be faster than vertical migration in zero field (20 mm in ≤14 days). Overall, our experiments demonstrate the existence of a magnetotactic advantage supporting vertical migration inside sediment, despite poor magnetic alignment. This advantage seems an essential condition for survival of displaced MCC cells, while MBV is able to migrate without a magnetic field and even against the direction imposed by polar M-A.

3. Is magnetotaxis used within the optimal living range?

Zero-field resilience experiments with undisturbed sediment show that MBV and MCC populations can survive without magnetotaxis if they were originally located within their preferred living range. Total MCC cell counts did not change significantly in zero field, while MBV decreased initially by 50% and remained stable at this level for 6 months. This result is unexpected when considering that displaced MBV cells have a much higher capability of migrating back to their preferred living depth in comparison to MCC. The apparent contradiction between experimental results with displaced and undisplaced MTB populations can be explained if magnetotaxis is somehow continuously used by MBV, even within the preferred living range, while it plays a minor role in case of undisturbed MCC populations. For example, MBV can take advantage from magnetotaxis if it needs to shuttle between oxic and anoxic sediment levels for satisfying different metabolic requirements, as postulated by [27]. This mechanism, called 'redoxtaxis', has been introduced to explain some morphological peculiarities of MBV cells, such as large size and the presence of sulphur inclusions that would be filled when plunging deep in sediment and oxidized after moving into more superficial layers. Empty and filled inclusions can be clearly seen as lighter and darker areas in TEM micrographs, sometimes coexisting within the same cell (e.g. Figure 1d).

The redoxtaxis hypothesis is indirectly supported by our experiments in several ways. First, the bi-modal depth distribution of MBV is compatible with cells spending most time at depths where sulphur oxidation can take place (shallower peak at $\sim$5 mm), and depths where sulphur is accumulated (deeper peak at $\sim$16 mm). Second, vertical shuttling between these depths occur over macroscopic ranges ($\sim$11 mm) largely exceeding the $\sim$1 mm minimum distance required for magnetotaxis to become effective in case of <1% alignments with the Earth magnetic field. For comparison, microscopic displacements required by MTB living at a single preferred depth for contrasting slow sediment mixing occur over distances where magnetotaxis is less effective, explaining the insensitivity of undisturbed MCC populations to the lack of a magnetic field. On the other hand, incorrect field polarities trigger a runaway mechanism by which displaced cells tend to move in the wrong direction.

Conclusions

The magnetotactic advantage of two wild MTB types living in sediment – undefined round cocci and *M. bavaricum* – has been investigated with experiments aimed at testing the capability of (1) undisturbed populations to survive zero-field and field reversal conditions, and (2) displaced populations to migrate back to their preferred living depths under different field configurations. These experiments confirmed that a magnetotactic advantage exists for both MTB types if they are displaced from preferred living depths, for example by bioturbation, even if their mean alignment with the Earth magnetic field is <1%. On the other hand, many responses of the two MTB types to our experimental conditions were clearly distinct and partially unexpected on the basis of current polar M-A models. For example, MBV can migrate over macroscopic distances in zero field, thus relying exclusively on chemotaxis,

while this is impossible for MCC. Displaced MBV cells can even override wrongly oriented fields, although at cost of a population decrease.

The ability of MBV to rely only on chemotaxis for macroscopic displacements can be explained by facultative tactile responses of cells colliding with sediment particles. Such responses, as observed in hanging drop assays in a magnetic field, include swimming reversals lasting for variable amounts of time, which, in zero field, might produce the 'tumbles' required by magnetotaxis. It is important to realize that these reversals have been observed on cells with a clearly defined HO state, and should therefore not be confused with HO-LO transitions in polar M-A. This type of reversals is also required to explain the capability of displaced MBV populations to migrate towards their preferred living depth against the direction dictated by polar M-A. We tentatively explain these observations with the possibility for MBV to switch between polar and axial M-A. Cells with a defined HO/LO state would maintain a dominant magnetotaxis polarity for a certain period of time. If HO/LO conditions persist over this time, magnetotaxis polarity would be reversed, becoming de facto part of an axial magnetotactic mechanism. The delayed onset of this mechanism would enable MBV to profit from specific advantages of polar and axial magnetotaxis, i.e. the insensitivity of axial magnetotaxis to the direction of chemical gradients with respect to the magnetic field, and the insensitivity of polar magnetotaxis to small-scale gradient fluctuations in which cells using 'fast' axial M-A would easily become trapped. MCC do not possess this ability, and appear to rely only on polar magnetotaxis.

Another unexpected difference between MBV and MCC is observed when stable populations living in undisturbed sediment are kept in a zero-field environment for long periods of time. In this case, MBV populations decrease by $\sim$50%, while no changes are observed for MCC. This difference can only be explained by different mobility requirements during stationary conditions. As previously proposed [27], MBV would continuously shuttle between sediment layers with different chemical properties for satisfying metabolic requirements (redoxtaxis), while cocci might seek a single preferred living depth, becoming less motile when this condition is satisfied. In this case, MCC would use magnetotaxis only for overcoming macroscopic displacements produced by sediment bioturbation.

Results of this study suggests that magnetotaxis might be more complicated than originally implied by the M-A model, and that mechanical interactions with sediment particles could play an important role. Furthermore, different magnetotaxis 'styles' might represent a response to specific biological require-ments. In this context, direct MTB observations in aqueous media provide only a very limited insight into complex interactions between magneto-taxis, chemical sensing, and sediment heterogeneities.

Supporting Information

Table S1 Data. This table contains all experimental data underlying Figures 1–8.

Text S1 Detailed description of experimental settings. Section 1: hanging drop assays. Section 2: resilience experiments. Section 3: microcosm preparation for vertical migration experiments.

Video S1 Hanging drop observations of *M. bavaricum* and cocci. Hanging drop assay showing *M. bavaricum* and unspecified magnetotactic cocci accumulating along the right edge of a water drop. The local magnetic field generated by Helmholtz coils is parallel to the observation plane and points to the right, as indicated by the white line inscribed in a circle. Therefore, MTB cells accumulating on the right edge are north-seeking (NS). *M. bavaricum* is easily distinguishable from cocci because of its elongated shape and larger size. During the first two parts of the video, few *M. bavaricum* cells, already immobilized in the partially dried drop edge, are surrounded by highly motile cocci performing backward loops not observed in case of freely swimming cells. The third part of the video shows another region of the drop edge with more, still motile *M. bavaricum* cells. The last part of the video show *M. bavaricum* and cocci swimming away from and back to the drop edge after reversing the field direction twice.

Video S2 In-field tactile response of *M. bavaricum* when colliding with sediment particles. Hanging drop assay prepared with higher sediment concentration for observing interactions of *M. bavaricum* cells with particles. The magnetic field points to the right. Tactile responses can be seen on freely swimming cells coming from the left. Some cells briefly reverse their swimming direction upon colliding with sediment particles (e.g. large cell near the center at 00′03″, and large cell on the bottom after at 00′37″). Other cells engage a long-lasting series of swimming direction reversals, oscillating around a mean position in proximity of sediment particles. These behaviors are never observed in absence of sediment particles. The field direction is reversed at 00′46″. Most cells that have accumulated on the right drop edge cross the region occupied by sediment particles without changing their swimming behavior, as a demonstration that tactile responses are facultative. A large cells swimming to the left near the video bottom stops its motion abruptly after touching a sediment particle (00′47″).

Video S3 In-field and zero-field swimming behavior of *M. bavaricum* in a sediment-rich, aqueous environment. Hanging drop assay prepared with higher sediment concentration for observing the swimming behavior of *M. bavaricum* cells. The magnetic field points to the left until 00′09″, and zero-field conditions are established afterwards, with a brief interruption

during 00:49–56″. Frequent in-field swimming direction reversals are related to the presence of sediment and are not observed in pure water. After establishing zero-field conditions, cells swim at random directions along nearly straight path interrupted by sudden direction changes according to the modes described in [24]. Such direction changes have been never observed in in hanging drop assays with pure water [16]. Some events are highlighted in the following. 00′20–25″ center-top: cell swims upwards, enter a sediment cluster and exits after a ∼180° turn deduced from the fact that the cell is still moving forward (absence of cell wiggling characteristic for backward swimming). 00′26–27″ center-right: cell reverses flagellar rotation and swims backward (wiggling) after coming close to an isolated sediment particle. 00′27–30″ center-right: cell 'collides' with a sediment particle, rotates ∼45°CW, swims backwards for a short time, and finally rotates ∼45°CCW after reversing the swimming direction a second time. This event corresponds to a 'run-reverse-flick' pattern [24]. 00′28–33″ bottom-right: cell collides with a sediment particles and reverses swimming direction (∼00′30″). While swimming backwards, it collides with another swimming cell (∼00′31″). During this collision, the swimming direction is reversed a second time after short tumbling corresponding to a 'run-and-tumble' pattern [24]. 00′33–37″ top-right: cell starts to tumble chaotically just after passing by a couple of sediment particles. 00′42–46″ center to top-left: sequence of two run-reverse-flick patterns [24] triggered in proximity to sediment particles.

Acknowledgments

We thank the fire brigade of Prien am Chiemsee for assisting us with lake sediment sampling and the Ludwig-Maximilians University for supporting the pond infrastructure.

Author Contributions

Conceived and designed the experiments: XM RE NP XL. Performed the experiments: XM MH. Analyzed the data: XM RE. Contributed reagents/materials/analysis tools: MH. Contributed to the writing of the manuscript: XM RE NP MH XL.

References

1. Blakemore RP (1982) Magnetotactic bacteria. Annu Rev Microbiol 36: 217-238.
2. Faivre D, Schüler D (2008) Magnetotactic bacteria and magnetosomes. Chem Rev 108: 4875-4898.
3. Blakemore RP (1975) Magnetotactic bacteria. Science 190: 377-379.
4. Blakemore RP, Frankel RB, Kalmijn AJ (1980) South-seeking magnetotactic bacteria in the Southern Hemisphere. Nature 286: 384-385.
5. Frankel RB, Bazylinski DA, Johnson MS, Taylor BL (1997) Magneto-aerotaxis in marine coccoid bacteria. Biophys J 73: 994-1000.
6. Simmons SL, Bazylinski DA, Edwards KJ (2006) South-seeking magnetotactic bacteria in the Northern Hemisphere. Science 311: 371-374.
7. Zhang WJ, Chen C, Li Y, Song T, Wu LF (2010) Configuration of redox gradient determines magnetotactic polarity of the marine bacteria MO-1. Environ Microbiol Rep 2: 646-650.
8. Chen C, Ma Q, Jiang W, Song T (2011) Phototaxis in the magnetotactic bacterium Magnetospirillum magneticum strain AMB-1 is independent of magnetic fields. Appl Microbiol Biotechnol 90: 269-275.
9. Shapiro OH, Hatzenpichler R, Buckley DH, Zinder SH, Orphan VJ (2011) Multicellular photo-magnetotactic bacteria. Environ Microbiol Rep 3: 233-238.
10. Spormann AM, Wolfe RS (1984) Chemotactic, magnetotactic and tactile behaviour in a magnetic spirillum. FEMS Microbiol Lett 22: 171-177.
11. Bazylinski DA, Frankel RB (2004) Magnetosome formation in prokaryotes. Nat Rev Microbiol 2: 217-230.
12. Smith MJ, Sheehan PE, Perry LL, Connor KO, Csonka LN, et al. (2006) Quantifying the magnetic advantage in magnetotaxis. Biophys J 91: 1098-1107.
13. Petermann H, Bleil U (1993) Detection of live magnetotactic bacteria in South Atlantic deep-sea sediments. Earth Planet Sci Lett 117: 223-228.
14. Petersen N, Weiss DG, Hojatollah V (1989) Magnetic bacteria in lake sediments. In: Lowes FJ, Collinson DW., Parry JH, Runcorn SK, Tozer DC, et al., editors. Geomagnetism and paleomagnetism. Springer Netherlands. 231-241.
15. Flies CB, Jonkers HM, Beer D de, Bosselmann K, Böttcher ME, et al. (2005) Diversity and vertical distribution of magnetotactic bacteria along chemical gradients in freshwater microcosms. FEMS Microbiol Ecol 52: 185-195.
16. Jogler C, Niebler M, Lin W, Kube M, Wanner G, et al. (2010) Cultivation-independent characterization of "Candidatus Magnetobacterium bavaricum" via ultrastructural, geochemical, ecological and metagenomic methods. Environ Microbiol 12: 2466-2478.
17. Mao X, Egli R, Petersen N, Hanzlik M (2014) Magnetotaxis and acquisition of detrital remanent magnetization by magnetotactic bacteria in natural sediment: First experimental results and theory. Geochem Geophys Geosyst 15: 255-283.
18. Pan Y, Petersen N, Davila AF, Zhang L, Winklhofer M, et al. (2005) The detection of bacterial magnetite in recent sediments of Lake Chiemsee (southern Germany). Earth Planet Sci Lett 232: 109-123.
19. Spring S, Amann R, Ludwig W, Schleifer KH, van Gemerden H, et al. (1993) Dominating role of an unusual magnetotactic bacterium in the microaerobic zone of a freshwater sediment. Appl Environ Microbiol 59: 2397-2403.
20. Hanzlik M, Winklhofer M, Petersen N (1996) Spatial arrangement of chains of magnetosomes in magnetotactic bacteria. Earth Planet Sci Lett 145: 125-134.
21. Cox BL, Popa R, Bazylinski Da, Lanoil B, Douglas S, et al. (2002) Organization and elemental analysis of P-, S-, and Fe-rich inclusions in a population of freshwater magnetococci. Geomicrobiol J 19: 387-406.
22. Lin W, Pan Y (2009) Specific primers for the detection of freshwater alphaproteobacterial magnetotactic cocci. Int Microbiol 12: 237-242.

23. Kirschvink JL (1980) South seeking magnetotactic bacteria. J Exp Biol 86: 345–347.

24. Berg HC, Brown D (1972) Chemotaxis in Escherichia coli analysed by three-dimensional tracking. Nature 239: 500–504.

25. Taktikos J, Stark H, Zaburdaev V (2013) How the motility pattern of bacteria affects their dispersal and chemotaxis. PLoS One 8: e81936.

26. Steinberger B, Petersen N, Petermann H, Weiss DG (1994) Movement of magnetic bacteria in time-varying magnetic fields. J Fluid Mech 273: 189–211.

27. Spring S, Schulze R, Schleifer K (2000) Identification and characterization of ecologically significant prokaryotes in the sediment of freshwater lakes?: molecular and cultivation studies. FEMS Microbiol Rev 24: 573–590.

18

Isolate-Dependent Growth, Virulence, and Cell Wall Composition in the Human Pathogen *Aspergillus fumigatus*

Nansalmaa Amarsaikhan, Evan M. O'Dea, Angar Tsoggerel, Henry Owegi, Jordan Gillenwater, Steven P. Templeton*

Department of Microbiology and Immunology, Indiana University School of Medicine – Terre Haute, Terre Haute, Indiana, United States of America

Abstract

The ubiquitous fungal pathogen *Aspergillus fumigatus* is a mediator of allergic sensitization and invasive disease in susceptible individuals. The significant genetic and phenotypic variability between and among clinical and environmental isolates are important considerations in host-pathogen studies of *A. fumigatus*-mediated disease. We observed decreased radial growth, rate of germination, and ability to establish colony growth in a single environmental isolate of *A. fumigatus*, Af5517, when compared to other clinical and environmental isolates. Af5517 also exhibited increased hyphal diameter and cell wall β-glucan and chitin content, with chitin most significantly increased. Morbidity, mortality, lung fungal burden, and tissue pathology were decreased in neutropenic Af5517-infected mice when compared to the clinical isolate Af293. Our results support previous findings that suggest a correlation between *in vitro* growth rates and *in vivo* virulence, and we propose that changes in cell wall composition may contribute to this phenotype.

Editor: Robert A. Cramer, Geisel School of Medicine at Dartmouth, United States of America

Funding: This research was supported by IUSM-TH start-up funds. The funders had no role in study design, data collection and analysis, decision to publish, or preparation of the manuscript.

Competing Interests: The authors have declared that no competing interests exist.

* E-mail: sptemple@iupui.edu

Introduction

Aspergillus fumigatus is a ubiquitous filamentous mold that is associated with pulmonary pathology in patients suffering from asthma, cystic fibrosis, and immune deficiencies [1]. In otherwise healthy individuals, inhalation of *A. fumigatus* conidia (asexual spores) has been associated with allergic sensitization and hypersensitivity pneumonitis [2,3]. In patients with chronic lung inflammatory diseases such as asthma or cystic fibrosis, inhalation of *A. fumigatus* can lead to allergic bronchopulmonary aspergillosis (ABPA), which is marked by fungal persistence in the airways and increased inflammatory responses. However, the most severe disease occurs in neutropenic individuals or patients treated with immune suppressive drugs after hematopoietic stem cell or organ transplantation. These patients are susceptible to development of invasive aspergillosis (IA), a serious infection associated with a high mortality rate [1,4].

Studies that attempt to identify virulence factors of *A. fumigatus* may be confounded by the extensive genetic and phenotypic variability observed between fungal isolates [5]. Sampling of health care centers reported a large diversity among clinical and environmental isolates in patients and in areas associated with patient care; in some instances changes in the environmental isolates that were sampled were seen over several months at the same location [6–11]. Although isolates may exhibit variability, only individual strains were able to be isolated from patients with aspergillosis [12]. Not surprisingly, when studied in experimental models, clinical isolates with higher *in vitro* growth rates exhibited increased virulence in mice when compared to slower growing isolates [13] or environmental isolates [14,15]. Therefore, there is a correlation between isolate virulence and *in vitro* growth rates, although specific phenotypic differences that may play a role in this association have yet to be closely examined.

Through targeted mutation of *A. fumigatus* genes, numerous virulence factors have been identified [1,16–19]. These include genes involved in thermotolerant growth, cell wall integrity, secretion of toxic metabolites, and the fungal response to environmental stress. To maintain a barrier of protection from the external environment, the cell wall of *A. fumigatus* contains α and β-glucans, chitin, galactomannan, melanin, and rodlet hydrophobins [19–21]. Not all of the genes that encode cell wall components are required for virulence in experimental invasive aspergillosis. For instance, deletion of the α-glucan encoding *ags1* or *ags2* had no effect on virulence, while mutation of *ags3* increased fungal disease [22]. Chitin, a polymer of N-acetylglucosamine that is covalently linked to β-glucan, is encoded by at least seven chitin synthase (*chs*) genes in *A. fumigatus* [23]. Deletion of individual *chs* genes did not alter fungal virulence in mice, though a double *chsC/G* mutant exhibited decreased growth and virulence [24]. Thus, fungal chitin synthesis is marked by redundancy, indicating the importance of this component to the growth and survival of *A. fumigatus*.

In this study, we examined phenotypic differences between two clinical and two environmental isolates of *A. fumigatus*. The two clinical strains, Af293 and Af13073, and one of the environmental strains (Af164), were similar with respect to *in vitro* radial growth, rate of germination, ability to establish colony growth, and cell wall chitin and β-glucan content. However, the environmental isolate Af5517 exhibited decreased radial, colony formation, and rate of germination along with increased hyphal diameter and cell wall chitin and β-glucan. Despite these differences, Af5517 was able to induce invasive aspergillosis in neutropenic mice, though with reduced virulence, lung inflammation, and *in vivo* fungal growth when compared to Af293. Thus, phenotypic differences may partly explain differences in virulence observed between clinical and environmental isolates of *A. fumigatus*.

Results

Decreased Radial Growth Rate and Colony Formation by the Environmental Isolate Af5517

We were interested in the phenotypic differences between clinical and environmental isolates of *A. fumigatus*. Results of a previous study indicated that the *in vitro* growth rates of *A. fumigatus* isolates exhibited significant variation [13]. For our study, we screened two clinical and two environmental isolates of *A. fumigatus* for phenotypic differences (Table 1). The radial growth of each isolate over the course of 10–12 days at ambient (22°C) and physiological (37°C) temperatures was compared. The two clinical isolates, Af293 and Af13073 exhibited steady radial growth at both temperatures, as did the environmental isolate Af164 (Figure 1A, B). In contrast, isolate Af5517 exhibited a significantly reduced growth rate in comparison to all other isolates, reaching a colony diameter that was reduced by half by 10 days after inoculation. After 24 hours in liquid culture, Af5517 conidia formed smaller, yet denser hyphal aggregates when compared to the other isolates (Figure 1C, top panels). Furthermore, Af5517 and Af164 displayed more hyphal vacuoles and an increase in hyphal diameter (Figures 1C and 1D). Despite the differences in radial growth and hyphal morphology, Af5517 hyphae did not exhibit altered biomass accumulation (Figure 1E).

We also compared the ability of each isolate to initiate conidial swelling and germination and support colony growth by agar plate dilution and quantification of resulting colonies. By flow cytometric analysis of forward scatter (size), we observed that the fold increase in size (i.e. conidial swelling) of each isolate was equivalent after 5 hours of incubation in AMM at 37°C (Figures 2A and 2B). In contrast, by visual examination of germling formation, germination of Af5517 was reduced after 8 and 10 hours (Figure 2C). The ability of Af5517 conidia to support colony growth on solid agar was also reduced (Figure 2D). These results suggest that the ability of Af5517 conidia to form germlings and establish colony growth on solid media is defective. Furthermore, these differences do not appear to be due to decreased viability, as early conidial swelling appeared to be equivalent, the rate of germination was equivalent at 16 hours, and an increased inoculum of conidia did not result in increased radial growth in isolate Af5517 (Figure S1).

Altered Cell Wall Composition in Isolate Af5517

Previous published reports have demonstrated altered growth upon mutation of cell wall components of *A. fumigatus*. More specifically, β-glucan and chitin contribute to the structural rigidity of the fungal cell wall, and changes in these components might affect radial growth [25]. In order to determine if the relative composition of β-glucan and chitin in the cell wall was different between isolates, we performed dot blot assays using a β-glucan-specific antibody or a chitin-binding probe [26]. In our assays, these reagents proved to be highly specific, as no cross-reactivity with other cell wall components was observed with the chitin

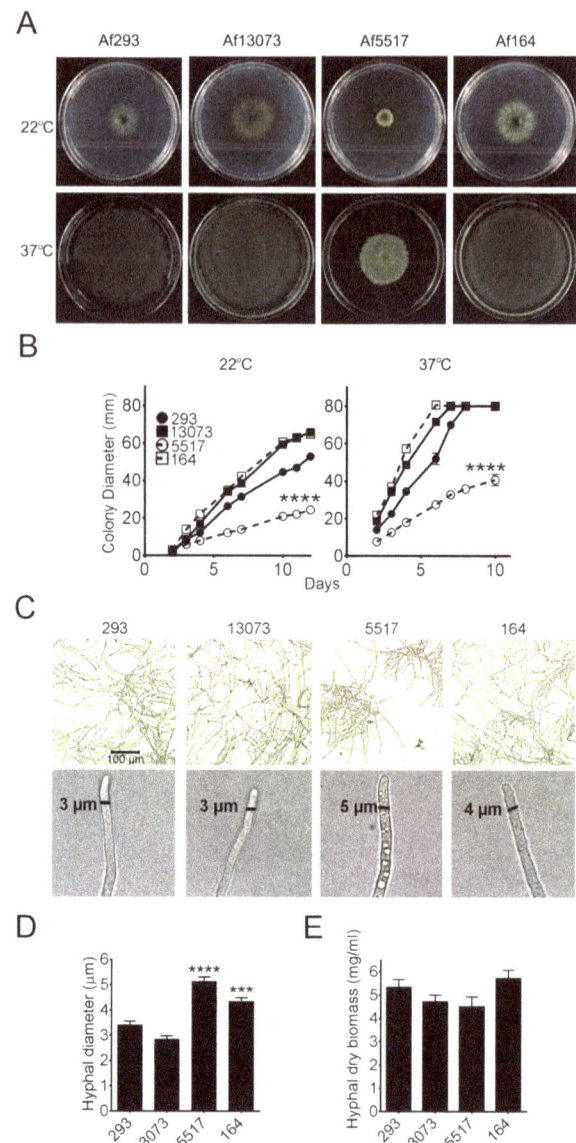

Figure 1. Decreased radial growth of *A. fumigatus* isolate Af5517. (**A**) Representative *Aspergillus* minimal media (AMM) plates 10 days after inoculation with 100 conidia of the indicated *A. fumigatus* isolates in triplicate. Plates were incubated at either 22°C or 37°C, as indicated. (**B**) Colony diameter during this incubation, measured daily. Clinical isolates are depicted with filled symbols connected by solid lines, while environmental isolates are open symbols connected by dotted lines. Growth of Af5517 was significantly different from all isolates after 4 days of growth. (**C**) Top panels, hyphal morphology of isolates after 24 hours growth in liquid culture at 37°C without shaking. Bottom panels, hyphal width was measured using SPOT Basic Software. (**D**) Summary of hyphal diameter measurements (n = 20–22/group). The diameters of Af5517 and Af164 hyphae were significantly increased when compared to Af293 and Af13073, and Af5517 diameter was increased compared to Af164 (p<0.01). (**E**) Hyphal mass accumulation of isolates after 24 hours growth in liquid culture with shaking. Data depict a summary of two experiments (n = 6/group). Each panel displayed is representative of two experiments. ***p<0.001, ****p< 0.0001.

binding probe or anti-β-glucan (Figure S2). When compared to the other isolates, the amount of chitin in mycelial extracts of Af5517 was markedly increased (Figure 3A, C). The relative amount of β-

Table 1. List of *A. fumigatus* isolates used in this study.

Source Name	Study Name	Source	Repository
Af293	Af293	Clinical	Fungal Genetics Stock Center
ATCC13073	Af13073	Clinical	American Type Culture Collection
NRRL5517	Af5517	Environmental	U.S. Agricultural Research Service
NRRL164	Af164	Environmental	U.S. Agricultural Research Service

glucan was also increased in Af5517 mycelia (Figure 3B, D). Thus, the cell wall composition in Af5517 is altered when compared to other isolates.

In dormant conidia, inner cell wall components such as chitin and β-glucan are masked by an inert hydrophobic rodlet layer [27]. Upon swelling and germination, this layer is degraded and

the underlying carbohydrate layers are exposed [27,28]. Since chitin and β-glucan are immune modulatory [23,29], increased exposure of these molecules may affect the ability of host cells to clear swelling or germinating conidia. Therefore, we stained conidia of each isolate with a chitin-specific probe, wheat germ agglutinin (WGA), or a β-(1,3)-glucan-specific antibody. We

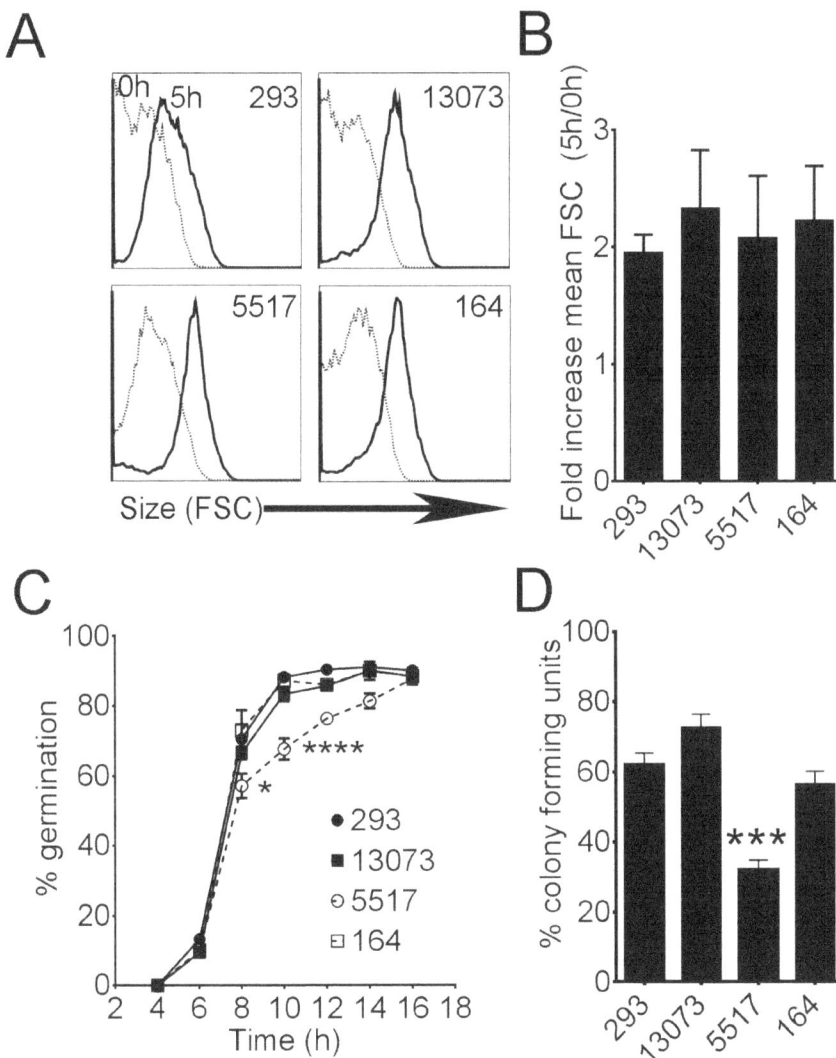

Figure 2. Decreased rate of germination and ability to establish colony growth by *A. fumigatus* isolate Af5517. (A,B) Flow cytometric analysis of conidial swelling in *A. fumigatus* isolates during 5 hours of incubation at 37°C. **(A)** Representative histograms from three experiments of forward scatter (FSC) from each isolate at 0 and 5 hours. **(B)** Increase in size (Conidial swelling) = FSC 5 h/FSC 0 h. Data are a summary of three experiments (n = 5). **(C)** Temporal quantification of germling formation by microscopic analysis. Summary of two experiments (n = 6). **(D)** Percent colony growth (CFU = colony forming unit), averaged from inoculations of 1000, 100, and 10 conidia, each in triplicate. Graphed data are a summary of two experiments. Decreased ability of Af5517 to establish colony growth was significantly different from all other isolates. ***p<0.001.

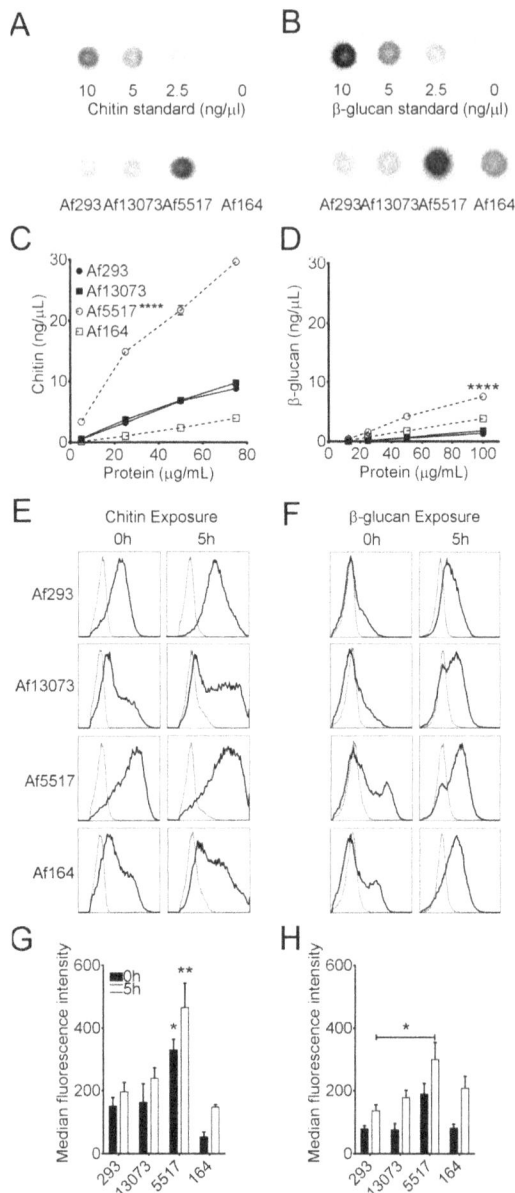

Figure 3. Increased cell wall chitin and β-glucan in *A. fumigatus* isolate Af5517. (**A,B**) Representative dot-blot images of processed *A. fumigatus* mycelial extracts and chitin (shrimp shell chitin) (**A**) or β-glucan (curdlan) (**B**) standards, probed with chitin binding probe or anti-(1,3)-β-glucan, respectively. Blots of mycelial extracts depict 75 μg (**A**) and 100 μg (**B**) of total protein of each isolates. (**C,D**) Chitin (**C**) or β-glucan (**D**) vs. total fungal protein in mycelial extracts as determined by dot blot assay. (**C**) Chitin was significantly increased compared to other isolates at 25, 50, or 75 μg of total protein, while β-glucan (**D**) was significant at 50 and 100 μg total protein. ****p<0.0001. (**A–D**) Panels are representative of two experiments. (**E,F**) Representative flow cytometric histograms of dormant (0 h) or swollen (5 h at 37°C) conidia of each isolate stained with the chitin binding wheat germ agglutinin (WGA, panel E) or anti-(1,3)-β-glucan (**F**). Negative controls are unstained conidia (**E**) or goat anti-mouse alexa-fluor 488 only (**F**) and depicted as dotted histograms. (**G,H**) Median fluorescence intensities of WGA (**G**) or anti-β-glucan (**H**) stained conidia after 0 or 5 hours incubation at 37°C. (**G**) Chitin exposure in Af5517 conidia was significantly increased in comparison will all other isolates after 0 and 5 hours incubation. *p<0.05. **p<0.01. Panels are a summary of three experiments.

incubated these conidia in AMM for 5 hours, and compared the surface exposure of chitin and β-glucan by flow cytometry. After 0 and 5 hours incubation, Af5517 conidia displayed a higher level of chitin exposure when compared to the other isolates (Figures 3E, G). β-glucan exposure was more modestly increased in Af5517 conidia, and only with statistical significance after 5 hours and only in comparison to Af293 (Figures 3F, 3H). Thus, both chitin and β-glucan levels and surface exposure are increased in isolate Af5517, with a more marked increase in chitin.

Decreased Lung Inflammation, Fungal Burden, and Virulence of Af5517 in Neutropenic Mice

A previous study reported that the *in vitro* growth rates of isolates of *A. fumigatus* were proportional with the severity of disease in a mouse model of invasive aspergillosis [13]. Since the *in vitro* radial growth and colony formation of Af5517 were decreased in comparison with other isolates, we hypothesized that the virulence would also be decreased. Because the other isolates were phenotypically similar, and to limit the use of experimental animals, we therefore chose to more closely examine the virulence of Af5517 by infecting mice with doses of 2×10^6 or 5×10^6 Af293 or Af5517 conidia. Af293 was chosen for comparison because it has been well-characterized in genomic studies and is commonly used in murine models of infection and immunity [30–32]. Furthermore, in preliminary studies, we did not detect significant differences in virulence between Af293 and the other isolates used in this study, Af13073 and Af164 (data not shown). When compared to Af5517, Af293-infected mice exhibited severe disease with no survival at the 5×10^6 dose, while at the 2×10^6 dose mortality was slightly decreased (Figure 4A). In contrast, only mice infected with the higher dose of Af5517 had the potential to become moribund. The symptoms of clinical disease over the course of infection were also more severe in Af293-infected mice at both doses (Figure 4C, D). Furthermore, the lung fungal burden at day 3 post-infection was markedly lower in Af5517-infected mice (Figure 4B).

In addition to the above factors, lung inflammation and *in vivo* fungal growth were also compared in histological sections from neutropenic mice infected with both isolates. Mice infected with Af293 displayed larger foci of inflammation and fungal growth that extended further into the lung parenchyma, whereas Af5517 infection and resultant inflammation was more bronchiocentric (Figure 5A). When lung fungal growth was quantified in Gomori's Methanamine Silver (GMS) stained sections, areas of fungal growth were significantly larger in Af293-infected mice with both doses tested (Figure 5B). Therefore, Af5517 displayed a less virulent phenotype in neutropenic mice than Af293.

Discussion

We observed that decreased radial growth of a single isolate, Af5517, was correlated with decreased virulence in a mouse model of invasive pulmonary aspergillosis. Although Af5517 conidia displayed a reduced ability to establish colony growth, our results suggest this is not due to decreased conidial viability, but rather due to limited and/or delayed growth after germination. We observed that conidial swelling was equivalent between isolates, and formation of germlings in Af5517 conidia appeared defective after 8 hours in liquid culture, reaching equivalent levels by 16 hours. Thus, the ability of *A. fumigatus* conidia to establish colony growth is not necessarily a direct measure of conidial viability.

In this study, we observed increased β-glucan and chitin content in the cell wall of the environmental isolate Af5517 when compared to the clinical isolates Af293 and Af13073 and the

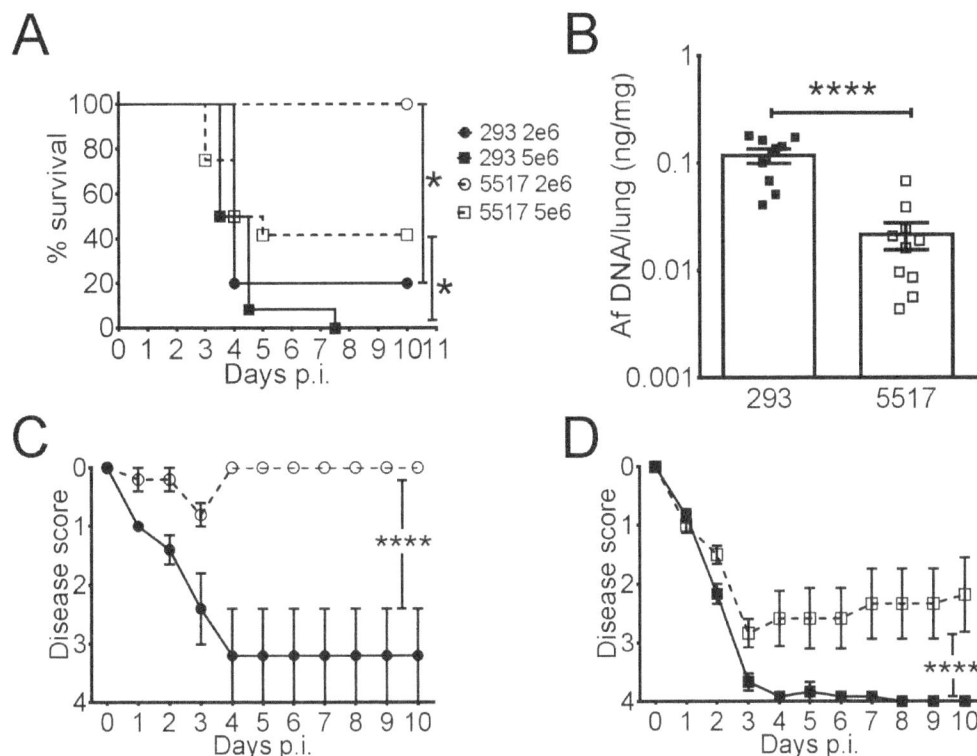

Figure 4. *A. fumigatus* isolate Af5517 is less virulent than Af293 in an animal model of pulmonary IA. Neutropenic BALB/c mice were infected with 2×10^6 (circles) or 5×10^6 (squares) conidia of isolates Af293 (closed symbols) or Af5517 (open symbols). (**A**) Survival curves depict death or moribund status for experimental animals over the course of the experiment. (**B**) Fungal burden of mice infected with 5×10^6 conidia of Af293 (closed symbols) or Af5517 (open symbols) was determined by quantification of fungal 18S rDNA. (**C,D**) Disease scores of mice infected with 2×10^6 (**C**) or 5×10^6 (**D**) conidia. Mice were scored daily for progression of disease as described in Materials and Methods. Graphed data depicts the summary of two experiments with 5–8 mice per group in panels A, C, and D. *p$<$0.05. ****p$<$0.0001.

environmental isolate Af164. In *A. fumigatus*, chitin synthesis is controlled by at least 7 genes that appear to be redundant, though there is evidence that several of these genes control different aspects of cell-wall synthesis during growth and germination [23]. In some instances, β-glucan and/or chitin synthesis may be affected by regulatory networks that are activated in response to environmental stress. These stresses include contact with antifungal drugs or growth in hypoxic conditions [33,34]. Other studies have shown that disruption of fungal genes such as the N-glycosylation-regulating gene Afstt3 increased cell-wall chitin via activation of the unfolded protein response (UPR), which is an ER stress-response [35]. It is interesting that both β-glucan and chitin content were increased in Af5517, since experimental evidence suggests that in some instances these components are reciprocally regulated. For example, chitin was increased in fungi grown in the presence of the β-glucan synthesis inhibitor caspofungin, while β-glucan was increased in the presence of the chitin synthesis-inhibiting nikkomycin Z [34]. Combined with our results, it appears that although reciprocal regulation is possible, it is not imperative. It is possible that in the isolate Af5517, multiple stress response regulatory genes may be constitutively activated, and this has resulted in alteration of cell wall content. Although chitin is a durable polymer of N-acetylglucosamine that possesses a high tensile strength that helps protect fungi from environmental stress [23], it is possible that an abundance of this rigid cell wall component could prove detrimental to fungal growth. Interestingly, Af5517 exhibited the largest hyphal diameter of all of the isolates examined, and it is likely that the altered cell wall composition accounts for this increase. Furthermore, our results

suggest that optimal chitin expression may prevent abnormal growth rates exhibited by Af5517.

In addition to their roles as structural components of the fungal cell wall, β-glucan and chitin differentially stimulate host immune responses. Of these components, the response to β-glucan has been more extensively studied, though initial studies often used the fungal preparation zymosan, which also contains mannan and other immune-stimulating components [36]. Fungal β-glucan is recognized by the C-type lectin receptor Dectin-1 after degradation of the rodlet layer during germination results in nascent surface exposure [27,29,37,38]. In our study, surface exposure of both chitin and β-glucan were increased in swollen Af5517 conidia. This altered expression could result in increased clearance of germinating Af5517 conidia, and thus might account for some of the observed decrease in virulence in Af5517-infected mice. Dectin-1 stimulation promotes phagocytosis, killing, and secretion of TNFα and CXCL2 in response to fungal particles [29,39]. In murine pulmonary responses to *A. fumigatus*, dectin-1 mediated TNF-α production and provided protection from infection [40,41]. Under hypoxic conditions, β-glucan and chitin expression were increased by *A. fumigatus*, resulting in increased macrophage and neutrophil fungal killing that was partly dependent on dectin-1 [33]. In contrast to studies of β-glucan, much less is known about chitin recognition by the immune system. A distinct chitin-binding receptor has yet to be identified, although TLR9, NOD2, and the mannose receptor were shown to interact with and be necessary for chitin-mediated macrophage IL-10 secretion [42]. Furthermore, chitin particles and chitin in fungal extracts mediated lung eosinophil recruitment in mice [43–45]. Eosinophils are effector

A

B

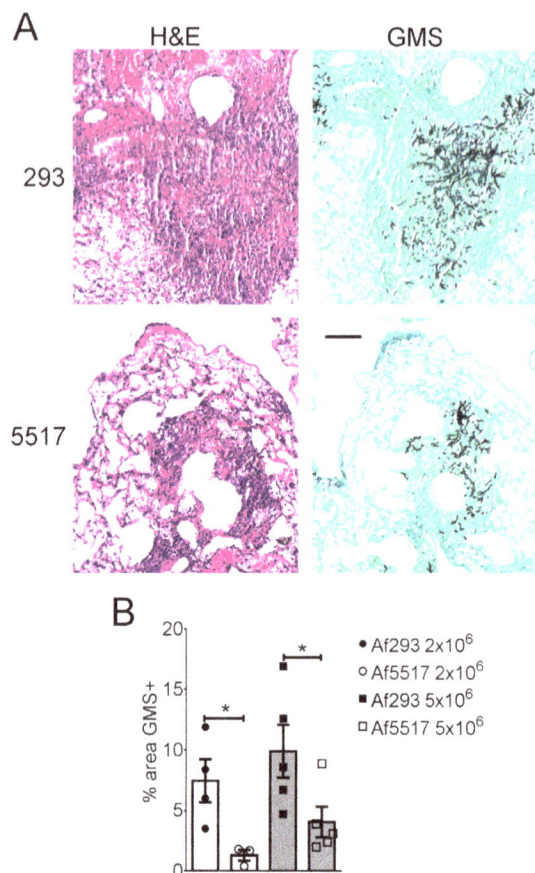

Figure 5. Histopathology of Af293 and Af5517 infection. Neutropenic BALB/c mice were infected with Af293 or Af5517 and sacrificed after 3 days for lung histological analysis. (**A**) Hematoxylin and Eosin (H&E)-stained sections (left panels) depict lung inflammation in mice infected with 5×10^6 conidia. Adjacent Gomori's Methanamine Silver (GMS)-stained sections (right panels) show areas of fungal growth. The black bar (bottom right panel) is equivalent to 100 μm. (**B**) GMS staining representing fungal growth was quantified in sections from mice infected with 2×10^6 or 5×10^6 conidia, with the mean of four representative fields displayed for each sample. *p<0.05.

cells associated with allergy and infection with helminths [46], intestinal parasites that, like fungi, express chitin [23]. In response to aspiration of Af5517 conidia, we have observed increased chitin-mediated airway eosinophil recruitment (O'Dea, E.M., Amarsaikhan, N., Li, H., Downey, J., Steele, E., Van Dyken, S.J., Locksley, R.M., Templeton, S.P. In press). Thus, the role of chitin expression and increased eosinophil recruitment in immune responses to *A. fumigatus* infection are the focus of our current and future investigation.

Although our study identified decreased growth in Af5517, other phenotypic differences between isolates could also contribute to decreased virulence and lung fungal burden. Numerous genes control the ability of *A. fumigatus* to support growth at physiological temperature, maintain cell wall integrity, process and uptake available nutrients, and respond to environmental stress [19]. Furthermore, several secreted toxins and fungal metabolites of *A. fumigatus* may interfere with immune-mediated host clearance, and these factors could be differentially expressed among clinical and environmental isolates [18]. In addition to the immune stimulatory chitin and β-glucan, expression of immune suppressive cell wall components such as galactomannan, melanin, and galactosami-

nogalactan might also differ between *A. fumigatus* isolates [47–49]. Results of a recent study suggest that immune responses to wild-type and mutant *A. fumigatus* strains also exhibit considerable variability [50]. It will thus be of interest to determine if these differences are driven by differential expression of immune-modulating cell wall components or other factors required for the growth and metabolism of *A. fumigatus* within host tissue. In order to completely understand the mechanisms of virulence and immune modulation by *A. fumigatus* isolates, extensive comparative analysis of genomic and proteomic differences between fungal isolates will likely be required.

In summary, we observed a decreased *in vitro* growth rate, rate of germination, and decreased colony formation along with increased hyphal diameter characterized by increased chitin and β-glucan in a single environmental isolate, Af5517, in comparison to other clinical and environmental isolates. In addition to these phenotypic differences, the virulence of Af5517 was also decreased in neutropenic mice. When considered together, the results of this and other studies support a correlation between *in vitro* growth rates and *in vivo* virulence [13–15]. Furthermore, alteration of cell-wall composition might be a contributing factor to this phenotype. However, more isolates that concurrently display slower growth, altered cell wall composition, and reduced virulence will need to be identified in order to conclusively define the relationships between these factors. In future studies, comparative analysis of fungal gene expression in these isolates will be necessary to identify candidate pathways responsible for associated changes in structure, growth, and virulence of this ubiquitous human pathogen.

Materials and Methods

Comparison of Morphology, Radial Growth, and Rates of Germination in *A. fumigatus* Isolates

Fungal isolates were obtained from the specified sources (Table 1). Fungi were grown on malt extract agar (MEA) plates at room temperature (RT) unless specified otherwise, and fungal conidia were isolated from plates using glass beads (BioSpec), and further separated from the beads by adding sterile PBS as previously described [51]. To compare radial growth, conidia were collected in sterile water, counted on a hemacytometer, and then diluted to 100 conidia/μL. A volume of 1 μL was pipetted onto the center of *Aspergillus* minimal media agar plates [52] and radial growth was measured daily for thirteen days. For microscopic examination of hyphal morphology, conidia were inoculated on coverslips immersed in 2 ml AMM overnight at 37°C. Coverslips were removed and washed with PBS and inverted on glass slides for microscopic analysis. To compare rates of germination, a previously described method was used, with slight modifications [53]. Briefly, conidia were incubated in 5×10^6 conidia/ml in AMM at 37°C with shaking at 200 rpm. Sampling was performed at 4, 6, 8, 10, 12, 14, and 16 hours and observed under light microscopy at 400x magnification. 24 hour-germinated samples were lyophilized for total biomass measurement or were analyzed for comparison of hyphal morphology by microscopy. Germination rates were reported as percent of germ tube forming conidia per random 100 conidia count. To compare the ability of fungal isolates to establish colony growth, conidia were collected in sterile water with 0.05% Tween 20, very lightly sonicated to disperse clumps, counted on a hemacytometer, and diluted to three concentrations: 1000, 100, and 10 conidia/mL. A volume of 1 mL of each suspension was inoculated in triplicate on MEA plates, and colonies were counted after a 72 hour incubation at 37°C. For the described experiments, all samples were prepared and data acquired in triplicate.

Comparison of Cell Wall Chitin and β-Glucan Content in A. fumigatus Isolates

To compare cell wall chitin and β-glucan content using a previously described dot blot method [44], isolates were cultured in malt extract broth for 5 days at room temperature with shaking at 100rpm. The hyphae were then lyophilized, bead-beaten, and suspended in sterile water with 0.05% Tween 20. Fungal protein concentrations were standardized using Pierce BCA Protein Assay kit (Fisher), and then the samples were diluted into four separate concentrations to ensure results within measurable range. Each sample was lightly sonicated to disperse clumps, then 1 μL was immediately dot-blotted in triplicate onto a nitrocellulose membrane (GE Healthcare). Shrimp shell chitin (Sigma-Aldrich) and curdlan (Wako) were used as positive controls. The membranes were dried at room temperature (RT), blocked in Tris buffered saline with 0.05% Tween 20 (TBST) with 5% nonfat dry milk, then incubated for 16–24 hours at 4°C with gentle rocking in TBST with 1% nonfat dry milk and either anti-(1–3)-β-glucan antibody (Biosupplies) or a probe containing a chitin-binding domain conjugated to FITC (expression clone provided by Yinhua Zhang, New England Biologicals) [54]. After washing with TBST, the membranes were incubated for 45 min at RT with gentle rocking in either HRP anti-mouse IgG (Jackson ImmunoResearch) or HRP anti-fluorescein (Invitrogen). After a final wash in TBST, membranes were developed with SuperSignal West Femto Substrate ECL kit (Fisher) and images were captured using a ChemiDoc-IT imaging system and analyzed with VisionWorks software (UVP). To quantify chitin and β-glucan exposure using a similar method that was previously described [55], conidia were kept dormant or allowed to germinate for 5 hours in AMM with gentle rocking, then stained with a 1:500 dilution of the chitin-binding wheat germ agglutinin-APC (Invitrogen) or 1:250 dilution of anti-β-glucan (Biosupplies) for 30 minutes on ice in the dark. For β-glucan detection, a 1:500 dilution of a secondary goat-anti mouse Alexa-Fluor 488 (Invitrogen) was added after washing for a second 30 minute incubation. Stained conidia were analyzed for surface staining or changes in forward scatter (size) on a Guava EasyCyte 8 HT flow cytometer (EMD Millipore).

Comparison of Fungal Virulence

BALB/c mice were obtained from Harlan Laboratories and bred in an AALAC accredited animal facility. Mice were used in experiments at 7–10 weeks of age. All animal procedures were approved by the Indiana State University Animal Care and Use Committee. For examination of virulence, mice involuntarily aspirated 50 uL sterile PBS containing 2×10^6 or 5×10^6 conidia of either Af293 or Af5517. Relative to this aspiration, neutrophils were depleted via IP injection of 0.5 mg anti-mouse-Ly-6G antibody (clone 1A8; BioXCell) one day before and after. Following aspiration, mice were scored daily for morbidity and mortality up to ten days. Morbidity was scored from 0 to 4 as follows: 0) healthy 1) minimal disease (e.g. ruffled fur), 2) moderate disease (e.g. ungroomed, hunched), 3) severe disease (e.g. severely hunched, changes in eye color, low motility), and 4) moribund/ deceased. Mice that became moribund and received a score of 4 were sacrificed. For histological comparison, mice were sacrificed 3 days post-infection, and lungs were perfused with 5 ml of saline followed by 5 ml of 10% formalin buffered saline, followed by inflation using a tracheal catheter with 1 ml of formalin buffered saline. Lungs were then removed and allowed to fix overnight in formalin buffered saline, followed by tissue embedding processing and staining of sections with hematoxylin and eosin and Gomori's methanamine silver stains. All histological sample processing and staining was performed by the Terre Haute Regional Hospital pathology laboratory

Fungal Burden Assay

To quantify fungal burden, mice were sacrificed three days after infection and lungs were collected and rapidly frozen in liquid nitrogen. Genomic DNA was extracted from 50–100 mg freeze dried, homogenized whole lung tissue using a previously described DNA extraction buffer for *Aspergillus* nucleic acids with subsequent phenol/chloroform extraction. A total of 500 ng genomic DNA was used for quantitative PCR to determine the fungal DNA content [56,57]. A qPCR fungal burden assay was performed to determine the amount of fungal 18S rDNA in lung extracts with 18S rDNA primer and probe sets with a modified probe quencher (5′-/56-FAM/AGC CAG CGG/ZEN/CCC GCA AAT G/ 3IABkFQ/-3′) [56,58]. A standard curve was prepared from Af293 genomic DNA and used to calculate the concentration of fungal DNA in each sample, with uninfected mouse lungs analyzed to confirm absence of contamination. Samples were amplified in duplicate with at least five technical replicates. The qPCR reaction was performed in Agilent Mx3005P with MxPro Software (Agilent Technologies). Ct values were used to calculate the corresponding fungal DNA content in the lung tissue and the fungal burden was reported as ng fungal DNA per mg of total lung DNA.

Data Analysis

Data analysis and resulting graphs were performed and prepared with Prism software (GraphPad). For statistical comparison of two groups, unpaired t tests were used. For multiple groups, one- or two-way ANOVA were used with Tukey's or Holm-Sidak's test for multiple comparisons, respectively. To compare survival or disease scores between groups, the Mantel-Cox log-rank test was used. The hyphal diameter of each isolate was measured using SPOT Basic Software (Diagnostic Instruments, Inc). For quantification of fungal growth on histological sections, the mean of four representative fields at 100x magnification of GMS+ staining for each sample was calculated using ImageJ software (National Institutes of Health).

Supporting Information

Figure S1 Radial growth of Af293 and Af5517 with increased inocula of Af5517 conidia. AMM plates were centrally inoculated with the indicated isolate and inoculum, and allowed to grow for 7 days, with the resulting diameter of growth measured. Data displayed are representative of two experiments.

Figure S2 Specificity of anti-β-glucan antibody and chitin binding probe for cell wall components of *Aspergillus fumigatus*. Shrimp shell chitin (Ch), curdlan (βg), and locust bean gum galactomannan (Gm) were blotted together on three separate membranes that were probed with chitin binding probe, anti-β-glucan, and ConA, respectively. Data displayed are representative of samples blotted in triplicate that gave similar results.

Acknowledgments

The authors would like to thank Hongtao Li for technical assistance and Joe Lewis for animal care.

Author Contributions

Conceived and designed the experiments: ST NA. Performed the experiments: NA EO AT HO JG. Analyzed the data: ST NA EO. Wrote the paper: ST NA EO.

References

1. Hohl TM, Feldmesser M (2007) *Aspergillus fumigatus*: principles of pathogenesis and host defense. Eukaryotic cell 6: 1953–1963.
2. Greenberger PA (2004) Mold-induced hypersensitivity pneumonitis. Allergy and Asthma Proceedings 25: 219–223.
3. Templeton SP, Buskirk AD, Green BJ, Beezhold DH, Schmechel D (2010) Murine models of airway fungal exposure and allergic sensitization. Medical Mycology 48: 217–228.
4. Grahl N, Puttikamonkul S, Macdonald JM, Gamcsik MP, Ngo LY, et al. (2011) In vivo Hypoxia and a Fungal Alcohol Dehydrogenase Influence the Pathogenesis of Invasive Pulmonary Aspergillosis. PLoS Pathogens 7: e1002145.
5. de Valk HA, Klaassen CH, Meis JF (2008) Molecular typing of Aspergillus species. Mycoses 51: 463–476.
6. Araujo R, Amorim A, Gusmao L (2010) Genetic diversity of Aspergillus fumigatus in indoor hospital environments. Med Mycol 48: 832–838.
7. Balajee SA, de Valk HA, Lasker BA, Meis JF, Klaassen CH (2008) Utility of a microsatellite assay for identifying clonally related outbreak isolates of Aspergillus fumigatus. J Microbiol Methods 73: 252–256.
8. Chazalet V, Debeaupuis JP, Sarfati J, Lortholary J, Ribaud P, et al. (1998) Molecular typing of environmental and patient isolates of Aspergillus fumigatus from various hospital settings. J Clin Microbiol 36: 1494–1500.
9. de Valk HA, Klaassen CH, Yntema JB, Hebestreit A, Seidler M, et al. (2009) Molecular typing and colonization patterns of Aspergillus fumigatus in patients with cystic fibrosis. J Cyst Fibros 8: 110–114.
10. Leenders AC, van Belkum A, Behrendt M, Luijendijk A, Verbrugh HA (1999) Density and molecular epidemiology of Aspergillus in air and relationship to outbreaks of Aspergillus infection. J Clin Microbiol 37: 1752–1757.
11. Vanhee LM, Symoens F, Jacobsen MD, Nelis HJ, Coenye T (2009) Comparison of multiple typing methods for Aspergillus fumigatus. Clin Microbiol Infect 15: 643–650.
12. Girardin H, Sarfati J, Traore F, Dupouy Camet J, Derouin F, et al. (1994) Molecular epidemiology of nosocomial invasive aspergillosis. J Clin Microbiol 32: 684–690.
13. Paisley D, Robson GD, Denning DW (2005) Correlation between in vitro growth rate and in vivo virulence in Aspergillus fumigatus. Med Mycol 43: 397–401.
14. Mondon P, De Champs C, Donadille A, Ambroise-Thomas P, Grillot R (1996) Variation in virulence of Aspergillus fumigatus strains in a murine model of invasive pulmonary aspergillosis. J Med Microbiol 45: 186–191.
15. Aufauvre-Brown A, Brown JS, Holden DW (1998) Comparison of virulence between clinical and environmental isolates of Aspergillus fumigatus. Eur J Clin Microbiol Infect Dis 17: 778–780.
16. Bhabhra R, Askew DS (2005) Thermotolerance and virulence of *Aspergillus fumigatus*: role of the fungal nucleolus. Medical Mycology 43 Suppl 1: S87–93.
17. Latgé JP (1999) *Aspergillus fumigatus* and aspergillosis. Clinical Microbiology Reviews 12: 310–350.
18. Tomee JF, Kauffman HF (2000) Putative virulence factors of *Aspergillus fumigatus*. Clinical and Experimental Allergy 30: 476–484.
19. Askew DS (2008) Aspergillus fumigatus: virulence genes in a street-smart mold. Current Opinion in Microbiology 11: 331–337.
20. Latge JP, Mouyna I, Tekaia F, Beauvais A, Debeaupuis JP, et al. (2005) Specific molecular features in the organization and biosynthesis of the cell wall of Aspergillus fumigatus. Med Mycol 43 Suppl 1: S15–22.
21. Thau N, Monod M, Crestani B, Rolland C, Tronchin G, et al. (1994) rodletless mutants of Aspergillus fumigatus. Infect Immun 62: 4380–4388.
22. Maubon D, Park S, Tanguy M, Huerre M, Schmitt C, et al. (2006) AGS3, an alpha (1–3) glucan synthase gene family member of Aspergillus fumigatus, modulates mycelium growth in the lung of experimentally infected mice. Fungal Genetics and Biology 43: 366–375.
23. Lenardon MD, Munro CA, Gow NA (2010) Chitin synthesis and fungal pathogenesis. Curr Opin Microbiol 13: 416–423.
24. Mellado E, Aufauvre-Brown A, Gow NA, Holden DW (1996) The Aspergillus fumigatus chsC and chsG genes encode class III chitin synthases with different functions. Mol Microbiol 20: 667–679.
25. Borgia PT, Dodge CL (1992) Characterization of Aspergillus nidulans mutants deficient in cell wall chitin or glucan. Journal of Bacteriology 174: 377–383.
26. Watanabe T, Ito Y, Yamada T, Hashimoto M, Sekine S, et al. (1994) The roles of the C-terminal domain and type III domains of chitinase A1 from Bacillus circulans WL-12 in chitin degradation. Journal of Bacteriology 176: 4465–4472.
27. Aimanianda V, Bayry J, Bozza S, Kniemeyer O, Perruccio K, et al. (2009) Surface hydrophobin prevents immune recognition of airborne fungal spores. Nature 460: 1117–1121.
28. Dague E, Alsteens D, Latge JP, Dufrene YF (2008) High-resolution cell surface dynamics of germinating Aspergillus fumigatus conidia. Biophys J 94: 656–660.
29. Brown GD (2006) Dectin-1: a signalling non-TLR pattern-recognition receptor. Nature Reviews: Immunology 6: 33–43.
30. Nierman WC, Pain A, Anderson MJ, Wortman JR, Kim HS, et al. (2005) Genomic sequence of the pathogenic and allergenic filamentous fungus Aspergillus fumigatus. Nature 438: 1151–1156.
31. Leal SM Jr, Cowden S, Hsia YC, Ghannoum MA, Momany M, et al. (2010) Distinct roles for Dectin-1 and TLR4 in the pathogenesis of Aspergillus fumigatus keratitis. PLoS Pathog 6: e1000976.
32. Warn PA, Sharp A, Morrissey G, Denning DW (2010) Activity of aminocandin (IP960; HMR3270) compared with amphotericin B, itraconazole, caspofungin and micafungin in neutropenic murine models of disseminated infection caused by itraconazole-susceptible and -resistant strains of Aspergillus fumigatus. Int J Antimicrob Agents 35: 146–151.
33. Shepardson KM, Ngo LY, Aimanianda V, Latge JP, Barker BM, et al. (2013) Hypoxia enhances innate immune activation to Aspergillus fumigatus through cell wall modulation. Microbes Infect 15: 259–269.
34. Verwer PE, van Duijn ML, Tavakol M, Bakker-Woudenberg IA, van de Sande WW (2012) Reshuffling of Aspergillus fumigatus cell wall components chitin and beta-glucan under the influence of caspofungin or nikkomycin Z alone or in combination. Antimicrob Agents Chemother 56: 1595–1598.
35. Li K, Ouyang H, Lu Y, Liang J, Wilson IB, et al. (2011) Repression of N-glycosylation triggers the unfolded protein response (UPR) and overexpression of cell wall protein and chitin in Aspergillus fumigatus. Microbiology 157: 1968–1979.
36. Sato M, Sano H, Iwaki D, Kudo K, Konishi M, et al. (2003) Direct binding of Toll-like receptor 2 to zymosan, and zymosan-induced NF-kappa B activation and TNF-alpha secretion are down-regulated by lung collectin surfactant protein A. J Immunol 171: 417–425.
37. Hohl TM, Van Epps HL, Rivera A, Morgan LA, Chen PL, et al. (2005) *Aspergillus fumigatus* triggers inflammatory responses by stage-specific beta-glucan display. PLoS Pathogens 1: e30.
38. Steele C, Rapaka RR, Metz A, Pop SM, Williams DL, et al. (2005) The beta-glucan receptor dectin-1 recognizes specific morphologies of Aspergillus fumigatus. PLoS Pathog 1: e42.
39. Brown GD, Herre J, Williams DL, Willment JA, Marshall AS, et al. (2003) Dectin-1 mediates the biological effects of beta-glucans. Journal of Experimental Medicine 197: 1119–1124.
40. Werner JL, Metz AE, Horn D, Schoeb TR, Hewitt MM, et al. (2009) Requisite role for the dectin-1 beta-glucan receptor in pulmonary defense against Aspergillus fumigatus. J Immunol 182: 4938–4946.
41. Faro-Trindade I, Willment JA, Kerrigan AM, Redelinghuys P, Hadebe S, et al. (2012) Characterisation of innate fungal recognition in the lung. PLoS One 7: e35675.
42. Wagener J, Malireddi RK, Lenardon MD, Koberle M, Vautier S, et al. (2014) Fungal Chitin Dampens Inflammation through IL-10 Induction Mediated by NOD2 and TLR9 Activation. PLoS Pathog 10: e1004050.
43. Reese TA, Liang HE, Tager AM, Luster AD, Van Rooijen N, et al. (2007) Chitin induces accumulation in tissue of innate immune cells associated with allergy. Nature 447: 92–96.
44. Van Dyken SJ, Garcia D, Porter P, Huang X, Quinlan PJ, et al. (2011) Fungal chitin from asthma-associated home environments induces eosinophilic lung infiltration. Journal of Immunology 187: 2261–2267.
45. Van Dyken SJ, Mohapatra A, Nussbaum JC, Molofsky AB, Thornton EE, et al. (2014) Chitin activates parallel immune modules that direct distinct inflammatory responses via innate lymphoid type 2 and gammadelta T cells. Immunity 40: 414–424.
46. Rothenberg ME, Hogan SP (2006) The eosinophil. Annu Rev Immunol 24: 147–174.
47. Chai LY, Vonk AG, Kullberg BJ, Verweij PE, Verschueren I, et al. (2011) Aspergillus fumigatus cell wall components differentially modulate host TLR2 and TLR4 responses. Microbes Infect 13: 151–159.
48. Fontaine T, Delangle A, Simenel C, Coddeville B, van Vliet SJ, et al. (2011) Galactosaminogalactan, a new immunosuppressive polysaccharide of Aspergillus fumigatus. PLoS Pathog 7: e1002372.
49. Scharf DH, Heinekamp T, Brakhage AA (2014) Human and plant fungal pathogens: the role of secondary metabolites. PLoS Pathog 10: e1003859.
50. Rizzetto L, Giovannini G, Bromley M, Bowyer P, Romani L, et al. (2013) Strain dependent variation of immune responses to A. fumigatus: definition of pathogenic species. PLoS One 8: e56651.
51. Templeton SP, Buskirk AD, Law B, Green BJ, Beezhold DH (2011) Role of germination in murine airway CD8+ T-cell responses to Aspergillus conidia. PLoS One 6: e18777.
52. Cove DJ (1966) The induction and repression of nitrate reductase in the fungus Aspergillus nidulans. Biochim Biophys Acta 113: 51–56.
53. Cramer RA Jr, Perfect BZ, Pinchai N, Park S, Perlin DS, et al. (2008) Calcineurin target CrzA regulates conidial germination, hyphal growth, and pathogenesis of Aspergillus fumigatus. Eukaryot Cell 7: 1085–1097.

54. Maduzia LL, Yu E, Zhang Y (2011) Caenorhabditis elegans galectins LEC-6 and LEC-10 interact with similar glycoconjugates in the intestine. Journal of Biological Chemistry 286: 4371–4381.

55. Levdansky E, Kashi O, Sharon H, Shadkchan Y, Osherov N (2010) The Aspergillus fumigatus cspA gene encoding a repeat-rich cell wall protein is important for normal conidial cell wall architecture and interaction with host cells. Eukaryot Cell 9: 1403–1415.

56. Bowman JC, Abruzzo GK, Anderson JW, Flattery AM, Gill CJ, et al. (2001) Quantitative PCR assay to measure Aspergillus fumigatus burden in a murine model of disseminated aspergillosis: demonstration of efficacy of caspofungin acetate. Antimicrob Agents Chemother 45: 3474–3481.

57. Gessner MA, Werner JL, Lilly LM, Nelson MP, Metz AE, et al. (2012) Dectin-1-dependent interleukin-22 contributes to early innate lung defense against Aspergillus fumigatus. Infect Immun 80: 410–417.

58. Li H, Barker BM, Grahl N, Puttikamonkul S, Bell JD, et al. (2011) The small GTPase RacA mediates intracellular reactive oxygen species production, polarized growth, and virulence in the human fungal pathogen Aspergillus fumigatus. Eukaryot Cell 10: 174–186.

Permissions

The contributors of this book come from diverse backgrounds, making this book a truly international effort. This book will bring forth new frontiers with its revolutionizing research information and detailed analysis of the nascent developments around the world.

We would like to thank all the contributing authors for lending their expertise to make the book truly unique. They have played a crucial role in the development of this book. Without their invaluable contributions this book wouldn't have been possible. They have made vital efforts to compile up to date information on the varied aspects of this subject to make this book a valuable addition to the collection of many professionals and students.

This book was conceptualized with the vision of imparting up-to-date information and advanced data in this field. To ensure the same, a matchless editorial board was set up. Every individual on the board went through rigorous rounds of assessment to prove their worth. After which they invested a large part of their time researching and compiling the most relevant data for our readers.

The editorial board has been involved in producing this book since its inception. They have spent rigorous hours researching and exploring the diverse topics which have resulted in the successful publishing of this book. They have passed on their knowledge of decades through this book. To expedite this challenging task, the publisher supported the team at every step. A small team of assistant editors was also appointed to further simplify the editing procedure and attain best results for the readers.

Apart from the editorial board, the designing team has also invested a significant amount of their time in understanding the subject and creating the most relevant covers. They scrutinized every image to scout for the most suitable representation of the subject and create an appropriate cover for the book.

The publishing team has been an ardent support to the editorial, designing and production team. Their endless efforts to recruit the best for this project, has resulted in the accomplishment of this book. They are a veteran in the field of academics and their pool of knowledge is as vast as their experience in printing. Their expertise and guidance has proved useful at every step. Their uncompromising quality standards have made this book an exceptional effort. Their encouragement from time to time has been an inspiration for everyone.

The publisher and the editorial board hope that this book will prove to be a valuable piece of knowledge for researchers, students, practitioners and scholars across the globe.

List of Contributors

Mohammad S. Hossain, Sampath Ramachandiran and Edmund K. Waller
Department of Hematology and Medical Oncology, Division of Stem Cell and Bone Marrow Transplantation, Winship Cancer Institute, Emory University School of Medicine, Atlanta, Georgia, United States of America

Andrew T. Gewirtz
Department of Biology, Georgia State University, Atlanta, Georgia, United States of America

Swapna Bhat, Tye O. Boynton, Dan Pham and Lawrence J. Shimkets
Department of Microbiology, University of Georgia, Athens, Georgia, United States of America

Adelfia Talà, Pietro Alifano and Salvatore Maurizio Tredici
Dipartimento di Scienze e Tecnologie Biologiche ed Ambientali, Università del Salento, Lecce, Italy

Domenico Delle Side, Giovanni Buccolieri, Luciano Velardi, Vincenzo Nassisi and Fabio Paladini
Dipartimento di Matematica e Fisica "Ennio De Giorgi", Università del Salento INFN – Lecce, Lecce, Italy

Mario De Stefano
Dipartimento di Scienze Ambientali, Seconda Universita` di Napoli, Caserta, Italy

Sonica Sondhi, Prince Sharma, Shilpa Saini and Naveen Gupta
Department of Microbiology, BMS Block, Panjab University, Chandigarh, India

Neena Puri
Department of Industrial Microbiology, Guru Nanak Khalsa College, Yamunanagar, Haryana, India

Satoshi Yamamoto, Hirotoshi Sato, Kohmei Kadowaki and Hirokazu Toju
Graduate School of Human and Environmental Studies, Kyoto University, Kyoto, Japan

Akifumi S. Tanabe
National Research Institute of Fisheries Science, Fisheries Research Agency, Yokohama, Japan

Amane Hidaka
Network center of Forest and Grassland Survey in Monitoring Sites 1000 Project, Japan Wildlife Research Center, c/o Filed Science Center for Northern Biosphere, Hokkaido University, Tomakomai, Japan

Stefano Romano and Vladimir Bondarev
Max Planck Institute for Marine Microbiology, Bremen, Germany

Thorsten Dittmar
Research Group for Marine Geochemistry, Institute for Chemistry and Biology of the Marine Environment (ICBM), University of Oldenburg, Oldenburg, Germany

Ralf J. M. Weber and Mark R. Viant
School of Biosciences, University of Birmingham, Birmingham, United Kingdom

Heide N. Schulz-Vogt
Department of Biological Oceanography, Leibniz-Institute for Baltic Sea Research Warnemuende (IOW), Rostock, Germany

Yan Chen, Stephen B. Melville and David L. Popham
Department of Biological Sciences, Virginia Tech, Blacksburg, Virginia, United States of America

W. Keith Ray and Richard F. Helm
Department of Biochemistry, Virginia Tech, Blacksburg, Virginia, United States of America

Teresa Kolle and Andrew F Bent
Department of Plant Pathology, University of Wisconsin – Madison, Madison, Wisconsin, United States of America

Martha E. Trujillo, Rodrigo Bacigalupe, Patricia Benito and Raúl Riesco
Departamento de Microbiología y Genética, Edificio Departamental, Campus Miguel de Unamuno, Universidad de Salamanca, Salamanca, Spain

Petar Pujic and Philippe Normand
Université Lyon 1,Université de Lyon, CNRS-UMR5557 Ecologie Microbienne, Villeurbanne, France

Yasuhiro Igarashi
Biotechnology Research Center, Toyama Prefectural University, Kurokawa, Imizu, Toyama, Japan

Claudine Médigue
Genoscope, CNRS-UMR 8030, Atelier de Ge´nomique Comparative, Evry, France

Alina Nescerecka
Department of Water Engineering and Technology, Riga Technical University, Riga, Latvia

Department of Environmental Microbiology, Eawag, Swiss Federal Institute for Aquatic Science and Technology, Dübendorf, Switzerland

Janis Rubulis and Talis Juhna
Department of Water Engineering and Technology, Riga Technical University, Riga, Latvia

Marius Vital and Frederik Hammes
Department of Environmental Microbiology, Eawag, Swiss Federal Institute for Aquatic Science and Technology, Dübendorf, Switzerland

Yaíma L. Lightfoot, Tao Yang, Bikash Sahay, Mojgan Zadeh and Mansour Mohamadzadeh
Department of Infectious Diseases and Pathology, University of Florida, Gainesville, Florida, United States of America
Division of Gastroenterology, Hepatology and Nutrition, Department of Medicine, University of Florida, Gainesville, Florida, United States of America

Sam X. Cheng
Division of Gastroenterology, Department of Pediatrics, University of Florida, Gainesville, Florida, United States of America

Gary P. Wang
Division of Infectious Diseases and Global Medicine, Department of Medicine, University of Florida, Gainesville, Florida, United States of America

Jennifer L. Owen
Department of Physiological Sciences, College of Veterinary Medicine, University of Florida, Gainesville, Florida, United States of America

Hsin-Hou Chang and Der-Shan Sun
Department of Molecular Biology and Human Genetics, Tzu-Chi University, Hualien, Taiwan,

Ya-Wen Chiang and Ting-Kai Lin
Department of Molecular Biology and Human Genetics, Tzu-Chi University, Hualien, Taiwan,
Institute of Medical Sciences, Tzu-Chi University, Hualien, Taiwan,

Guan-Ling Lin and You-Yen Lin
Institute of Medical Sciences, Tzu-Chi University, Hualien, Taiwan,

Jyh-Hwa Kau
Department of Microbiology and Immunology, National Defense Medical Center, Taipei, Taiwan

Hsin-Hsien Huang and Hui-Ling Hsu
Institute of Preventive Medicine, National Defense Medical Center, Taipei, Taiwan

Jen-Hung Wang
Department of Medical Research, Tzu Chi General Hospital, Hualien, Taiwan

Lumeng Ye, Falk Hildebrand, Jozef Dingemans and Pierre Cornelis
Department of Bioengineering Sciences, Research group Microbiology, Vrije Universiteit Brussel and VIB Structural Biology Brussels, Brussels, Belgium,

Steven Ballet and George Laus
Chemistry Department, Vrije Universiteit Brussel, Pleinlaan 2, 1050 Brussels, Belgium

Sandra Matthijs
Institut de Recherches Microbiologiques - Wiame, Campus du CERIA, Brussels, Belgium

Roeland Berendsen
Plant-Microbe Interactions, Utrecht University, Utrecht, The Netherlands

Marco Plomp and Alexander J. Malkin
Biosciences and Biotechnology Division, Physical and Life Sciences Directorate, Lawrence Livermore National Laboratory, Livermore, California, United States of America

Alicia Monroe Carroll and Peter Setlow
Department of Molecular Biology and Biophysics, University of Connecticut Health Center, Farmington, Connecticut, United States of America

Robert J. Huber, Michael A. Myre and Susan L. Cotman
Center for Human Genetic Research, Massachusetts General Hospital, Harvard Medical School, Boston, Massachusetts, United States of America

Ben Ryall, Marta Carrara, James E. A. Zlosnik, Xiaoyun Lee, Zhen Wong, Kathryn E. Lougheed and Huw D. Williams
Department of Life Sciences, Faculty of Natural Sciences, Imperial College London, Sir Alexander Fleming Building, London, United Kingdom

Volker Behrends
Department of Life Sciences, Faculty of Natural Sciences, Imperial College London, Sir Alexander Fleming Building, London, United Kingdom
Department of Surgery and Cancer, Faculty of Medicine, Imperial College London, Sir Alexander Fleming Building, London, United Kingdom

Xuegang Mao
College of Geographical Sciences, Fujian Normal University, Fuzhou, China

Department of Earth and Environmental Sciences, Ludwig-Maximilians University, Munich, Germany

Ramon Egli
Central institute for Meteorology and Geodynamics, Vienna, Austria

Nikolai Petersen
Department of Earth and Environmental Sciences, Ludwig-Maximilians University, Munich, Germany

Marianne Hanzlik
Chemistry Department, Munich Technical University, Munich, Germany

Xiuming Liu
College of Geographical Sciences, Fujian Normal University, Fuzhou, China
Department of Environment and Geography, Macquarie University, Sydney, Australia

Nansalmaa Amarsaikhan, Evan M. O'Dea, Angar Tsoggerel, Henry Owegi, Jordan Gillenwater and Steven P. Templeton
Department of Microbiology and Immunology, Indiana University School of Medicine – Terre Haute, Terre Haute, Indiana, United States of America

Index

A

Adenosine Tri-phosphate (atp), 106, 108, 111, 115
Allergic Sensitization, 211, 218
Allo-bmt Recipients, 1, 3, 7-8, 11
Amino Acid Limitation, 13
Antagonistic Activity, 141, 146, 150-151
Anthrax Disease, 23
Anthrax Lethal Toxin (lt),, 129
Anti-cmv Immunity, 1
Antibiotic Treatments, 129
Aspergillus Fumigatus, 211, 217-219

B

Bacillus Anthracis, 23, 30, 117, 127-129, 139-140, 157
Bacillus Spores, 155, 169
Bacillus Subtilis, 30, 34, 39, 64-65, 73-74, 104, 155, 169-170
Bacillus Tequilensis Sn4, 31-32, 35, 39
Bacterial Endospores, 65
Barrier Dysfunction, 117, 119, 121
Biological Instability, 106-107, 109, 111-112, 114
Bone Marrow Transplantation, 1, 12

C

Carbon Storage Organelles, 13
Cell Concentrations, 106, 110-114, 172, 202
Cell Wall Composition, 211-213, 216
Chlorinated Drinking Water, 106-107, 109, 114
Coat Structure, 155, 158, 160-162, 165-166, 168-169
Cystic Fibrosis (cf), 186
Cytotoxic Nk Cells, 1, 4, 8

D

Decontamination Efficacy, 23, 29
Defense Signaling Activation, 75
Determining Mechanisms, 155
Dictyostelium Discoideum, 171, 184-185
Distribution Network, 106-107, 109-112, 114-115
Diverse Clades, 41
Dysbiosis, 117, 119-120, 122, 127

E

Ectomycorrhizal, 41-42, 44-45, 47, 49, 51-53
Endophytic Actinobacterium, 87
Endophytic Life Style, 87
Environmental Resistance, 155
Erythrocytic Mobilization, 129, 137

Exo-metabolome, 54-55, 59-63
Extracellular Nature, 31

F

Fatty Acids, 13-17, 19-21, 56
Flow Cytometric (fcm), 106
Fungus-fungus Interactions, 41-42, 51

G

Gastrointestinal (gi) Anthrax, 117
Genome Features, 87, 89
Genome Sequence Analysis, 89, 141, 143
Germinating Spores, 65, 156, 170
Germination Proteins, 65-66, 69, 72-74
Graftvs-host Disease (gvhd), 1
Granulocyte Colony-stimulating Factor (g-csf), 129-130

H

Host-plant Roots, 41
Human Pathogen, 211, 216

I

Immune Suppression, 117, 126-127
Inorganic Compounds, 31
Interaction Sites, 75, 77-80, 84
Invasive Disease, 211

J

Juvenile Ncl (jncl), 171

L

Leucinerich Repeat (lrr), 75
Lipid Bodies, 13-14, 16-20
Lt-mediated Pathogenesis, 129
Lupinus Angustifolius, 87, 101, 103, 105

M

Magneto-chemotaxis, 197
Magnetotactic Bacteria (mtb), 197
Mass Spectrometry, 54-55, 60, 63-67, 74, 142-143, 146, 150, 185
Membrane Lipids, 13, 17, 19-20
Metal Tolerant Laccase, 31
Microbe-associated Molecular Pattern (mamp), 75
Micromonospora Lupini Strain, 87
Microorganisms, 30, 54, 61-62, 64, 87, 89-91, 97, 100, 102, 105-107, 116, 150, 170

Mitogen-activated Protein Kinases (mapks), 117

Monomeric Protein, 31, 36

Mortality, 1, 4, 11, 65, 117, 119, 125-126, 129-130, 132, 134, 137, 139, 211, 214, 217

Mucoid Switch, 186-187, 194

Myxococcus Xanthus, 13, 21-22

N

Neuronal Ceroid Lipofuscinoses (ncl), 171

Nitrogen-fixing Nodules, 87

Nutrient Germinants, 65-66, 71

O

Oceanic Dissolved Organic Matter (dom), 54

P

Pattern-recognition Receptors (prrs), 75

Phosphate Limitation, 54-55, 57, 60-62, 64

Physicochemical Properties, 155

Pleiotropic Effects, 76, 171, 184, 194

Pseudomonas Aeruginosa, 64, 76, 99, 104, 141, 150-151, 153-154, 186, 195-196

Pseudomonas Putida, 64, 141, 143, 146, 150, 153

Pseudomonas Sp. Pathogens, 141

Pseudovibrio Sp. Fo-beg1, 54, 56-57, 59-63

Pyoverdine Siderophore, 141, 153

Q

Quercus Species, 41-43, 45, 49

Quorum Sensing Systems, 186-187, 189, 193-194

R

Receptor-like Kinases (rlks), 75

Root-endophytic Fungi, 41, 52

S

Signaling Initiation, 75

Spatial Segregation, 41-44, 47, 49, 51

Spore Decontamination, 65

Spore Pathogenesis, 155

Sporicidal Disinfectants, 23

Sporulation, 13, 17-18, 22, 66, 73-74, 104, 155-157, 160, 167-170

Surrogate Spores, 23

T

Terrestrial Ecosystems, 41

Thermo-alkali-stable, 31-32, 36, 39

Toll-like Receptor 5 (tlr5), 1

Transposon Mutants, 141, 150

Triacylglycerides (tags), 13

U

Ultra-high Resolution, 54, 59-60, 62